Introduction to Quantum Algorithms

Pure and Applied
UNDERGRADUATE // TEXTS · 64

Introduction to Quantum Algorithms

Johannes A. Buchmann

AMERICAN
MATHEMATICAL
SOCIETY
Providence, Rhode Island USA

2020 *Mathematics Subject Classification.* Primary 11-xx, 68-xx, 81-xx, 94-xx.

For additional information and updates on this book, visit
www.ams.org/bookpages/amstext-64

Library of Congress Cataloging-in-Publication Data

Cataloging-in-Publication Data has been applied for by the AMS.
See http://www.loc.gov/publish/cip/.
DOI: https://doi.org/10.1090/amstext/64

Contents

Preface

The advent of quantum computing

Computing has become indispensable in all areas of society, economy, science, and engineering. As the scope and complexity of problems we aim to solve with computers expand, the demand for computing power continues to skyrocket. For decades, Moore's law has driven the exponential growth of computing resources, enabling faster and more capable machines. However, physical limitations loom on the horizon, challenging the continuation of this trend. Addressing the impending limitations of classical computing requires innovative solutions. New computing architectures and algorithms that can significantly outperform classical systems would be a game changer, promising the ability to tackle problems that are currently beyond reach.

Such an innovative concept is that of a quantum computer. It harnesses the principles of quantum mechanics. While in classical computing, bits can represent either a "0" or a "1", quantum bits or qubits can exist in superpositions of these values. This property allows quantum registers with several qubits to store and process exponentially more information than classical registers with an equivalent number of bits. However, as this book will demonstrate, harnessing this property of quantum computers presents significant challenges.

The concept of quantum computing began to take shape in the early 1980s, thanks to the visionary work of several researchers. The mathematician and physicist Yuri Manin [**Man80**], [**Man99**] pondered whether computers based on the principles of quantum mechanics could challenge the Church-Turing thesis and go beyond the capabilities of Turing machines. Physicist Paul Benioff [**Ben80**] proposed the model of a quantum Turing machine, laying the theoretical foundation for quantum computation. Physics Nobel laureate Richard Feynman [**Fey82**] recognized the potential of quantum computers to simulate physics experiments involving quantum effects. Quantum algorithms became known to virtually everyone in the world through the groundbreaking

work of Peter Shor [**Sho94**] on quantum polynomial time factoring and discrete logarithm algorithms. Shor's work alarmed the world as it revealed the vulnerability of
all known public-key cryptography, one of the fundamental pillars of cybersecurity, to
quantum computer attacks.

Another early advancement in quantum computing that garnered significant at-
tention was Lov Grover's algorithm [**Gro96**], offering a quadratic speedup for unstructured search problems. This breakthrough further fueled the growing interest in quantum computing. Grover's algorithm captured widespread interest because of its ability
to solve a very generic problem, making it useful across a wide range of applications.

In the decades following these early developments, many more quantum algorithms have been discovered. An example is the HHL algorithm [**HHL08**], which can
be used to find properties of solutions of large sparse linear systems with certain properties, providing an exponential speedup over classical solvers like Gauss elimination.
Since linear algebra is one of the most important tools in all areas of science and engineering, the HHL algorithm has wide applications, including machine learning, which
is one of the most significant techniques in computer science today.

This progress should not deceive us, as the development of quantum algorithms
remains a significant challenge. Sometimes, there is the impression that quantum computing allows all computations to be parallelized and significantly accelerated. However, that is not the case. In reality, each new quantum algorithm requires a unique
idea. Consequently, such algorithms can currently only accelerate a few computation problems. Moreover, only very few of these improvements come with exponential
speedups.

All the algorithms mentioned in this book are designed for universal, gate-based
quantum computers, which are the most widely recognized and extensively researched
type of quantum computers. In addition to universal quantum computers, there are
more specialized types of quantum computers, such as quantum annealers and quantum simulators. Quantum annealers utilize annealing techniques to solve optimization problems by finding the lowest-energy state of a physical system. On the other
hand, quantum simulators are specifically designed to simulate quantum systems and
study quantum phenomena. However, this book focuses on universal quantum computers due to their versatility and because they are the most interesting from a computer science perspective.

The goal of the book

The book is well suited for readers with a basic understanding of calculus, linear algebra, algebra, and probability theory, as well as algorithms and complexity as taught in
basic university courses on these subjects. However, recognizing that the readers may
have diverse backgrounds in computer science, mathematics, physics, and engineering, my goal is to make the book self-contained and rigorous from the perspectives of all
these subjects. Thus, even readers with little knowledge of algorithms and complexity
can acquire this by studying Chapter 1. Similarly, for those lacking knowledge of certain topics in linear algebra, algebra, or probability theory, the appendix and Chapter 2
can compensate. Furthermore, the physics introduced in Chapter 3 is sufficient for the

subsequent chapters. I also adopted this approach in writing the book due to my own experience. Despite having degrees in mathematics and physics and being a computer science professor for over 30 years, I found myself needing to refresh my memory on several required concepts and to learn new material. Therefore, my objective is to make the presentation understandable with a minimum of prerequisites, ensuring clarity for both myself and the readers.

My approach of covering all the details will lead to the situation that some readers may already possess knowledge covered in the introductory chapters. However, even they are likely to encounter new and vital information in these chapters, essential for understanding quantum algorithms. For example, Chapter 1 gives an introduction to the theory of reversible computation, which is not typically part of the standard computer science education. Chapter 2 introduces mathematicians to the Dirac notation, commonly used by physicists. Chapter 3 further expands the understanding of physicists by applying the quantum mechanics postulates to quantum gates and circuits. Therefore, I encourage those with prior knowledge to read these sections, taking note of the notation used in the book and of unfamiliar results. This is vital for grasping the intricacies of my explanation of quantum algorithms.

The structure of the book

As described, the initial three chapters lay the groundwork required to dive into quantum computing.

Chapter 1 presents the theoretical aspects of classical computation relevant to understanding quantum computing. It begins with classical algorithms, particularly focusing on probabilistic algorithms and their complexity analysis. This is pivotal because quantum algorithms, such as Shor's algorithm, can be seen as probabilistic algorithms with quantum circuits as subroutines. Therefore, the analysis of probabilistic algorithms effortlessly extends to the analysis of quantum algorithms. Furthermore, the classical complexity classes elucidated in this chapter serve as a blueprint for defining quantum complexity classes. The second part of this chapter is dedicated to the computational model of classical Boolean circuits, specifically examining the theory of reversible circuits, which bears a direct connection to their quantum counterparts. This serves as a foundation for designing basic quantum algorithms using quantum circuits. The chapter establishes that every Boolean function can be computed by a Boolean circuit and covers universal sets of classical gates, uniform families of circuits, and their correlation with classical algorithms and their complexity. Moreover, it demonstrates that every Boolean circuit can be transformed into a reversible circuit — an essential result later applied to show that every Boolean function can be computed by a quantum circuit.

Chapter 2 explores the theory of finite-dimensional Hilbert spaces, which forms the mathematical framework for modeling the physics of quantum computing. We note, however, that finite-dimensional Hilbert spaces may fall short of modeling general quantum mechanics. The chapter begins by establishing the necessary foundations, including essential concepts like inner products, orthogonality, linear maps, and their adjoints. It also introduces the Schur decomposition theorem, which provides valuable tools for subsequent discussions. Moving forward, the chapter familiarizes the reader with significant operators in quantum mechanics, such as Hermitian, unitary, and normal operators. Of particular significance is the spectral theorem, a fundamental result that offers profound insights into these operators and their characteristics. The consequences of the spectral theorem are also explored to enrich the reader's understanding. Furthermore, the chapter delves into the concept of tensor products of finite-dimensional Hilbert spaces, a crucial notion in quantum computing. The discussion culminates with an elucidation of the Schmidt decomposition theorem, which plays a pivotal role in characterizing the entanglement of quantum systems.

Chapter 3 constitutes the third foundational pillar of quantum computing required in this book, encompassing the essential background of quantum mechanics. This chapter introduces the relevant quantum mechanics postulates. To illustrate their relevance, the chapter applies these postulates to introduce fundamental concepts of quantum computing, including quantum bits, registers, gates, and circuits. Simple examples of quantum computation are provided to enhance the reader's understanding of the connection between the postulates and quantum algorithms. In addition, the chapter provides the foundation for the geometric interpretation of quantum computation. It achieves this by establishing the correspondence between states of individual quantum bits and points on the Bloch sphere, a pivotal concept in quantum computing visualization. Moreover, the chapter presents an alternative description of the relevant quantum mechanics using density operators. This approach enables the modeling of the behavior of components within composed quantum systems.

The foundational groundwork laid out in the initial three chapters, including the domains of computer science, mathematics, and physics, sets the stage for a comprehensive exploration of quantum algorithms in Chapter 4. This chapter embarks on this transformative journey by shedding light on pivotal quantum gates, which serve as the fundamental constituents of quantum circuits. We start by introducing single-qubit gates, demonstrating that their operations can be perceived as rotations within three-dimensional real space. Subsequently, we delve into the realm of multiple-qubit operators, with a particular focus on controlled operators. In addition, this chapter familiarizes readers with the significance of ancillary and erasure gates, which play a vital role in the augmentation and removal of quantum bits. Leveraging analogous outcomes from classical reversible circuits, the chapter shows that every Boolean function can be realized through a quantum circuit. In contrast to the classical scenario, the quantum case does not adhere to the notion that a finite set of quantum gates suffices to implement any quantum operator. Instead, finite sets of quantum gates are presented, enabling the approximation of all quantum circuits. Lastly, the chapter ushers in the concept of quantum complexity theory, using the analogy between classical probabilistic algorithms and quantum algorithms. It introduces the complexity class BQP, which stands for bounded-error quantum polynomial time.

The following four chapters focus on specific quantum algorithms.

Chapter 5 introduces early algorithms designed to illustrate the quantum computing advantage. We begin with the Deutsch algorithm, as presented in David Deutsch's seminal paper from 1985 [**Deu85**], and its generalization by David Deutsch and Richard

Jozsa in 1992 [**DJ92**]. Subsequently, we explore Daniel R. Simon's algorithm, proposed in 1994 [**Sim94**], which demonstrated a quadratic speedup compared to the best-known classical algorithm for its specific problem. Through the explanations of these algorithms, two key principles emerge, elucidating why quantum computing can surpass classical computing. The first principle is quantum parallelism, which capitalizes on the ability of quantum registers to exist in superpositions of states, allowing for simultaneous computation of multiple possibilities. The second principle is quantum interference, empowering quantum algorithms to concentrate probability amplitudes on correct answers while suppressing incorrect ones. This phenomenon increases the likelihood of obtaining the desired solution to a computational problem. Also, the important phase-kickback method is explained.

Chapter 6 introduces the most famous quantum algorithms, namely Peter Shor's algorithms for factoring integers and computing discrete logarithms [**Sho94**]. The chapter commences by presenting an overview of Shor's integer factorization algorithm, serving as a road map for the subsequent concepts introduced and their utilization within this algorithm. Next, the key ingredient of Shor's algorithms is presented: the Quantum Fourier Transform. A detailed explanation illustrates how this operator and its inverse can be efficiently implemented using simple quantum gates. The Quantum Fourier Transform is then employed to solve the problem of quantum phase estimation. By utilizing quantum phase estimation, a polynomial time quantum algorithm for computing the order of an integer modulo another positive integer is developed, leading to a polynomial time quantum algorithm for integer factorization. Moreover, the chapter reveals how quantum phase estimation enables the computation of discrete logarithms modulo positive integers in polynomial time.

Chapter 7 explores another significant quantum algorithm with numerous applications: the Grover search algorithm from 1996. This algorithm offers a quadratic advantage when searching for a specific target item in an unstructured set. The chapter delves into the concept of amplitude amplification, a powerful technique that plays a pivotal role in Grover's algorithm. Moreover, the chapter introduces the quantum counting algorithm, a notable contribution by Gilles Brassard, Peter Høyer, and Alain Tapp in 1998 [**BHT98**]. This algorithm utilizes Grover's algorithm in combination with quantum phase estimation to count the number of solutions for a given search problem.

Chapter 8 provides an overview of the HHL algorithm, which focuses on computing properties of solutions for large sparse linear systems over the complex numbers. The primary purpose of this chapter is to demonstrate how the ideas of previously introduced algorithms in this book enable the design of such an advanced algorithm.

The book also features an extensive appendix that serves as a valuable resource for readers. It introduces the mathematical notation used throughout the book and covers essential concepts, including results that might not commonly be part of standard mathematical university education. Appendix A begins by exploring fundamental mathematical concepts, including groups, rings, and fields. It then delves into the necessary topics in number theory, such as the greatest common divisor and its computation, prime factor decomposition, and continued fractions. In addition, this part of the appendix lists essential trigonometric identities and inequalities that play a crucial role in the main part of the book. Appendix B focuses on linear algebra. Its first part briefly reviews important concepts and results. The second part covers the concept of tensor products, which is of significant importance in quantum computing and is typically not included in introductory courses in linear algebra. Lastly, Appendix C contains the required notions and results from probability theory. This knowledge is essential for the analysis of probabilistic and quantum algorithms.

What is not covered

The main focus of this book is on the theory of quantum algorithms, rather than the practical aspects of their implementation. However, it is essential to acknowledge two significant aspects related to realizing practical quantum computers, which our presentation does not cover.

One such aspect is quantum error correction, which plays a crucial role in quantum computing. Quantum systems are highly sensitive to their environment, leading to decoherence, which can introduce errors in quantum computations. Quantum error correction techniques are designed to protect quantum computing from the harmful effects of decoherence and errors, making quantum algorithms more reliable and robust.

Another critical aspect is the physical implementation of quantum computers. Various approaches are being explored to build qubits and quantum gates for practical quantum computers. These approaches include using superconducting circuits, trapped ions, photonic qubits, topological qubits, and quantum dots, among others. Each of these approaches has its advantages and challenges, and researchers are actively working to develop scalable and efficient quantum computing technologies.

An overview of the main results of quantum computing, including quantum error correction and quantum computer architectures, can be found in the Wikipedia article "Timeline of quantum computing" [**Aut23**].

For instructors

This book is suitable for self-study. It is also intended and has been used for teaching introductory courses on quantum algorithms. My recommendation to instructors for such a course is as follows: If most of the participants are already familiar with the required basics of algorithms and complexity, linear algebra, algebra, and probability theory, the course should cover Chapters 3, 4, 5, 6, 7, and 8 in this order, exploring different aspects of quantum algorithms. Individual students lacking some basic knowledge can familiarize themselves with those topics using the detailed explanations in the respective parts of the book. If the majority of the participants in the course is unfamiliar with certain basic topics, the instructor may want to briefly cover them either in a preliminary lecture or when they are used during the course.

Depending on the instructor's intentions and the available time, the course may focus more on theoretical explanations and proofs or on the practical aspects of how quantum algorithms work. In both situations, students who desire more background than is covered in the course can supplement their knowledge through self-study of the corresponding book parts.

Acknowledgements

I express my sincere gratitude to the following individuals who have been instrumental in supporting me throughout the process of writing this book. Their invaluable advice, discussions, and comments have played a pivotal role in shaping the content and quality of this work: Gernot Alber, Gerhard Birkl, Jintai Ding, Samed Düzlü, Fritz Eisenbrand, Marc Fischlin, Mika Göös, Iryna Gurevich, Matthieu Nicolas Haeberle, Taketo Imaizumi, Michael Jacobson, Nigam Jigyasa, Norbert Lutkenhaus, Alastair Kay, Juliane Krämer, Gen Kimura, Michele Mosca, Jigyasa Nigam, Rom Pinchasi, Ahamad-Reza Sadeghi, Masahide Sasaki, Alexander Sauer, Florian Schwarz, Tsuyoshi Takagi, Shusaku Uemura, Thomas Walther, Yuntao Wang, and Ho Yun. Their dedication to sharing their expertise and knowledge has been truly invaluable, and I am deeply grateful for their willingness to engage in insightful discussions and provide constructive feedback throughout this journey.

I also extend my heartfelt gratitude to Ina Mette, the responsible person at AMS. Her belief in the potential of this book and her continuous encouragement to pursue this project have been instrumental in its realization. I am deeply grateful for her unwavering support and guidance throughout the writing process. I am also grateful to Arlene O'Sean and her team at AMS for their great job in carefully proofreading the book and making its appearance so nice.

I learned a lot from the books on quantum computing by Michael A. Nielsen and Isaac L. Chuang [**NC16**] and by Phillip Kaye, Raymond Laflamme, and Michele Mosca [**KLM06**].

The writing of this book would not have been possible without the invaluable contributions of several open-source LaTeX packages, which greatly facilitated the presentation of complex concepts. I extend my gratitude to the creators and maintainers of these packages:

- The powerful TikZ library[1] and its extension circuitikz[2] were instrumental in visualizing circuits and diagrams throughout the book.
- I used the open source TikZ code for illustrating the right-hand rule,[3]
- For the clear representation of quantum circuits, I relied on quantikz.[4]
- To illustrate quantum states on Bloch spheres, I used the blochsphere package.[5]
- The packages algorithm and algorithmpseudocode[6] were indispensable in presenting algorithms in a structured and easily understandable format.
- For handling the Dirac notation with ease, I benefited from the physics package.[7]

---

[1]https://tikz.net/
[2]https://ctan.org/pkg/circuitikz
[3]https://tikz.net/righthand_rule/
[4]https://ctan.org/pkg/quantikz
[5]https://ctan.org/pkg/blochsphere
[6]https://www.overleaf.com/learn/latex/Algorithms
[7]https://www.ctan.org/pkg/physics

I am sincerely grateful to the open-source community for making these and many more tools available, enhancing the quality of this work and simplifying its creation.

Finally, I would like to acknowledge the support provided by ChatGPT[8] in improving many formulations of my presentation. As I am not a native speaker of the English language, this assistance was of great help.

---

[8]https://chat.openai.com/

Chapter 1

Classical Computation

Quantum algorithms outperform the best-known classical algorithms in numerous computational tasks. To establish and demonstrate these advancements, we rely on essential mathematical frameworks. These frameworks aid our understanding of classical and quantum computation, enable us to address questions about problem solubility on both classical and quantum computers, and assess the computational efficiency of these algorithms.

This chapter presents models fundamental to classical computation, providing the basis for the corresponding models in quantum computation. We begin by explaining a model for classical deterministic and probabilistic algorithms. It forms the basis for reviewing key concepts and results in classical complexity theory. The part on the analysis of probabilistic algorithms is particularly important, as it plays a crucial role in later chapters where we discuss quantum algorithms which are probabilistic in nature. The methods used for analyzing probabilistic algorithms, including success probability amplification, will be directly applicable to the analysis of quantum algorithms.

Moving on from the algorithmic model, we introduce the classical circuit model of computation, with a particular emphasis on reversible circuits. These reversible circuits can be readily transformed into quantum circuits, providing the essential insight that quantum computing is Turing complete.

1.1. Deterministic algorithms

In the realm of computer algorithms, there exist various formal models. The most famous is the model of Turing machines. It was invented in 1936 by the mathematician Alan Turing. The Turing machine model has mathematical rigor. But its connection with algorithms implemented using real-world programming languages is difficult to understand and describe. Hence, we will introduce a model of computation that strikes a balance between a formal description and a representation resembling real-world computing. A very good and comprehensive presentation of computer algorithms that uses similar modeling is the book by Thomas H. Cormen, Charles E. Leiserson, Ronald L. Rivest, and Clifford Stein [**CLRS22**].

1.1.1. Basics. To explain our model, we introduce some basic concepts and results. We begin by defining alphabets.

Definition 1.1.1. An *alphabet* is a finite nonempty set.

Example 1.1.2. The simplest alphabet is the *unary alphabet* $\{I\}$, which contains only the symbol I. The most commonly used alphabet in computer science is the *binary alphabet* $\{0, 1\}$, where each element is referred to as a *bit*. Other commonly used alphabets in computer science include the set $\{0, 1, 2, 3, 4, 5, 6, 7, 8, 9\}$ of *decimal digits*, the set $\{0, 1, 2, 3, 4, 5, 6, 7, 8, 9, A, B, C, D, E, F\}$ of *hexadecimal digits*, and the *Latin alphabet* $\mathcal{R} = \{a, \ldots, z, A, \ldots, Z, \_\}$ that includes lowercase and uppercase Latin letters, as well as the space symbol $\_$.

Next, we introduce words over alphabets.

Definition 1.1.3. Let Σ be an alphabet.

(1) By Σ^* we denote the set of all finite sequences over Σ including the empty sequence which is denoted by ( ).

(2) The elements of Σ^* are called *words* or *strings* over Σ.

(3) If w is a word over Σ, then its elements are called the *characters* of w and $|w|$ denotes the length of w.

(4) The set of words over Σ of length $n \in \mathbb{N}_0$ is denoted by S^n.

(5) If $n \in \mathbb{N}_0$ and $\vec{s} \in \Sigma^n$, then we write $\vec{s} = s_0 s_1 \cdots s_{n-1}$, $\vec{s} = $ "$s_0 \cdots s_{k-1}$", $\vec{s} = (s_0 s_1 \cdots s_{n-1})$, or $\vec{s} = (s_0, s_1, \ldots, s_{n-1})$. We also use other numberings of the characters. For example, we may start the numbering with 1.

Example 1.1.4. The sequence III is a word over the alphabet $\{I\}$. The sequence $(0, 1, 0)$ is a word over the alphabets of bits, decimal digits, and hexadecimal digits. The sequence $(0, 1, 2)$ is a word over the alphabets of decimal digits and hexadecimal digits. The sequence $(0, 1, A)$ is a word over the alphabet of hexadecimal digits. All these words have length 3. Finally "another_failure" is a word over the alphabet of Roman letters.

Next, we define encodings.

Definition 1.1.5. An *encoding* of a set S with respect to an alphabet Σ is an injective map $e : S \to \Sigma^*$.

Example 1.1.6. A simple encoding of the set $\mathbb{N}_0$ of nonnegative integers is

$$(1.1.1) \qquad\qquad e : \mathbb{N}_0 \to \{I\}^*, \quad a \mapsto I^a = \underbrace{I \cdots I}_{a \text{ times}}.$$

For $a \in \mathbb{N}_0$ we call $e(a) = I^a$ the *unary representation* of a. Its length is a.

As an important way of encoding nonnegative integers, we introduce their binary representation.

Proposition 1.1.7. *Let $a \in \mathbb{N}$. Then there is a uniquely determined sequence $\vec{b} = (b_0, \ldots, b_{n-1}) \in \{0,1\}^*$ with $b_0 = 1$ such that*

$$(1.1.2) \qquad a = \sum_{i=0}^{n-1} b_i 2^{n-i-1}.$$

The length of this sequence is $n = \lfloor \log_2 a \rfloor + 1$.

Exercise 1.1.8. Prove Proposition 1.1.7.

Definition 1.1.9. Let $a \in \mathbb{N}$.

(1) The sequence $\vec{b}$ from Proposition 1.1.7 is called the *binary expansion* or *binary representation* of a.

(2) The positive integer $n = \lfloor \log_2 a \rfloor + 1$ is called the *binary length* or *bit length* of a. It is denoted by bitLength(a).

(3) The *binary expansion* or *binary representation* of 0 is defined to be (0). The *binary length* or *bit length* of 0 is set to 1.

Example 1.1.10. The binary expansion of 7 is 111 since $7 = 2^2 + 2^1 + 2^0$. The binary length of 7 is 3.

Exercise 1.1.11. Determine the binary expansion and the binary length of 251.

As we have seen, every finite sequence of bits that starts with the bit '1' is assigned to a uniquely determined positive integer as its binary expansion. Now, we also assign uniquely determined nonnegative integers to all sequences in $\{0,1\}^*$ including those that start with the bit 0 using the following definition.

Definition 1.1.12. For all $n \in \mathbb{N}$ and $\vec{b} = (b_0, \ldots, b_{n-1}) \in \{0,1\}^*$ we set

$$(1.1.3) \qquad \text{stringToInt}(\vec{b}) = \sum_{i=0}^{n-1} b_i 2^{n-i-1}$$

and call this value the *integer represented by $\vec{b}$*.

Note that nonnegative integers are represented by infinitely many strings in $\{0,1\}^*$. Specifically, they are represented by all strings that result from prepending a string consisting only of zeros to their binary expansions. Also, the number 0 is represented by all finite sequences of the bit "0".

Exercise 1.1.13. Use Proposition 1.1.7 to show that the map (1.1.3) is a bijection.

1.1.2. Data types. Algorithms perform operations on objects of specific data types, such as bits or integers. In this context, a *data type* refers to a nonempty set, and the individual elements within this set are called *data type objects*.

In the presented model, the *elementary data types* consist of the following sets:

- the set $\{0,1\}$ of bits,
- the set $\mathbb{Z}$ of integers,
- the set $\mathcal{R}$ of lowercase and uppercase Latin letters, including the space symbol ␣.

Table 1.1.1. Elementary data types and their sizes.

Data type	Element	Size
$\{0, 1\}$	b	$\text{size}(b) = \text{O}(1)$
$\mathcal{R}$	x	$\text{size}(x) = \text{O}(1)$
$\mathbb{Z}$	a	$\text{size}(a) = \text{O}(\text{bitLength}(|a|))$

Another common data type is that of a *floating point number*, which represents an approximation to a real number. However, analyzing algorithms that use this data type is more complex, as it requires considering error propagation. For our purposes, we do not need to take this data type into account.

Objects of these elementary data types are represented and stored using bits. The encoding of these objects can be defined in a straightforward manner for both bits and Roman characters. Integers are encoded using the binary expansion of their absolute value along with the indication of their sign. The *size* of these encodings refers to the number of bits used. For a data type object a, the size of its encoding is denoted by size a. It may vary depending on the specific computing platform, programming language, or other relevant factors. However, detailed discussions regarding these specific implementations are beyond the scope of this book, and we present only a summary of the different sizes of the elementary data types in Table 1.1.1.

In our model, advanced data types include vectors and matrices over some data type, which are described in Sections B.1 and B.4. For example, $(1, 2, 3)$ is an integer vector. Similarly, $\begin{pmatrix} 0 & 1 \\ 1 & 0 \end{pmatrix}$ is a bit matrix. Vectors and matrices are also encoded by bits, using the encoding method of their data type. The encodings of vectors and matrices possess the following properties. For any $k \in \mathbb{N}$, the size of the encoding of a vector $\vec{s} = (s_0, \ldots, s_{k-1})$ over some data type satisfies

$$(1.1.4) \qquad \text{size}(\vec{s}) = \text{O}\left(\sum_{i=0}^{k-1} \text{size } s_i\right).$$

Similarly, for any $k, l \in \mathbb{N}$, the size of the encoding of a matrix $M = (m_{i,j})_{i \in \mathbb{Z}_k, j \in \mathbb{Z}_l}$ over some data type satisfies

$$(1.1.5) \qquad \text{size}(M) = \text{O}\left(\sum_{i \in \mathbb{Z}_k, j \in \mathbb{Z}_l} \text{size } m_{i,j}\right).$$

Here we have used $\mathbb{Z}_k = \{0, \ldots, k - 1\}$.

1.1.3. Variables. A *variable* is a symbol that serves as a reference to a memory unit with the capability of storing an element of a specific data type. Once defined, the data type of a variable remains fixed. The *value* of a variable corresponds to the content stored in its associated memory cell. Thus, the size of a variable is determined by the size of its value. This concept is illustrated in Figure 1.1.1, where the variable a represents a memory cell containing the integer 19, while the variable b represents a memory cell containing the word "bit". Therefore, the value of a is 19, and the value of b is "bit".

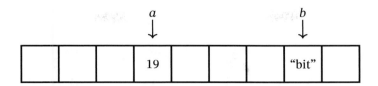

Figure 1.1.1. The variables a and b represent memory cells that contain 19 and "bit", respectively.

In most programming languages, variables must be explicitly declared before they can be used. This declaration serves three essential purposes: first, it specifies the data type of the variable; second, it reserves the necessary memory space to store the variable's value; and third, it assigns a name to the variable for easy reference within the code. However, to streamline the presentation of algorithms and to keep the focus on the core logic, we omit explicit variable declarations and assume that the data type and name of the variables are apparent from the context in which they are used.

1.1.4. Instructions. *Instructions* serve as fundamental building blocks of algorithms and play a central role in defining their functionality. In this section, we will introduce the instructions, also known as *statements*, that algorithms in our model can execute, and we will provide upper bounds on their running times and memory size requirements. We will use the terms "instructions" and "statements" interchangeably to refer to these essential algorithmic components.

Unless otherwise specified, we measure the running times in *bit operations*. By bit operations, we refer to all operations that a computer can perform on a single or two bits. For example, a collection of bit operations commonly available on most computers is shown in Table 1.1.3. The concept is that all operations are executed using these bit operations, and the number of such bit operations is counted. This concept is justified by the fact that, as we will see in Section 1.5.3, all operations can be implemented using the operators NOT, AND, and OR. Analogously, we measure the space requirement by counting the number of memory units used during the computation.

The basic instruction is the *assign instruction*:

$$(1.1.6) \qquad a \leftarrow b.$$

It sets the value of a variable a of a certain data type to an element b of this data type or to the value of a variable b of the same data type as a. The running time and space requirement for this operation is $O(\text{size } b)$.

Algorithms may also assign the result of an operation to a variable. An example of such an instruction is

$$(1.1.7) \qquad a \leftarrow b + 3.$$

The right-hand side of this assignment is the arithmetic expression $b + 3$. When such an instruction is executed, the arithmetic expression is evaluated first. In the example,

Table 1.1.2. Permitted operations on integers, their running times, and space requirements for operands of size $O(n)$.

Operation	Operands	Result	Time	Space
absolute value	$a \in \mathbb{Z}$	$\lvert a \rvert$	$O(n)$	$O(n)$
add	$a, b \in \mathbb{Z}$	$a + b$	$O(n)$	$O(n)$
subtract	$a, b \in \mathbb{Z}$	$a - b$	$O(n)$	$O(n)$
multiply	$a, b \in \mathbb{Z}$	$a * b$	$O(n^2)$	$O(n)$
divide	$a, b \in \mathbb{Z}, b \neq 0$	$\lfloor a/b \rfloor$	$O(n^2)$	$O(n)$
remainder	$a, b \in \mathbb{Z}, b \neq 0$	$a \bmod b$	$O(n^2)$	$O(n)$
equal	$a, b \in \mathbb{Z}$	$a = b$	$O(n)$	$O(n)$
less than	$a, b \in \mathbb{Z}$	$a < b$	$O(n)$	$O(n)$
less than or equal to	$a, b \in \mathbb{Z}$	$a \leq b$	$O(n)$	$O(n)$
square root	$a \in \mathbb{Z}$	$\lfloor \sqrt{a} \rfloor$	$O(n^2)$	$O(n)$
floor	$a, b \in \mathbb{Z}, b \neq 0$	$\lfloor a/b \rfloor$	$O(n^2)$	$O(n)$
ceiling	$a, b \in \mathbb{Z}, b \neq 0$	$\lceil a/b \rceil$	$O(n^2)$	$O(n)$
next integer	$a, b \in \mathbb{Z}, b \neq 0$	$\lfloor a/b \rceil$	$O(n^2)$	$O(n)$
bit length	$a \in \mathbb{N}_0$	$\mathrm{bitLength}(a)$	$O(\mathrm{size}\, a)$	$O(\mathrm{size}\, a)$
string to int	$\vec{s} \in \{0,1\}^*$	$\mathrm{stringToInt}(\vec{s})$	$O(\mathrm{size}\, \vec{s})$	$O(\mathrm{size}\, \vec{s})$

it depends on the value of the variable b. Then the result is assigned to a. It is permitted that the variable on the left side also appears on the right side. For instance, the instruction

$$(1.1.8) \qquad\qquad\qquad c \leftarrow c + 1$$

increments a counter c by 1.

Next, we present the operations that may be used in expressions on the right side of an assign instruction. The permitted operations on integers are listed in Table 1.1.2, including their time and space requirements. The results of the operations absolute value, floor, ceiling, next integer, add, subtract, multiply, divide, and remainder are integers. The results of the comparisons are the bits 0 or 1 where 1 stands for "true" and 0 stands for "false". For a description and analysis of these algorithms see [**AHU74**] and [**Knu82**].

In most programming languages, only integers with limited bit lengths are available, such as 64-bit integers. When there is a need to work with integers of arbitrary length, specialized algorithms are required to handle operations on such numbers. To simplify our descriptions, we assume that operations on integers of arbitrary length are available as basic operations. These operations can be realized using the running time and memory space as listed in Table 1.1.2 but there exist much more efficient algorithms for integer multiplication and division with remainder. For instance, a highly efficient integer multiplication algorithm developed by David Harvey and Joris van der Hoeven [**HvdH21**] has running time $O(n \log n)$ for n-bit operands. Additionally, it is known that for any integer multiplication algorithm with a running time of $M(n)$, there exist division with remainder and square root algorithms with a running time of $O(M(n))$ (see [**AHU74**, Theorem 8.5]). However, in practice, these faster algorithms

Table 1.1.3. Permitted logic operations.

Name	Logic operator
AND	$\wedge$
OR	$\vee$
NOT	$\neg$
NAND	$\uparrow$
NOR	$\downarrow$
XOR	$\oplus$

Table 1.1.4. Functions implemented by the permitted logic operations.

a	b	$a \wedge b$	$a \vee b$	$\neg a$	$a \uparrow b$	$a \downarrow b$	$a \oplus b$
0	0	0	0	1	1	1	0
0	1	0	1	1	1	0	1
1	0	0	1	0	1	0	1
1	1	1	1	0	0	0	0

are only advantageous for handling very large numbers. For more typical integer sizes, classical algorithms may still be more efficient in terms of practical performance.

On the bits 0 and 1 our algorithms can perform the logic operations that are listed in Table 1.1.3. They implement the functions shown in Table 1.1.4. All permitted logic operations run in time $O(1)$ and require space $O(1)$.

Exercise 1.1.14. Show that $\oplus$ is addition in $\mathbb{Z}_2$ and $\wedge$ is multiplication in $\mathbb{Z}_2$.

Algorithms may also use the branch statements **for**, **while**, **repeat**, and **if**. They initiate the execution of a sequence of instructions if some branch condition is satisfied. In the analysis of algorithms, we will assume that the time and space required to execute an algorithm part that uses a branch instruction is the time and space required to evaluate the branch condition and the corresponding sequence of instructions, possibly several times. Branch instructions together with the corresponding instruction sequence are referred to as *loops*.

We now provide more detailed explanations of the branch instructions using the examples shown in Figures 1.1.2 and 1.1.3 utilizing pseudocode that is further described in Section 1.1.5.

A **for** statement appears at the beginning of an instruction sequence and is ended by an **end for** statement. This instruction sequence is executed for all values of a specified variable, as indicated in the **for** statement. In the **for** loop in Figure 1.1.2, the variable is i, and the instruction $p \leftarrow 2p$ is executed for all i from $i = 1$ to $i = e$. After i iterations of this instruction, the value of p is 2^i. So after completion of the **for** loop, the value of p is 2^e.

Also, **while** statements appear at the beginning of an instruction sequence after which there is an **end while** statement. The instruction sequence is executed as long as the condition in the **while** statement is true. For instance, the **while** loop in Figure

$$p \leftarrow 1 \qquad\qquad i \leftarrow 1 \qquad\qquad i \leftarrow 0$$

<table>
<tr><td>$p \leftarrow 1$</td><td>$i \leftarrow 1$</td><td>$i \leftarrow 0$</td></tr>
<tr><td>for $i = 1$ to e do</td><td>$p \leftarrow 1$</td><td>$p \leftarrow 1$</td></tr>
<tr><td>$\quad p \leftarrow 2p$</td><td>while $i \leq e$ do</td><td>repeat</td></tr>
<tr><td>end for</td><td>$\quad p \leftarrow 2p$</td><td>$\quad p \leftarrow 2p$</td></tr>
<tr><td></td><td>$\quad i \leftarrow i + 1$</td><td>$\quad i \leftarrow i + 1$</td></tr>
<tr><td></td><td>end while</td><td>until $i = e$</td></tr>
</table>

Figure 1.1.2. for, while, and **repeat** loops that compute 2^e.

1.1.2 also computes 2^e. For this, the counting variable i is initialized to 1 and the variable p is initially set to 1. Before each round of the **while** loop, the logic expression $i \leq e$ is evaluated. If it is true, then the instruction sequence in the **while** loop is executed. In the example, p is set to $2p$ and the counting variable i is increased by 1. After the kth iteration of the **while** loop, the value of p is 2^k and the counting variable is $i = k$. Hence, after the eth iteration of the **while** loop we have $p = 2^e$ and $i = e + 1$. So the **while** condition is violated and the computation continues with the first instruction following the **while** loop.

Next, **repeat** statements are also followed by an instruction sequence that is ended by an **until** statement that contains a condition. If this condition is satisfied, the computation continues with the first instruction after the **until** statement. Otherwise, the instruction sequence is executed again. In the example in Figure 1.1.2, the instruction $p \leftarrow 2p$ is executed until the counting variable i is equal to e. Note that the instruction sequence is executed at least once. Therefore, this **repeat** loop cannot compute 2^0.

Exercise 1.1.15. Find **for, while,** and **repeat** loops that compute the integer represented by a bit sequence $s_0 s_1 \cdots s_{n-1} \in \{0, 1\}^n$ where $n \in \mathbb{N}$.

Now we explain **if** statements. The three different ways to use **if** are shown in Figure 1.1.3. Such a statement is followed by a sequence of instructions that is ended by an **end if** statement. The instruction sequence may be interrupted by an **else** statement or by one or more **else if** statements. The code segment on the left side of Figure 1.1.3 checks whether $a < 0$ is true, in which case the instruction $a \leftarrow -a$ is executed. Otherwise, the computation continues with the instruction following the **end if** statement. This code segment computes the absolute value of a because a is set to $-a$ if a is negative and otherwise remains unchanged. The code segment in the middle of Figure 1.1.3 checks whether a is divisible by 11. If so, the variable s is set to 1 and otherwise to 0. Finally, the code segment on the right side of Figure 1.1.3 first checks if $a > 0$ in which case the variable s is set to 1. Next, if $a = 0$, s is set to 0. Finally, s is set to -1 if $a < 0$. The result is the sign s of a.

A computation terminates when the **return** instruction is used. This instruction makes the result available for further use and takes the form

$$\text{(1.1.9)} \qquad\qquad\qquad \textbf{return}(a)$$

where a is an element or variable of some data type or a sequence of such objects. The time and space requirement for this operation is $O(S)$, where S represents the sum of the sizes of the objects in the **return** statement.

```
if a < 0 then        if a mod 11 = 0     if a > 0 then
    a ← −a           then                    s ← 1
end if                   s ← 1           else if a = 0
                     else                then
                         s ← 0               s ← 0
                     end if              else if a < 0
                                         then
                                             s ← −1
                                         end if
```

Figure 1.1.3. if statements.

Algorithms can also return the result of expressions. For instance, a **return** instruction may be of the form

$$(1.1.10) \qquad \textbf{return}(2 * a + b).$$

In this case, the expression is evaluated first, and then the result is returned. The time and space requirements of this instruction are $O(t)$ and $O(s)$, respectively, where t and s denote the time and space required to evaluate the corresponding expression.

Finally, a call to a subroutine is also considered a valid instruction in our model. It takes the form

$$(1.1.11) \qquad a \leftarrow A(b)$$

where A represents an algorithm, as elaborated in the subsequent section. For example, such an instruction may be expressed as

$$(1.1.12) \qquad b \leftarrow \text{power}(a, e).$$

Here, the call $\text{power}(a, e)$ invokes the subroutine, returning the result of a^e, where a and e are both nonnegative integers.

The time and space requirements associated with a subroutine call are $O(t)$ and $O(s)$, respectively, where t and s represent the time and space required by the subroutine to execute. Section 1.4 provides an explanation of how these requirements are determined.

1.1.5. Definition of deterministic algorithms. In this section, we explain the concept of a deterministic algorithm and illustrate how such an algorithm can be presented using pseudocode. Pseudocode serves as a high-level depiction of an algorithm's logic and sequence of steps, with the aim of ensuring ease of comprehension for human readers. To simplify the discussion, we will just talk about an "algorithm" instead of a "deterministic algorithm". In contrast, in the next section, we introduce probabilistic algorithms.

As mentioned earlier, our model of computation is a compromise between a formal description and a representation that resembles real algorithms written in some programming language. Therefore, in this section, we do not use formal definitions.

But in the next sections, we formally define the properties of algorithms. These definitions can be made precise if a formal model of computation is used, such as the Turing machine model.

We illustrate our algorithm model using the Euclidean algorithm. The corresponding pseudocode is shown in Algorithm 1.1.16.

Algorithm 1.1.16. Euclidean algorithm

Input: $a, b \in \mathbb{Z}$
Output: $\gcd(a, b)$

  1: $\gcd(a, b)$
  2:      $a \leftarrow |a|$
  3:      $b \leftarrow |b|$
  4:      **while** $b \neq 0$ **do**
  5:          $r \leftarrow a \bmod b$
  6:          $a \leftarrow b$
  7:          $b \leftarrow r$
  8:      **end while**
  9:      **return** a
 10: **end**

An algorithm A has the following components:

(1) An **Input** statement. It specifies a finite number of *input variables*, their data types, and the permitted values of these variables. The set of all permitted input value tuples is referred to as Input(A).

(2) An **Output** statement. For every $a \in$ Input(A) it specifies properties that may depend on a and makes a return value a correct output. The set of all correct outputs for input a is denoted by Output(A, a).

(3) An algorithm name followed by the sequence of input variables which is used when A is called by other algorithms as a subroutine.

(4) A finite sequence of instructions that ends with **end**.

For example, in the Euclidean algorithm, the **Input** statement specifies that there are two integer input variables a and b which may take any integer value. According to the **Output** statement, the algorithm returns $\gcd(a, b)$. The name of the algorithm and the sequence of input variables can be seen in line 1 of the pseudocode: $\gcd(a, b)$. It is followed by 8 instructions. The last line of the pseudocode is **end**.

Let A be an algorithm that has k input variables $v_0, \ldots, v_{k-1}$ of data types $D_0, \ldots, D_{k-1}$, respectively. Then the set Input(A) of all allowed input values of A satisfies Input(A) $\subset D_0 \times \cdots \times D_{k-1}$. For the Euclidean algorithm, we have $k = 2, D_0 = D_1 = \mathbb{Z}$, and Input($A$) $= \mathbb{Z} \times \mathbb{Z}$. Running A with input $a = (a_0, \ldots, a_{k-1}) \in$ Input(A) means the following. A assigns a_i to the input variable v_i for all $i \in \mathbb{Z}_k$. Then it executes its sequence of instructions. This process is called the *run* of A with input a. We list a few requirements that every deterministic algorithm must satisfy.

(1) Each run of the algorithm with a permitted input carries out a **return** instruction. This means that the algorithm terminates on any input $a \in$ Input(A).

(2) When the algorithm performs a **return** instruction, the return value is correct; i.e., it has the property specified in the **Output** statement.

(3) Executing the **return** instruction is the only way the algorithm can terminate. This means that after executing a statement that is not a **return** instruction there is always a next instruction that the algorithm carries out.

Example 1.1.17. We describe the run of the Euclidean algorithm with input $(a, b) = (100, 35)$. The instructions in lines 2 and 3 replace a and b by their absolute values. For the chosen input, they have no effect. Since $b = 35$, the **while** condition is satisfied. Hence, the Euclidean algorithm executes $r \leftarrow 100 \bmod 35 = 30$, $a \leftarrow b = 35$, and $b \leftarrow r = 30$. After this, the **while** condition is still satisfied since $b = 30$. So the Euclidean algorithm executes $r \leftarrow 35 \bmod 30 = 5$, $a \leftarrow b = 30$, and $b \leftarrow r = 5$. Also, after this iteration of the **while** loop, the **while** condition is still satisfied since $b = 5$. The Euclidean algorithm executes $r \leftarrow 30 \bmod 5 = 0$, $a \leftarrow b = 5$, and $b \leftarrow r = 0$. Now, the **while** condition is violated. So the **while** loop is no longer executed. Instead, the **return** instruction following **end while** is carried out. This means that the algorithm returns 5 which is gcd$(100, 35)$.

We model the run of an algorithm A on an input $a \in$ Input(A) as a sequence of *states*. A state describes the situation of the algorithm immediately before the execution of an instruction. It includes the contents of the memory cells corresponding to the variables in the algorithm and the instruction to be executed next. For example, consider the run of the Euclidean algorithm with input $(100, 35)$. Tables 1.1.5 and 1.1.6 show the first and final states of this run.

If the instruction in a state is not the **return** statement, then the algorithm performs this instruction. This may change the value of the variables. Then the algorithm enters the next state since in our model termination is only possible when the **return** instruction is carried out. This next state is uniquely determined by the previous state. So, the input $a \in$ Input(A) uniquely determines the execution of the algorithm and its return value, which is referred to by $A(a)$. This explains the name "deterministic

Table 1.1.5. Beginning of the run of the Euclidean algorithm with input $(100, 35)$.

State#	Memory contents			Next instruction
	a	b	r	
1	100	35		$a \leftarrow |a|$
2	100	35		$b \leftarrow |b|$
3	100	35		**while** $b \neq 0$ **do**
4	100	35		$r \leftarrow a \bmod b$
5	100	35	30	$a \leftarrow b$
6	35	35	30	$b \leftarrow r$
7	35	30	30	**end while**
8	35	30	30	**while** $b \neq 0$ **do**

Table 1.1.6. End of the run of the Euclidean algorithm with input $(100, 35)$.

State#	Memory contents			Next instruction
	a	b	r	
1	30	5	5	**while** $b \neq 0$ **do**
2	30	5	5	$r \leftarrow a \bmod b$
3	30	5	0	$a \leftarrow b$
4	5	5	0	$b \leftarrow r$
5	5	0	0	**end while**
6	5	0	0	**while** $b \neq 0$ **do**
7	5	0	0	**return** a

algorithm". For instance, consider State 3 in Table 1.1.5. The value of b is 35. So the **while** condition is satisfied. The execution of the **while** instruction does not change the values of a, b, or r and causes the next instruction to be $r \leftarrow a \bmod b$. So State 4 is uniquely determined by State 3.

Since we require deterministic algorithms to always terminate, the same state cannot occur repeatedly in an algorithm run. Otherwise, the algorithm would enter an infinite loop. In other words, the states in algorithm runs are pairwise different.

It is important to prove the *correctness of an algorithm*. This means that on input of any $a \in \text{Input}(A)$ the algorithm terminates and its output has the specified properties. In Example 1.1.18, we present the correctness proof of the Euclidean algorithm.

Example 1.1.18. We prove the correctness of the Euclidean algorithm. First, note that after b is replaced by its absolute value, the sequence of values of b is strictly decreasing since starting from the second b, any such value is the remainder of a division by the previous value of b. So at some point, we must have $b = 0$ which means that the algorithm terminates. Next, as Exercise 1.1.19 shows, the value of $\gcd(a, b)$ in line 4 is always the same. But when the algorithm terminates, we have $b = 0$ and therefore $\gcd(a, b) = \gcd(a, 0) = a$. The fact that $\gcd(a, b)$ does not change is called an *algorithm invariant*. Such invariants are frequently used in correctness proofs of algorithms.

Exercise 1.1.19. Show that in line 4 of the Euclidean algorithm, the value of $\gcd(a, b)$ is always the same.

As a further example, we present a deterministic factoring algorithm. Recall that a *composite integer* is an integer a that can be written as $a = bc$ where b, c are integers and b is a *proper divisor* of a; i.e., $a \bmod b = 0$ and $0 < |b| < |a|$. The goal of a factoring algorithm is to find a proper divisor of a given integer a if a is composite. Algorithm 1.1.21 is such an algorithm. It is based on the fact that any composite integer a has a proper divisor b with $1 < b \leq \sqrt{|a|}$. This is proved in Exercise 1.1.20. The input of the algorithm is an integer $a > 1$. It enumerates all integers b with $1 < b \leq \sqrt{a}$ and checks whether b is a divisor of a. If no divisor is found, then a is proved to be a prime number in which case the algorithm returns 0.

Exercise 1.1.20. Show that every composite integer a has a proper divisor b such that $1 < b \leq \sqrt{|a|}$.

Algorithm 1.1.21. A deterministic factoring algorithm

Input: $a \in \mathbb{Z}_{>1}$
Output: A proper divisor b of a if a is composite, or 0 if a is a prime number

1:   detFactor(a)
2:      **for** all $b = 2, \ldots, \lfloor \sqrt{a} \rfloor$ **do**
3:         **if** $a \bmod b = 0$ **then**
4:            **return** b
5:         **end if**
6:      **end for**
7:      **return** 0
8:   **end**

Exercise 1.1.22. Let $a = 35$. Determine the first three and the last three states of the run of Algorithm 1.1.21.

1.1.6. Decision algorithms. In classical complexity theory, *decision algorithms* play an important role. Such an algorithm decides whether a string $s \in \{0, 1\}^*$ belongs to a subset L of $\{0, 1\}$ which is called a *language*. The input of a decision algorithm is a string $\vec{s}$ in $\{0, 1\}^*$. The output is 0 or 1, where 1 means that the input $\vec{s}$ belongs to L and 0 means that $\vec{s}$ belongs to the complement of L in $\{0, 1\}^*$. We also say that the algorithm *decides the language L*.

Example 1.1.23. Algorithm 1.1.24 decides the language L that consists of all strings in $\vec{s} \in \{0, 1\}^*$ representing composite integers. It works like Algorithm 1.1.21 except that the output corresponding to a composite number is 1 and the output is 0 if the input string $\vec{s}$ is (), represents 0, 1, or a prime number.

There is a close connection between decision and more general algorithms. For example, as shown in Example 1.4.21, an algorithm that decides whether an integer has a proper divisor below a given bound can be transformed into an integer factoring algorithm with almost the same efficiency. This can be generalized to many algorithmic problems.

Algorithm 1.1.24. Compositeness decision algorithm

Input: $\vec{s} \in \{0, 1\}^*$
Output: 1 if stringToInt($\vec{s}$) is composite and 0 otherwise

1:   decideComp($\vec{s}$)
2:      $a \leftarrow$ stringToInt($\vec{s}$)
3:      **for** all $b = 2, \ldots, \lfloor \sqrt{a} \rfloor$ **do**
4:         **if** $a \bmod b = 0$ **then**
5:            **return** 1
6:         **end if**
7:      **end for**
8:      **return** 0
9:   **end**

1.1.7. Time and space complexity. Let A be an algorithm. Its efficiency depends on the time complexity and the memory requirements of A which we discuss in this section.

Definition 1.1.25. (1) The *running time* or *time complexity* of A for a particular input $a \in \text{Input}(A)$ is the sum of the time required for reading the input a which is $O(\text{size}(a))$ and the running times of the instructions executed during the algorithm run with input a.

(2) The *worst-case running time* or *worst-case time complexity* of A is the function

$$(1.1.13) \qquad\qquad \text{wTime}_A : \mathbb{N} \to \mathbb{R}_{\geq 0}$$

that sends a positive integer n which is the size of an input of A to the maximum running time of A over all inputs of size n. If n is not the size of an input of a, then we set $\text{wTime}_A(n) = 0$.

Next, we define the space complexity of A.

Definition 1.1.26. (1) The *space complexity* of A for a particular input a is the total amount of memory space that is used in the algorithm run with input a.

(2) The *worst-case space complexity* of A is the function

$$(1.1.14) \qquad\qquad \text{wSpace}_A : \mathbb{N} \to \mathbb{R}_{\geq 0}$$

that sends a positive integer n which is the size of an input of A to the maximum space complexity of A over all inputs of size n. If n is not the size of an input of a, then we set $\text{wSpace}_A(n) = 0$.

Using the Definitions 1.1.25 and 1.1.26, we define the asymptotic time and space complexity of deterministic algorithms.

Definition 1.1.27. Let $f : \mathbb{N} \to \mathbb{R}_{>0}$ be a function. We say that A has *asymptotic worst-case running time or space complexity* $O(f)$ if $\text{wTime}_A = O(f)$ or $\text{wSpace}_A = O(f)$, respectively. The words "asymptotic" and "worst-case" may also be omitted.

It is common to use special names for certain time and space complexities. Several of these names are listed in Table 1.1.7.

Table 1.1.7. Asymptotic time and space complexity names.

Name	Time or space complexity
constant	$O(1)$
logarithmic	$O(\log n)$
linear	$O(n)$
quasilinear	$O(n(\log n)^c)$ for some $c \in \mathbb{N}$
quadratic	$O(n^2)$
cubic	$O(n^3)$
polynomial	$O(n^c)$ for some $c \in \mathbb{N}$
subexponential	$O(2^{n^\varepsilon})$ for all $\varepsilon \in \mathbb{R}_{>0}$
exponential	$O(2^{n^c})$ for some $c \in \mathbb{N}$

Exercise 1.1.28. Show that quasilinear complexity can also be written as $n^{1+o(1)}$, polynomial complexity as $n^{O(1)}$ or $2^{O(\log n)}$, subexponential complexity as $2^{o(n)}$, and exponential complexity as $2^{n^{O(1)}}$.

Example 1.1.29. We analyze the time and space complexity of the Euclidean Algorithm 1.1.16. Let $a, b \in \mathbb{Z}$ be the input of the algorithm, and let n be the maximum of $\mathrm{size}(a)$ and $\mathrm{size}(b)$. The time to read the input a, b is $O(n)$. After the operations in lines 2 and 3 we have $a, b \geq 0$. The time and space complexity of these instructions is $O(n)$. If $b = 0$, then the **while** loop is not executed and a is returned, which takes time $O(n)$. If $b \neq 0$ and $a \leq b$, then after the first iteration of the **while** loop, we have $b < a$. It follows from Exercise 1.1.30 that the total number of executions of the **while** loop is $O(n)$. Also, by this exercise, the size of the operands used in the executions of the **while** loop is $O(n)$. So, the running time of each iteration is $O(n^2)$ and the space requirement is $O(n)$. This shows that the worst-case running time of the Euclidean algorithm is $O(n^3)$ and the worst-case space complexity is $O(n)$. Thus, the Euclidean algorithm has cubic running time. Using more complicated arguments, it can even be shown that this algorithm has quadratic running time (see Theorem 1.10.5 in [**Buc04**]).

What is the practical relevance of worst-case running times when comparing algorithms? Let us take two algorithms, A and A', both designed to solve the same problem, such as computing the greatest common divisors. It is essential to note that if algorithm A has a smaller asymptotic running time than algorithm A', it does not automatically make A superior to A' in practice. This comparison only indicates that A is faster than A' for inputs greater than a certain length. However, this input length can be so large that it becomes irrelevant for most real-world use cases.

For example, in [**AHU74**] it is shown that for any integer multiplication algorithm with a worst-case time complexity of $M(n)$, there exists a gcd algorithm with a worst-case time complexity of $O(M(n) \log(n))$. Additionally, [**HvdH21**] presents an integer multiplication algorithm with a worst-case running time of $O(n \log n)$. As a result, there is a corresponding gcd algorithm with a worst-case running time of $O(n \log^2 n)$. However, this improved complexity only outperforms the $O(n^2)$ algorithm for very large integers, which may not occur in most common input sizes.

Exercise 1.1.30. Let $a, b \in \mathbb{N}$, $a > b$, be the input of the Euclidean algorithm. Let $r_0 = a$ and $r_1 = b$. Denote by k the number of iterations of the **while** loop executed in the algorithm and denote by $r_2, r_3, \ldots, r_{k+1}$ the sequence of remainders r which are computed in line 5 of the Euclidean algorithm. Prove that the sequence $(r_i)_{0 \leq i \leq k+1}$ is strictly decreasing and that $r_{i+2} < r_i/2$ for all $i \in \mathbb{Z}_k$. Conclude that $k = O(\mathrm{size}\, a)$.

Example 1.1.31. We determine the worst-case time and space complexity of the Deterministic Factoring Algorithm 1.1.21. Let $n = \mathrm{bitLength}\, a$. The number of iterations of the **for** loop in this algorithm is $O(2^{n/2})$. Each iteration of the **for** loop requires time $O(n^2)$ and space $O(n)$. Hence, the worst-case time complexity of Algorithm 1.1.21 is $O(n^2 2^{n/2}) = 2^{O(n)}$ and the worst-case space complexity is $O(n)$. So, the algorithm has exponential running time and linear space complexity.

1.2. Probabilistic algorithms

Quantum algorithms are inherently probabilistic; i.e., the output for a given input is not uniquely determined but follows a probability distribution on the possible outputs. This section discusses classical probabilistic algorithms. In many cases, they are much more efficient than their deterministic counterparts. An example is the probabilistic solution of the Deutsch-Jozsa problem in Section 5.3.1.

1.2.1. Definition of probabilistic algorithms. A probabilistic algorithm has the same four components as a deterministic algorithm: the **Input** and **Output** statement, the algorithm name followed by the sequence of input variables, and a sequence of instructions that is ended by **end**. In addition, states, runs, and complexities of probabilistic algorithms are defined analogously to their deterministic counterparts. The differences between the two types of algorithms are now listed. For this, let A be a probabilistic algorithm.

(1) The probabilistic algorithm A may call the subroutine coinToss. It returns 0 or 1, both with probability $\frac{1}{2}$.

(2) The probabilistic algorithm A may call other probabilistic algorithms subroutines if they satisfy the following condition. Given a permitted input, they terminate and return one of finitely many possible outputs according to a probability distribution.

(3) The run of A on input of some $a \in \text{Input}(A)$ may depend on a and the return values of the probabilistic subroutines called during the run of the algorithm. Therefore, in contrast to deterministic algorithms, this run may not be uniquely determined by a.

(4) A may not terminate, since termination may depend on certain return values of some probabilistic subroutine that may never occur.

(5) Let $a \in \text{Input}(A)$ and suppose that A terminates on input of a with output o. Then o may not be uniquely determined by a, but it may also depend on the return values of the probabilistic subroutine calls during the run of A. Also, we may have $o \in \text{Output}(A, a)$, $o =$ "Failure", which indicates that the algorithm did not find a correct output or o has neither of these properties.

(6) Due to the special meaning of the return value "Failure", it must never be a correct output.

So an important difference between deterministic and probabilistic algorithms is that the latter are not required to always return correct outputs. As we shall see, correct outputs occur with a certain *success probability*. We will show in Proposition 1.3.5 that the second condition in (2) follows from the property that the probabilistic subroutine always terminates.

We present two examples of probabilistic algorithms that may be used as subroutines in probabilistic algorithms.

Example 1.2.1. On input of $k \in \mathbb{N}$, Algorithm 1.2.2 returns a random bit string of length k with a uniform distribution. Also, on the same input, Algorithm 1.2.3 returns a random integer a with bit length at most k with the uniform distribution.

Algorithm 1.2.2. Selecting a uniformly distributed random bit string of fixed length

Input: $k \in \mathbb{N}$
Output: $s \in \{0, 1\}^k$
1:    randomString(k)
2:      **for** $i = 0$ to $k - 1$ **do**
3:        $s_i \leftarrow$ coinToss
4:      **end for**
5:      **return** $\vec{s} = (s_0, \ldots, s_{k-1})$
6: **end**

Algorithm 1.2.3. Selecting a uniformly distributed random positive integer of bounded bitlength

Input: $k \in \mathbb{N}$
Output: $b \in \mathbb{N}_0$ with bitLength$(b) \leq k$
1:    randomInt(k)
2:      $\vec{s} \leftarrow$ randomString(k)
3:      $b \leftarrow$ stringToInt($\vec{s}$)
4:      **return** b
5: **end**

We also use the following terminology. If a probabilistic algorithm A returns an output from Output(A, a) for a specific input $a \in$ Input(A), we refer to the corresponding algorithm run as a "success". Otherwise, it is considered a "failure". Notably, if the output of A on input a is "Failure", it indicates that the algorithm did not find a correct output. However, if the output is not "Failure", it is not immediately evident whether the result is correct or not, that is, whether the algorithm run was a success or not. This must be checked by other means.

A probabilistic algorithm that, upon termination, always returns a correct result or "Failure" is called "error-free".

Example 1.2.4. We present an algorithm that implements the *Fermat test* to determine whether a positive integer is composite. Given an input $a \in \mathbb{N}_{>1}$, the algorithm randomly selects $b \in \mathbb{Z}_a$ with the uniform distribution. If the condition

$$(1.2.1) \qquad 1 < \gcd(a, b) < a \vee b^{a-1} \not\equiv 1 \bmod a$$

holds, the algorithm returns 1; otherwise, it returns 0. Fermat's Little Theorem (see [**Buc04**, Theorem 2.11.1]) guarantees that condition (1.2.1) implies the compositeness of a. However, it is essential to note that the converse may not be true, since a could be a *Carmichael number*. Carmichael numbers are composite numbers that satisfy $b^{a-1} \equiv 1 \bmod a$ for all $b \in \mathbb{Z}_a^*$. For example, 561, 1105, and 1729 are the first three Carmichael numbers. Moreover, as shown in [**AGP94**], there are infinitely many Carmichael numbers. Since Carmichael numbers are composite, the Fermat test will return 0 for these inputs, making the algorithm non-error-free.

Exercise 1.2.5. (1) Write pseudocode for the Fermat test described in Example 1.2.4.

(2) Find a composite number a such that on input of a the algorithm of Example 1.2.4 sometimes returns 0 and sometimes 1.

There are the following two types of probabilistic algorithms.

(1) *Monte Carlo algorithms.* They always terminate but may not always be successful.

(2) *Las Vegas algorithms.* They may not terminate but when they terminate, they are successful.

Algorithms 1.2.2 and 1.2.3 are examples of Monte Carlo algorithms. We will see in Proposition 1.3.5 that Monte Carlo algorithms are exactly the possible subroutines of probabilistic algorithms; i.e., they terminate on every permitted input and return one of finitely many possible outputs according to a probability distribution. Another example of a Monte Carlo algorithm is the following.

Example 1.2.6. Algorithm 1.2.8 is an error-free Monte Carlo factoring algorithm that is based on the fact that a composite integer $a \in \mathbb{N}$ has a proper divisor $b \in \mathbb{N}$ such that

(1.2.2) $\mathrm{bitLength}(b) \leq \mathrm{m}(a) = \lceil (\mathrm{bitLength}\, a)/2 \rceil.$

This is shown in Exercise 1.2.7. On input of $a \in \mathbb{Z}_{>1}$, Algorithm 1.2.8 computes the integer b represented by a uniformly distributed random bit string of length $\mathrm{m}(a)$. The algorithm returns b if this number is a proper divisor of a. Then the algorithm run was successful. Otherwise, it returns "Failure" which means that the algorithm did not find a proper divisor of a. The algorithm always terminates since it tests only a single b but it may not always be successful.

Exercise 1.2.7. Show that every composite number $a \in \mathbb{N}$ has a proper divisor with bitlength at most $\mathrm{m}(a)$. Also, show that $\mathrm{m}(a)$ can be computed in linear time.

Algorithm 1.2.8. Monte Carlo factoring algorithm

Input: $a \in \mathbb{N}_{>1}$
Output: A proper divisor $b \in \mathbb{N}$ of a
  1: mcFactor(a)
  2:     $b \leftarrow \mathrm{randomInt}(\mathrm{m}(a))$
  3:     **if** $1 < b < a \wedge a \bmod b = 0$ **then**
  4:         **return** b
  5:     **end if**
  6:     **return** "Failure"
  7: **end**

Example 1.2.9. Algorithm 1.2.10 is a Las Vegas factoring algorithm which calls mcFactor until a proper divisor of a is found. This may take forever. But if the algorithm terminates, then it is successful.

Algorithm 1.2.10. Las Vegas factoring algorithm

Input: $a \in \mathbb{N}_{>1}$
Output: A proper divisor $b \in \mathbb{N}$ of a
  1: lvFactor(a)
  2:    **repeat**
  3:       $b \leftarrow$ mcFactor(a)
  4:    **until** $b \neq$ "Failure"
  5:    **return** b
  6: **end**

The approach used in Algorithm 1.2.10 can be extended to create a more general version, allowing any error-free Monte Carlo algorithm A to be transformed into a Las Vegas algorithm. This transformation is achieved through Algorithm 1.2.11. When given an input $a \in \text{Input}(A)$, this algorithm repeatedly executes $A(a)$ until a successful outcome is obtained. As this algorithm is akin to performing a Bernoulli experiment, we refer to it as the *Bernoulli algorithm* associated with A.

Algorithm 1.2.11. Bernoulli algorithm associated with an error-free Monte Carlo algorithm A

Input: $a \in \text{Input}(A)$
Output: $b \in \text{Output}(A, a)$
  1: bernoulli$_A$(a)
  2:    $b \leftarrow$ "Failure"
  3:    **while** $b =$ "Failure" **do**
  4:       $b \leftarrow A(a)$
  5:    **end while**
  6:    **return** b
  7: **end**

On the other hand, every Las Vegas algorithm can indeed be transformed into an error-free Monte Carlo algorithm. This conversion entails monitoring the number of calls made to the probabilistic subroutines while the algorithm runs. The algorithm terminates if the Las Vegas algorithm produces a successful outcome or if the count of subroutine calls exceeds a predetermined threshold value, which may vary depending on the specific input of the algorithm. In the event of success, the algorithm returns the output of the Las Vegas algorithm. However, if the threshold is surpassed, it returns the result "Failure."

Exercise 1.2.12. Change Algorithm 1.2.10 to an error-free Monte Carlo algorithm that, on input of $a \in \mathbb{Z}_{>1}$, performs at most bitLength(a) coin tosses.

Exercise 1.2.13. Change Algorithm 1.2.10 to an error-free Monte Carlo algorithm that, on input of $a \in \mathbb{Z}_{>1}$, performs at most bitLength(a) coin tosses.

1.2.2. Probabilistic decision algorithms. We now introduce *probabilistic decision algorithms*. As deterministic decision algorithms, their purpose is to decide the membership of $\vec{s} \in \{0,1\}^*$ in a language $L \subset \{0,1\}^*$. Such an algorithm always returns 1 or 0. It satisfies $\text{Output}(A,\vec{s}) = \{1\}$ for all $\vec{s} \in L$ and $\text{Output}(A,\vec{s}) = \{0\}$ for all $\vec{s} \in \{0,1\}^* \setminus L$. However, recall that runs of probabilistic decision algorithms do not have to be successful. So, the algorithm may return 0 if $\vec{s} \in L$ and 1 if $\vec{s} \in \{0,1\}^* \setminus L$.

There are three different types of probabilistic decision algorithms. To define them, let A be a probabilistic algorithm that decides a language L.

(1) A is called *true-biased* if it never returns *false positives*. So, if on input of $\vec{s} \in \{0,1\}^*$ the algorithm returns 1, then $\vec{s} \in L$.

(2) A is called *false-biased* if it never returns *false negatives*. So, if at the input of $\vec{s} \in \{0,1\}^*$ the algorithm returns 0, then $\vec{s} \notin L$.

(3) If A is true-biased or false-biased, then it is also called an *algorithm with one-sided error*.

(4) A is called an *algorithm with two-sided error* if it can return false positives and false negatives.

Note that a false-biased algorithm can always be transformed into a true-biased algorithm. We only need to replace the language to be decided by its complement in $\{0,1\}^*$ and change the outputs 0 and 1.

Example 1.2.14. Consider Algorithm 1.2.15 that decides whether or not the integer that corresponds to a string in $\{0,1\}^*$ is composite or not. On the input of $\vec{s} \in \{0,1\}^*$, the algorithm computes the corresponding integer a and calls mcFactor. If this subroutine returns a proper divisor of a, then the algorithm returns 1. Otherwise, it returns 0. This is a true-biased Monte Carlo decision algorithm. If it returns 1, then $\vec{s}$ represents a composite integer. But if it returns 0, then the integer represented by $\vec{s}$ may or may not be composite.

Algorithm 1.2.15. True-biased Monte Carlo compositeness decision algorithm

Input: $\vec{s} \in \{0,1\}^*$
Output: 1 if $\text{stringToInt}(\vec{s})$ is composite and 0 otherwise
  1: mcComposite$(\vec{s})$
  2:     $a \leftarrow \text{stringToInt}(\vec{s})$
  3:     $b \leftarrow \text{mcFactor}(a)$
  4:     **if** $b \in \mathbb{N}$ **then**
  5:        **return** 1
  6:     **else**
  7:        **return** 0
  8:     **end if**
  9: **end**

Example 1.2.16. Algorithm 1.2.17 is a somewhat artificial example of a Monte Carlo decision algorithm with two-sided error. On input of $\vec{s} \in \{0,1\}^*$ it computes $a = \text{stringToInt}(\vec{s})$. Then it tosses a coin, calls mcFactor, and returns 1 if the coin toss gives 1 or mcFactor(a) returns a proper divisor of a. Otherwise, it returns 0. The algorithm returns a false negative answer if a is composite and coinToss and mcFactor both return 0. Also, it returns a false positive answer if a is a prime number and coinToss gives 1.

Algorithm 1.2.17. Monte Carlo compositeness decision algorithm with two-sided error

---

Input: $\vec{s} \in \{0,1\}^*$
Output: 1 if stringToInt$(\vec{s})$ is composite and 0 otherwise
  1: mcComposite2(a)
  2:    $a \leftarrow$ stringToInt$(\vec{s})$
  3:    $c \leftarrow$ coinToss
  4:    $b \leftarrow$ mcFactor(a)
  5:    **if** $c = 1 \vee b \in \mathbb{N}$ **then**
  6:       **return** 1
  7:    **else**
  8:       **return** 0
  9:    **end if**
 10: **end**

1.3. Analysis of probabilistic algorithms

This section discusses the analysis of the time complexity and success probability of probabilistic algorithms.

1.3.1. A discrete probability space. Our first goal is to define a discrete probability space that is the basis of the analyses. In this section, A denotes a probabilistic algorithm. We first introduce some notation.

Consider a run R of A with input $a \in \text{Input}(A)$ and let $l \in \mathbb{N}_0 \cup \{\infty\}$ be the number of probabilistic subroutine calls in R. For instance, in Algorithm 1.2.2 we have Input$(A) = \mathbb{N}$ and for $a \in \mathbb{N}$ it holds that $l = a$. In contrast, in Algorithm 1.2.10, the number l of probabilistic subroutine calls may be infinite.

For all $k \in \mathbb{N}$, $k \leq l$, let a_k be the input of the kth probabilistic subroutine call in R if this subroutine requires an input, let r_k be its output, and let p_k be the probability that on input of a_k the output r_k occurs. These quantities are well-defined since we require that probabilistic algorithms may only use probabilistic subroutines that on any input terminate and return one of finitely many possible outputs according to some probability distribution. For example, for the probabilistic subroutine coinToss there is no input, the output is 0 or 1, and the probability of both outputs is $\frac{1}{2}$. We call $\vec{r} = (r_k)_{k \leq l}$ the *random sequence of the run R*. So the random sequence of a run of randomInt with input $a \in \mathbb{N}$ is in $\{0,1\}^a$.

We denote the set of all random sequences of runs of A with input a by Rand(A, a) and the set of finite strings in Rand(A, a) by FRand(A, a). So for $A =$ randomInt and $a \in \mathbb{N}$ we have Rand$(A, a) =$ FRand$(A, a) = \{0,1\}^a$. We note that for any $a \in \text{Input}(A)$, each $\vec{r} \in \text{Rand}(A, a)$ is the random sequence of exactly one run of A. We call it the *run of A corresponding to* $\vec{r}$. This run terminates if and only if $\vec{r} \in \text{FRand}(A, a)$ in which case we write $A(a, \vec{r})$ for the return value of this run.

Finally, let $k \in \mathbb{N}_0$, $k \leq l$, and let $\vec{r} = (r_0, \ldots, r_{k-1})$ be a prefix of a random sequence of a run of A with input a. Also, for $0 \leq i < k$ denote by p_i the probability for the return value r_i to occur. Then we set

$$(1.3.1) \qquad\qquad \text{Pr}_{A,a}(\vec{r}) = \prod_{i=0}^{k-1} p_i.$$

This is the probability that $\vec{r}$ occurs as the prefix of the random sequence of a run of A with input a. For instance, if $A = \text{randomInt}$ and $a \in \mathbb{N}$, then for all $k \in \mathbb{N}_0$ with $k \leq a$ and all $\vec{r} \in \{0, 1\}^k$, we have $\text{Pr}_{A,a}(\vec{r}) = \frac{1}{2^k}$.

Exercise 1.3.1. Determine $\text{Rand}(A, a)$, $\text{FRand}(A, a)$, and $\text{Pr}_{A,a}$ for $A = \text{lvFactor}$ specified in Algorithm 1.2.10 and $a \in \text{Input}(A) = \mathbb{N}_{>1}$.

The next lemma allows the definition of the probability distribution that we are looking for.

Lemma 1.3.2. *Let $a \in \text{Input}(A)$. The (possibly infinite) sum*

$$(1.3.2) \qquad\qquad \sum_{\vec{r} \in \text{FRand}(A,a)} \text{Pr}(\vec{r})$$

converges, its limit is in the interval $[0, 1]$, and it is independent of the ordering of the terms in the sum.

Proof. First, note the following. If the sum in (1.3.2) is convergent, then it is absolute convergent since the terms in the sum are nonnegative. So Theorem C.1.4 implies that its limit is independent of the ordering of the terms in the sum.

To prove the convergence of the sum, set

$$(1.3.3) \qquad\qquad t_k = \sum_{\vec{r} \in \text{FRand}, |\vec{r}| \leq k} \text{Pr}_{A,a}(\vec{r})$$

for all $k \in \mathbb{N}_0$. Then the sum in (1.3.2) is convergent if and only if the sequence (t_k) converges. For $k \in \mathbb{N}_0$ let Rand_k be the set of all prefixes of length at most k of sequences in $\text{Rand}(A, a)$. We will prove below that

$$(1.3.4) \qquad\qquad \sum_{\vec{r} \in \text{Rand}_k} \text{Pr}_{A,a}(\vec{r}) = 1$$

for all $k \in \mathbb{N}_0$. This implies

$$(1.3.5) \qquad\qquad t_k \leq \sum_{\vec{r} \in \text{Rand}_k} \text{Pr}_{A,a}(\vec{r}) = 1$$

for all $k \in \mathbb{N}_0$. Since the elements of the sequence (t_k) are nondecreasing this proves the convergence of (t_k) and thus of the infinite sum (1.3.2).

We will now prove (1.3.4) by induction on k. Since Rand_0 only contains the empty sequence, (1.3.4) holds for $k = 0$. For the inductive step, assume that $k \in \mathbb{N}_0$ and that (1.3.4) holds for k. Denote by Rand'_k the set of all sequences of length at most k in $\text{Rand}(A, a)$ and denote by Rand''_k the set of sequences of length k that are proper prefixes of strings in $\text{Rand}(A, a)$. For $\vec{r} \in \text{Rand}_k''$ let $m(\vec{r})$ be the number of possible outputs of the $(k + 1)$st call of a probabilistic subroutine when the sequence of return values of the previous calls was $\vec{r}$, let $r_i(\vec{r})$ be the ith of these outputs, and let $p_i(\vec{r})$ be its probability. These quantities exist by the definition of probabilistic algorithms. Then we have

(1.3.6) $\text{Rand}_{k+1} = \text{Rand}_k' \cup \{\vec{r} \| r_i(\vec{r}) : \vec{r} \in \text{Rand}_k'' \text{ and } 1 \le i \le m(\vec{r})\}.$

Also, we have

(1.3.7)
$$\sum_{i=1}^{m(\vec{r})} p_i(\vec{r}) = 1$$

for all $\vec{r} \in \text{Rand}_k''$. This implies

(1.3.8)
$$\begin{aligned}
\sum_{\vec{r} \in \text{Rand}_{k+1}} \text{Pr}_{A,a}(\vec{r}) &= \sum_{\vec{r} \in \text{Rand}_k'} \text{Pr}_{A,a}(\vec{r}) + \sum_{\vec{r} \in \text{Rand}_k''} \sum_{i=1}^{m(\vec{r})} \text{Pr}_{A,a}(\vec{r} \| s_i(\vec{r})) \\
&= \sum_{\vec{r} \in \text{Rand}_k'} \text{Pr}_{A,a}(\vec{r}) + \sum_{\vec{r} \in \text{Rand}_k''} \text{Pr}_{A,a}(\vec{r}) \sum_{i=1}^{m(\vec{r})} p_i(\vec{r}) \\
&= \sum_{\vec{r} \in \text{Rand}_k'} \text{Pr}_{A,a}(\vec{r}) + \sum_{\vec{r} \in \text{Rand}_k''} \text{Pr}_{A,a}(\vec{r}) \\
&= \sum_{\vec{r} \in \text{Rand}_k} \text{Pr}_{A,a}(\vec{r}) = 1. \qquad \square
\end{aligned}$$

Lemma 1.3.2 allows the definition of the probability distribution that we are looking for. This is done in the following proposition.

Proposition 1.3.3. *For every $a \in \text{Input}(A)$ we define*

(1.3.9) $$\text{Pr}_{A,a}(\infty) = 1 - \sum_{\vec{r} \in \text{FRand}(A,a)} \text{Pr}_{A,a}(\vec{r}).$$

Then $(\text{FRand}(A, a) \cup \{\infty\}, \text{Pr}_{A,a})$ *is a discrete probability space. Also, if* $\text{Pr}_{A,a}(\infty) = 0$, *then* $(\text{FRand}(A, a), \text{Pr}_{A,a})$ *is a discrete probability space.*

Exercise 1.3.4. Prove Proposition 1.3.3.

For $a \in \text{Input}(a)$ and $\vec{r} \in \text{FRand}(A, a)$, the value $\text{Pr}_{A,a}(\vec{r})$ is the probability that $\vec{r}$ is the random sequence of a run of A with input a. Also, $\text{Pr}_{A,a}(\infty)$ is the probability that on input of a, the algorithm A does not terminate.

An important type of algorithms A that satisfy $\text{Pr}_{A,a}(\infty) = 0$ for all $a \in \text{Input}(a)$ is Monte Carlo algorithms. We now show that they are exactly the probabilistic algorithms that, according to the specification in Section 1.2.1, can be called by probabilistic algorithms as subroutines.

Proposition 1.3.5. *Let A be a Monte Carlo algorithm and let $a \in$ Input(A). Then the following hold.*

(1) *The running time of A on input of a is bounded by some $k \in \mathbb{N}$ that may depend on a.*

(2) *On input of a, algorithm A returns one of finitely many possible outputs according to a probability distribution.*

Proof. We first show that the length of all $\vec{r} \in$ FRand(A, a) is bounded by some $k \in \mathbb{N}$. This shows that there are only finitely many possible runs of A on input of a which implies the first assertion.

Assume that no such upper bound exists. We inductively construct prefixes $\vec{r}_k = (r_0, \ldots, r_k)$, $k \in \mathbb{N}_0$, of an infinite sequence $\vec{r} = (r_0, r_1, \ldots)$ that are also prefixes of arbitrarily long strings in Rand(A, a); that is, for all $k \in \mathbb{N}_0$ and $l \in \mathbb{N}$ the sequence $\vec{r}_k$ is a prefix of a sequence in Rand(A, a) of length at least l. Then $\vec{r}$ is an infinite sequence in Rand(A, a) that contradicts the assumption that A is a Monte Carlo algorithm.

For the base case, we set $\vec{r}_0 = ()$. This is a prefix of all strings in Rand(A, a) that, by our assumption, may be arbitrarily long. For the inductive step, assume that $k \in \mathbb{N}$ and that we have constructed $\vec{r}_{k-1} = (r_0, \ldots, r_{k-1})$. By the definition of probabilistic algorithms, there are finitely many possibilities to select r_k in such a way that the sequence $(r_0, \ldots, r_k)$ is the prefix of a string in Rand(A, a). For at least one of these choices, this sequence is a prefix of arbitrarily long strings in Rand(A, a) because, by the induction hypothesis, $\vec{r}_{k-1}$ has this property. We select such an r_k and this concludes the inductive construction and the proof of the first assertion.

Together with Proposition 1.3.3, the first assertion of the proposition implies the second one. $\square$

1.3.2. The success probability of Monte Carlo algorithms. In addition to the running time, the probability of success is also crucial for the efficiency of a Monte Carlo algorithm. We now define this probability.

Definition 1.3.6. Let A be a Monte Carlo algorithm, let $a \in$ Input(A), and denote by Rand$_{\text{succ}}(A, a)$ the set of all $\vec{r} \in$ Rand(A, a) such that the run of A associated with $\vec{r}$ is successful; i.e., $A(a, \vec{r}) \in$ Output(A, a). Then we set

$$(1.3.10) \qquad p_A(a) = \sum_{\vec{r} \in \text{Rand}_{\text{succ}}(A,a)} \text{Pr}_{A,a}(\vec{r})$$

and call this quantity the *success probability of A on input of a*. Also, the value

$$(1.3.11) \qquad q_A(a) = 1 - p_A(a)$$

is called the *failure probability of A on input of a*.

Exercise 1.3.7. Prove that for all $a \in$ Input(A), the sum in (1.3.10) is convergent and its limit is independent of the ordering of the terms in the sum.

Example 1.3.8. Let $A =$ mcFactor specified in Algorithm 1.2.8 and let $a \in$ Input$(A) = \mathbb{N}_{>1}$. Then Rand$_{\text{succ}}(A, a)$ is the set of all sequences (b) where b is a proper divisor of

a of bitlength at most $m(a)$. By Exercise 1.2.7, this set is not empty. Therefore, the success probability $p_A(a)$ of A on input of a is at least $1/2^{m(a)}$.

We can use the definition of the success probability to show that Bernoulli algorithms terminate with probability 1.

Proposition 1.3.9. *Let A be a Bernoulli algorithm. Then we have* $\mathrm{Pr}_{A,a}(\infty) = 0$ *for all* $a \in \mathrm{Input}(A)$.

Proof. Denote by A' the error-free Monte Carlo algorithm used in A. Let $a \in \mathrm{Input}(A')$. Then $\mathrm{FRand}(A, a)$ consists of all strings $\vec{r} = \vec{r}_1 || \cdots || \vec{r}_k$ where $k \in \mathbb{N}$, $\vec{r}_i \in \mathrm{Rand}(A', a)$ for $1 \le i \le k$, $A'(a, \vec{r}_i) = $ "Failure" for $1 \le i < k$, and $A'(a, \vec{r}_k) \ne $ "Failure". So we obtain

$$(1.3.12) \qquad \sum_{\vec{r} \in \mathrm{FRand}(A,a)} \mathrm{Pr}_{A,a}(s) = p_{A'}(a) \sum_{k=0}^{\infty} (1 - p_{A'}(a))^k = \frac{p_{A'}(a)}{p_{A'}(a)} = 1.$$

This implies the assertion. $\qquad\qquad\qquad\qquad\qquad\qquad\qquad\qquad\qquad\qquad\square$

1.3.3. Expected running time. The probability space defined in Proposition 1.3.3 also allows the definition of the expected running time of random algorithms.

Definition 1.3.10. Let A be a probabilistic algorithm and let $a \in \mathrm{Input}(A)$ such that $\mathrm{Pr}_{A,a}(\infty) = 0$. Then the *expected running time of A on input of a* is defined as the expectation of the random variable $\mathrm{time}_{A,a}$ that sends $\vec{r} \in \mathrm{FRand}(A, a)$ to the running time of the algorithm run associated with $\vec{r}$. It is denoted by $\mathrm{eTime}_A(a)$. So we have

$$(1.3.13) \qquad \mathrm{eTime}_A(a) = \sum_{\vec{r} \in \mathrm{FRand}(A,a)} \mathrm{time}_{A,a}(\vec{r}) \mathrm{Pr}_{A,a}(\vec{r}).$$

Example 1.3.11. Let $A = \mathrm{mcFactor}$ which is specified in Algorithm 1.2.8 and let $a \in \mathrm{Input}(A)$. Then $\mathrm{FRand}(A, a)$ is the set of all one-element sequences (b), where b is an integer that can be represented by a bit string of length $m(a)$. So $|\mathrm{FRand}(A, a)| \le 2^{m(a)}$. Also, by Proposition 1.3.5 we have $\mathrm{Pr}_{A,a}(\infty) = 0$. So $\mathrm{eTime}_A(a)$ is defined. Since each run of A on input a has running time $\mathrm{O}(\mathrm{size}^2 a)$, we have

$$(1.3.14) \qquad \mathrm{eTime}_A(a) = \mathrm{O}\left(\mathrm{size}^2 a \sum_{\vec{r} \in \mathrm{FRand}(A,a)} \frac{1}{2^{m(a)}} \right) = \mathrm{O}(\mathrm{size}^2 a).$$

So the expected running time is quadratic. However, the success probability of A on input of a may be as small as $1/2^{m(a)}$. As we will see in Section 1.3.4, this success probability can be amplified by repeatedly calling A. But in order to obtain success probability $\ge \frac{2}{3}$, exponential expected running time is required.

The next proposition determines the expected running time of Bernoulli algorithms.

Proposition 1.3.12. *Let A be an error-free Monte Carlo algorithm, let $a \in \mathrm{Input}(A)$, and let t be an upper bound on the running time of A with input of a. Then the expected running time of* $\mathrm{bernoulli}_A(a)$ *specified in Algorithm 1.2.11 is* $\mathrm{O}(t/p_A(a))$.

Proof. We use the fact that for all $c \in \mathbb{R}$ with $0 \leq c < 1$ we have

$$(1.3.15) \qquad \sum_{i=0}^{\infty} kc^k = \frac{c}{(1-c)^2}.$$

So the expected number of calls of A until $\text{bernoulli}_A(a)$ is successful is

$$(1.3.16) \qquad p_A(a) \sum_{k=0}^{\infty} kq_A(a)^k = \frac{p_A(a)}{p_A(a)^2} = \frac{1}{p_A(a)}.$$

The statement about the expected running time is an immediate consequence of this result. $\qquad\qquad\square$

Example 1.3.13. Proposition 1.3.12 allows the analysis of lvFactor specified in Algorithm 1.2.10. Let $n \in \mathbb{N}$ and let $a \in \mathbb{N}_{>1}$ be an input of size n. It follows from Example 1.3.8 that the success probability of mcFactor(a) is at least $1/2^{m(a)} \geq 1/2^{n/2+1}$. Also, the worst-case running time of mcFactor(a) is $O(n^2)$. It therefore follows from Proposition 1.3.12 that the expected running time of mcFactor(a) is $O(n^2 2^{n/2})$. So the expected running time is exponential which shows that this probabilistic algorithm has no advantage over the deterministic Algorithm 1.1.21.

1.3.4. Amplifying success probabilities. In Example 1.3.8, we have seen that the success probability of mcFactor specified in Algorithm 1.2.8 with input $a \in \mathbb{N}_{>1}$ is at least $1/2^{m(a)}$. We now explain how this success probability and the success probability of every error-free Monte Carlo algorithm A can be amplified by repeatedly calling it with the same input. Algorithm 1.3.14 implements this idea.

Algorithm 1.3.14. Repeated application of an error-free Monte Carlo algorithm

---

Input: $a \in \text{Input}(A)$ for an error-free Monte Carlo algorithm A used as a subroutine, $k \in \mathbb{N}$
Output: $b \in \text{Output}(A, a)$
1: $\text{repeat}_A(a, k)$
2:      **for** $i = 1$ to k **do**
3:          $b \leftarrow A(a)$
4:          **if** $b \neq$ "Failure" **then**
5:              **return** b
6:          **end if**
7:      **end for**
8:      **return** "Failure"
9: **end**

Definition 1.3.15. Let $a \in \text{Input}(A)$. We denote the success probability of $\text{repeat}_A(a, k)$ by $p_A(a, k)$ and the failure probability of this call by $q_A(a, k) = 1 - p_A(a, k)$.

The next proposition shows that $q_A(a, k)$ decreases exponentially in k.

Proposition 1.3.16. *Let $k \in \mathbb{N}$, and let $a \in \text{Input}(A)$ with $p_A(a) < 1$. Then we have the following:*

$$(1.3.17) \qquad e^{-kp_A(a)/q_A(a)} \leq q_A(a, k) \leq e^{-kp_A(a)}.$$

Proof. Write $p = p_A(a)$ and $q = q_A(a) = 1 - p$. Then we have

(1.3.18)
$$q_A(a, k) = q^k.$$

Since we assume that $q > 0$, it follows from [**Abr72**, (4.1.33) and (4.1.36)] that

(1.3.19)
$$1 - \frac{1}{q} \leq \log q \leq q - 1.$$

This implies

(1.3.20)
$$k \log q \leq k(q - 1) = -kp$$

and

(1.3.21)
$$k \log q \geq k\left(1 - \frac{1}{q}\right) = k\frac{q-1}{q} = -k\frac{p}{q}.$$

So (1.3.20) and (1.3.21) imply

(1.3.22)
$$e^{-kp/q} \leq q^k \leq e^{-kp}$$

as asserted. $\qquad\qquad\qquad\qquad\qquad\qquad\qquad\qquad\qquad\qquad\qquad\qquad\qquad\qquad\square$

The next corollary shows how to choose k in order to obtain a desired success probability. It also gives a lower bound for k that corresponds to a given success probability.

Corollary 1.3.17. *Let $a \in \mathrm{Input}(A)$ with $p_A(a) > 0$ and let $\varepsilon \in \mathbb{R}$ with $0 < \varepsilon \leq 1$.*

(1) *If $k \geq \log(1/\varepsilon)/p_A(a)$, then $p_A(a, k) \geq 1 - \varepsilon$.*

(2) *If $p_A(a, k) \geq 1 - \varepsilon$, then $k \geq \log(1/\varepsilon)q_A(a)/p_A(a)$.*

Exercise 1.3.18. Prove Corollary 1.3.17.

Example 1.3.19. Consider $A = \mathsf{mcFactor}$ as specified in Algorithm 1.2.8. In Example 1.3.8, we have seen that $p_A(a) \geq 1/2^{m(a)} > 0$ for all $a \in \mathbb{Z}_{>1}$. So repeat_A can be used to amplify this probability. For example, if we choose $\varepsilon = 1/3$ and $k \geq (\log 3)2^{m(a)} \geq \log(1/\varepsilon)/p_A(a)$, then Corollary 1.3.17 implies $p_A(a, k) \geq \frac{2}{3}$. Since

$$m(a) \geq \mathrm{bitLength}(a)/2,$$

this number k of calls to A is exponential in size a. Therefore, again, this algorithm does not give an asymptotic advantage over the deterministic Algorithm 1.1.21.

We can also amplify the success probability of decision algorithms with errors. Consider a true-biased decision algorithm A that decides a language L. We can modify this algorithm to make it an error-free Monte Carlo algorithm. For this, we set $\mathrm{Output}(A, \vec{s}) = \{1\}$ for all $\vec{s} \in L$, $\mathrm{Output}(A, \vec{s}) = \emptyset$ for all $\vec{s} \in \{0, 1\}^* \setminus L$ and we replace the return value 0 by "Failure". So, the success probability of A can be amplified using Algorithm 1.3.14. Analogously, the success probability of false-biased decision algorithms can be amplified.

Next, we consider a Monte Carlo decision algorithm A with two-sided error that decides a language L. Such an algorithm never gives certainty about whether an input $\vec{s} \in \{0, 1\}^*$ belongs to L or not. However, the probability of success can be increased by using a majority vote. To do this, we run the algorithm k times with input $\vec{s}$ for some $k \in \mathbb{N}$ and count the number of positive responses 1 and the number of negative answers 0 and return 1 or 0 depending on which answer has the majority. This is done in Algorithm 1.3.20.

Algorithm 1.3.20. Success probability amplifier for a Monte Carlo decision algorithm A with two-sided error

---

Input: $\vec{s} \in \{0, 1\}^*$, $k \in \mathbb{N}$
Output: 1 if $\vec{s} \in L$ and 0 if $\vec{s} \in \{0, 1\}^* \setminus L$ where L is the language decided by the Monte Carlo decision algorithm A that is used as a subroutine

```
 1:  majorityVote_A(s, k)
 2:     l = 0
 3:     for i = 1 to k do
 4:         l ← l + A(s)
 5:     end for
 6:     if l > k/2 then
 7:         return 1
 8:     else
 9:         return 0
10:     end if
11:  end
```

We will show that under certain conditions, Algorithm 1.3.20 amplifies the success probability of decision algorithms with two-sided error. For this, we need the following definition.

Definition 1.3.21. Assume that a Monte Carlo algorithm A decides a language L, let $\vec{s} \in L$, and let $b \in \{0, 1\}$. Then we write $\Pr(A(s) = b)$ for the probability that on input of $\vec{s}$ the algorithm A returns b.

Proposition 1.3.22. *Let A be a Monte Carlo algorithm that decides a language L, let $\vec{s} \in L$, and let $\varepsilon \in \mathbb{R}_{>0}$ such that $\Pr(A(\vec{s}) = 1) \geq \frac{1}{2} + \varepsilon$. Then for all $k \in \mathbb{N}$ we have*

$$(1.3.23) \qquad \Pr(\text{majorityVote}_A(\vec{s}, k) = 1) > 1 - e^{-2k\varepsilon^2}.$$

Proof. Let $k \in \mathbb{N}$. We prove the assertion by showing that

$$(1.3.24) \qquad q = \Pr(\text{majorityVote}_A(\vec{s}, k) = 0) < e^{-2k\varepsilon^2}.$$

Consider a run of majorityVote_A with input $\vec{s}, k$. Denote by $\vec{r} \in \{0, 1\}^k$ the random output sequence corresponding to this run. Then A returns 0 if and only if at most $k/2$ entries in $\vec{r}$ are 1. The probability for such an $\vec{r}$ to occur is at most

$$(1.3.25) \qquad \left(\frac{1}{2} - \varepsilon\right)^{\frac{k}{2}} \left(\frac{1}{2} + \varepsilon\right)^{\frac{k}{2}} = \frac{(1 - 4\varepsilon^2)^{\frac{k}{2}}}{2^k}.$$

Since the number of such sequences $\vec{r}$ is at most 2^k and because by [**Abr72**, (4.2.30)] we have $1 - x < e^{-x}$ for all $x > -1$, it follows that

$$(1.3.26) \qquad q \leq (1 - 4\varepsilon^2)^{\frac{k}{2}} < e^{-2k\varepsilon^2}.$$

This concludes the proof. $\square$

Example 1.3.23. Let $A = \mathsf{mcComposite2}$ specified in Algorithm 1.2.17 which is the Monte Carlo compositeness test with two-sided error and let $\vec{s} \in \{0,1\}^*$ such that $a = \mathsf{stringToInt}(\vec{s})$ is composite. The call $A(\vec{s})$ returns 1 if the first coin toss gives 1 which happens with probability $\frac{1}{2}$ or $\mathsf{mcFactor}(a)$ returns a proper divisor of $a = \mathsf{stringToInt}(\vec{s})$ which occurs with probability $\geq \frac{1}{2^m}$ where $m = \mathsf{m}(a)$. Therefore, we have

$$(1.3.27) \qquad \Pr(A(\vec{s}) = 1) \geq 1 - \frac{1}{2}\left(1 - \frac{1}{2}^m\right) = \frac{1}{2} + \frac{1}{2^{m+1}}.$$

So in Proposition 1.3.22 we can set $\varepsilon = \frac{1}{2^{m+1}}$ and obtain

$$(1.3.28) \qquad \Pr(\mathsf{majorityVote}_A(\vec{s}, k) = 1) \geq 1 - e^{-2k\varepsilon^2} = 1 - e^{-\frac{k}{2^{2m+1}}}$$

for all $k \in \mathbb{N}$.

Exercise 1.3.24. Use the result in Example 1.4.7 to determine k such that

$$\Pr(\mathsf{majorityVote}_A(\vec{s}, k) = 1) \geq \frac{2}{3}.$$

1.4. Complexity theory

Classical complexity theory allows us to assess the efficiency of algorithms and the difficulty of solving computational problems. In this section, we present important notions and results of this theory. They are required as a basis for quantum complexity theory.

1.4.1. Computational problems.
An important question in complexity theory is: How efficiently can a computational problem be solved? In cryptography, for example, it is crucial to know how quickly an encryption system can be broken. That is why we start by defining computational problems

Definition 1.4.1. A *computational problem* is a triplet $\mathrm{CP} = (I, O, R)$ where I is a subset of the Cartesian product of finitely many data types where I and O are subsets of Cartesian products of finitely many data types.

Also, $R \subset I \times O$ such that for all $a \in I$ there is $b \in O$ with $(a, b) \in R$.

Definition 1.4.2. Let $\mathrm{CP} = (I, O, R)$ be a computational problem.

(1) The elements of I are called the *instances* of CP.

(2) If $(a, b) \in R$, then b is called a *solution* of the problem instance a.

Example 1.4.3. By the *square root problem* we mean the triplet $(\mathbb{N}, \mathbb{Z}, R)$ where $R = \{(a, b) \in \mathbb{N} \times \mathbb{Z} : b^2 = a \text{ or } b = 0 \text{ if } a \text{ is not a square in } \mathbb{N}\}$. An instance of the square root problem is 4. It has the two solutions -2 and 2. Another instance is 2. It has the solution 0 that indicates that 2 is not a square in $\mathbb{N}$. We can also define this problem differently by only allowing problem instances that are squares.

We define what it means that an algorithm solves a computational problem.

Definition 1.4.4. Let $CP = (I, O, R)$ be a computational problem.

(1) We say that a deterministic algorithm A *solves* CP if $I \subset \text{Input}(A)$ and on input of a problem instance $a \in I$ the algorithm returns a solution of a.

(2) We say that a Monte Carlo algorithm A solves CP if $I \subset \text{Input}(A)$ and on input of $a \in I$ a successful run of A returns a solution of a.

(3) We say that a Las Vegas algorithm solves CP if $I \subset \text{Input}(A)$ and on input of $a \in I$ the algorithm either terminates and returns a solution of a or does not terminate.

Example 1.4.5. The *gcd problem* is the triplet $CP = (\mathbb{Z}^2, \mathbb{N}_0, R)$ where $R = \{(a, b, c) \in \mathbb{Z}^3 : c = \gcd(a, b)\}$. An instance of this problem is $(100, 35)$. The unique solution of this instance is 5. Also, the Euclidean algorithm solves this problem.

Exercise 1.4.6. Find a deterministic algorithm that solves the square root problem from Example 1.4.3 in polynomial time.

Example 1.4.7. By the *integer factorization problem* we mean the triplet $(C, \mathbb{N}, R)$ where C is the set of all positive composite integers and

$$R = \{(a, b) \in C \times \mathbb{N} : b \text{ is a proper divisor of } a\}.$$

Algorithms 1.1.21, 1.2.8, and 1.2.10 are deterministic, Monte Carlo, and Las Vegas algorithms that solve the integer factorization problem.

1.4.2. Complexity of computational problems. We define the complexity of computational problems.

Definition 1.4.8. Let CP be a computational problem and let $f : \mathbb{N} \to \mathbb{R}_{>0}$ be a function

(1) We say that CP *can be solved in (deterministic) time* $O(f)$ or *has time complexity* $O(f)$ if there is a deterministic algorithm that solves CP and has running time $O(f)$.

(2) We say that CP can be solved in *(deterministic) linear, quasilinear, quadratic, cubic, polynomial, subexponential,* or *exponential time* or has this time complexity if there is a deterministic algorithm that solves CP and has the respective time complexity.

(3) The corresponding space complexities are defined analogously.

Example 1.4.9. As seen in Example 1.1.29, the gcd problem from Example 1.4.5 can be solved in deterministic time $O(n^3)$. As noted in this example, the gcd problem can even be solved in deterministic time $O(n^2)$ or $O(n \log^2 n)$ and linear space. Thus this problem can be solved in polynomial time or, more precisely, cubic, quadratic, or even quasilinear time.

Example 1.4.10. As seen in Example 1.1.31, the integer factorization problem can be solved in deterministic exponential time and linear space.

Now we introduce the corresponding probabilistic complexity notions.

Definition 1.4.11. Let CP be a computational problem and let $f : \mathbb{N} \to \mathbb{R}_{>0}$ be a function.

(1) We say that CP *can be solved in probabilistic time* $O(f)$ if there is a Monte Carlo algorithm that solves CP and has running time $O(f)$ and success probability $\geq \frac{2}{3}$.

(2) We say that CP can be solved in *probabilistic linear, quasilinear, quadratic, cubic, polynomial, subexponential,* or *exponential time* if there is a Monte Carlo algorithm with the respective running time that solves CP and has success probability $\geq \frac{2}{3}$.

Exercise 1.4.12 shows that the value $\frac{2}{3}$ in Definition 1.4.11 may be replaced by any real number in $\left]\frac{1}{2}, 1\right]$ without changing the complexity of the computational problem.

Exercise 1.4.12. Let CP be a computational problem, let $f : \mathbb{N} \to \mathbb{R}_{>0}$ be a function, and let $p \in \left]\frac{1}{2}, 1\right]$. Use Proposition 1.3.22 to show that CP can be solved in probabilistic time $O(f)$ if and only if there is a Monte Carlo algorithm that solves CP in time $O(f)$ and has success probability $\geq p$.

Example 1.4.13. It follows from Example 1.3.19 that Algorithm 1.3.14 with subroutine $A = \mathsf{mcFactor}(a)$ and $k = \lceil (\log 3) 2^{\mathsf{m}(a)} \rceil$ is an error-free Monte Carlo algorithm that solves the integer factorization problem in probabilistic exponential time. We note that this problem can even be solved in probabilistic subexponential time (see [**LP92**], [**BLP93**]) but no classical polynomial time algorithm for this problem is known.

1.4.3. Complexity classes. In this section, we delve into the definition of complexity classes, which serve to group languages that satisfy specific complexity conditions. The foundation of this concept was laid in the early 1970s. Over the years, complexity theory has witnessed the introduction of numerous complexity classes, and extensive research has been conducted to study their interrelationships. For the scope of this discussion, we will focus on a select few complexity classes that hold relevance to our context.

We begin with the definition of the most basic complexity classes.

Definition 1.4.14. Let $f : \mathbb{N} \to \mathbb{R}_{>0}$ be a function.

(1) The *complexity class* DTIME(f) is the set of all languages L for which there is a deterministic algorithm that decides L and has time complexity $O(f)$.

(2) The *complexity class* DSPACE(f) is the set of all languages L for which there is a deterministic algorithm that decides L and has space complexity $O(f)$.

We also define the following more concrete complexity classes.

Definition 1.4.15. (1) The *complexity class* P is the set of all languages L for which there is a deterministic polynomial time algorithm which decides L.

(2) The *complexity class* PSPACE is the set of all languages L for which there is a deterministic polynomial space algorithm which decides L.

(3) The complexity class EXPTIME is the set of all languages L for which there is a deterministic exponential time algorithm which decides L.

Exercise 1.4.16. Consider the language L of all strings that correspond to squares in $\mathbb{N}$. Show that L is in P.

Example 1.4.17. As shown in 2002 by Manindra Agrawal, Neeraj Kayal, and Nitin Saxena [**AKS04**], the language L of all bit strings that correspond to composite integers is in P. Therefore, it can be decided in polynomial time whether a positive integer is a prime number composite. However, if the algorithm of Agrawal, Kayal, and Saxena finds that a positive integer is composite, it does not give a proper divisor of this number. Finding such a divisor appears to be a much harder problem (see Example 1.4.13).

We also define two probabilistic complexity classes.

Definition 1.4.18. (1) The *complexity class* PP is the set of all languages L for which there is a polynomial time Monte Carlo algorithm A which decides L and satisfies $\Pr(A(s) = 1) > \frac{1}{2}$ for all $s \in L$ and $\Pr(A(s) = 0) > \frac{1}{2}$ for all $s \in \{0, 1\}^* \setminus L$.

(2) The *complexity class* BPP is the set of all languages L for which there is a polynomial time Monte Carlo algorithm A which decides L and satisfies $\Pr(A(s) = 1) \geq \frac{2}{3}$ for all $s \in L$ and $\Pr(A(s) = 0) \geq \frac{2}{3}$ for all $s \in \{0, 1\}^* \setminus L$.

We note that the constant $\frac{2}{3}$ in Definition 1.4.18 can be replaced by any other constant p with $\frac{1}{2} < p \leq 11$. This flexibility is established by Proposition 1.3.22, which asserts that when commencing with a success probability exceeding 1/2, Algorithm 1.3.20 can be employed to obtain a success probability arbitrarily close to 1.

Finally, we introduce the complexity class NP. We begin with a motivating example.

Example 1.4.19. In 1742, the German mathematician Christian Goldbach wrote a letter to Leonard Euler in which he proposed the conjecture that every even integer ≥ 4 is the sum of two odd prime numbers. For instance, we have $4 = 1 + 3, 6 = 3 + 3$, and $8 = 3 + 5$. To this day, this conjecture has neither been proven nor disproved by a counterexample. In order to find such a counterexample, one would have to find an even positive integer which is not the sum of two primes.

To frame this problem as a decision problem, we identify positive integers with their binary expansions and consider the *Goldbach language* L comprising all integers which are the sum of two odd prime numbers. So the Goldbach conjecture states that L is the set of all even integers ≥ 4. A proof of the Goldbach conjecture would imply that $L \in$ P since deciding membership of L would mean deciding whether an integer is ≥ 4 is even. But L is not known to be in P. However, we can verify in polynomial time that $a \in L$ if we are given a prime number p such that $a - p$ is also a prime number. For this, we apply the polynomial time primality test to p and $a - p$ that was mentioned in Example 1.4.17. The prime number p is called a *certificate* for the Goldbach language membership of a.

There are many other languages L that have a property analogous to that of the Goldbach language presented in Example 1.4.19. Abstractly speaking, this property is the following. For $\vec{s} \in \{0, 1\}^*$ it may be hard to decide whether $\vec{s} \in L$. But for each $\vec{s} \in L$ there is a *certificate* t which allows us to verify in polynomial time in $|\vec{s}|$ that $\vec{s} \in L$. For the Goldbach language, the certificate is the prime number p such that $a - p$ is a prime number. The set of all languages with this property is denoted by NP, which stands for *nondeterministic polynomial time*. This name comes from another NP modeling that we do not discuss here (see [**LP98**]). Here is a formal definition of NP.

Definition 1.4.20. (1) The *complexity class* NP is the set of all languages L with the following properties.
  (a) There is a deterministic polynomial time algorithm A with $\text{Input}(A) = \{0, 1\}^*$ $\times \{0, 1\}^*$ such that $A(\vec{s}, \vec{t}) = 1$ implies $\vec{s} \in L$ for all $\vec{s}, \vec{t} \in \Sigma^*$.
  (b) There is $c \in \mathbb{N}$ that may depend on L so that for all $\vec{s} \in L$ there is $\vec{t} \in \{0, 1\}^*$ with $|\vec{t}| \leq |\vec{s}|^c$ and $A(\vec{s}, \vec{t}) = 1$.
  If $\vec{s} \in L$ and $\vec{t} \in \{0, 1\}^*$ such that $A(\vec{s}, \vec{t}) = 1$, then $\vec{t}$ is called a *certificate* for the membership of $\vec{s}$ in L.

(2) The *complexity class* Co-NP is the set of all languages L such that $\{0, 1\}^* \setminus L \in$ NP.

One of the big open research problems in computer science is finding out whether P is equal to NP. It is one of the seven Millennium Prize Problems. They are well-known mathematical problems that were selected by the Clay Mathematics Institute in the year 2000. The Clay Institute has pledged a US\$1 million prize for the correct solution to any of the problems.

The complexity theory that we have explained so far only refers to solving language decision problems but not to more general computational problems such as finding proper divisors of composite integers. But, as illustrated in the next example, there is a close connection between these two problem classes.

Example 1.4.21. Consider the set

$$(1.4.1) \qquad L = \{(a, c) \in \mathbb{N}^2 : c \leq a, a \text{ has a proper divisor } \leq c\}.$$

By identifying the elements (a, c) of L with a linear length string representation in size a, we consider L as a language, that is, a subset of $\{0, 1\}^*$. We will now show that L is in P if and only if the integer factorization problem from Example 1.4.7 can be solved in polynomial time.

Suppose that A is a polynomial time algorithm that on input of a composite $a \in \mathbb{N}$ finds a proper divisor of a. Then the following algorithm A' decides L in polynomial time. On input of $a \in \mathbb{N}$, the algorithm checks whether $a = 1$ or a is a prime number. In both cases, A' returns 0. As explained in Example 1.4.17, this test can be carried out in polynomial time in size a. If a is neither 1 nor a prime number, A' invokes A and finds a proper divisor $b \in \mathbb{N}$ of a. If $b \leq c$, then A' returns 1. Otherwise, this procedure is applied to b and a/b and so on until a sufficiently small divisor is found or until it is clear that all prime divisors of the input are greater than c. Exercise 1.4.22 shows that this algorithm has polynomial running time.

Conversely, let A be an algorithm that decides L in polynomial time. We present a polynomial time algorithm A' that uses A as a subroutine and finds a proper divisor of a composite integer $a \in \mathbb{N}$. It uses two integers $l, u \in \mathbb{N}$ with $1 \leq u \leq v \leq a$. They define an interval $[u, v]$ such that inside this interval there is a proper divisor of a. During the execution of the algorithm, the interval shrinks exponentially. After $O(\text{size } a)$ iterations, we have $u = v$. Then the algorithm returns $b = u$. To achieve this, the algorithm initially sets $u \leftarrow 2$ and $v \leftarrow a-1$. Since a is composite, the interval $[u, v]$ contains a proper divisor of a, but not the interval $[1, u - 1]$. While $u < v$, algorithm A' repeats the following steps. It determines $m = \lfloor \frac{v-u}{2} \rfloor$ and calls $A(a, u + m)$. If the return value is 1, then A' sets v to $u + m$. So $[u, v]$ contains a proper divisor of a, but $[1, u - 1]$ does not. If the return value is 0, then A' sets u to $u + m + 1$. Again $[u, v]$ contains a proper divisor of a but $[1, u - 1]$ does not. If after this step we have $u = v$, then the algorithm returns $b = u$. It is a proper divisor of a since $[u, v]$ contains such a divisor. Since in each iteration of this **while** loop the interval $[u, v]$ is roughly cut in half and the initial length of the interval is $a - 2$, the number of iterations is $O(\text{size } a)$. Also, because A is a polynomial time algorithm, it follows that A' runs in polynomial time.

Exercise 1.4.22. Write pseudocode for the algorithms sketched in Example 1.4.21 and analyze them.

The method explained in Example 1.4.21 can be generalized to all algorithmic problems for which the solution length is polynomially bounded in the instance length. This is shown in Exercise 1.4.23.

Exercise 1.4.23. Consider a computational problem $\text{CP} = (I, O, R)$ with the following property. There is $c \in \mathbb{N}$ such that for all $a \in I$ and all solutions b of a we have $\text{size } b \leq (\text{size } a)^c$. Define a language L that can be decided in polynomial time if and only if the computational problem can be solved in polynomial time.

The next theorem describes the relation between the complexity classes that we have introduced. This theorem is also illustrated in Figure 1.4.1.

Theorem 1.4.24. *We have* $\text{P} \subset \text{NP} \subset \text{PSPACE} \subset \text{EXPTIME}$ *and* $\text{P} \subset \text{BPP} \subset \text{PP} \subset \text{PSPACE}$.

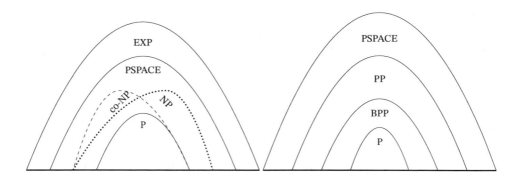

Figure 1.4.1. Relation between deterministic and probabilistic complexity classes.

Proof. Since our computational model is only semiformal, we can only sketch the proofs of the inclusions. But with the ideas presented below, the proofs can be carried out in any of the formal models of computation, for instance the Turing machine model.

We clearly have $P \subset NP$.

To prove $NP \subset PSPACE$, let $L \in NP$. Also, let A be an algorithm and let $c \in \mathbb{N}$ be a constant with the properties from Definition 1.4.20. Then this algorithm can be transformed as follows into an algorithm A' that decides L in polynomial space. The input set of A' is $\{0, 1\}^*$. On input of $\vec{s} \in \{0, 1\}^*$ the modified algorithm runs the algorithm A with all possible certificates $\vec{t} \in \{0, 1\}^*$ such that $|\vec{t}| \leq |\vec{s}|^c$. It returns 1 if A returns 1 for one of these certificates and 0 otherwise. It follows from Definition 1.4.20 that A' decides L. Also, since A is a polynomial time algorithm and because $|\vec{t}| \leq |\vec{s}|^c$, it follows that A' has polynomial space complexity.

Next, we show that $PSPACE \subset EXPTIME$. Let $L \in PSPACE$ and let A be an algorithm with polynomial space complexity that decides L. So there is a constant $c \in \mathbb{N}$ such that on input of $\vec{s} \in \{0, 1\}^*$ the size of the memory used by the algorithm is at most $|\vec{s}|^c$. Therefore, the number of states of the algorithm run with input $\vec{s}$ is $O(2^{n^c})$ since the number of instructions that the algorithm may use is a constant. This implies that the algorithm has exponential running time.

Now we turn to probabilistic algorithms. Clearly, we have $P \subset BPP \subset PP$.

To see that $PP \subset PSPACE$, let L be a language in PP and let A be a Monte Carlo algorithm that decides L as described in Definition 1.4.18. Using A, we construct an algorithm A' with polynomial space complexity that decides L. Let $\vec{s} \in \{0, 1\}^*$. Since A has polynomial running time, there is $c \in \mathbb{N}$ such that the number of calls of probabilistic subroutines in a run of A on input of $\vec{s}$ is at most $|\vec{s}|^c$. On input of a, the algorithm A' runs algorithm A with all random sequences corresponding to runs of A with input $\vec{s}$. Their number is bounded by $\leq |\vec{s}|^c$. If the majority of these runs returns 1, the return value of A' is 1. Otherwise, A' returns 0. It follows from the definition of the complexity class PP that A' decides L. Also, A' has polynomial space complexity. $\qquad\square$

We note that the relation between BPP and NP is unknown.

1.5. The circuit model

The second computational model that we present is that of Boolean circuits. It is the basis for the circuit model for quantum computation.

1.5.1. Logic gates. Logic gates are fundamental components of Boolean circuits. A *logic gate* is a device that implements a *Boolean function* $\{0, 1\}^n \to \{0, 1\}^m$, where m and n are natural numbers. The availability of specific gates depends on the computing platform being used. In this context, we will focus solely on the gates with $m = 1$. Table 1.5.1 presents commonly used logic gates and their corresponding implemented functions are listed in Table 1.1.4. These gates can be realized using various technologies, such as diodes or transistors acting as electronic switches. Additionally, they can

Table 1.5.1. List of important logic gates.

Name	Logic operator	Circuit symbol
AND	$\wedge$	D
OR	$\vee$	D
NOT	$\neg$	$\triangleright$
NAND	$\uparrow$	D
NOR	$\downarrow$	D
XOR	$\oplus$	D

also be constructed using alternative technologies such as vacuum tubes, electromagnetic relays, or even mechanical elements.

1.5.2. Boolean circuits. Next, we define Boolean circuits.

Definition 1.5.1. A *Boolean circuit* is a tuple $C = (V, E, G, L)$ where (V, E) is a directed acyclic graph, G is a set of logic gates, and

$$L : V \to \{\mathsf{I}, \mathsf{O}, 0, 1\} \cup G$$

is a map that labels the elements of V which are called *vertices* or *nodes*. A node labeled I is called an *input node*. A node labeled O is called an *output node*. A node labeled 0 or 1 is called a *constant node*. All other nodes are called *gates*. Also, the circuit C satisfies the following conditions.

(1) Input nodes and constant nodes have indegree 0.

(2) There is an ordering $I \to \mathbb{Z}_{|I|}$ on the set I of input nodes.

(3) The output nodes have indegree 1 and outdegree 0.

(4) There is an ordering $O \to \mathbb{Z}_{|O|}$ on the set O of output nodes.

(5) Let $g \in G$ and assume that g implements a function with k inputs and l outputs. Then its indegree is k and its outdegree is l. Also, there is an ordering $I(g) \to \mathbb{Z}_k$ on the set $I(g)$ of incoming edges of g and there is an analogous ordering on the outgoing edges of g.

Boolean circuits are also referred to as *logic circuits* or simply as *circuits*. We introduce a few important notions for Boolean circuits.

Definition 1.5.2. Let C be a Boolean circuit.

(1) The *depth* of a node v of C is the maximum length of a path from an input node or a constant node to v.

(2) The *depth* of C is the maximum depth of all nodes of C. It is denoted by $\mathrm{depth}(C)$.

(3) The *size* of C is the number of nodes of C. It is denoted by $|C|$.

Example 1.5.3. Figure 1.5.1 shows two examples of Boolean circuits. The first implements NAND using one AND and one NOT gate. The second implements XOR using one NAND, one OR, and one AND gate.

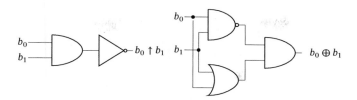

Figure 1.5.1. Circuit implementations of NAND and XOR.

Exercise 1.5.4. Verify that the circuits in Figure 1.5.1 implement NAND and XOR.

In the second circuit in Figure 1.5.1, the input nodes have outdegree 2. This is represented by a fanout symbol $\top$. Fanout operations are used in circuits in order to increase the outdegree of logic gates. When we describe the simulation of circuits by reversible circuits in Section 1.7, we will consider fanout symbols as gates.

Next, we define the functions that are computed by circuits. Let $C = (V, E, G, L)$ be a circuit with n input nodes and m output nodes. To simplify the description, we assume that all gates in G implement functions $\{0, 1\}^l \to \{0, 1\}$ for some $l \in \mathbb{N}$. The generalization to arbitrary gates is straightforward.

The circuit C computes a function

$$(1.5.1) \qquad f : \{0, 1\}^n \to \{0, 1\}^m.$$

To specify this function, we let $\vec{b} = (b_0, \ldots, b_{n-1}) \in \{0, 1\}^n$ and construct $f(\vec{b})$. For this, we use a *value function*

$$(1.5.2) \qquad B : V \to \{0, 1\}$$

which we define by induction on the depths of the nodes in V. For the base case, we specify the following.

(1) For constant nodes v labeled 0 or 1 we set $B(v)$ to 0 or 1, respectively.

(2) Let v_i be the input nodes of C for $0 \le i < n$. Then we set $B(v_i) = b_i, 0 \le i < n$. Here, we use the ordering on the input nodes to assign a bit b_i to an input node v_i.

For the inductive step, let K be the depth of C and let k be a positive integer with $0 < k \le K$. Assume that $B(v)$ has been defined for all nodes v of depth less than k. Let v be a node of depth k. We define $B(v)$ as follows. Since the depth of the node v is greater than 0, it is either a gate or an output node. Assume that v is a gate. Let

$$(1.5.3) \qquad g : \{0, 1\}^l \to \{0, 1\}$$

be the Boolean function implemented by this gate. Then g has l incoming edges and, by definition, there is an ordering $(e_0, \ldots, e_{l-1})$ on these edges. Denote by $u_0, \ldots, u_{l-1}$ the nodes in the circuit such that e_i is an outgoing edge of u_i for $0 \le i < l$. Then the nodes $u_0, \ldots, u_{l-1}$ have depth less than k. Therefore, the values $B(u_i), 0 \le i < l$, are already defined. We set

$$(1.5.4) \qquad B(v) = g(B(u_0), \ldots, B(u_{n-1})).$$

Assume that v is an output node. By definition, it has indegree 1. Let u be the node in V from which there exists an edge to v. Then we define

$$(1.5.5) \qquad\qquad\qquad B(v) = B(u).$$

Finally, let $(y_0, \ldots, y_m)$ be the ordered sequence of output nodes of C. Then we set

$$(1.5.6) \qquad\qquad\qquad f(\vec{b}) = (B(y_0), \ldots, B(y_{m-1})).$$

Examples of circuits and the functions that they implement can be seen in Figure 1.5.1.

Exercise 1.5.5. Define the function computed by a circuit that uses logic gates with more than one output gate.

1.5.3. Universal sets of gates. Which gates do we really need in circuits? In this section we show that very few suffice. We start with a definition.

Definition 1.5.6. A set G of logic gates is called *universal for classical computation* if for all $m, n \in N$ and every function $f : \{0,1\}^n \to \{0,1\}^m$ there is a circuit that only uses gates from G and computes f.

Now we present a very simple set of logic gates that is universal for classical computation.

Theorem 1.5.7. *The set* $\{\mathsf{NOT}, \mathsf{AND}, \mathsf{OR}\}$ *is universal for classical computation.*

Proof. It suffices to prove the theorem for Boolean functions

$$(1.5.7) \qquad\qquad\qquad f : \{0,1\}^n \to \{0,1\}.$$

If the function f has more components, we can use circuits that implement the individual components to construct a circuit C that computes f. For this, we proceed as follows. Let $f = (f_0, \ldots, f_l)$ with Boolean functions $f_i : \{0,1\}^n \to \{0,1\}$ for $0 \le i < m$. For all $i \in \mathbb{Z}_m$ let C_i be a circuit that computes f_i. We use fanout operations in order to make the constant and input nodes available for all circuits C_i. Then $(o_0, \ldots, o_{l-1})$ are used as the output nodes of C where o_i is the output node of C_i.

We now prove the assertion for functions as in (1.5.7) by induction on n. For the base case, let $n = 1$. There are four Boolean functions $f : \{0,1\} \to \{0,1\}$, namely the following:

$$(1.5.8) \qquad\qquad b \mapsto 0, \quad b \mapsto 1, \quad b \mapsto b, \quad b_0 \mapsto \neg b.$$

Implementations of these functions by circuits that use only the NOT, AND, and OR gates are shown in Figure 1.5.2.

Now let $n > 0$ and assume that all functions $\{0,1\}^{n-1} \to \{0,1\}$ can be implemented by circuits that use only the NOT, AND, and OR gates. Let $f : \{0,1\}^n \to \{0,1\}$. For $b \in \{0,1\}$ define the functions

$$(1.5.9) \qquad f_b : \{0,1\}^{n-1} \to \{0,1\}, \quad (b_0, \ldots, b_{n-2}) \mapsto f(b_0, \ldots, b_{n-2}, b).$$

Then for every $(b_0, \ldots, b_{n-2}, b) \in \{0,1\}^n$ we can write

$$(1.5.10) \qquad f(b_0, \ldots, b_{n-2}, b) = (f_0(b_0, \ldots, b_{n-2}) \wedge \neg b) \vee (f_1(b_0, \ldots, b_{n-2}) \wedge b).$$

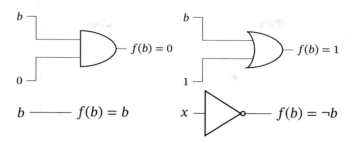

Figure 1.5.2. Base case of the induction proof in Theorem 1.5.7: circuits that compute the four functions $f : \{0,1\} \to \{0,1\}$.

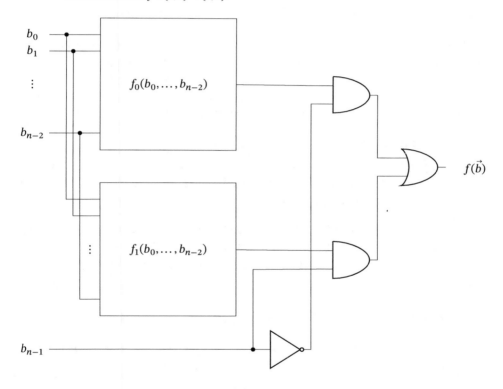

Figure 1.5.3. Inductive step in the proof of Theorem 1.5.7.

By the induction hypothesis, there exist circuits that implement f_0 and f_1 and use only the gates NOT, OR, and AND. Therefore, the circuit in Figure 1.5.3 that uses the circuits for f_0 and f_1 implements the function f. $\qquad\square$

From Theorem 1.5.7 we obtain the following important corollary.

Corollary 1.5.8. *For all* $m, n \in \mathbb{N}$ *and every function* $f : \{0,1\}^n \to \{0,1\}^m$ *there is a circuit that computes* f.

Next, we present an even smaller universal set of gates.

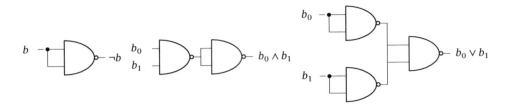

Figure 1.5.4. Implementation of NOT, AND, and OR by NAND.

Theorem 1.5.9. *The gate set {NAND} is universal for classical computing.*

Proof. By Theorem 1.5.7 it suffices to show that the gates NOT, AND, and OR can be implemented by a circuit that uses only the NAND gate. This is shown in Figure 1.5.4.

$\square$

Exercise 1.5.10. Show that the set {NOR} is universal for classical computation.

Finally, we define the complexity of Boolean functions.

Definition 1.5.11. Let G be a universal set of logic gates and let f be a Boolean function. The *circuit size complexity* of f with respect to G is the minimum size of any circuit that computes f. It is also called the *circuit complexity of f*.

If it is clear which universal set of logic gates we refer to, we simply speak about the circuit size complexity of a Boolean function.

We note that there is also the notion of circuit depth complexity which is not required in our context.

1.6. Circuit families and circuit complexity

In this section we discuss the notion of circuit families which is then used to introduce circuit complexity theory.

1.6.1. Circuit families. Individual circuits cannot compute functions $f : \{0,1\}^* \to \{0,1\}^*$ or decide languages since their input length is fixed. To solve these more general problems, we need families of circuits. We fix a finite universal set of logical gates and assume from now on that all circuits are constructed using these gates. Such universal sets are presented in Theorems 1.5.7 and 1.5.9.

Definition 1.6.1. A *family of circuits* or *circuit family* is a sequence $(C_n)_{n \in \mathbb{N}}$ of circuits such that the circuit C_n has n input nodes for all $n \in \mathbb{N}$.

Next, we describe what it means for a circuit family to compute a function, solve a computational problem, or decide a language. In doing so, we must take into account that circuits have a fixed output length. Example 1.6.2 illustrates how to deal with this.

Example 1.6.2. Consider the function $f : \{0,1\}^* \to \{0,1\}^*$, $a \mapsto f(a) = a^2$ where we identify the elements of $\{0,1\}^*$ with the integers they represent. How can we implement this function using a circuit family? For $n \in \mathbb{N}$, let

$$(1.6.1) \qquad f_n : \{0,1\}^n \to \{0,1\}^*, \quad a \mapsto a^2.$$

In order to implement f_n as a circuit, we must use representations of the function values that have the same length for all inputs of length n. So we prepend an appropriate number of zeros to the binary representations to ensure that they all have the same length $2n$. For example, for $n = 2$ we write $f_2(00) = 0000$, $f_2(01) = 0001$, $f_2(10) = 0100$, $f_2(11) = 1001$. In the same way, circuits C_n can be constructed that implement f_n for all $n \in \mathbb{N}$. We note that the binary expansion of the function values $\neq 0$ can be obtained from the function values represented by bit strings of length n by deleting the leading zeros.

The idea of Example 1.6.2 can be used for all functions $f : \{0,1\}^* \to \{0,1\}^*$ whose function values are encoded either by 0 or by bit strings in $\{0,1\}^*$ starting with bit 1. This encoding can be easily obtained from any encoding by prefixing the representations of the function values different from 0 with 1. So without loss of generality, we only consider functions that satisfy

$$(1.6.2) \qquad |\vec{s}| = |\vec{s}'| \Rightarrow |f(\vec{s})| = |f(\vec{s}')| \quad \text{for all } \vec{s}, \vec{s}' \in \{0,1\}^*.$$

Definition 1.6.3. Let $f : \{0,1\}^* \to \{0,1\}^*$ be a function that satisfies (1.6.2) and let $C = (C_n)_{n \in \mathbb{N}}$ be a circuit family. We say that C computes f if for all $n \in \mathbb{N}$, the circuit C_n computes the function $f_n : \{0,1\}^n \to \{0,1\}^*, \vec{s} \mapsto f(\vec{s})$.

We also define what it means for a circuit family to solve a computational problem $\mathrm{CP} = (I, O, R)$. Analogous to (1.6.2), we may, without loss of generality, assume that the solutions of all instances of a fixed length also have a fixed length. This means that

$$(1.6.3) \qquad |\vec{s}| = |\vec{s}'| \Rightarrow |\vec{t}| = |\vec{t}'| \quad \text{for all } (\vec{s}, \vec{t}), (\vec{s}', \vec{t}') \in R.$$

We assume in the following that the encodings of computational problems have this property.

Definition 1.6.4. Let $\mathrm{CP} = (I, O, R)$ be a computational problem and let $C = (C_n)_{n \in \mathbb{N}}$ be a circuit family. We say that C solves CP if for all $n \in \mathbb{N}$ on input of $a \in \{0,1\}^n \cap I$ the circuit C_n computes a solution b of a.

Finally, we define what it means for a circuit family to decide a language.

Definition 1.6.5. Let L be a language, and let $C = (C_n)_{n \in \mathbb{N}}$ be a circuit family. We say that C decides L if for all $n \in \mathbb{N}$ on input of $\vec{s} \in \{0,1\}^n$ the circuit C_n returns 1 if $\vec{s} \in L$ and 0 otherwise.

From Corollary 1.5.8 we obtain the following result.

Theorem 1.6.6. *For all functions* $f : \{0,1\}^* \to \{0,1\}^*$, *computational problems* CP, *and languages L there is a circuit family that computes f, solves* CP, *or decides L.*

Theorem 1.6.6 demonstrates that circuit families are more powerful than algorithms in terms of computation. It is known that certain functions $f : \{0, 1\}^* \to \{0, 1\}^*$ cannot be computed by algorithms (see [**Dav82**]). However, as this theorem shows, for all such functions, there exists a circuit family that can compute them. This is possible because an individual circuit can be designed for each input length. The next section introduces a more limited concept of circuit families that possesses capabilities equivalent to the concept of algorithms.

1.6.2. Uniform circuit families. We now introduce uniform circuit families and obtain a computational model that corresponds to the algorithmic one. For the next definition, we assume that we have fixed some encoding of circuits by bit strings. Following [**Wat09**] we require the following.

(1) The encoding is *sensible*: every circuit is encoded by at least one bit string, and every bit string encodes at most one quantum circuit.

(2) The encoding is *efficient*: there is $c \in \mathbb{N}$ such that every circuit C has an encoding of length at least size C and at most $(\text{size } C)^c$.

(3) Information about the structure of a circuit is computable in polynomial time from an encoding of the circuit.

"Structure information" means, for example, information about what the input nodes, the gates, and the output nodes are and how these nodes are connected.

We define uniform circuit families.

Definition 1.6.7. A circuit family $C = (C_n)$ is called *uniform* if there is a deterministic algorithm which on input of I^n, $n \in \mathbb{N}$, outputs the encoding of C_n.

After the next definition, we explain why the input of the algorithm in Definition 1.6.7 is I^n and not simply n. Here, we remark the following: It can be shown that uniform circuit families are *Turing complete*, meaning that their computing power is equivalent to that of Turing machines. This is important because the *Turing-Church thesis* states that Turing machines are the most powerful computing devices imaginable. This means that a function $f : \{0, 1\}^* \to \{0, 1\}^*$ is computable by a human being following an algorithm, ignoring resource limitations, if and only if it is computable by a Turing machine. In today's computer science, the Turing-Church thesis is still considered to be true.

Now we define the P-uniform circuit families.

Definition 1.6.8. A circuit family $C = (C_n)$ is called *P-uniform* if there is a deterministic polynomial time algorithm which on input of I^n, $n \in \mathbb{N}$, outputs the encoding of C_n.

Why is the input of the algorithm represented in unary encoding I^n of n, rather than the binary expansion of this number? The key reason is that the algorithm running time should be polynomial in n. The running time of algorithms is typically measured as a function of the length of their input. Therefore, the input length of this algorithm must be proportional to n. If we were to use the binary encoding of n, its length would only be of the order $\log n$, which is much smaller than n. As a result, the algorithm running time would not be polynomial in n. On the other hand, the unary encoding I^n has a length proportional to n, making it suitable for ensuring polynomial running time in terms of the input size.

1.6.3. Circuit complexity. Now we define the size complexity of circuit families and different circuit complexity classes.

Definition 1.6.9. Let $C = (C_n)_{n \in \mathbb{N}}$ be a circuit family and let $f : \mathbb{N} \to \mathbb{N}$ be a function.

(1) The *size complexity* of C is the function $\mathbb{N} \to \mathbb{N}, n \mapsto |C_n|$.

(2) The complexity class $\mathrm{SIZE}(f)$ is the set of all languages that can be decided by a P-uniform circuit family with size complexity $O(f)$.

The next theorem establishes a connection between algorithmic and circuit complexity classes.

Theorem 1.6.10. *Let* $f : \mathbb{N} \to \mathbb{N}$. *Then* $\mathrm{DTIME}(f) \subset \mathrm{SIZE}(f \log f)$.

For the proof of this theorem, see [**Vol99**] and [**AB09**]. It is beyond the scope of this book.

From Theorem 1.6.10 we obtain the following corollary which characterizes the complexity class P in terms of polynomial size uniform circuit families.

Corollary 1.6.11. *A language L is in* P *if and only if L is in* $\mathrm{SIZE}(n^c)$ *for some $c \in \mathbb{N}$.*

1.7. Reversible circuits

This section explores reversible circuits, which play a crucial role in quantum computing, as we will see in Section 4.7. Reversible circuits can be easily converted into quantum circuits by substituting classical reversible gates with their quantum equivalents. A significant objective of this section is to demonstrate that any circuit can be emulated by a reversible circuit. This, in turn, implies that there exists a quantum circuit for every Boolean function, capable of computing that function.

1.7.1. Basics. We define reversible gates and circuits.

Definition 1.7.1. *A reversible gate or circuit is a logic gate or circuit that implements an invertible function $f : \{0,1\}^n \to \{0,1\}^n$ for some $n \in \mathbb{N}$, respectively.*

The only reversible gate that we have seen so far is the NOT gate. All other gates in Table 1.5.1 are not reversible.

An important reversible gate with two input nodes is the *controlled not* gate which is denoted by CNOT. It applies the NOT operation to a *target bit t* if a *control bit c* is 1. Otherwise, the target bit remains unchanged. Therefore, the target qubit becomes $c \oplus t$. The control bit is never changed. Two variants of CNOT are shown in Figure 1.7.1. In the left CNOT gate, the first bit is the control, and the second bit is the target. In the right CNOT gate, the roles of the bits are reversed. A circuit implementation of the left CNOT gate using one XOR gate is shown in Figure 1.7.2. Figure 1.7.3 presents two more CNOT variants. They flip the target bit t conditioned on the control bit c being 0.

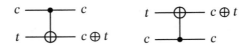

Figure 1.7.1. CNOT gates that change the target bit t conditioned on the control bit c being 1.

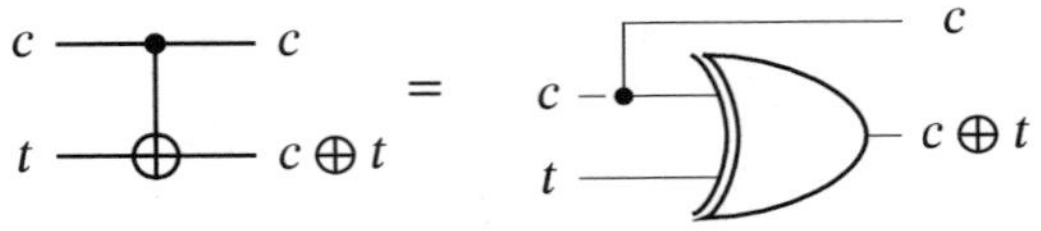

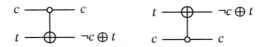

Figure 1.7.2. Circuit implementation of a CNOT gate using one XOR gate.

Figure 1.7.3. CNOT gates that change the target bit t conditioned on the control bit c being 0.

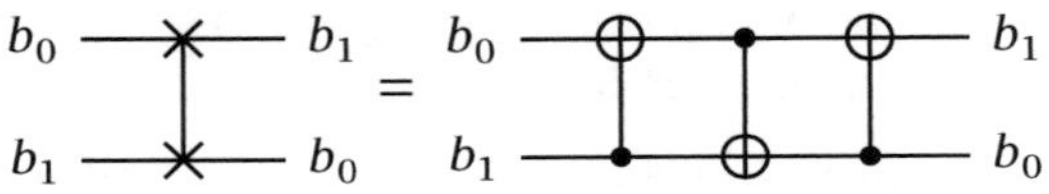

Figure 1.7.4. The SWAP gate and its implementation using three CNOT gates.

Another important gate is the SWAP gate. On input of a pair (b_0, b_1) of bits it returns (b_1, b_0). This gate is shown in Figure 1.7.4 together with an implementation that uses only three CNOT gates.

Next, we show that we can implement every permutation of the entries of an n-bit string by a reversible circuit that uses at most $n - 1$ SWAP gates.

Proposition 1.7.2. *Let* $n \in \mathbb{N}$ *and let* $\pi \in S_n$. *Then the map*

$$(1.7.1) \qquad f_\pi : \{0, 1\}^n \to \{0, 1\}^n, \quad (b_0, \dots, b_{n-1}) \mapsto (b_{\pi(0)}, \dots b_{\pi(n-1)})$$

can be implemented by a circuit that uses at most $n - 1$ *SWAP or at most* $3n$ *CNOT gates.*

Proof. The proposition follows Theorem A.4.25 which states that π is the product of at most $n - 1$ transpositions. □

Example 1.7.3. Consider the permutation

$$(1.7.2) \qquad\qquad\qquad \pi = \begin{pmatrix} 0 & 1 & 2 & 3 \\ 1 & 3 & 0 & 2 \end{pmatrix}.$$

We have $\pi = (2, 1) \circ (1, 3) \circ (0, 2)$. So the circuit in Figure 1.7.5 implements π.

We now introduce the *Toffoli gate* which was proposed in 1980 by Tomaso Toffoli and is shown in Figure 1.7.6. It implements the bijection

$$(1.7.3) \qquad\qquad \{0, 1\}^3 \to \{0, 1\}^3, \quad (c_0, c_1, t) \mapsto (c_0, c_1, c_0 \wedge c_1 \oplus t).$$

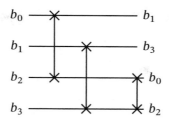

Figure 1.7.5. Implementation of the permutation π in (1.7.2).

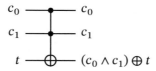

Figure 1.7.6. A Toffoli or CCNOT gate.

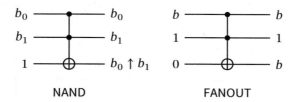

Figure 1.7.7. Reversible circuits that implement the NAND and FANOUT gates.

This gate leaves the *control bits* c_0 and c_1 unchanged and modifies the *target bit* t conditioned on both control bits c_0 and c_1 being 1. Toffoli gates are also called CCNOT gates: a NOT operation controlled by two control bits.

The Toffoli gate has the important property that it allows implementations of the NAND and fanout operations. This is shown in Figure 1.7.7. As we will see in Section 1.7.2, this property implies that Toffoli gates can be used to transform every Boolean circuit into a reversible circuit.

Exercise 1.7.4. Verify that the circuits in Figure 1.7.7 are reversible and implement the NAND and the FANOUT operation, respectively.

Another gate that can be used to make every circuit reversible is the *Fredkin gate*. It was introduced by Edward Fredkin in 1969. It implements the bijection

$$(1.7.4) \quad \begin{aligned} &\{0,1\}^3 \to \{0,1\}^3, \\ &(c, t_0, t_1) \mapsto (c, (\neg c \wedge t_0) \vee (c \wedge t_1), (c \wedge t_0) \vee (\neg c \wedge t_1)). \end{aligned}$$

This function does not change the *control bit* c, swaps the *target bits* b_1 and b_2 if the control bit c is 1, and leaves them unchanged otherwise (see Exercise 1.7.5). Because of this property, the Fredkin gate is also called the *controlled swap* gate and is denoted by CSWAP: a swap controlled by one control bit.

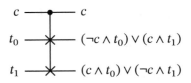

Figure 1.7.8. A Fredkin or CSWAP gate.

Exercise 1.7.5. Determine the truth tables of the Toffoli and the Fredkin gates and use them to verify that they implement the functions in (1.7.3) and (1.7.4). Also, verify that the two functions are bijections.

Exercise 1.7.6. (1) Find an implementation of the Toffoli gate that uses only NOT, AND, and OR gates.

(2) Find an implementation of the Fredkin gate that uses only NOT, AND, and OR gates.

(3) Find implementations of the NAND and fanout operations that use only Fredkin gates.

1.7.2. Construction of reversible circuits. In this section, we provide an illustrative example that demonstrates the construction of reversible circuits using reversible gates. This construction will be subsequently adopted in Section 3.3.4 and further formalized in Definition 4.7.1 for generating quantum circuits. The example is shown in Figure 1.7.9.

The circuit implements a bijection $f : \{0,1\}^4 \to \{0,1\}^4$. As shown in Figure 3.3.5, it can be written as $f = f_2 \circ f_1 \circ f_0$ where $f_0, f_1, f_2 : \{0,1\}^4 \to \{0,1\}^4$ are bijections. Each of these functions is obtained by applying invertible gates to certain bits and applying the identity function $I : \{0,1\} \to \{0,1\}$ to the remaining bits. The functions are

$$(1.7.5) \qquad f_0 = (\text{NOT}, I, I, I), \quad f_1 = (\text{CCNOT}, I), \quad f_2 = (I, \text{CSWAP}).$$

The circuit operates on the input $(0,1,0,0)$ as follows:

$$(1.7.6) \qquad (0,1,0,0) \underset{f_0}{\mapsto} (1,1,0,0) \underset{f_1}{\mapsto} (1,1,1,0) \underset{f_2}{\mapsto} (1,1,0,1).$$

Exercise 1.7.7. Determine $f(\vec{x})$ for the function f implemented by the circuit in Figure 1.7.9 for all $\vec{x} \in \{0,1\}^4$.

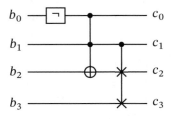

Figure 1.7.9. Reversible circuit.

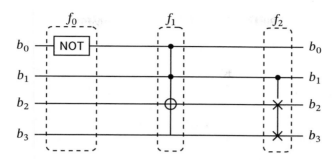

Figure 1.7.10. The functions f_0, f_1, f_2 corresponding to the reversible circuit from Figure 1.7.9.

This construction can be easily extended to circuits that handle inputs of any size. Moreover, the construction enables the solution of the subsequent exercise.

Exercise 1.7.8. Show that any circuit that uses only reversible gates is reversible.

1.7.3. Every function can be computed by a reversible circuit. In this section, we show that every Boolean function $f : \{0, 1\}^n \rightarrow \{0, 1\}^m$ with $m, n \in \mathbb{N}$ can be computed by a reversible circuit that uses only Toffoli gates. Since, in general, the function f is not reversible, this cannot mean that f is implementable by a reversible circuit. It means that there is an invertible circuit such that the function h implemented by it allows us to obtain $f(\vec{x})$ for every $\vec{x} \in \{0, 1\}^n$ very easily. Theorems 1.7.10 and 1.7.12 will make this statement precise.

To simplify our description, we consider fanout operations also as gates which we write as FANOUT. Before we state the first important result of this section, we introduce some more notation.

Definition 1.7.9. (1) For a circuit C we denote by $|C|_F$ the number of gates including FANOUT gates that it uses.

(2) For a Boolean function f denote by $|f|_F$ the minimum value of $|C|_F$ over all circuits C that implement f and use only NAND and FANOUT gates.

The idea of the construction of a reversible circuit that implements f is to start from a circuit C that implements f and uses only NAND and FANOUT gates. Such a circuit exists by Theorem 1.5.9. Then all NAND and FANOUT gates in C are replaced by their reversible counterparts shown in Figure 1.7.7.

Theorem 1.7.10. *For all Boolean functions $f : \{0, 1\}^n \rightarrow \{0, 1\}^m$, $m, n \in \mathbb{N}$, there is $p \in \mathbb{N}_0$, $p \leq 2|f|_F$, a reversible circuit C_r of size $|f|_F$ that uses only Toffoli gates, $\vec{a} \in \{0, 1\}^p$, and a function $g : \{0, 1\}^n \rightarrow \{0, 1\}^{n+p-m}$ such that C_r implements a function*

$$(1.7.7) \qquad h : \{0, 1\}^n \times \{0, 1\}^p \rightarrow \{0, 1\}^m \times \{0, 1\}^{n+p-m}$$

with

$$(1.7.8) \qquad h(\vec{x}, \vec{a}) = (f(\vec{x}), g(\vec{x}))$$

for all $\vec{x} \in \{0,1\}^n$. The bits in $\vec{a}$ are called ancilla bits. *The functional value $g(\vec{x})$ is called* garbage.

Proof. We prove the theorem by induction on $k = |f|_F$.

For the base case, let $f : \{0,1\}^n \to \{0,1\}^m$, $m, n \in \mathbb{N}$, with $|f|_F = 0$ and let C be a circuit that implements f with $|C|_F = 0$. Then C has only input and output nodes. Input nodes have indegree 0 and outdegree 1, while output nodes have indegree 1 and outdegree 0. Therefore, we have $n = m$ and the function f permutes the input bits and is therefore a bijection. So $p = 0$, $C_r = C$, $\vec{a} = ()$, and $g : \{0,1\}^n \to \{0,1\}^0$, $\vec{x} \mapsto ()$ have the asserted properties.

For the inductive step, let $f : \{0,1\}^n \to \{0,1\}^m$, $m, n \in \mathbb{N}$, with $k = |f|_F > 0$ and assume that for every Boolean function f' with $|f'|_F < k$ the assertion of Theorem 1.7.10 holds. Let C be a circuit that implements the function f, uses only NAND and FANOUT gates, and satisfies $|C|_F = k$.

Since $k > 0$, it follows that C contains at least one gate, which is a NAND or a FANOUT gate. This implies that C has at least one of the following properties.

(1) There is a FANOUT gate in C whose outgoing edges are incoming edges of two output nodes y_i and y_j of C where $i, j \in \mathbb{Z}_m$, $i \neq j$.

(2) There is a NAND gate in C whose outgoing edge is the incoming edge of an output node y_i of C where $i \in \mathbb{Z}_m$.

Assume that C has the first property. To construct C_r we proceed as follows. Remove the FANOUT gate and the corresponding output nodes from C. Connect the incoming edge of the removed FANOUT gate to a new output node. Since FANOUT gates do not change the input, we also denote it by y_i. The resulting circuit is denoted by C'. An example for C and C' is shown in Figure 1.7.11. In this example we have $n = 1$, $m = 2$, $i = 0$, $j = 1$, and the function implemented by C is $f(x_0) = (x_0, x_0)$.

Let $f' : \{0,1\}^n \to \{0,1\}^{m-1}$ be the function implemented by C'. In Figure 1.7.11 we have $f'(x_0) = x_0$. Since a FANOUT gate was removed from C to obtain C', we have $|f'|_F < k = |f|_F$. Apply the induction hypothesis to f' and obtain p', C'_r, $\vec{a}'$, and g' as described in Theorem 1.7.10. In the example in Figure 1.7.11 the circuit C' is reversible. So, we can choose $C'_r = C'$. Note that $|C'_r| = 0 = |f|_F - 1$ which means that $p' = 0$, $\vec{a}' = ()$, and $g' : \{0,1\} \to \{0,1\}^0$, $(b) \mapsto ()$ have the required properties.

We construct the reversible circuit C_r. We set $p = p' + 2$ and add two ancilla bits a_{p-2} and a_{p-1} to C'_r. Additionally, we add a Toffoli gate to C'_r that replaces the

$$
\begin{array}{cc}
x_0 \;\bullet\!\!-\; y_0 = x_0 & \\
\quad\;\;\Big\lfloor\!\!\!-\; y_1 = x_0 & \qquad x_0 \;\text{——}\; y_0 = x_0 \\
\end{array}
$$

$$
\qquad\quad C \qquad\qquad\qquad\qquad\qquad C'
$$

Figure 1.7.11. $C'_r = C'$ is obtained by removing the fanout node from C.

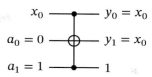

Figure 1.7.12. Construction of C_r from C'_r in Figure 1.7.11.

removed FANOUT gate. The first input of this gate is the output bit y_i of C'_r, and the second and third inputs are the new ancilla bits $a_{p-2} = 0$ and $a_{p-1} = 1$. As shown in Exercise 1.7.4, the output of the Toffoli gate is $(y_i, y_i, 1)$. The first two output edges of the Toffoli gate are connected to two output nodes that are in the same position as the removed output nodes of the removed FANOUT gate. Then $p = p' + 2$, C_r, $\vec{a} = \vec{a}'||(0, 1)$, and $g(\vec{x}) = g'(\vec{x})||(1)$ have the required properties. We also note that $|C_r| = |C'_r| + 1 = |C'|_F + 1 = |C|_F$ and $p = p' + 2 \leq 2|C'|_F + 2 = 2|C|_F$.

Figure 1.7.12 shows how this construction works for the example in Figure 1.7.11. There, the circuit C_r is simply the Toffoli gate that implements the fanout operation and we have $p = 2 = p' + 2$, $g(x_0) = 1$, and $|C_r| = 1 = |f|_F$.

Now suppose that C has the second property; i.e., there is a NAND gate whose outgoing edge is connected to an output node y_i of C for some $i \in \mathbb{Z}_m$. Remove this NAND gate and the corresponding output node y_i from C. Add two new output gates y'_i and y'_m to C and connect the incoming edges of the removed NAND to y'_i and y'_m. Denote by C' the resulting circuit and by f' the function implemented by C'. Then we have $|f'|_F = k - 1$. In the example shown in Figure 1.7.13 we have $n = 2$, $m = 1$, $y_0 = f(x_0, x_1) = x_0 \wedge x_1$, and $(y'_0, y'_1) = f'(x_0, x_1) = (x_0, x_1)$.

Apply the induction hypothesis to f' and obtain p', C'_r, $\vec{a}'$, and g' as described in the assertion of Theorem 1.7.10. In the example in Figure 1.7.13 we can set $C'_r = C'$ because C' is reversible. So we have $p' = 0$, $\vec{a}' = ()$, and $g' : \{0, 1\}^2 \to \{0, 1\}^0$, $\vec{b} \mapsto ()$. The reversible circuit C_r is obtained from C'_r as follows. We set $p = p' + 1$ and add one ancilla bit $a_{p-1} = 1$. In addition, we add a Toffoli gate that replaces the removed NAND gate. Its first input is $a_{p-1} = 1$. The two other inputs are y'_i and y'_m. The corresponding output gates are removed. Then the output of the Toffoli gate is $(y'_i \uparrow y'_m, y'_i, y'_m)$. The first outgoing edge of the Toffoli gate is connected to a new output gate y_i. The two other outgoing edges are connected to two new garbage output gates. So we have $g(\vec{x}) = g'(\vec{x})||(y'_i, y'_m)$. We note that $|C_r| = |C'_r| + 1 = |C'|_F + 1 = |C|_F$ and $p = p' + 1 \leq 2|C'|_F + 1 < 2|C'|_F + 2 = 2|C|_F$.

Figure 1.7.14 shows how this construction works for the example in Figure 1.7.12. In this example, a_0 is the first, x_0 is the second, and x_1 to the third input of the Toffoli gate. So the circuit C_r is a simple modification of the Toffoli gate that implements the

$$x_0 \qquad x_1 \qquad y_1 = x_0 \uparrow x_1 \qquad\qquad x_0 \longrightarrow y'_0 = x_0$$
$$x_1 \longrightarrow y'_1 = x_1$$

Figure 1.7.13. $C'_r = C'$ is obtained by removing the NAND gate from C.

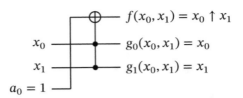

Figure 1.7.14. Construction of C_r from C_r' in Figure 1.7.13.

NAND gate and we have $p = 1 = p' + 1$, $g(x_0, x_1) = (x_0, x_1)$, and $|C_r| = 1 = |f|_F$. This concludes the proof.   □

Exercise 1.7.11. State and prove a theorem analogous to Theorem 1.7.10 where Fredkin gates are used instead of Toffoli gates.

When we use the construction from the proof of Theorem 1.7.10 to construct quantum circuits, then the garbage may be problematic. Therefore, we need the following theorem whose proof uses the so-called *uncompute trick*.

Theorem 1.7.12. *For all Boolean functions $f : \{0, 1\}^n \to \{0, 1\}^m$, $m, n \in \mathbb{N}$, there is $p \in \mathbb{N}_0$, $p \le 2|f|_F$, a reversible circuit D_r with $|D_r| = O(|f|_F)$ that uses only Toffoli, NOT, and CNOT gates such that D_r implements a function*

$$(1.7.9) \qquad h : \{0, 1\}^n \times \{0, 1\}^{n+p} \times \{0, 1\}^m \to \{0, 1\}^n \times \{0, 1\}^{n+p} \times \{0, 1\}^m$$

with

$$(1.7.10) \qquad\qquad h(\vec{x}, \vec{0}, \vec{y}) = (\vec{x}, \vec{0}, \vec{y} \oplus f(\vec{x}))$$

for all $\vec{x} \in \{0, 1\}^n$ and $\vec{y} \in \{0, 1\}^m$.

Proof. Let $f : \{0, 1\}^n \to \{0, 1\}^m$, $m, n \in \mathbb{N}$. Let p, C_r, $\vec{a}$, and g be as in Theorem 1.7.10. We construct the circuit D_r from C_r. This construction is illustrated in Figure 1.7.15 for C_r from Figure 1.7.14.

D_r has a total of $2n + p + m$ input nodes. The initial sequence consists of the first n nodes, represented as $\vec{x} = (x_0, \ldots, x_{n-1})$, followed by $\vec{x}' = (x'_0, \ldots, x'_{n-1})$. Subsequently, we include sequences of p ancillary input nodes, denoted as $\vec{a}' = (a'_0, \ldots, a_{p-1})$, and m input nodes, $\vec{y} = (y_0, \ldots, y_m)$. In the example shown in Figure 1.7.15, where $n = 2$, $m = p = 1$, we append two input nodes after x_1, set a_0 to 0, and introduce the input node y_0.

The circuit D_r applies a bitwise CNOT to $\vec{x}$ and $\vec{x}'$. If $\vec{x}' = \vec{0}$, then this operation copies $\vec{x}$ to $\vec{x}'$. In D_r, there is also a NOT gate after each ancilla input node whose value in $\vec{a}$ is 1. These NOT gates change an ancillary bit vector $\vec{0}$ of length p to $\vec{a}$. In the example, we have $a_0 = 1$. Therefore, a NOT gate is inserted behind the input node a_0.

Now, the reversible circuit C_r is applied to the input $\vec{x}' \| \vec{a}'$. This does not change $\vec{x}$ and $\vec{y}$. The circuit D_r then copies $f(\vec{x})$ to $\vec{y}$ using bitwise CNOT. In the example, C_r produces the bit string $(f(x_0, x_1), g_0(x_0, x_1), g_1(x_0, x_1))$ where $f(x_0, x_1) = x_0 \uparrow x_1$. In addition, a CNOT gate is required to copy $f(x_0, x_1)$ to y_0.

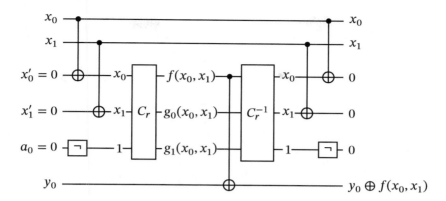

Figure 1.7.15. Construction of D_r from Theorem 1.7.12.

Finally, the *uncompute trick* is used. The inverse circuit C_r^{-1} is applied to the bits with indices $n \ldots n + p - 1$. This gives $(\vec{x}, \vec{a})$. Since the first n input nodes have not changed their values, CNOT gates can be used to change $\vec{x}'$ back to $\vec{0}$. In addition, applying NOT gates to the appropriate ancilla bits maps $\vec{a}$ to $\vec{0}$. In the example, two CNOT gates are used to obtain $\vec{x}' = \vec{0}$. Also, one NOT gate is required to change $a_0 = 1$ to $a_0 = 0$. The assertion about the size of D_r is verified in Exercise 1.7.13. □

Exercise 1.7.13. Show that the circuit D_r from Theorem 1.7.12 satisfies $|D_r| = O(|f|_F)$ and determine an appropriate O-constant.

Exercise 1.7.14. Construct a reversible circuit that computes the function $f(b_0, b_1) = b_0 \downarrow b_1$.

Chapter 2

Hilbert Spaces

Hilbert spaces, named after the mathematician David Hilbert, serve as the fundamental mathematical framework for quantum mechanics. In the context of quantum computing, finite-dimensional complex Hilbert spaces prove to be sufficient. These spaces are complex vector spaces equipped with an inner product, known as state spaces in the context of quantum computing. They provide a powerful mathematical representation of the quantum mechanics that is required for understanding quantum algorithms, in particular quantum states, their evolution, and measurement. In our presentation, we introduce and use the bra-ket notation. It is also referred to as Dirac notation due to its originator, physicist Paul Dirac, in the 1930s, and plays a central role in quantum mechanics. Its primary purpose is to simplify the presentation of the mathematical framework of Hilbert spaces, making it more accessible and concise.

It is crucial to emphasize the following point: in the realm of general quantum mechanics, which models the behavior of particles at the atomic and subatomic levels, finite-dimensional Hilbert spaces frequently fall short when addressing numerous aspects of the theory. As we confront progressively more complex and intricate quantum systems, the limitations inherent in finite-dimensional spaces become increasingly apparent. This expansion of complexity necessitates a departure from the confines of linear algebra, leading quantum theory into the profound domains of mathematical analysis

Appendix B presents general linear algebra, which forms the foundation for this chapter. Our objective here is to introduce the concept of finite-dimensional Hilbert spaces and explore their notable properties. We employ the widely used Dirac notation from physics, which proves to be elegant, and we select our examples from the realm of quantum computing. After the introductory sections, this chapter delves into vital concepts crucial for our discussion of quantum mechanics and quantum algorithms. We explore various special operators, such as Hermitian, unitary, and normal operators, involutions, and projections, which play pivotal roles in modeling quantum computation. Of particular significance are the spectral theorem, essential, for example, when stating the quantum mechanical postulates, and the Schmidt decomposition theorem, enabling the characterization of a fundamental quantum mechanical phenomenon: entanglement.

Throughout this chapter, k is a positive integer and $\mathbb{H}$ denotes a complex vector space of dimension k.

2.1. Kets and state spaces

In this section, we introduce the ket notation which is used to denote elements of the complex vector space $\mathbb{H}$. We illustrate this notation by applying it to the so-called state spaces, which provide a convenient way of modeling the states of quantum computers. Additionally, we demonstrate how elements of $\mathbb{H}$ can be represented as vectors in the complex vector space $\mathbb{C}^k$.

2.1.1. Kets. As explained in Appendix B, elements of vector spaces are called vectors and are denoted by $\vec{v}$ for some character v. In quantum physics, the following notation is used.

Definition 2.1.1. Every element of $\mathbb{H}$ is called a *ket*. Such a ket is denoted by $|\varphi\rangle$ for some character φ and is pronounced "ket-φ". The character φ may be replaced by any other character, number, or even word.

It is important to note that the sum of two kets $|\varphi\rangle, |\psi\rangle \in \mathbb{H}$ is written as $|\varphi\rangle + |\psi\rangle$ and not as $|\varphi + \psi\rangle$. In the same way, all expressions that contain several kets are written by keeping the $|\,\rangle$ notation for all kets. In the next section, we will present examples of the ket notation.

2.1.2. State spaces. We introduce the finite-dimensional complex vector spaces that will be central for modeling quantum computing. Let $n \in \mathbb{N}$. Consider the lexico-graphically ordered sequence

$$(2.1.1) \qquad\qquad B_n = \left(|\vec{b}\rangle \right)_{\vec{b} \in \{0,1\}^n}.$$

For example, if $n = 2$, then this sequence is

$$(2.1.2) \qquad\qquad B_2 = (|00\rangle, |01\rangle, |10\rangle, |11\rangle).$$

We define $\mathbb{H}_n$ as the set of all formal linear combinations of the elements of B_n with complex coefficients; that is,

$$(2.1.3) \qquad \mathbb{H}_n = \sum_{\vec{b} \in \{0,1\}^n} \mathbb{C} |\vec{b}\rangle = \left\{ \sum_{\vec{b} \in \{0,1\}^n} \alpha_{\vec{b}} |\vec{b}\rangle : \alpha_{\vec{b}} \in \mathbb{C} \text{ for all } \vec{b} \in \{0,1\}^n \right\}.$$

Note that the first sum in (2.1.3) is direct. On $\mathbb{H}_n$, we define componentwise addition and multiplication with complex scalars as follows. If

$$(2.1.4) \qquad |\varphi\rangle = \sum_{\vec{b} \in \{0,1\}^n} \alpha_{\vec{b}} |\vec{b}\rangle, \quad |\psi\rangle = \sum_{\vec{b} \in \{0,1\}^n} \beta_{\vec{b}} |\vec{b}\rangle$$

where $\alpha_{\vec{b}}, \beta_{\vec{b}} \in \mathbb{C}$ for all $\vec{b} \in \{0,1\}^n$, then we set

(2.1.5)
$$|\varphi\rangle + |\psi\rangle = \sum_{\vec{b} \in \{0,1\}^n} (\alpha_{\vec{b}} + \beta_{\vec{b}}) |\vec{b}\rangle.$$

Also, if $\gamma \in \mathbb{C}$, then we set

(2.1.6)
$$\gamma |\varphi\rangle = \sum_{\vec{b} \in \{0,1\}^n} \gamma \alpha_{\vec{b}} |\vec{b}\rangle.$$

Then $\mathbb{H}_n$ is a complex vector space, and the sequence $\left(|\vec{b}\rangle\right)_{\vec{b} \in \{0,1\}^n}$ is a basis of $\mathbb{H}_n$. Therefore, the dimension of $\mathbb{H}_n$ is 2^n. This vector space will be used in Section 3.1.5 to describe the states of n-qubit registers. This explains the following definition.

Definition 2.1.2. (1) The 2^n-dimensional complex vector space $\mathbb{H}_n$ from (2.1.3) equipped with addition and scalar multiplication as defined in (2.1.5) and (2.1.6) is called the *n-qubit state space*. In particular, $\mathbb{H}_1$ is called the *single-qubit state space*.

(2) The lexicographically ordered sequence $\left(|\vec{b}\rangle\right)_{\vec{b} \in \{0,1\}^n}$ is called the *computational basis* of $\mathbb{H}_n$.

Example 2.1.3. In classical computing, the bits 0 and 1 are used. In quantum computing, these bits are replaced by *quantum bits* or *qubits*. The *state* of a qubit is an element in the single-qubit state space $\mathbb{H}_1$. This will be explained in more detail in Section 3.1.2. The computational basis of $\mathbb{H}_1$ is $B = (|0\rangle, |1\rangle)$. Another basis of $\mathbb{H}_1$ is

(2.1.7)
$$(|x_+\rangle, |x_-\rangle) = \left(\frac{|0\rangle + |1\rangle}{\sqrt{2}}, \frac{|0\rangle - |1\rangle}{\sqrt{2}} \right).$$

Here the symbols x_+ and x_- are used to denote the basis elements. This basis will play a role in later sections.

Exercise 2.1.4. Show that $(|x_+\rangle, |x_-\rangle)$ is a basis of the single-qubit state space $\mathbb{H}_1$.

Example 2.1.5. We describe an alternative representation of the computational basis elements of the n-qubit state space $\mathbb{H}_n$. For this, we use the map

(2.1.8)
$$\text{stringToInt} : \{0,1\}^n \to \mathbb{Z}_{2^n}, \quad \vec{b} = (b_0 \cdots b_{n-1}) \mapsto \sum_{i=0}^{n-1} b_i 2^{n-i-1}$$

which was introduced in Definition 1.1.12. Also, in Exercise 1.1.13 it was shown that this map is a bijection. Using this bijection, we identify the bit vectors in $\{0,1\}^n$ with the integers in $\mathbb{Z}_{2^n}$. For example, the bit vector (010) is identified with the integer $0 \cdot 2^2 + 1 \cdot 2^1 + 0 \cdot 2^0 = 2$.

So we can write the computational basis of $\mathbb{H}_n$ as $(|b\rangle_n)_{b \in \mathbb{Z}_{2^n}}$, where the index n indicates that the number in the ket is considered as an n-bit string. For instance, the computational basis of $\mathbb{H}_2$ is $(|0\rangle_2, |1\rangle_2, |2\rangle_2, |3\rangle_2)$. We also obtain the following alternative representation of $\mathbb{H}_n$:

(2.1.9)
$$\mathbb{H}_n = \left\{ \sum_{b=0}^{2^n-1} \alpha_b |b\rangle_n : \alpha_b \in \mathbb{C} \text{ for all } b \in \mathbb{Z}_{2^n} \right\}.$$

2.1.3. Vector representations. Let $B = (|b_0\rangle, \ldots, |b_{k-1}\rangle)$ be any basis of $\mathbb{H}$. In Definition B.6.6 we have assigned to each $|\varphi\rangle \in \mathbb{H}$ its coefficient vector

$$(2.1.10) \qquad |\varphi\rangle_B = (\alpha_0, \ldots, \alpha_{k-1}) \in \mathbb{C}^k$$

with respect to B. It is the uniquely determined vector in $\mathbb{C}^k$ with

$$(2.1.11) \qquad |\varphi\rangle = B |\varphi\rangle_B = \sum_{i=0}^{k-1} \alpha_i |b_i\rangle.$$

Theorem B.6.7 states that the map

$$(2.1.12) \qquad \mathbb{H} \to \mathbb{C}^k, \quad |\varphi\rangle \mapsto |\varphi\rangle_B$$

is an isomorphism of $\mathbb{C}$-vector spaces. Using this isomorphism, we identify kets in $\mathbb{H}$ with vectors in $\mathbb{C}^k$ which is useful in many contexts.

Example 2.1.6. The definition of the single qubit state space $\mathbb{H}_1$ and Exercise 2.1.4 tell us that $B = (|0\rangle, |1\rangle)$ and $C = (|x_+\rangle, |x_-\rangle)$ from (2.1.7) are bases of $\mathbb{H}_1$. Let

$$(2.1.13) \qquad |\varphi\rangle = |0\rangle + i |1\rangle.$$

Then the coefficient vector of $|\varphi\rangle$ with respect to the basis B is

$$(2.1.14) \qquad |\varphi\rangle_B = (1, i).$$

Also, we have

$$(2.1.15) \qquad \begin{aligned} |\varphi\rangle &= |0\rangle + i |1\rangle \\ &= \frac{1+i}{\sqrt{2}} |x_+\rangle + \frac{1-i}{\sqrt{2}} |x_-\rangle. \end{aligned}$$

Hence, the coefficient vector of $|\varphi\rangle$ with respect to the basis C is

$$(2.1.16) \qquad |\varphi\rangle_C = \left(\frac{1+i}{\sqrt{2}}, \frac{1-i}{\sqrt{2}} \right).$$

2.2. Inner products

In this section, we introduce and discuss inner products on $\mathbb{H}$.

2.2.1. Basics. Inner products on $\mathbb{H}$ are maps $\mathbb{H} \times \mathbb{H} \to \mathbb{C}$ with certain properties which are explained in Definition 2.2.1. We will write them as

$$(2.2.1) \qquad \langle \cdot | \cdot \rangle : \mathbb{H} \times \mathbb{H} \to \mathbb{C}, \quad (|\varphi\rangle, |\psi\rangle) \mapsto \langle \varphi | \psi \rangle$$

and use the following simplifying notation. Let

$$(2.2.2) \qquad |\varphi\rangle = \sum_{i=0}^{m-1} \alpha_i |\varphi_i\rangle, \quad |\psi\rangle = \sum_{i=0}^{n-1} \beta_i |\psi_i\rangle$$

with $m, n \in \mathbb{N}$, $\alpha_i \in \mathbb{C}$, $|\varphi_i\rangle \in \mathbb{H}$ for all $i \in \mathbb{Z}_m$, and $\beta_i \in \mathbb{C}$, $|\psi_i\rangle \in \mathbb{H}$ for all $i \in \mathbb{Z}_n$. Then we write

$$(2.2.3) \qquad \langle \varphi | \psi \rangle = \left(\sum_{i=0}^{m-1} \alpha_i \langle \varphi_i | \right) \left(\sum_{i=0}^{n-1} \beta_i |\psi_i\rangle \right).$$

So, we change each ket $|\varphi_i\rangle$ in the left argument to a so-called *bra* $\langle\varphi_i|$ (see Section 2.2.3 for an explanation) with the same symbol inside and omit the outer $\langle\rangle$. Using this notation, we now define inner products.

Definition 2.2.1. An *inner product on* $\mathbb{H}$ is a map

$$(2.2.4) \qquad \langle\cdot|\cdot\rangle : \mathbb{H} \times \mathbb{H} \to \mathbb{C}, \quad (|\varphi\rangle, |\psi\rangle) \mapsto \langle\varphi|\psi\rangle$$

that satisfies the following three conditions for all kets $|\xi\rangle, |\varphi\rangle, |\psi\rangle \in \mathbb{H}$ and all scalars $\alpha \in \mathbb{C}$.

(1) *Linearity in the second argument*: $\langle\xi|(|\varphi\rangle + |\psi\rangle) = \langle\xi|\varphi\rangle + \langle\xi|\psi\rangle$ and $\langle\varphi|(\alpha|\psi\rangle) = \alpha\langle\varphi|\psi\rangle$.

(2) *Conjugate symmetry*: $\langle\psi|\varphi\rangle = \overline{\langle\varphi|\psi\rangle}$. This property is also called *Hermitian symmetry* or *conjugate commutativity*. It implies that $\langle\varphi|\varphi\rangle$ is a real number.

(3) *Positive definiteness*: $\langle\varphi|\varphi\rangle \geq 0$ and $\langle\varphi|\varphi\rangle = 0$ if and only if $|\varphi\rangle = 0$. This property is also called *positivity*.

Inner products on real vector spaces are defined analogously, but the conjugate symmetry condition becomes a symmetry condition. Note that the definition of inner products does not require $\mathbb{H}$ to be finite dimensional.

Exercise 2.2.2. Show that for all $|\varphi\rangle \in \mathbb{H}$ the inner product $\langle\varphi|\varphi\rangle$ is a real number.

We present three important properties of inner products.

Proposition 2.2.3. *Let* $\langle\cdot|\cdot\rangle$ *be an inner product on* $\mathbb{H}$. *Then for all* $\alpha \in \mathbb{C}, |\xi\rangle, |\varphi\rangle, |\psi\rangle \in \mathbb{H}$, *and the zero element* $\vec{0} \in \mathbb{H}$ *the following hold.*

(1) $\langle\vec{0}|\varphi\rangle = \langle\varphi|\vec{0}\rangle = 0$.

(2) $(\langle\xi| + \langle\varphi|)|\psi\rangle = \langle\xi|\psi\rangle + \langle\varphi|\xi\rangle$ *and* $(\alpha\langle\varphi|)|\psi\rangle = \overline{\alpha}\langle\varphi|\psi\rangle$.

(3) $(\langle\varphi| + \langle\psi|)(|\varphi\rangle + |\psi\rangle) = \langle\varphi|\varphi\rangle + 2\Re\langle\varphi|\psi\rangle + \langle\psi|\psi\rangle$.

The second property in Proposition 2.2.3 is called *sesquilinearity* or *conjugate linearity* of the inner product in the first argument.

Exercise 2.2.4. Prove Proposition 2.2.3.

Using the linearity of the inner product in the second argument and the conjugate linearity in the first argument we obtain the distributive law

$$(2.2.5) \qquad \left(\sum_{i=0}^{m-1} \alpha_i \langle\varphi_i|\right)\left(\sum_{j=0}^{n-1} \beta_j |\psi_j\rangle\right) = \sum_{i=0}^{m-1}\sum_{j=0}^{n-1} \overline{\alpha_i}\beta_j\langle\varphi_i|\psi_j\rangle$$

where $m, n \in \mathbb{N}$, $\alpha_i \in \mathbb{C}$, $|\varphi_i\rangle \in \mathbb{H}$ for all $i \in \mathbb{Z}_m$, and $\beta_j \in \mathbb{C}$, $|\psi_j\rangle \in \mathbb{H}$ for all $j \in \mathbb{Z}_n$.

The next proposition is useful in many contexts.

Proposition 2.2.5. *If* $\langle\cdot|\cdot\rangle$ *is an inner product on* $\mathbb{H}$ *and if* $\mathbb{H}'$ *is a linear subspace in* $\mathbb{H}$, *then the restriction* $\langle\cdot|\cdot\rangle_{\mathbb{H}'}$ *of* $\langle\cdot|\cdot\rangle$ *to* $\mathbb{H}'$ *is an inner product on* $\mathbb{H}'$.

Exercise 2.2.6. Prove Proposition 2.2.5.

2.2.2. Construction of inner products. We construct inner products on $\mathbb{H}$ and begin with the case $\mathbb{H} = \mathbb{C}^k$. For the construction we define the dual of vectors in $\mathbb{C}^k$. In Section B.4.1 we identify every vector $\vec{v} = (v_0, \dots, v_{k-1}) \in \mathbb{C}^k$ with the matrix

$$(2.2.6) \qquad \begin{pmatrix} v_0 \\ v_1 \\ \vdots \\ v_{k-1} \end{pmatrix} \in \mathbb{C}^{(k,1)}$$

that has $\vec{v}$ as its only column vector. This is used in the following definition.

Definition 2.2.7. Let $k \in \mathbb{N}$ and let $\vec{v} = (v_0, \dots, v_{k-1}) \in \mathbb{C}^k$. Then we define the *dual* $\vec{v}^*$ of $\vec{v}$ as the complex conjugate and transpose of the matrix in $\mathbb{C}^{(1,k)}$ with $\vec{v}$ as its only row vector; that is,

$$(2.2.7) \qquad \vec{v}^* = \overline{\vec{v}^{\mathrm{T}}} = (\overline{v_0}, \dots, \overline{v_{k-1}}).$$

The definition of matrix multiplication allows multiplying the dual $\vec{v}^*$ of a vector $\vec{v} = (v_0, \dots, v_{k-1}) \in \mathbb{C}^k$ with another vector $\vec{w} = (w_0, \dots, w_{k-1}) \in \mathbb{C}^k$. The result is

$$(2.2.8) \qquad \vec{v}^* \vec{w} = (\overline{v_0}, \dots, \overline{v_{k-1}}) \begin{pmatrix} w_0 \\ \vdots \\ w_{k-1} \end{pmatrix} = \sum_{i=0}^{k-1} \overline{v_i} w_i.$$

Example 2.2.8. Let $k = 2$, $\vec{v} = (1, i)$, and $\vec{w} = (i, 1)$. Then we have $\vec{v}^* = (1, -i)$ and $\vec{v}^* \vec{w} = i - i = 0$.

Now we introduce the standard Hermitian inner product on $\mathbb{C}^k$.

Theorem 2.2.9. *The map*

$$(2.2.9) \qquad \langle \cdot | \cdot \rangle : \mathbb{C}^k \times \mathbb{C}^k \to \mathbb{C}, \quad (\vec{v}, \vec{w}) \mapsto \langle \vec{v} | \vec{w} \rangle = \vec{v}^* \vec{w}$$

is an inner product on the complex vector space $\mathbb{C}^k$. It is called the Hermitian inner product *on $\mathbb{C}^k$.*

We will always write the standard Hermitian inner product on $\mathbb{C}^k$ as $\langle \cdot | \cdot \rangle$.

Exercise 2.2.10. Prove Theorem 2.2.9.

We use Theorem 2.2.9 to construct further inner products on $\mathbb{H}$.

Corollary 2.2.11. *Let B be a basis of $\mathbb{H}$. Then the map*

$$(2.2.10) \qquad \langle \cdot | \cdot \rangle_B : \mathbb{H} \times \mathbb{H} \to \mathbb{C}, \quad (|\varphi\rangle, |\psi\rangle) \mapsto \langle \varphi | \psi \rangle_B = |\varphi\rangle_B^* |\psi\rangle_B$$

is an inner product on $\mathbb{H}$. It is called the Hermitian inner product *on $\mathbb{H}$ with respect to the basis B.*

Exercise 2.2.12. Prove Corollary 2.2.11.

Next, we show that the Hermitian inner product with respect to a basis B of $\mathbb{H}$ can be used to determine the coefficient vectors of kets in $\mathbb{H}$ with respect to B.

Proposition 2.2.13. *Let $B = (|b_0\rangle, \ldots, |b_{k-1}\rangle)$ be a basis of $\mathbb{H}$. Then the following hold.*

(1) *For all $i, j \in \mathbb{Z}_k$ we have*

(2.2.11) $$\langle b_i | b_j \rangle_B = \delta_{i,j}.$$

(2) *For all $|\varphi\rangle \in \mathbb{H}$ we have*

(2.2.12) $$|\varphi\rangle = \sum_{i=0}^{k-1} \langle b_i | \varphi \rangle_B |b_i\rangle.$$

Exercise 2.2.14. Prove Proposition 2.2.13.

2.2.3. Bras. We introduce and discuss the bra notation that will further simplify the presentation of the theory of Hilbert spaces. We assume that $\langle \cdot | \cdot \rangle$ is an inner product on $\mathbb{H}$. In the next definition, the dual of $\mathbb{H}$ is used, which — as explained in Section B.3.2 — is $\mathbb{H}^* = \mathrm{End}(\mathbb{H}, \mathbb{C})$.

Definition 2.2.15. Every element of the dual $\mathbb{H}^*$ of $\mathbb{H}$ is called a *bra*. Such a bra is denoted by $\langle \varphi |$ for some character φ and is pronounced "bra-φ". The character φ may be replaced by any other character, number, or even word.

The next theorem associates with every ket in $\mathbb{H}$ a bra in $\mathbb{H}^*$.

Theorem 2.2.16. (1) *For all $|\varphi\rangle \in \mathbb{H}$ the map*

(2.2.13) $$\langle \varphi | : \mathbb{H} \to \mathbb{C}, \quad |\psi\rangle \mapsto \langle \varphi | \psi \rangle$$

is in the dual $\mathbb{H}^*$ of $\mathbb{H}$. It is called the dual of $|\varphi\rangle$.

(2) *The map*

(2.2.14) $$\mathbb{H} \mapsto \mathbb{H}^*, \quad |\varphi\rangle \mapsto \langle \varphi |$$

is a conjugate linear bijection; that is, for all $|\varphi\rangle, |\psi\rangle \in \mathbb{H}$ and $\alpha \in \mathbb{C}$ the following hold. For $|\xi\rangle = |\varphi\rangle + |\psi\rangle$ we have $\langle \xi | = \langle \varphi | + \langle \psi |$ and for $|\xi\rangle = \alpha |\varphi\rangle$ we have $\langle \xi | = \overline{\alpha} \langle \varphi |$.

Proof. It follows from the linearity in the second argument of the inner product that for all $|\varphi\rangle \in \mathbb{H}$ the map $\langle \varphi |$ from (2.2.13) is in $\mathbb{H}^*$. Furthermore, due to the positivity of the inner product, the map (2.2.14) is injective. Hence, Theorem B.6.18 implies that this map is a bijection. Its conjugate linearity follows from Proposition 2.2.3. $\square$

Theorem 2.2.16 shows that all elements in $\mathbb{H}^*$ can be written as $\langle \varphi |$ for some uniquely determined $|\varphi\rangle \in \mathbb{H}$ and vice versa. Therefore, we will always use the same character, number, or word inside $|\rangle$ and $\langle |$ to denote the kets and bras that correspond to each other. The construction of $\langle \varphi |$ from $|\varphi\rangle \in \mathbb{H}$ will be explained in Proposition 2.2.37.

The bra notation is quite elegant since for every $|\varphi\rangle, |\psi\rangle \in \mathbb{H}$ the image of $|\psi\rangle \in \mathbb{H}$ under $\langle \varphi |$ is obtained by "gluing" the two expressions $\langle \varphi |$ and $|\psi\rangle$ together, giving $\langle \varphi | \psi \rangle$. We also obtain the following interpretation of the notation introduced in (2.2.3): the inner product of a linear combination of kets with another linear combination of kets is obtained by gluing together the linear combination of bras corresponding to the first linear combination of kets with the second linear combination of kets. The linear combinations are written in parentheses. Also, the distributive law from (2.2.5) holds for bras and kets.

Exercise 2.2.17. Determine the images of the computational basis states of $\mathbb{H}_1$ under $\langle x_+|$ and $\langle x_-|$.

2.2.4. Hilbert spaces. We define finite-dimensional Hilbert spaces. To define general Hilbert spaces, the notion of a *complete metric space* is required, which is beyond the scope of this book.

Definition 2.2.18. A finite-dimensional Hilbert space is a pair $(V, \langle \cdot | \cdot \rangle)$ where V is a finite-dimensional real or complex vector space and $\langle \cdot | \cdot \rangle$ is an inner product on V.

By Corollary 2.2.11, there is an inner product on every finite-dimensional complex vector space. Hence, every finite-dimensional complex vector space can be made a Hilbert space by choosing such an inner product on it. Also, if $(\mathbb{H}, \langle \cdot | \cdot \rangle)$ is a finite-dimensional complex Hilbert space and if $\mathbb{H}'$ is a linear subspace of $\mathbb{H}$, then it follows from Proposition 2.2.5 that $(\mathbb{H}', \langle \cdot | \cdot \rangle_{\mathbb{H}'})$ is also a finite-dimensional complex Hilbert space where $\langle \cdot | \cdot \rangle_{\mathbb{H}'}$ denotes the restriction of the inner product $\langle \cdot | \cdot \rangle$ to $\mathbb{H}'$.

State spaces are of particular importance in the discussion of quantum algorithms. We now define inner products on these spaces so that they become Hilbert spaces.

Definition 2.2.19. Let $n \in \mathbb{N}$. Then we denote by $\langle \cdot | \cdot \rangle$ the inner product on $\mathbb{H}_n$ with respect to the computational basis of $\mathbb{H}_n$. We also write $\mathbb{H}_n$ for the Hilbert space $(\mathbb{H}_n \langle \cdot | \cdot \rangle)$.

We now discuss some examples of finite-dimensional complex Hilbert spaces.

Example 2.2.20. Consider the single-qubit state space $\mathbb{H}_1$. We determine the inner product of

$$(2.2.15) \qquad |\varphi\rangle = \alpha |0\rangle + \beta |1\rangle \quad \text{and} \quad |\psi\rangle = \gamma |0\rangle + \delta |1\rangle$$

where $\alpha, \beta, \gamma, \delta \in \mathbb{C}$. Using (2.2.5) and Proposition 2.2.13, we obtain the following:

$$(2.2.16) \qquad \langle \varphi | \psi \rangle = \overline{\alpha} \gamma + \overline{\beta} \delta.$$

Next, we recall that $C = (|x_+\rangle, |x_-\rangle)$ introduced in (2.1.7) is another basis of $\mathbb{H}_1$. So $(\mathbb{H}_1, \langle \cdot | \cdot \rangle_C)$ is also a Hilbert space. The representation of $|\varphi\rangle$ and $|\psi\rangle$ with respect to C is

$$(2.2.17) \qquad |\varphi\rangle = \frac{\alpha + \beta}{\sqrt{2}} |x_+\rangle + \frac{\alpha - \beta}{\sqrt{2}} |x_-\rangle, \quad |\psi\rangle = \frac{\gamma + \delta}{\sqrt{2}} |x_+\rangle + \frac{\gamma - \delta}{\sqrt{2}} |x_-\rangle.$$

This implies

$$
\begin{aligned}
(2.2.18) \qquad \langle \varphi | \psi \rangle_C &= \frac{1}{2}((\overline{\alpha} + \overline{\beta})(\gamma + \delta) + (\overline{\alpha} - \overline{\beta})(\gamma - \delta)) \\
&= \frac{1}{2}(\overline{\alpha}\gamma + \overline{\alpha}\delta + \overline{\beta}\gamma + \overline{\beta}\delta + \overline{\alpha}\gamma - \overline{\alpha}\delta - \overline{\beta}\gamma + \overline{\beta}\delta) \\
&= \overline{\alpha}\gamma + \overline{\beta}\delta.
\end{aligned}
$$

We see that

$$(2.2.19) \qquad \langle \varphi | \psi \rangle_B = \langle \varphi | \psi \rangle_C$$

for all $|\varphi\rangle, |\psi\rangle \in \mathbb{H}_1$. Hence, the Hilbert spaces $(\mathbb{H}, \langle \cdot | \cdot \rangle_B)$ and $(\mathbb{H}, \langle \cdot | \cdot \rangle_C)$ are identical.

Exercise 2.2.21. Find a basis C of $\mathbb{H}_1$ such that the Hilbert spaces $(\mathbb{H}_1, \langle \cdot | \cdot \rangle)$ and $(\mathbb{H}_1, \langle \cdot | \cdot \rangle_C)$ are different.

2.2.5. Norm. Another important notion is the norm on a Hilbert space which we define now.

Definition 2.2.22. A *norm* on $\mathbb{H}$ is a function $f : \mathbb{H} \to \mathbb{R}, |\varphi\rangle \mapsto f |\varphi\rangle$ which for all $|\varphi\rangle, |\psi\rangle \in \mathbb{H}$ and all $\alpha \in \mathbb{C}$ satisfies the following conditions.

(1) *Triangle inequality:* $f(|\varphi\rangle + |\psi\rangle) \leq f |\varphi\rangle + f |\psi\rangle$.

(2) *Absolute homogeneity:* $f(\alpha |\varphi\rangle) = |\alpha| f |\varphi\rangle$.

(3) *Positive definiteness:* $f |\varphi\rangle \geq 0$ and $f |\varphi\rangle = 0$ if and only $|\varphi\rangle = 0$.

Exercise 2.2.23. Verify that the map

$$(2.2.20) \qquad \mathbb{C} \to \mathbb{C}, \quad \alpha = \mathfrak{R}\alpha + i\mathfrak{I}\alpha \mapsto |\alpha| = \sqrt{(\mathfrak{R}\alpha)^2 + (\mathfrak{I}\alpha)^2}$$

is a norm on $\mathbb{C}$.

Exercise 2.2.24. Show that the map

$$(2.2.21) \qquad |\cdot| : \mathbb{C}^k \to \mathbb{C}, \quad \vec{v} = (v_0, \ldots, v_{k-1}) \mapsto |\vec{v}| = \sum_{i=0}^{k-1} |v_i|$$

is a norm on $\mathbb{C}^k$.

We show how to construct a norm from an inner product on $\mathbb{H}$. For this, we let $\langle \cdot | \cdot \rangle$ be such an inner product. We refer to the Hilbert space $(\mathbb{H}, \langle \cdot | \cdot \rangle)$ simply by $\mathbb{H}$. But we must keep in mind which inner product we have chosen on $\mathbb{H}$ since changing the inner product changes the norm.

Proposition 2.2.25. *The map*

$$(2.2.22) \qquad \|\cdot\| : \mathbb{H} \to \mathbb{R}, \quad |\varphi\rangle \mapsto \|\varphi\| = \sqrt{\langle \varphi | \varphi \rangle}$$

is a norm on $\mathbb{H}$ that satisfies the Cauchy-Schwarz inequality

$$(2.2.23) \qquad |\langle \varphi | \psi \rangle| \leq \|\varphi\| \|\psi\|$$

for all $|\varphi\rangle, |\psi\rangle \in \mathbb{H}$.

Proof. We start by proving the Cauchy-Schwarz inequality. Let $|\varphi\rangle, |\psi\rangle \in \mathbb{H}$ and $\alpha \in \mathbb{C}$. For $x \in \mathbb{R}$ let

$$(2.2.24) \qquad \begin{aligned} p(x) &= ((\langle \varphi| - x\langle \psi|)(|\varphi\rangle - x |\psi\rangle)) \\ &= x^2 \langle \psi | \psi \rangle - (\langle \varphi | \psi \rangle + \langle \psi | \varphi \rangle)x + \langle \varphi | \varphi \rangle. \end{aligned}$$

Since $\langle \varphi | \psi \rangle + \langle \psi | \varphi \rangle = 2\mathfrak{R}\langle \varphi | \psi \rangle$ is a real number, it follows that the coefficients of $p(x)$, considered as a quadratic polynomial, are real numbers. The discriminant of this polynomial (see Exercise A.4.49) is

$$(2.2.25) \qquad \Delta(p) = (\langle \varphi | \psi \rangle + \langle \psi | \varphi \rangle)^2 - 4\langle \varphi | \varphi \rangle \langle \psi | \psi \rangle.$$

This equation and the conjugate symmetry of the inner product imply

$$
\begin{aligned}
\Delta(p) &= (\langle\varphi|\psi\rangle + \langle\psi|\varphi\rangle)^2 - 4\langle\varphi|\varphi\rangle\langle\psi|\psi\rangle \\
&= |\langle\varphi|\psi\rangle + \langle\psi|\varphi\rangle|^2 - 4\langle\varphi|\varphi\rangle\langle\psi|\psi\rangle \\
&\leq (|\langle\varphi|\psi\rangle| + |\langle\psi|\varphi\rangle|)^2 - 4\langle\varphi|\varphi\rangle\langle\psi|\psi\rangle \\
&= 4(|\langle\varphi|\psi\rangle|^2 - \langle\varphi|\varphi\rangle\langle\psi|\psi\rangle).
\end{aligned}
$$

(2.2.26)

But $p(x)$ is nonnegative for all $x \in \mathbb{R}$. Therefore, this polynomial can have at most one real root, which means that its discriminant is nonpositive. So (2.2.26) implies the Cauchy-Schwarz inequality.

Now we apply the Cauchy-Schwarz inequality and obtain

$$
\begin{aligned}
\||\varphi\rangle - |\psi\rangle\|^2 &= ((\langle\varphi| - \langle\psi|)(|\varphi\rangle - |\psi\rangle)) \\
&= \|\varphi\|^2 - \langle\varphi|\psi\rangle - \langle\psi|\varphi\rangle + \|\psi\|^2 \\
&\leq \|\varphi\|^2 - 2\|\varphi\|\|\psi\| + \|\psi\|^2 \\
&= (\|\varphi\| - \|\psi\|)^2.
\end{aligned}
$$

(2.2.27)

This implies the triangle inequality.

Next, we let $\alpha \in \mathbb{C}$. Then, the linearity in the second element and the conjugate linearity in the first argument of the inner product imply

(2.2.28) $$\|\alpha\,|\varphi\rangle\|^2 = (\alpha\,|\varphi\rangle)(\alpha\,|\varphi\rangle) = |\alpha|^2\langle\varphi|\varphi\rangle = |\alpha|^2\|\varphi\|^2.$$

This implies the absolute homogeneity of the norm.

Finally, the positive definiteness of $\|\cdot\|$ immediately follows from the positive definiteness of the inner product. $\qquad\square$

Definition 2.2.26. The norm $\|\cdot\| \to \mathbb{C}$, $|\varphi\rangle \mapsto \|\varphi\| = \sqrt{\langle\varphi|\varphi\rangle}$ defined in Proposition 2.2.25 is called the *Euclidean norm* on the Hilbert space $\mathbb{H}$. It depends on the inner product on $\mathbb{H}$. For $|\varphi\rangle \in \mathbb{H}$ we also refer to $\|\varphi\|$ as the *length* of $|\varphi\rangle$.

2.2.6. Orthogonality. We discuss the important concept of orthogonality. For this, we fix an inner product $\langle\cdot|\cdot\rangle$ on $\mathbb{H}$.

Definition 2.2.27. (1) Two kets in $\mathbb{H}$ are called *orthogonal to each other* if their inner product is zero.

(2) Two subsets S_0 and S_1 of $\mathbb{H}$ are called *orthogonal to each other* if all $|\phi_0\rangle \in S_0$ and $|\varphi_1\rangle \in S_1$ are orthogonal to each other.

(3) A sequence B in $\mathbb{H}$ is called *orthogonal* if any two different elements of B are orthogonal.

(4) A sequence B in $\mathbb{H}$ is called *orthonormal* if it is orthogonal and all its elements have Euclidean norm 1.

(5) By an *orthogonal or orthonormal basis* of $\mathbb{H}$ we mean a basis of $\mathbb{H}$ that is orthogonal or orthonormal as a sequence, respectively.

Example 2.2.28. The empty sequence () in $\mathbb{H}$ is orthonormal since it has no elements and, therefore, all statements about all of its elements are true.

Exercise 2.2.29. Let B be a basis of $\mathbb{H}$ and consider the Hilbert space $(\mathbb{H}, \langle \cdot | \cdot \rangle_B)$. Show that the basis B is an orthonormal basis in this Hilbert space.

Exercise 2.2.30. Show that the basis $(|x_+\rangle, |x_-\rangle)$ of $\mathbb{H}_1$ is orthonormal.

The next theorem presents the *Gram-Schmidt procedure* that constructs an orthogonal basis from any basis of $\mathbb{H}$.

Theorem 2.2.31 (Gram-Schmidt procedure). *Let $C = (|c_0\rangle, \ldots, |c_{k-1}\rangle)$ be a basis of $\mathbb{H}$. Set*

$$(2.2.29) \qquad\qquad |b_0\rangle = |c_0\rangle$$

and for $1 \leq j < k$ let

$$(2.2.30) \qquad\qquad |b_j\rangle = |c_j\rangle - \sum_{i=0}^{j-1} \frac{\langle b_i | c_j \rangle}{\langle b_i | b_i \rangle} |b_i\rangle.$$

Then $(|b_0\rangle, \ldots, |b_{k-1}\rangle)$ is an orthogonal basis of $\mathbb{H}$ and for $0 \leq j < k$ we have

$$(2.2.31) \qquad\qquad \mathrm{Span}\{|b_0\rangle, \ldots, |b_j\rangle\} = \mathrm{Span}\{|c_0\rangle, \ldots, |c_j\rangle\}.$$

Proof. We prove the assertion by induction on the dimension k of $\mathbb{H}$. If $k = 1$, then $(|b_0\rangle) = (|c_0\rangle)$ is an orthogonal basis of $\mathbb{H}$. Assume that $k > 1$ and that the assertion holds for $k-1$. Set $\mathbb{H}' = \mathrm{Span}\{|c_0\rangle, \ldots, |c_{k-2}\rangle\}$. The induction hypothesis then implies that $B' = (|b_0\rangle, \ldots, |b_{k-2}\rangle)$ is an orthogonal basis of $\mathbb{H}'$ and (2.2.31) holds for $0 \leq j \leq k-2$. Furthermore, the definition of $|b_{k-1}\rangle$ in (2.2.30) implies (2.2.31) for $j = k-1$. It remains to show that $\langle b_j | b_{k-1} \rangle = 0$ for $0 \leq j \leq k-2$. So, let $j \in \{0, \ldots, k-2\}$. Then the linearity in the second argument of the inner product and the orthogonality of the sequence B' imply

$$
\begin{aligned}
(2.2.32) \qquad \langle b_j | b_{k-1} \rangle &= \langle b_j | \left(|c_{k-1}\rangle - \sum_{i=0}^{k-2} \frac{\langle b_i | c_{k-1} \rangle}{\langle b_i | b_i \rangle} |b_i\rangle \right) \\
&= \langle b_j | c_{k-1} \rangle - \langle b_j | \sum_{i=0}^{k-2} \frac{\langle b_i | c_{k-1} \rangle}{\langle b_i | b_i \rangle} |b_i\rangle \\
&= \langle b_j | c_{k-1} \rangle - \sum_{i=0}^{k-2} \frac{\langle b_i | c_{k-1} \rangle}{\langle b_i | b_i \rangle} \langle b_j | b_i \rangle \\
&= \langle b_j | c_{k-1} \rangle - \langle b_j | c_{k-1} \rangle = 0.
\end{aligned}
$$

This concludes the proof. $\qquad\qquad\qquad\qquad\qquad\qquad\qquad\qquad\qquad\qquad\qquad\square$

The process of constructing the orthogonal basis $B = (|b_0\rangle, \ldots, |b_{k-1}\rangle)$ from the basis $C = (|c_0\rangle, \ldots, |c_{k-1}\rangle)$ of $\mathbb{H}$ presented in Theorem 2.2.31 is referred to as the *Gram-Schmidt orthogonalization of C*. We also call the resulting orthogonal basis B the *Gram-Schmidt orthogonalization* of the basis C.

Example 2.2.32. Consider the basis $(|c_0\rangle, |c_1\rangle) = (|0\rangle, |0\rangle + |1\rangle)$ of the single-qubit state space $\mathbb{H}_1$. It is not orthogonal since

(2.2.33) $$\langle c_0 | c_1 \rangle = \langle 0 | (|0\rangle + |1\rangle) = \langle 0 | 0 \rangle + \langle 0 | 1 \rangle = 1.$$

We apply Gram-Schmidt orthogonalization to this basis. We obtain

(2.2.34) $$|b_0\rangle = |c_0\rangle = |0\rangle$$

and

(2.2.35)
$$|b_1\rangle = |c_1\rangle - \frac{\langle b_0 | c_1 \rangle}{\langle b_0 | c_0 \rangle} |b_0\rangle = (|0\rangle + |1\rangle) - \frac{\langle 0 | (|0\rangle + |1\rangle)}{\langle 0 | 0 \rangle} |0\rangle$$
$$= (|0\rangle + |1\rangle) - (\langle 0 | 0 \rangle + \langle 0 | 1 \rangle) |0\rangle = (|0\rangle + |1\rangle) - |0\rangle = |1\rangle.$$

Hence, the Gram-Schmidt orthogonalization of the basis $(|0\rangle, |0\rangle + |1\rangle)$ of $\mathbb{H}_1$ gives $(|0\rangle, |1\rangle)$.

From Theorem 2.2.31 we obtain the following theorem.

Theorem 2.2.33. *Every orthogonal or orthonormal sequence in $\mathbb{H}$ is linearly independent and can be extended to an orthogonal or orthonormal basis of $\mathbb{H}$, respectively.*

Proof. Let $l \in \mathbb{N}$ and let $C' = (|c_0\rangle, \ldots |c_{l-1}\rangle)$ be an orthogonal sequence in $\mathbb{H}$. By Theorem B.7.2 it can be extended to a basis C of $\mathbb{H}$. Let $(|b_0\rangle, \ldots, |b_{k-1}\rangle)$ be the Gram-Schmidt orthogonalization of C. We show by induction on l that

(2.2.36) $$(|b_0\rangle, \ldots, |b_{l-1}\rangle) = (|c_0\rangle, \ldots, |c_{l-1}\rangle).$$

For $l = 1$ the assertion follows from (2.2.29). Now let $l \in \mathbb{N}$, $1 < l \leq k$, and assume that

(2.2.37) $$(|b_0\rangle, \ldots, |b_{l-2}\rangle) = (|c_0\rangle, \ldots, |c_{l-2}\rangle).$$

Then we have

(2.2.38)
$$|b_{l-1}\rangle = |c_{l-1}\rangle - \sum_{i=0}^{l-2} \frac{\langle b_i | c_{l-1} \rangle}{\langle b_i | b_i \rangle} |b_i\rangle$$
$$= |c_{l-1}\rangle - \sum_{i=0}^{l-2} \frac{\langle c_i | c_{l-1} \rangle}{\langle c_i | c_i \rangle} |c_i\rangle = |c_{l-1}\rangle.$$

Hence, $(|b_l\rangle, \ldots, |b_{k-1}\rangle)$ extends C' to an orthogonal basis of $\mathbb{H}$.

In order to extend an orthonormal sequence to an orthonormal basis, one first extends this sequence to an orthogonal basis and then normalizes the appended elements. □

Theorem 2.2.33 implies the following result.

Corollary 2.2.34. *Every finite-dimensional Hilbert space has an orthonormal basis*

Example 2.2.35. On $\mathbb{H}_1$ we use the Hermitian inner product with respect to the computational basis $(|0\rangle, |1\rangle)$. Then $|x_+\rangle = \frac{|0\rangle + |1\rangle}{\sqrt{2}}$ has length 1. We use Gram-Schmidt orthogonalization to append $B = (|x_+\rangle)$ to an orthonormal basis of $\mathbb{H}_1$. First, we note that $((|b_0\rangle, |c_1\rangle) = (|x_+\rangle, |0\rangle)$ is a basis of $\mathbb{H}_1$ which is not orthogonal. Gram-Schmidt orthogonalization gives

(2.2.39)
$$|b_1\rangle = |c_1\rangle - \frac{\langle b_0|c_1\rangle}{\langle b_0|b_0\rangle}|b_0\rangle = \frac{|0\rangle - |1\rangle}{2}.$$

Since $\||b_1\rangle\| = 1/\sqrt{2}$, this gives the orthonormal basis $(|b_0\rangle, \sqrt{2}|b_1\rangle) = (|x_+\rangle, |x_-\rangle)$ of $\mathbb{H}_1$

Corollary 2.2.34 implies the following result, which shows that all inner products on $\mathbb{H}$ are Hermitian inner products with respect to some basis of $\mathbb{H}$.

Proposition 2.2.36. *If B is an orthonormal basis of $\mathbb{H}$, then $\langle \cdot | \cdot \rangle$ is the Hermitian inner product on $\mathbb{H}$ with respect to B.*

Proof. Let $|\varphi\rangle, |\psi\rangle \in \mathbb{H}$, and let $|\varphi\rangle_B = (\alpha_0, \ldots, \alpha_{k-1})$, $|\psi\rangle_B = (\beta_0, \ldots, \beta_{k-1})$. Then we have

(2.2.40)
$$\langle \varphi, \psi \rangle = \left(\sum_{i=0}^{k-1} \alpha_i \langle b_i| \right)\left(\sum_{j=0}^{k-1} \beta_j |b_i\rangle\right) = \sum_{i,j=0}^{k-1} \overline{\alpha_i}\beta_j \langle b_i, b_j\rangle$$
$$= \sum_{i,j=0}^{k-1} \overline{\alpha_i}\beta_j \delta_{i,j} = \sum_{i=0}^{k-1} \overline{\alpha_i}\beta_i = |\varphi\rangle_B^* |\psi\rangle_B$$

as asserted. $\square$

By Theorem 2.2.16, we have $\mathbb{H}^* = \{\langle\varphi| : |\varphi\rangle \in \mathbb{H}\}$. The next proposition shows how $|\varphi\rangle$ is constructed from $\varphi \in \mathbb{H}^*$.

Proposition 2.2.37. *Let $B = (|b_0\rangle, \ldots, |b_{k-1}\rangle)$ be an orthonormal basis of $\mathbb{H}$ and let $\langle\varphi| \in \mathbb{H}^*$. Then we have*

(2.2.41)
$$|\varphi\rangle = \sum_{i=0}^{k-1} \langle\varphi|b_i\rangle |b_i\rangle.$$

Exercise 2.2.38. Verify Proposition 2.2.37.

Example 2.2.39. Consider the map

(2.2.42)
$$f : \mathbb{H}_2 \to \mathbb{H}_2, \quad \sum_{i=0}^{3} \alpha_i |i\rangle_2 \mapsto 2\alpha_0 + \alpha_3$$

where $\alpha_i \in \mathbb{C}$ for all $i \in \mathbb{Z}_4$. It is easy to verify that $f \in \mathbb{H}^*$. From Proposition 2.2.37 we know that for

(2.2.43)
$$|\varphi\rangle = \sum_{i=0}^{3} f |i\rangle_2 |i\rangle_2 = 2|0\rangle_2 + |3\rangle_2$$

we have $f = \langle\varphi|$.

2.2.7. Orthogonal complements. We define orthogonal complements of subsets of $\mathbb{H}$ and discuss their properties.

Proposition 2.2.40. *Let $S \subset \mathbb{H}$. Then the set*

$$(2.2.44) \qquad S^{\perp} = \{|\varphi\rangle \in \mathbb{H} : \langle \psi, \varphi \rangle = 0 \text{ for every } |\psi\rangle \in S\}$$

is a linear subspace of $\mathbb{H}$. It is called the orthogonal complement of S. If $|\varphi\rangle \in \mathbb{H}$, then we write $|\varphi\rangle^{\perp}$ for $\{|\varphi\rangle\}^{\perp}$ and call this subspace the orthogonal complement of $|\varphi\rangle$.

Example 2.2.41. We determine the orthogonal complement $|0\rangle^{\perp}$ of $|0\rangle$ in $\mathbb{H}_1$. Let $|\psi\rangle \in \mathbb{H}_1$,

$$(2.2.45) \qquad |\psi\rangle = \alpha |0\rangle + \beta |1\rangle$$

with $\alpha, \beta \in \mathbb{C}$. Then $|\psi\rangle \in |0\rangle^{\perp}$ if and only if $0 = \langle \psi|0\rangle = \alpha\langle 0|0\rangle + \beta\langle 1|0\rangle = \alpha$. This implies that

$$(2.2.46) \qquad |0\rangle^{\perp} = \mathbb{C}\,|1\rangle.$$

Proposition 2.2.42. *Let $\mathbb{H}(0), \mathbb{H}(1)$ be linear subspaces of $\mathbb{H}$. Then the following hold.*

(1) $(\mathbb{H}(0)^{\perp})^{\perp} = \mathbb{H}(0)$.

(2) $\mathbb{H}$ *is the direct sum of $\mathbb{H}(0)$ and $\mathbb{H}(0)^{\perp}$ and $\dim \mathbb{H}(0) + \dim \mathbb{H}(0)^{\perp} = \dim \mathbb{H}$.*

(3) *If B_0 is an orthonormal basis of $\mathbb{H}(0)$ and B_1 is an orthonormal basis of $\mathbb{H}^{\perp}$, then $B_0 \parallel B_1$ is an orthonormal basis of $\mathbb{H}$.*

(4) *If $\mathbb{H} = \mathbb{H}(0) + \mathbb{H}(1)$ and $\mathbb{H}(0)$ and $\mathbb{H}(1)$ are orthogonal to each other, then $\mathbb{H}(1) = \mathbb{H}(0)^{\perp}$.*

Exercise 2.2.43. Prove Proposition 2.2.42.

Using Proposition 2.2.42 we can prove the following more general statement.

Proposition 2.2.44. *Let $l \in \mathbb{N}$ and let $\mathbb{H}(0), \dots, \mathbb{H}(l-1)$ be subspaces of $\mathbb{H}$. Then the following hold.*

(1) *If $\mathbb{H}(0), \dots, \mathbb{H}(l-1)$ are pairwise orthogonal to each other, then their sum is direct.*

(2) *The subspaces $\mathbb{H}(0), \dots, \mathbb{H}(l-1)$ are pairwise orthogonal to each other if and only if there are orthonormal bases $B_0, \dots, B_{l-1}$ of $\mathbb{H}(0), \dots, \mathbb{H}(l-1)$, respectively, such that $B = B_0 \parallel \cdots \parallel B_{l-1}$ is an orthonormal basis of $\mathbb{H}(0) + \cdots + \mathbb{H}(l-1)$.*

Proof. Without loss of generality, we assume that the sum of the subspaces $\mathbb{H}(i)$ is $\mathbb{H}$.

We begin by proving the first assertion. For $i \in \mathbb{Z}_l$ let $|\varphi_i\rangle \in \mathbb{H}(i)$ such that

$$(2.2.47) \qquad \sum_{i=0}^{l-1} |\varphi_i\rangle = 0.$$

Since the subspaces $\mathbb{H}(i)$ are pairwise orthogonal, for all $j \in \mathbb{Z}_l$ we have

$$(2.2.48) \qquad 0 = \left\langle \varphi_j \left| \sum_{i=0}^{l-1} |\varphi_i\rangle \right. \right\rangle = \sum_{i=0}^{l-1} \langle \varphi_j|\varphi_i\rangle = \langle \varphi_j|\varphi_j\rangle$$

and therefore $|\varphi_j\rangle = 0$.

Next, we turn to the second assertion. Assume that the subspaces $\mathbb{H}(i)$ are pairwise orthogonal. We prove the existence of bases B_i with the asserted properties by induction on l. For $l = 1$, we can choose an orthonormal basis B_0 of $\mathbb{H}(0)$ that exists by Corollary 2.2.34. Assume that $l > 1$ and that the assertion holds for $l - 1$. According to the induction hypothesis, there are orthonormal bases $B_0, \ldots, B_{l-2}$ of $\mathbb{H}(0), \ldots, \mathbb{H}(l-2)$, respectively, such that $B' = B_0 \parallel \cdots \parallel B_{l-2}$ is an orthonormal basis of the sum $\mathbb{H}'$ of these subspaces. It follows from Proposition 2.2.42 that $\mathbb{H}(l-1) = (\mathbb{H}')^\perp$ and there is an orthonormal basis B_{l-1} of $\mathbb{H}(l-1)$ such that $B' \parallel B_{l-1} = \mathbb{H}$.

To prove the converse of the second assertion, assume that there are orthonormal bases B_i of $\mathbb{H}(i)$ such that their concatenation is an orthonormal basis of $\mathbb{H}$. It is then easy to verify that the subspaces are pairwise orthogonal to each other. $\square$

2.3. Linear maps

In this section, we study linear maps between Hilbert spaces. They are also called *linear operators*. We let $\mathbb{H}$, $\mathbb{H}'$, $\mathbb{H}''$ be Hilbert spaces of dimension k, l, and m, respectively, and denote the inner product on these Hilbert spaces by $\langle \cdot | \cdot \rangle$.

2.3.1. Matrix representations.
Matrix representations of homomorphisms between vector spaces are discussed in Section B.6.3. We briefly summarize this concept. Let $B = (|b_0\rangle, \ldots, |b_{k-1}\rangle)$ be a basis of $\mathbb{H}$ and let $C = (|c_0\rangle, \ldots, |c_{l-1}\rangle)$ be a basis of $\mathbb{H}'$. Then the representation matrix of $f \in \mathrm{Hom}(\mathbb{H}, \mathbb{H}')$ with respect to these bases is the matrix in $\mathbb{C}^{(l,k)}$ whose column vectors are the coefficient vectors of $f |b_0\rangle, \ldots, f |b_{k-1}\rangle$ with respect to the basis C; that is,

$$(2.3.1) \qquad \mathrm{Mat}_{B,C}(f) = ((f |b_0\rangle)_C, \ldots, (f |b_{k-1}\rangle)_C) \in \mathbb{C}^{(l,k)}.$$

Theorem B.6.10 states that the map

$$(2.3.2) \qquad \mathrm{Hom}(\mathbb{H}, \mathbb{H}') \to \mathbb{C}^{(l,k)}, \quad f \mapsto \mathrm{Mat}_{B,C}(f)$$

is an isomorphism of $\mathbb{C}$-vector spaces. Its inverse is

$$(2.3.3) \qquad \mathbb{C}^{(l,k)} \to \mathrm{Hom}(\mathbb{H}, \mathbb{H}'), \quad A \mapsto f_{A,B,C}$$

where

$$(2.3.4) \qquad f_{A,B,C} : \mathbb{H} \to \mathbb{H}', \quad |\psi\rangle \mapsto CA |\psi\rangle_B.$$

If $\mathbb{H} = \mathbb{H}'$ and $B = C$, then we write $\mathrm{Mat}_B(f)$ for $\mathrm{Mat}_{B,B}(f)$ and $f_{A,B}$ for $f_{A,B,B}$. Also, if D is a basis of $\mathbb{H}''$, if $f \in \mathrm{Hom}(\mathbb{H}, \mathbb{H}')$ and $g \in \mathrm{Hom}(\mathbb{H}', \mathbb{H}'')$, then

$$(2.3.5) \qquad M_{B,D}(g \circ f) = M_{B,C}(f) M_{C,D}(g).$$

Example 2.3.1. The *Pauli X operator* on $\mathbb{H}_1$ is defined as

$$(2.3.6) \qquad X : \mathbb{H}_1 \to \mathbb{H}_1, \quad \alpha |0\rangle + \beta |1\rangle \mapsto \beta |0\rangle + \alpha |1\rangle.$$

It swaps the vectors of the computational basis $B = (|0\rangle, |1\rangle)$. Hence, its matrix representation with respect to B is

$$(2.3.7) \qquad \mathrm{Mat}_B(X) = \begin{pmatrix} 0 & 1 \\ 1 & 0 \end{pmatrix}.$$

We also determine the matrix representation of the Pauli X operator with respect to the basis

$$(2.3.8) \qquad C = (|x_+\rangle, |x_-\rangle) = \left(\frac{|0\rangle + |1\rangle}{\sqrt{2}}, \frac{|0\rangle - |1\rangle}{\sqrt{2}} \right)$$

of $\mathbb{H}_1$. We note that

$$(2.3.9) \qquad X|x_+\rangle = \frac{X|0\rangle + X|1\rangle}{\sqrt{2}} = \frac{|1\rangle + |0\rangle}{\sqrt{2}} = |x_+\rangle$$

and

$$(2.3.10) \qquad X|x_-\rangle = \frac{X|0\rangle - X|1\rangle}{\sqrt{2}} = \frac{|1\rangle - |0\rangle}{\sqrt{2}} = -|x_-\rangle.$$

Hence, we have

$$(2.3.11) \qquad \mathrm{Mat}_C(X) = \begin{pmatrix} 1 & 0 \\ 0 & -1 \end{pmatrix}.$$

Note that this matrix is different from $\mathrm{Mat}_B(X)$.

Example 2.3.2. The *Pauli Z operator*

$$(2.3.12) \qquad\qquad Z : \mathbb{H}_1 \to \mathbb{H}_1$$

has the representation matrix

$$(2.3.13) \qquad A = \mathrm{Mat}_B(Z) = \begin{pmatrix} 1 & 0 \\ 0 & -1 \end{pmatrix}$$

with respect to the computational basis of $\mathbb{H}_1$. Note that this matrix is equal to $\mathrm{Mat}_C(X)$ from (2.3.11). So the Pauli Z operator is

$$(2.3.14) \qquad Z = f_{A,B} : \mathbb{H}_1 \to \mathbb{H}_1, \quad \alpha|0\rangle + \beta|1\rangle \mapsto \alpha|0\rangle - \beta|1\rangle.$$

Exercise 2.3.3. (1) Determine the matrix representation of the *Pauli Y operator*

$$(2.3.15) \qquad Y : \mathbb{H}_1 \to \mathbb{H}_1, \quad \alpha|0\rangle + \beta|1\rangle \mapsto -i\beta|0\rangle + i\alpha|1\rangle$$

with respect to the computational basis of $\mathbb{H}_1$.

(2) Determine the matrix representations of the Pauli Y and Z operators with respect to the basis $C = (|x_-\rangle, |x_+\rangle)$ from (2.3.8).

Exercise 2.3.4. (1) Find the matrix representation of the *Hadamard operator*

$$(2.3.16) \qquad H : \mathbb{H}_1 \to \mathbb{H}_1, \quad \alpha|0\rangle + \beta|1\rangle \mapsto \alpha|x_+\rangle + \beta|x_-\rangle$$

with respect to the computational basis of $\mathbb{H}_1$.

(2) Use the matrix representations of the operators H, X, Y, and Z to show that

$$(2.3.17) \qquad HXH = Z, \quad HYH = -Y, \quad HZH = X.$$

Since (2.3.2) is an isomorphism of $\mathbb{C}$-vector spaces, every map $f \in \mathrm{Hom}(\mathbb{H}, \mathbb{H}')$ is uniquely determined by its representation matrix $\mathrm{Mat}_{B,C}(f)$. Therefore, we can define linear maps in $\mathrm{Hom}(\mathbb{H}, \mathbb{H}')$ by their action on the elements of a basis of $\mathbb{H}$. In particular, operators on state spaces $\mathbb{H}_n$ will typically be described by their effect on the computational basis elements. For instance, using this representation, the Hadamard operator from (2.3.16) can written as

$$(2.3.18) \qquad H : \mathbb{H}_1 \to \mathbb{H}_1, \quad |0\rangle \mapsto |x_+\rangle, |1\rangle \mapsto |x_-\rangle.$$

Finally, we introduce a further simplification of the notation. For this, we let $T \in \text{Hom}(\mathbb{H}, \mathbb{H}')$, $|\varphi\rangle \in \mathbb{H}$, and $|\psi\rangle \in \mathbb{H}'$. Then applying the operator $\langle\varphi|$ to $T|\psi\rangle \in \mathbb{H}$ has the same effect as applying the composite operator $\langle\varphi| \circ T$ to $|\psi\rangle \in \mathbb{H}'$. Therefore, we write

$$(2.3.19) \qquad \langle\varphi|T|\psi\rangle = \langle\varphi|\,(T\,|\psi\rangle) = \langle\varphi| \circ T\,|\psi\rangle.$$

We can use this notation to describe the representation matrices of homomorphisms between Hilbert spaces.

Proposition 2.3.5. *Let $B = (|b_0\rangle, \ldots, |b_{k-1}\rangle)$ and $C = (|c_0\rangle, \ldots, |c_{l-1}\rangle)$ be orthonormal bases of $\mathbb{H}$ and $\mathbb{H}'$, respectively. Then the matrix representation of a linear map $T \in \text{Hom}(\mathbb{H}, \mathbb{H}')$ with respect to these bases is*

$$(2.3.20) \qquad \text{Mat}_{B,C}(T) = ((\langle c_i|T|b_j\rangle)_{i\in\mathbb{Z}_l, j\in\mathbb{Z}_k} \in \mathbb{C}^{(l,k)}.$$

Proof. Write $\text{Mat}_{B,C}(T) = (\alpha_{i,j})$. Then for all $i \in \mathbb{Z}_k$ and $j \in \mathbb{Z}_l$ the linearity of the inner product in the second argument and the fact that $\langle b_i|b_j\rangle = \delta_{i,j}$ for all $i, j \in \mathbb{Z}_k$ imply

$$(2.3.21) \qquad \langle c_i|T|b_j\rangle = \langle c_i|\left(\sum_{m=0}^{k-1} \alpha_{m,j}\,|c_m\rangle\right) = \sum_{m=0}^{k-1} \alpha_{m,j}\langle c_i|c_m\rangle = \alpha_{i,j}. \qquad \square$$

Example 2.3.6. Denote by $\langle\cdot|\cdot\rangle$ the Hermitian inner product on $\mathbb{H}_1$ with respect to $B = (|0\rangle, |1\rangle)$. By Proposition 2.3.5, the representation matrix of the Pauli Z operator with respect to B is

$$(2.3.22) \qquad \text{Mat}_B(Z) = \begin{pmatrix} \langle 0|Z|0\rangle & \langle 0|Z|1\rangle \\ \langle 1|Z|0\rangle & \langle 1|Z|1\rangle \end{pmatrix} = \begin{pmatrix} \langle 0|0\rangle & -\langle 0|1\rangle \\ \langle 1|0\rangle & -\langle 1|1\rangle \end{pmatrix} = \begin{pmatrix} 1 & 0 \\ 0 & -1 \end{pmatrix}.$$

2.3.2. Adjoints. In this section, we introduce adjoints of matrices over $\mathbb{C}$ and of linear maps between finite-dimensional Hilbert spaces and discuss their properties. We start by defining the adjoints of complex matrices.

Definition 2.3.7. The *adjoint* of a matrix $A \in \mathbb{C}^{(k,l)}$ is $A^* = \overline{A^{\text{T}}} \in \mathbb{C}^{(l,k)}$.

The adjoint of a matrix over $\mathbb{C}$ is also called its *Hermitian adjoint*, *Hermitian conjugate*, or *Hermitian transpose*.

The notation A^* for the adjoint of a matrix A over $\mathbb{C}$ is in agreement with Definition 2.2.7 where the dual of a complex vector is specified as the matrix that has the conjugate of this vector as its only row.

Example 2.3.8. Consider the matrix

$$(2.3.23) \qquad A = \begin{pmatrix} 1 & i & 1+i \\ 1-i & i & 1 \end{pmatrix} \in \mathbb{C}^{(2,3)}.$$

Its adjoint is

$$(2.3.24) \qquad A^* = \begin{pmatrix} 1 & 1+i \\ -i & -i \\ 1-i & 1 \end{pmatrix} \in \mathbb{C}^{(3,2)}.$$

Here are some important properties of the adjoint of matrices.

Proposition 2.3.9. *Let $A \in \mathbb{C}^{(k,l)}$, $B \in \mathbb{C}^{(m,n)}$, and $\alpha \in \mathbb{C}$. Then we have*

$$(2.3.25) \qquad (A^*)^* = A,$$

$$(2.3.26) \qquad (A + B)^* = A^* + B^* \ \textit{if} \ m = k \ \textit{and} \ n = l,$$

$$(2.3.27) \qquad (\alpha A)^* = \overline{\alpha} A^*,$$

$$(2.3.28) \qquad \mathrm{rank}(A) = \mathrm{rank}(A^*),$$

$$(2.3.29) \qquad (AB)^* = B^* A^* \ \textit{if} \ l = m.$$

Exercise 2.3.10. Prove Proposition 2.3.9.

The next proposition characterizes the adjoints of matrices.

Proposition 2.3.11. *Let $A \in \mathbb{C}^{(k,l)}$. Then for all $\vec{v} \in \mathbb{C}^k$ and all $\vec{w} \in \mathbb{C}^l$ the adjoint A^* of A satisfies*

$$(2.3.30) \qquad \left\langle \vec{v} \middle| A\vec{w} \right\rangle = \left\langle A^* \vec{v} \middle| \vec{w} \right\rangle$$

and A^* is the only matrix in $\mathbb{C}^{(l,k)}$ with this property.

Proof. Let $\vec{v} \in \mathbb{C}^k$, $\vec{w} \in \mathbb{C}^l$. Then we have

$$
\begin{aligned}
\langle A^* \vec{v} | \vec{w} \rangle &= (A^* \vec{v})^* \vec{w} & \text{by (2.2.9),} \\
&= \vec{v}^* \left(A^* \right)^* \vec{w} & \text{by (2.3.29),} \\
&= \vec{v}^* A \vec{w} & \text{by (2.3.25),} \\
&= \langle \vec{v} | A \vec{w} \rangle & \text{by (2.2.9).}
\end{aligned}
$$

To show that A^* is the only matrix in $\mathbb{C}^{(l,k)}$ that satisfies (2.3.30), let $A' \in \mathbb{C}^{(l,k)}$ such that

$$(2.3.31) \qquad \langle \vec{v} | A \vec{w} \rangle = \langle A' \vec{v} | \vec{w} \rangle$$

for all $\vec{v} \in \mathbb{C}^k$ and $\vec{w} \in \mathbb{C}^l$. Denote by $\vec{e}_0, \ldots, \vec{e}_{k-1}$ and $\vec{f}_0, \ldots, \vec{f}_{l-1}$ the standard unit vectors of $\mathbb{C}^k$ and $\mathbb{C}^l$, respectively. Then for all $i \in \mathbb{Z}_k$ and $j \in \mathbb{Z}_l$ we have

$$
\begin{aligned}
\left\langle \vec{e}_i \middle| A \vec{f}_j \right\rangle &= \left\langle A' \vec{e}_i \middle| \vec{f}_j \right\rangle & \text{by (2.3.31),} \\
&= \left\langle \vec{e}_i \middle| (A')^* \vec{f}_j \right\rangle & \text{by (2.3.30) and (2.3.25).}
\end{aligned}
$$

So (2.3.25) implies $A^* = A'$. $\qquad \qquad \square$

From Proposition 2.3.11 we obtain the following result which allows us to define adjoints of linear operators on Hilbert spaces.

Proposition 2.3.12. *If $A \in \mathrm{Hom}(\mathbb{H}', \mathbb{H})$, then there is a uniquely determined operator $A^* \in \mathrm{Hom}(\mathbb{H}, \mathbb{H}')$ such that*

$$(2.3.32) \qquad \left\langle \varphi \middle| A \left| \psi \right\rangle \right\rangle = \left\langle A^* \left| \varphi \right\rangle \middle| \psi \right\rangle$$

for all $|\varphi\rangle \in \mathbb{H}$ and $|\psi\rangle \in \mathbb{H}'$. The operator A^* is called the adjoint *of A.*

Proof. Let $A \in \mathrm{Hom}(\mathbb{H}', \mathbb{H})$. Choose orthonormal bases B of $\mathbb{H}$ and C of $\mathbb{H}'$. Let A^* be the linear map in $\mathrm{Hom}(\mathbb{H}, \mathbb{H}')$ with representation matrix $(\mathrm{Mat}_{C,B}(A))^*$. It then follows from Proposition 2.3.11 that (2.3.32) holds for all $|\varphi\rangle \in \mathbb{H}$ and $|\psi\rangle \in \mathbb{H}'$. The uniqueness of A^* follows from the uniqueness of A^* in Proposition 2.3.11. □

Exercise 2.3.13. Verify that Proposition 2.3.9 also holds for linear operators.

Finally, we mention some properties of the adjoint of endomorphisms and square matrices over $\mathbb{C}$.

Proposition 2.3.14. *Let $A \in \mathbb{C}^{(k,k)}$ or $A \in \mathrm{End}(\mathbb{H})$. Then the following hold.*

(1) *The determinant, trace, characteristic polynomial, and eigenvalues of A^* are the complex conjugates of the determinant, trace, characteristic polynomial, and eigenvalues of A, respectively.*

(2) *If A is invertible, then A^* is invertible, and we have $(A^*)^{-1} = (A^{-1})^*$.*

Exercise 2.3.15. Prove Proposition 2.3.14.

2.3.3. The Hilbert-Schmidt inner product. We define an inner product on $\mathbb{C}^{(l,k)}$ and on $\mathrm{Hom}(\mathbb{H}, \mathbb{H}')$.

Proposition 2.3.16. *The map*

$$(2.3.33) \qquad \langle \cdot | \cdot \rangle : \mathbb{C}^{(l,k)} \times \mathbb{C}^{(l,k)} \to \mathbb{C}, \quad (A, B) \mapsto \langle A | B \rangle = \mathrm{tr}(A^* B)$$

is an inner product on $\mathbb{C}^{(l,k)}$. It is called the Hilbert-Schmidt *inner product on $\mathbb{C}^{(l,k)}$.*

Proof. The map $\mathbb{C}^{(l,k)} \to \mathbb{C}^{kl}$, which sends a matrix $A \in \mathbb{C}^{(l,k)}$ to the concatenation of its column vectors, is an isomorphism of $\mathbb{C}$-vector spaces. We use this map to identify the matrices in $\mathbb{C}^{(l,k)}$ with vectors in $\mathbb{C}^{kl}$. Using this identification, the map (2.3.33) is the standard Hermitian inner product on $\mathbb{C}^{(l,k)}$. □

Corollary 2.3.17. *The map*

$$(2.3.34) \qquad \langle \cdot | \cdot \rangle : \mathrm{Hom}(\mathbb{H}, \mathbb{H}') \times \mathrm{Hom}(\mathbb{H}, \mathbb{H}'), \quad (A, B) \mapsto \langle A | B \rangle = \mathrm{tr}(A^* B)$$

is an inner product on $\mathrm{Hom}(\mathbb{H}, \mathbb{H}')$. It is called the Hilbert-Schmidt *inner product on* $\mathrm{End}(\mathbb{H})$.

Equipped with the Hilbert-Schmidt inner product, the complex vector space $\mathrm{Hom}(\mathbb{H}, \mathbb{H}')$ becomes a Hilbert space.

We present another way of writing the Hilbert-Schmidt inner product.

Proposition 2.3.18. *Let $A, B \in \mathbb{C}^{(l,k)}$ and denote by $\vec{a}_0, \dots, \vec{a}_{k-1}$ the column vectors of A and by $\vec{b}_0, \dots, \vec{b}_{l-1}$ the column vectors of B. Then we have*

$$(2.3.35) \qquad\qquad \langle A|B \rangle = \sum_{i=0}^{l-1} \langle \vec{a}_i | \vec{b}_i \rangle.$$

Exercise 2.3.19. Prove Proposition 2.3.18.

Example 2.3.20. Let

$$(2.3.36) \qquad\qquad A = \begin{pmatrix} 2 & i \\ 3 & -1 \end{pmatrix} \quad \text{and} \quad B = \begin{pmatrix} i & 1 \\ 2 & 4 \end{pmatrix}.$$

Then we have

$$(2.3.37) \qquad \langle A|B \rangle = \operatorname{tr} A^* B = \operatorname{tr} \begin{pmatrix} 2 & 3 \\ -i & -1 \end{pmatrix} \begin{pmatrix} i & 1 \\ 2 & 4 \end{pmatrix} = \operatorname{tr} \begin{pmatrix} 2i+6 & 14 \\ -1 & -i-4 \end{pmatrix} = i + 2.$$

The norm induced by the Hilbert-Schmidt inner product is

$$(2.3.38) \qquad \mathbb{C}^{(l,k)} \to \mathbb{R}, \quad A = (a_{i,j}) \mapsto \|A\| = \operatorname{tr}(AA^*) = \sqrt{\sum_{i \in \mathbb{Z}_k, j \in \mathbb{Z}_l} |a_{i,j}|^2}.$$

Example 2.3.21. Let

$$(2.3.39) \qquad\qquad A = \begin{pmatrix} 1 & i \\ 1+i & 1-i \end{pmatrix}.$$

Then we have

$$(2.3.40) \qquad \|A\|^2 = |1|^2 + |i|^2 + |1+i|^2 + |1-i|^2 = 1 + 1 + 2 + 2 = 6.$$

2.4. Endomorphisms

In this section, we discuss endomorphisms of a Hilbert space $\mathbb{H}$ of finite dimension k with inner product $\langle \cdot | \cdot \rangle$ and their properties.

2.4.1. Basics.
The representation matrices of endomorphisms of $\mathbb{H}$ are the matrices in $\mathbb{C}^{(k,k)}$.

As examples we will use the Pauli operators X, Y, and Z on $\mathbb{H}_1$ that were already introduced in Examples 2.3.1 and 2.3.2 and Exercise 2.3.3. Their representation matrices with respect to the computational basis $(|0\rangle, |1\rangle)$ are the *Pauli matrices*

$$(2.4.1) \qquad X = \begin{pmatrix} 0 & 1 \\ 1 & 0 \end{pmatrix}, \quad Y = \begin{pmatrix} 0 & -i \\ i & 0 \end{pmatrix}, \quad Z = \begin{pmatrix} 1 & 0 \\ 0 & -1 \end{pmatrix}.$$

In the examples, we will also use the Hadamard operator on $\mathbb{H}_1$. Its representation matrix with respect to $(|0\rangle, |1\rangle)$ is the *Hadamard matrix*

$$(2.4.2) \qquad\qquad H = \frac{1}{\sqrt{2}} \begin{pmatrix} 1 & 1 \\ 1 & -1 \end{pmatrix}.$$

The next theorem presents formulas for the characteristic polynomial $p_A(x)$, the trace $\operatorname{tr}(A)$, and the determinant $\det(A)$ of endomorphisms A of $\mathbb{H}$ and matrices A in $\mathbb{C}^{(k,k)}$.

Proposition 2.4.1. *Let $A \in \mathbb{C}^{(k,k)}$ or $A \in \text{End}(\mathbb{H})$. Let Λ be the set of eigenvalues of A. For each $\lambda \in \Lambda$ let m_λ be its algebraic multiplicity. Then we have*

(2.4.3)
$$p_A(x) = \prod_{\lambda \in \Lambda}(x - \lambda)^{m_\lambda},$$

(2.4.4)
$$\text{tr}(A) = \sum_{\lambda \in \Lambda} m_\lambda \lambda,$$

(2.4.5)
$$\det(A) = \prod_{\lambda \in \Lambda} \lambda^{m_\lambda}.$$

Proof. Let $A \in \text{End}(\mathbb{H})$ or $A \in \mathbb{C}^{(k,k)}$. The first assertion follows from the fact that $\mathbb{C}$ is algebraically closed, which implies that $p_A(x)$ is a product of linear factors. The details are beyond the scope of this book. The other two assertions are derived from Proposition B.5.27. □

Example 2.4.2. The characteristic polynomial of the identity operator I_1 on $\mathbb{H}_1$ is

(2.4.6)
$$p_I(x) = \det(xI - I) = \det\begin{pmatrix} x-1 & 0 \\ 0 & x-1 \end{pmatrix} = (x-1)^2.$$

Hence, the only eigenvalue of I is 1. It has algebraic multiplicity 2 and we have $\text{tr}(I) = 1 + 1 = 2$ and $\det(I) = 1 \cdot 1 = 1$.

The characteristic polynomial of the Pauli X operator is

(2.4.7)
$$p_X(x) = \det(xI - X) = \det\begin{pmatrix} x & -1 \\ -1 & x \end{pmatrix} = (x-1)(x+1).$$

Hence, the eigenvalues of X are 1 and -1, both with algebraic multiplicity 1, and we have $\text{tr}(X) = 1 + (-1) = 0$ and $\det(X) = 1 \cdot (-1) = -1$.

Exercise 2.4.3. Use Proposition 2.4.1 to show that the Pauli Y and Z matrices have the eigenvalues 1 and -1, trace 0, and determinant -1.

Next, we note the following.

Theorem 2.4.4. *If all eigenvalues of $A \in \mathbb{C}^{(k,k)}$ have algebraic multiplicity 1, then A is diagonizable.*

Exercise 2.4.5. Prove Theorem 2.4.4.

Example 2.4.6. According to Example 2.4.2, the Pauli matrix X has the eigenvalues 1 and -1 and both have algebraic multiplicity 1. Therefore, by Theorem 2.4.4 this matrix is diagonalizable. In fact, by Exercise 2.3.4 we have

(2.4.8)
$$HXH = Z$$

with the Pauli matrix

(2.4.9)
$$Z = \begin{pmatrix} 1 & 0 \\ 0 & -1 \end{pmatrix}$$

which is a diagonal matrix and the *Hadamard matrix*

(2.4.10)
$$H = \frac{1}{\sqrt{2}} \begin{pmatrix} 1 & 1 \\ 1 & -1 \end{pmatrix}.$$

Since $H^{-1} = H$, it follows from (2.4.8) that

(2.4.11) $$H^{-1}XH = Z$$

which shows that X is diagonizable.

In order for a matrix $A \in \mathbb{C}^{(k,k)}$ to be diagonizable, it is not necessary that all of its eigenvalues have algebraic multiplicity 1. For example, as shown in Example 2.4.2, the diagonal identity matrix I has 1 as its single eigenvalue. So the question arises of whether all matrices in $\mathbb{C}^{(k,k)}$ are diagonizable. But this is not the case as the next example shows.

Example 2.4.7. Consider the matrix

(2.4.12) $$A = \begin{pmatrix} 1 & 1 \\ 0 & 1 \end{pmatrix}.$$

Its characteristic polynomial is $p_A(x) = (x-1)^2$. Hence 1 is the only eigenvalue of A and it has algebraic multiplicity 2. Also, the eigenspace of this eigenvector is the kernel of the matrix

(2.4.13) $$1 \cdot I - A = \begin{pmatrix} 0 & -1 \\ 0 & 0 \end{pmatrix}.$$

This matrix has rank 1 and therefore the dimension of this eigenspace is 1. So according to Theorem B.7.28, the matrix A is not diagonizable.

Definition 2.4.8. A matrix $A \in \mathbb{C}^{(k,k)}$ or an operator $A \in \text{End}(\mathbb{H})$ is called an *involution* if $A^2 = I_k$ or $A^2 = I_{\mathbb{H}}$, respectively.

Exercise 2.4.9. Prove that the Pauli matrices and operators are involutions.

2.4.2. Hermitian matrices and operators. In this section, we introduce and discuss Hermitian matrices and operators. They will be used in Section 3.4 to model quantum mechanical measurements.

Definition 2.4.10. A matrix $A \in \mathbb{C}^{(k,k)}$ or operator $A \in \text{End}(\mathbb{H})$ is called *Hermitian* or *self-adjoint* if $A = A^*$.

Hermitian matrices and operators are named after the French mathematician Charles Hermite who lived in the 19th century and made significant contributions to many areas of mathematics.

Example 2.4.11. The matrix

(2.4.14) $$A = \begin{pmatrix} 1 & i \\ -i & 1 \end{pmatrix}$$

is Hermitian since

(2.4.15) $$A^* = \overline{A^T} = \overline{\begin{pmatrix} 1 & -i \\ i & 1 \end{pmatrix}} = \begin{pmatrix} 1 & i \\ -i & 1 \end{pmatrix}.$$

Exercise 2.4.12. Show that the Pauli operators X, Y, and Z and the Hadamard operator H are Hermitian.

We present properties of Hermitian matrices and operators.

Proposition 2.4.13. (1) *The diagonal elements of Hermitian matrices are real numbers.*

(2) *The determinants, trace, and eigenvalues of Hermitian matrices or operators are real numbers.*

(3) *The inverse of an invertible Hermitian matrix or operator is Hermitian.*

(4) *The sum of two Hermitian matrices or operators is Hermitian.*

(5) *The product AB of two Hermitian matrices $A, B \in \mathbb{C}^{(k,k)}$ or operators $A, B \in \mathrm{End}(\mathbb{H})$ is Hermitian if and only if $AB = BA$.*

(6) *If $A, B \in \mathbb{C}^{(k,k)}$ or $A, B \in \mathrm{End}(\mathbb{H})$, then ABA is Hermitian.*

Exercise 2.4.14. Prove Proposition 2.4.13.

2.4.3. Unitary matrices and operators. In Section 3.3, unitary operators will be used to model the evolution of quantum systems over time. This section introduces these operators and presents their properties.

Definition 2.4.15. A matrix $U \in \mathbb{C}^{(k,k)}$ or operator $U \in \mathrm{End}(\mathbb{H})$ is called *unitary* if $U^*U = UU^* = I_k$ or $U^*U = UU^* = I_\mathbb{H}$, respectively

Example 2.4.16. Consider the Hadamard matrix

$$(2.4.16) \qquad H = \frac{1}{\sqrt{2}} \begin{pmatrix} 1 & 1 \\ 1 & -1 \end{pmatrix}.$$

We have

$$(2.4.17) \qquad HH^* = H^2 = \begin{pmatrix} 1 & 0 \\ 0 & 1 \end{pmatrix}.$$

Hence, H is unitary.

Exercise 2.4.17. Show that Hermitian matrices or operators are involutions if and only if they are unitary. Conclude that the Pauli operators X, Y, and Z and the Hadamard operator are unitary.

We prove a number of equivalent characterizations of unitary matrices.

Proposition 2.4.18. *Let $U \in \mathbb{C}^{(k,k)}$. Then the following statements are equivalent.*

(1) *U is unitary.*

(2) *U is invertible and $U^{-1} = U^*$.*

(3) *The columns of the matrix U form an orthonormal basis of $\mathbb{C}^k$.*

(4) *The rows of the matrix U form an orthonormal basis of $\mathbb{C}^k$.*

(5) *$\langle U\vec{v}, U\vec{w} \rangle = \langle \vec{v}, \vec{w} \rangle$ for all $\vec{v}, \vec{w} \in \mathbb{C}^k$*

(6) *$\left\| U\vec{v} \right\| = \left\| \vec{v} \right\|$ for all $\vec{v} \in \mathbb{C}^k$.*

Proof. Let $U \in \mathbb{C}^{(k,k)}$. Statements (1) and (2) are equivalent by the definition of unitary matrices and invertibility. Next, we note that $U^* = U^{-1}$ if and only if $UU^* = I_k$ which is equivalent to the sequence of row vectors of U being an orthonormal basis of $\mathbb{C}^k$. Also, the equivalence of the second and fourth property can be deduced from $U^*U = I_k$.

We show that statement (1) and statement (5) are equivalent. Let $U \in \mathbb{C}^{(k,k)}$ be unitary and let $\vec{v}, \vec{w} \in \mathbb{C}^k$. Then $U^*U = I_k$ and Proposition 2.3.11 imply $\langle U\vec{v}, U\vec{w} \rangle = \langle U^*U\vec{v}, \vec{w} \rangle = \langle \vec{v}, \vec{w} \rangle$. Conversely, assume that $\langle U\vec{v}, U\vec{w} \rangle = \langle \vec{v}, \vec{w} \rangle$ for all $\vec{v}, \vec{w} \in \mathbb{C}^k$. This implies

$$(2.4.18) \qquad \vec{e}_i^* U^* U \vec{e}_j = \langle U\vec{e}_i | U\vec{e}_j \rangle = \langle \vec{e}_i | \vec{e}_j \rangle = \delta_{i,j}$$

for all $i, j \in \mathbb{Z}_k$ where $\vec{e}_i$ is the ith standard unit vector in $\mathbb{C}^k$ for $0 \leq i < k$. This means that $U^*U = I_k$. So Corollary B.5.21 implies that U is invertible and $U^* = U^{-1}$; i.e., $U^*U = UU^* = I_k$.

Finally, we show that statements (1) and (6) are equivalent. Statement (6) follows immediately from statement (5) which is equivalent to statement (1). Conversely, assume that $\langle U\vec{v}, U\vec{v} \rangle = \langle \vec{v}, \vec{v} \rangle$ for all $\vec{v} \in \mathbb{C}^k$. We show that

$$(2.4.19) \qquad \vec{e}_i^* U^* U \vec{e}_j = \delta_{i,j}$$

for all $i, j \in \mathbb{Z}_k$. Then Corollary B.5.21 implies that U is invertible and $U^* = U^{-1}$; i.e., $U^*U = UU^* = I_k$. For all $i \in \mathbb{Z}_k$ we have

$$(2.4.20) \qquad \vec{e}_i^* U^* U \vec{e}_i = \langle U\vec{e}_i | U\vec{e}_i \rangle = \langle \vec{e}_i, \vec{e}_i \rangle = 1.$$

Next, let $i, j \in \mathbb{Z}_k$ and assume that $i \neq j$. Then we have

$$\begin{aligned}
2 &= \langle \vec{e}_i + \vec{e}_j | \vec{e}_i + \vec{e}_j \rangle \\
&= \langle U(\vec{e}_i + \vec{e}_j) | U(\vec{e}_i + \vec{e}_j) \rangle \\
&= \left\| U\vec{e}_i \right\|^2 + \langle U\vec{e}_i | U\vec{e}_j \rangle + \langle U\vec{e}_j | U\vec{e}_i \rangle + \left\| U\vec{e}_j \right\|^2 \\
&= 2 + 2\Re \vec{e}_i^* U^* U \vec{e}_j.
\end{aligned}$$

It follows that $\Re \vec{e}_i^* U^* U \vec{e}_j = 0$. Applying similar arguments to $\langle U(\vec{e}_i + i\vec{e}_j) | U(\vec{e}_i + i\vec{e}_j) \rangle$ it can be shown that $\Im \vec{e}_i^* U^* U \vec{e}_j = 0$. Therefore, (2.4.19) holds. □

Exercise 2.4.19. Show that permutation matrices are unitary.

Proposition 2.4.18 implies the following result.

Theorem 2.4.20.   (1) *The set of all unitary matrices in* $\mathbb{C}^{(k,k)}$ *is a subgroup of* $\mathrm{GL}(k, \mathbb{C})$. *It is denoted by* $\mathrm{U}(k)$ *and is called the* unitary group *of rank k.*

   (2) *The set of all unitary matrices of determinant 1 is a subgroup of* $\mathrm{U}(k)$. *It is called the* special unitary group *of rank k and is denoted by* $\mathrm{SU}(k)$.

Proof. Let $U, V \in \mathbb{C}^{(k,k)}$ be unitary. It then follows from Lemma B.5.23, Proposition 2.3.9, and Proposition 2.4.18 that $(UV)^{-1} = V^{-1}U^{-1} = V^*U^* = (UV)^*$. Also, I_k is unitary. Therefore, the set of unitary matrices is a subgroup of $\mathrm{GL}(k, \mathbb{C})$. Since the product of two matrices of determinant 1 has determinant 1, it follows that $\mathrm{SU}(k)$ is a subgroup of $\mathrm{U}(k)$. □

From Proposition 2.4.18 we also obtain characterizations of unitary operators on $\mathbb{H}$. To state them, we need the following definition.

Definition 2.4.21. Let $\mathbb{H}'$ be another Hilbert space with an inner product $\langle\cdot|\cdot\rangle$. A map $U \in \mathrm{Hom}(\mathbb{H}, \mathbb{H}')$ is called an *isometry* between $\mathbb{H}$ and $\mathbb{H}'$ if $\langle\varphi|\psi\rangle = \langle U\,|\varphi\rangle\,\big|\,U\,|\psi\rangle\rangle$ for all $|\varphi\rangle, |\psi\rangle \in \mathbb{H}$. .

Example 2.4.22. Let B be a basis of $\mathbb{H}$. Denote by $\langle\cdot|\cdot\rangle$ the standard Hermitian inner product on $\mathbb{C}^k$. So $(\mathbb{H}, \langle\cdot|\cdot\rangle_B)$ and $(\mathbb{C}^k, \langle\cdot|\cdot\rangle)$ are Hilbert spaces. By Corollary 2.2.11, the map

$$(2.4.21) \qquad \mathbb{H} \to \mathbb{C}^k, \quad |\varphi\rangle \mapsto |\varphi\rangle_B$$

is an isometry between these Hilbert spaces.

Finally, we characterize the set of all orthonormal bases of $\mathbb{H}$.

Corollary 2.4.23. *Let B be an orthonormal basis of $\mathbb{H}$. Then the set of all orthonormal bases of $\mathbb{H}$ is the coset $B\mathrm{U}(k)$ in $\mathrm{GL}(k, \mathbb{C})$.*

Exercise 2.4.24. Prove Corollary 2.4.23.

2.4.4. Outer products. Let $B = (|b_0\rangle, \dots, |b_{k-1}\rangle)$ be an orthonormal basis of $\mathbb{H}$. We define the outer product of elements of $\mathbb{H}$.

Definition 2.4.25. Let $|\varphi\rangle, |\psi\rangle \in \mathbb{H}$. Then the *outer product* of $|\varphi\rangle$ and $|\psi\rangle$ is the endomorphism

$$(2.4.22) \qquad |\varphi\rangle\langle\psi| : \mathbb{H} \to \mathbb{H}, \quad |\xi\rangle \mapsto |\varphi\rangle\langle\psi|\xi\rangle$$

of $\mathbb{H}$.

In formula (2.4.22) we deviate from the usual notation and write the scalar product of the complex number $\alpha = \langle\psi|\xi\rangle$ with $|\varphi\rangle \in \mathbb{H}$ as $|\varphi\rangle\,\alpha$ instead of $\alpha\,|\varphi\rangle$. This allows for a more intuitive notation.

Example 2.4.26. The computational basis of the single-qubit state space $\mathbb{H}_1$ is $(|0\rangle, |1\rangle)$. Examples of the outer products of kets in $\mathbb{H}_1$ are $|0\rangle\langle0|$, $|0\rangle\langle1|$, $|1\rangle\langle0|$, and $|1\rangle\langle1|$. Let $|\psi\rangle = \alpha\,|0\rangle + \beta\,|1\rangle \in \mathbb{H}_1$ with complex coefficients α and β. Then the images of this ket under the four outer products are

$$|0\rangle\langle0|\,(\alpha\,|0\rangle + \beta\,|1\rangle) = \alpha\,|0\rangle\langle0|0\rangle + \beta\,|0\rangle\langle0|1\rangle = \alpha\,|0\rangle,$$
$$|0\rangle\langle1|\,(\alpha\,|0\rangle + \beta\,|1\rangle) = \alpha\,|0\rangle\langle1|0\rangle + \beta\,|0\rangle\langle1|1\rangle = \beta\,|0\rangle,$$
$$|1\rangle\langle0|\,(\alpha\,|0\rangle + \beta\,|1\rangle) = \alpha\,|1\rangle\langle0|0\rangle + \beta\,|1\rangle\langle0|1\rangle = \alpha\,|1\rangle,$$
$$|1\rangle\langle1|\,(\alpha\,|0\rangle + \beta\,|1\rangle) = \alpha\,|1\rangle\langle1|0\rangle + \beta\,|1\rangle\langle1|1\rangle = \beta\,|1\rangle.$$

We present a more abstract interpretation of the outer product, which is useful for computations. We recall from Section B.3 that we can view each $|\varphi\rangle \in \mathbb{H}$ as the linear map

$$(2.4.23) \qquad \mathbb{C} \to \mathbb{H}, \quad \alpha \mapsto \alpha\,|\varphi\rangle$$

in $\mathrm{Hom}(\mathbb{C}, \mathbb{H})$. In this way, we obtain an isomorphism $\mathbb{H} \to \mathrm{Hom}(\mathbb{C}, \mathbb{H})$ of $\mathbb{C}$-vector spaces. Also, if we identify $|\varphi\rangle$ with the map in (2.4.23), then we can write

$$(2.4.24) \qquad\qquad |\varphi\rangle\langle\psi| = |\varphi\rangle \circ \langle\psi| .$$

We present a formula for the representation matrix of $|\varphi\rangle\langle\psi|$ with respect to the basis B where $|\varphi\rangle$ and $|\psi\rangle \in \mathbb{H}$. For this, we let

$$(2.4.25) \qquad\qquad |\varphi\rangle_B = (\alpha_0, \dots, \alpha_{k-1}), \quad |\psi\rangle_B = (\beta_0, \dots, \beta_{k-1}).$$

Then we have

$$(2.4.26) \qquad\qquad \mathrm{Mat}_B(|\varphi\rangle\langle\psi|) = |\varphi\rangle_B^* \, |\psi\rangle_B = (\alpha_i \overline{\beta_j})_{0 \le i,j < k}.$$

From (2.4.26) we obtain the following results by applying the rules of matrix multiplication and the formula for the trace.

Proposition 2.4.27. *Let* $|\varphi\rangle, |\psi\rangle, |\xi\rangle, |\chi\rangle \in \mathbb{H}$ *and let* $\alpha \in \mathbb{C}$. *Then the following hold.*

(1) $(|\varphi\rangle + |\psi\rangle)\langle\xi| = |\varphi\rangle\langle\xi| + |\psi\rangle\langle\xi|.$

(2) $|\varphi\rangle(\langle\psi| + \langle\xi|) = |\varphi\rangle\langle\psi| + |\varphi\rangle\langle\xi|.$

(3) $(\alpha\,|\varphi\rangle)\langle\psi| = \alpha\,|\varphi\rangle\langle\psi|.$

(4) $|\varphi\rangle(\alpha\,\langle\psi|) = \overline{\alpha}\,|\varphi\rangle\langle\psi|.$

(5) $(|\varphi\rangle\langle\psi|)^* = |\psi\rangle\langle\varphi|.$

(6) $\mathrm{tr}(|\varphi\rangle\langle\psi|) = \langle\psi|\varphi\rangle.$

(7) $|\varphi\rangle\langle\psi| \circ |\xi\rangle\langle\chi| = \langle\psi|\xi\rangle\,|\varphi\rangle\langle\chi|.$

(8) $\langle\varphi|\psi\rangle\langle\xi|\chi\rangle = \big\langle\varphi\big|(|\psi\rangle\langle\xi|)\big|\chi\big\rangle.$

Exercise 2.4.28. Prove Proposition 2.4.27.

In addition, the following proposition can be deduced from (2.4.26).

Proposition 2.4.29. *Let* A *be a linear operator on* $\mathbb{H}$ *and let* $|\varphi\rangle, |\psi\rangle \in \mathbb{H}$. *Then we have the following.*

(1) $A \circ |\varphi\rangle\langle\psi| = A\,|\varphi\rangle\langle\psi|.$

(2) $|\varphi\rangle\langle\psi| \circ A^* = |\varphi\rangle\,A\,|\psi\rangle.$

(3) $\mathrm{tr}\, A \circ |\varphi\rangle\langle\psi| = \mathrm{tr}\,|\varphi\rangle\langle\psi| \circ A^* = \langle\psi|A|\varphi\rangle.$

Exercise 2.4.30. Prove Proposition 2.4.29.

The outer product can be used to construct an orthonormal basis of $\mathrm{End}(\mathbb{H})$ as follows.

Proposition 2.4.31. *The sequence* $(|b_i\rangle\langle b_j|)_{i,j \in \mathbb{Z}_k}$ *is an orthonormal basis of the* $\mathbb{C}$-*algebra* $\mathrm{End}(\mathbb{H})$ *with respect to the Hilbert-Schmidt inner product. Furthermore, for any* $A \in \mathrm{End}(\mathbb{H})$ *we have*

$$(2.4.27) \qquad\qquad A = \sum_{i,j=0}^{k-1} \langle b_i|A|b_j\rangle\,|b_i\rangle\langle b_j| .$$

Proof. Let $i, j, u, v \in \mathbb{Z}_k$. Then Proposition 2.4.27 implies $\mathrm{tr}(|b_i\rangle\langle b_j| \circ |b_u\rangle\langle b_v|) = \langle b_j|b_u\rangle\langle b_i|b_v\rangle = \delta_{j,u}\delta_{i,v}$. Hence, the sequence $(|b_i\rangle\langle b_j|)$ is orthonormal. Since its length is n^2 which is the dimension of $\mathrm{End}(\mathbb{H})$ over $\mathbb{C}$, it is a basis of this $\mathbb{C}$-algebra. Also, (2.4.27) follows from (2.2.12). $\qquad\square$

From Proposition 2.4.31 we obtain the following results.

Corollary 2.4.32. *For any $A \in \mathrm{End}(\mathbb{H})$ and any $j \in \mathbb{Z}_k$ we have*

$$(2.4.28) \qquad A\,|b_j\rangle = \sum_{i=0}^{k-1}\langle b_i|A|b_j\rangle\,|b_i\rangle.$$

Corollary 2.4.33.

$$(2.4.29) \qquad I_{\mathbb{H}} = \sum_{i=0}^{k-1}|b_i\rangle\langle b_i|.$$

Exercise 2.4.34. Prove Corollaries 2.4.32 and 2.4.33.

2.4.5. Projections. In this section, we introduce projections and, in particular, orthogonal projections. Let $B = (|b_0\rangle, \ldots, |b_{k-1}\rangle)$ be an orthonormal basis of $\mathbb{H}$.

Definition 2.4.35. (1) An operator $P \in \mathrm{End}(\mathbb{H})$ is called a *projection* if $P^2 = P$.

(2) A projection $P \in \mathrm{End}(\mathbb{H})$ is called *orthogonal* if $P\,|\varphi\rangle$ and $|\varphi\rangle - P\,|\varphi\rangle$ are orthogonal to each other for all $|\varphi\rangle \in \mathbb{H}$.

A projection is sometimes also called a *projector*.

Example 2.4.36. Consider the map

$$(2.4.30) \qquad P : \mathbb{C}^2 \to \mathbb{C}^2, \quad (\alpha, \beta) \mapsto (-\beta, \beta).$$

This map is $\mathbb{C}$-linear with the representation matrix

$$(2.4.31) \qquad P = \begin{pmatrix} 0 & -1 \\ 0 & 1 \end{pmatrix}.$$

Also, it is a projection since

$$(2.4.32) \qquad P^2 = \begin{pmatrix} 0 & -1 \\ 0 & 1 \end{pmatrix} \cdot \begin{pmatrix} 0 & -1 \\ 0 & 1 \end{pmatrix} = \begin{pmatrix} 0 & -1 \\ 0 & 1 \end{pmatrix}.$$

But P is not an orthogonal projection. To see this, we note that $\langle P(1,2)|(1,2) - P(1,2)\rangle = \langle(-2,2)|(1,2) - (-2,2)\rangle = \langle(-2,2)|(3,0)\rangle = -6 \neq 0$.

Example 2.4.37. Consider the map

$$(2.4.33) \qquad P : \mathbb{C}^2 \to \mathbb{C}^2, \quad (\alpha, \beta) \mapsto (0, \beta).$$

This map is $\mathbb{C}$-linear with the representation matrix

$$(2.4.34) \qquad P = \begin{pmatrix} 0 & 0 \\ 0 & 1 \end{pmatrix}.$$

Also, it is a projection since

(2.4.35)
$$P^2 = \begin{pmatrix} 0 & 0 \\ 0 & 1 \end{pmatrix} \begin{pmatrix} 0 & 0 \\ 0 & 1 \end{pmatrix} = \begin{pmatrix} 0 & 0 \\ 0 & 1 \end{pmatrix}.$$

To show that P is an orthogonal projection, we note that $\langle P(\alpha, \beta) | (\alpha, \beta) - P(\alpha, \beta) \rangle = \langle (0, \beta) | (\alpha, 0) \rangle = 0$ for all $\alpha, \beta \in \mathbb{C}$.

Exercise 2.4.38. Show that for any orthogonal projection P on $\mathbb{H}$ and any $|\psi\rangle \in \mathbb{H}$ we have $\|P|\psi\rangle\| \leq \||\psi\rangle\|$.

Proposition 2.4.39. *Let $P \in \mathrm{End}(\mathbb{H})$. Then the following are true.*

(1) *If P is a projection, then P^* is a projection.*

(2) *If P is an orthogonal projection, then P^* is an orthogonal projection.*

Exercise 2.4.40. Prove Proposition 2.4.39.

We characterize orthogonal projections.

Proposition 2.4.41. *A projection $P \in \mathrm{End}(\mathbb{H})$ is orthogonal if and only if P is Hermitian.*

Exercise 2.4.42. Prove Proposition 2.4.41.

Example 2.4.43. Let $|\varphi\rangle \in \mathbb{H}$ with $\langle \varphi | \varphi \rangle = 1$. We claim that the map $P = |\varphi\rangle\langle\varphi|$ is an orthogonal projection. To see this, let $|\psi\rangle \in \mathbb{H}$. Then the linearity of the inner product in the second argument implies

$$\begin{aligned} P^2 |\psi\rangle &= P(P|\psi\rangle) = P(|\varphi\rangle\langle\varphi|\psi\rangle) \\ &= \langle\varphi|\psi\rangle P |\varphi\rangle = \langle\varphi|\psi\rangle\langle\varphi|\varphi\rangle |\varphi\rangle \\ &= |\varphi\rangle\langle\varphi|\psi\rangle = P|\psi\rangle. \end{aligned}$$

Also, we have

$$\left\langle P|\psi\rangle \,\big|\, |\psi\rangle - P|\psi\rangle \right\rangle = \overline{\langle\varphi|\psi\rangle}(\langle\varphi|\psi\rangle - \langle\varphi|\psi\rangle\langle\varphi|\varphi\rangle) = 0.$$

We generalize Example 2.4.43.

Proposition 2.4.44. *Let $l \in \mathbb{N}$, and let $\mathbb{H}(0), \ldots, \mathbb{H}(l-1)$ be linear subspaces of $\mathbb{H}$ which are orthogonal to each other such that $\mathbb{H} = \mathbb{H}(0) + \cdots + \mathbb{H}(l-1)$. Then the following hold.*

(1) *For $|\varphi\rangle \in \mathbb{H}$ let $|\varphi\rangle = \sum_{i=0}^{l-1} |\varphi(i)\rangle$ be the uniquely determined representation of $|\varphi\rangle$ as a sum of elements $|\varphi_i\rangle$ in $\mathbb{H}(i)$. Then, for all $i \in \mathbb{Z}_l$ the map*

(2.4.36)
$$P_i : \mathbb{H} \to \mathbb{H}(i), \quad |\varphi\rangle \mapsto |\varphi(i)\rangle$$

is an orthogonal projection. It is called the orthogonal projection of $\mathbb{H}$ onto $\mathbb{H}(i)$. Also, for $|\varphi\rangle \in \mathbb{H}$ the image $P_i |\varphi\rangle$ is called the orthogonal projection of $|\varphi\rangle$ onto $\mathbb{H}(i)$.

(2) *$P_0, \ldots, P_{l-1}$ are orthogonal to each other with respect to the Hilbert-Schmidt inner product.*

(3) *We have $\sum_{i=0}^{l-1} P_i = I_{\mathbb{H}}$.*

(4) *Let $B_0, \ldots, B_{l-1}$ be orthonormal bases of $\mathbb{H}(0), \ldots, \mathbb{H}(l-1)$, respectively, such that $B = B_0 \| \cdots \| B_{l-1}$ is an orthonormal basis of $\mathbb{H}$ which exists by Proposition 2.2.44. Then for all $i \in \mathbb{Z}_l$ we have*

$$(2.4.37) \qquad P_i = \sum_{|b\rangle \in B_i} |b\rangle \langle b|.$$

Proof. Let $i \in \mathbb{Z}_l$ and let $|\varphi\rangle \in \mathbb{H}$. The uniqueness of the representation of the elements of $\mathbb{H}$ as a sum of elements in the $\mathbb{H}(i)$ implies $P_i^2 \varphi = P_i(|\varphi(i)\rangle) = |\varphi(i)\rangle$. Also, the sequence (P_i) is orthogonal because of the orthogonality of the $\mathbb{H}(i)$. This proves the first assertion. Next, for $i, j \in \mathbb{Z}_l$ with $i \neq 0$ we have $P_i P_j = 0$. Therefore, the P_i are orthogonal to each other with respect to the Hilbert-Schmidt inner product. So the P_i are linearly independent by Corollary 2.2.34. The last assertion follows from Proposition 2.4.31. $\qquad \square$

Example 2.4.45. Recall that

$$(2.4.38) \qquad (|x_+\rangle, |x_-\rangle) = \left(\frac{|0\rangle + |1\rangle}{\sqrt{2}}, \frac{|0\rangle - |1\rangle}{\sqrt{2}} \right)$$

is an orthonormal basis of $\mathbb{H}_1$. The orthogonal projection of $|0\rangle$ onto $\mathbb{C}\,|x_+\rangle$ is

$$(2.4.39) \qquad |x_+\rangle \langle +|0\rangle = \frac{1}{\sqrt{2}}\,|x_+\rangle.$$

2.4.6. Schur decomposition. As we have seen in Section 2.4.1, not all matrices in $\mathbb{C}^{(k,k)}$ are diagonalizable. However, we can prove the following weaker result, which will allow us to prove the spectral theorem in the next section. It was first proved by the mathematician Issai Schur in the early 20th century.

Theorem 2.4.46 (Schur decomposition theorem). *Let $A \in \mathbb{C}^{(k,k)}$. Assume that A has the l distinct eigenvalues $\lambda_0, \ldots, \lambda_{l-1}$ with algebraic multiplicities $m_0, \ldots, m_{l-1}$. Then $k = \sum_{i=0}^{l-1} m_i$ and there is a unitary matrix $U \in \mathbb{C}^{(k,k)}$ and an upper triangular matrix T with diagonal*

$$(2.4.40) \qquad (\underbrace{\lambda_0, \ldots, \lambda_0}_{m_0}, \underbrace{\lambda_1, \ldots, \lambda_1}_{m_1}, \ldots, \underbrace{\lambda_{l-1}, \ldots, \lambda_{l-1}}_{m_{l-1}})$$

such that

$$(2.4.41) \qquad A = UTU^*.$$

Such a representation is called Schur decomposition *of A.*

Proof. We prove the assertion by induction on k and, in doing so, present an algorithm to construct a Schur decomposition of A. For $k = 1$ the assertion is true since in this case, the matrix A is in upper triangular form. So we can set $U = I_1$.

Let $k > 1$ and assume that the assertion holds for all $m < k$. Let $\vec{v}$ be an eigenvector associated with the eigenvalue λ_0 that exists by Proposition B.7.21. Assume, without loss of generality, that $\left\| \vec{v} \right\| = 1$. By Theorem 2.2.33, there is a matrix $X \in C^{(k,k-1)}$ such that the column vectors of the matrix

$$(2.4.42) \qquad (\vec{v}\,X)$$

form an orthonormal basis of $\mathbb{C}^k$. Proposition 2.4.18 implies that this matrix is unitary. So we have

$$
\begin{pmatrix} \vec{v}^* \\ X^* \end{pmatrix} A \begin{pmatrix} \vec{v} & X \end{pmatrix} = \begin{pmatrix} \vec{v}^* A \vec{v} & \vec{v}^* A X \\ X^* A \vec{v} & X^* A X \end{pmatrix}
$$

(2.4.43)

$$
= \begin{pmatrix} \lambda_0 \vec{v}^* \vec{v} & \vec{v}^* A X \\ \lambda_0 X^* \vec{v} & X^* A X \end{pmatrix} = \begin{pmatrix} \lambda_0 & \vec{v}^* A X \\ 0 & X^* A X \end{pmatrix}.
$$

The lower-left corner of this matrix is zero because all columns of X are orthogonal to $\vec{v}$. Also, $X^* A X$ is in $\mathbb{C}^{(k-1,k-1)}$ and since the matrix $(\vec{v} X)$ is unitary, we have $(\vec{v} X)^{-1} = (\vec{v} X)^*$. So by (2.4.43) and Proposition B.5.30 we have

(2.4.44) $$p_A(x) = (x - \lambda_0) p_{X^* A X}(x)$$

which implies

(2.4.45) $$p_{X^* A X}(x) = (x - \lambda_0)^{m_0 - 1} \prod_{i=1}^{l-1} (x - \lambda_i)^{m_i}.$$

By the induction hypothesis, there is a unitary matrix $Y \in \mathbb{C}^{(k-1,k-1)}$ such that

(2.4.46) $$Z = Y^* X^* A X Y$$

is an upper triangular matrix with diagonal

(2.4.47) $$(\underbrace{\lambda_0, \dots, \lambda_0}_{m_0 - 1}, \underbrace{\lambda_1, \dots, \lambda_1}_{m_1}, \dots, \underbrace{\lambda_{l-1}, \dots, \lambda_{l-1}}_{m_{l-1}}).$$

Define

(2.4.48) $$U = (\vec{v} X Y).$$

Then $U \in \mathbb{C}^{(k,k)}$ and this matrix is unitary because

(2.4.49) $$U^* U = \begin{pmatrix} \vec{v}^* \\ Y^* X^* \end{pmatrix} (\vec{v} X Y) = \begin{pmatrix} 1 & \vec{v}^* X Y \\ Y^* X^* \vec{v} & Y^* X^* X Y \end{pmatrix} = \begin{pmatrix} 1 & 0 \\ 0 & I_{k-1} \end{pmatrix}.$$

Also, we have

$$
U^* A U = \begin{pmatrix} \vec{v}^* \\ Y^* X^* \end{pmatrix} A (\vec{v} X Y)
$$

(2.4.50)

$$
= \begin{pmatrix} \vec{v}^* A \vec{v} & \vec{v}^* A X Y \\ Y^* X^* A \vec{v} & Y^* X^* A X Y \end{pmatrix}
$$

$$
= \begin{pmatrix} \lambda_0 & \vec{v}^* A X Y \\ 0 & Z \end{pmatrix}
$$

which is an upper triangular matrix. Denote it by T. The diagonal of the upper triangular matrix Z is shown in (2.4.47). Therefore, by (2.4.50) the diagonal of T is

(2.4.51) $$(\underbrace{\lambda_0, \dots, \lambda_0}_{m_0}, \underbrace{\lambda_1, \dots, \lambda_1}_{m_1}, \dots, \underbrace{\lambda_{l-1}, \dots, \lambda_{l-1}}_{m_{l-1}}).$$

So we have found the Schur decomposition $A = U T U^*$ of A. $\square$

Example 2.4.47. Consider the matrix

$$(2.4.52) \qquad A = \begin{pmatrix} 3 & -2 \\ 2 & -1 \end{pmatrix}.$$

Its characteristic polynomial is $p_A(x) = (x-1)^2$. So 1 is the only eigenvalue of A and its algebraic multiplicity is 2. However, its geometric multiplicity is only 1 since the matrix $I - A$ has rank 1. Therefore, A is not diagonalizable. We determine the Schur decomposition of A using the construction of the proof of Theorem 2.4.46. We have $l = 1$, $\lambda_0 = 1$, and $m_0 = 2$. The case $k = 1$ is trivial. So, let $k = 2$ and use the unitary eigenvector $\vec{v} = \frac{1}{\sqrt{2}}(1,1)$ of A. The only column vector of the matrix

$$(2.4.53) \qquad X = \frac{1}{\sqrt{2}} \begin{pmatrix} 1 \\ -1 \end{pmatrix}$$

appends $\vec{v}$ to an orthonormal basis of $\mathbb{C}^2$. So, the matrix from (2.4.42) is

$$(2.4.54) \qquad (\vec{v} \quad X) = \frac{1}{\sqrt{2}} \begin{pmatrix} 1 & 1 \\ 1 & -1 \end{pmatrix}.$$

Now we have

$$(2.4.55) \qquad \begin{pmatrix} \vec{v}^* \\ X^* \end{pmatrix} A (\vec{v} X) = \frac{1}{2} \begin{pmatrix} 1 & 1 \\ 1 & -1 \end{pmatrix} \begin{pmatrix} 3 & -2 \\ 2 & -1 \end{pmatrix} \begin{pmatrix} 1 & 1 \\ 1 & -1 \end{pmatrix} = \begin{pmatrix} 1 & 4 \\ 0 & 1 \end{pmatrix}.$$

This is already the upper triangular matrix that we are looking for. But let us see how the construction proceeds. The matrix X^*AX is (1). We can choose the unitary matrix $Y = (1)$. Then $Z = Y^*X^*AXY = (1)$. We set

$$(2.4.56) \qquad U = (\vec{v} \quad XY) = \frac{1}{\sqrt{2}} \begin{pmatrix} 1 & 1 \\ 1 & -1 \end{pmatrix}$$

and obtain

$$(2.4.57) \qquad U^*AU = T = \begin{pmatrix} 1 & 4 \\ 0 & 1 \end{pmatrix}.$$

So $A = UTU^*$ is a Schur decomposition of A.

2.4.7. The spectral theorem. The aim of this section is to introduce the renowned spectral theorem, which establishes the diagonalizability of normal matrices and the existence of an orthonormal basis consisting of eigenvectors for normal operators. The finite-dimensional version which we present here goes back to the early 20th century and is closely associated with the contributions of mathematicians such as David Hilbert. It assumes pivotal significance in the postulates of quantum mechanics presented in Chapter 3. As previously mentioned, the broader domain of quantum mechanics necessitates the infinite-dimensional analog, which can be attributed to mathematicians like John von Neumann and Hermann Weyl.

Definition 2.4.48. A matrix $A \in \mathbb{C}^{(k,k)}$ or operator $A \in \text{End}(\mathbb{H})$ is called *normal* if $A^*A = AA^*$.

The next proposition presents important examples of normal matrices and operators.

Proposition 2.4.49. *If $A \in \mathbb{C}^{(k,k)}$ or $A \in \mathrm{End}(V)$ is Hermitian or unitary, then A is normal.*

Exercise 2.4.50. Prove Proposition 2.4.49.

Here are some properties of normal matrices and operators.

Proposition 2.4.51.   (1) *The adjoint of a normal matrix or operator is normal.*

(2) *Every diagonal matrix in $\mathbb{C}^{(k,k)}$ is normal.*

(3) *A matrix in $\mathbb{C}^{(k,k)}$ that is both normal and upper triangular is a diagonal matrix.*

Proof. Let $A \in \mathbb{C}^{(k,k)}$ be a normal matrix. We apply (2.3.25) and obtain $(A^*)^*A^* = AA^* = A^*A = A^*(A^*)^*$. This proves the first assertion.

To show the second assertion, let $D \in \mathbb{C}^{(k,k)}$ be a diagonal matrix with diagonal $(\lambda_0, \ldots, \lambda_{k-1}) \in \mathbb{C}^k$. Then D^*D and DD^* are diagonal matrices with diagonal $(|\lambda_0|^2, \ldots, |\lambda_{k-1}|^2)$. Hence, we have $DD^* = D^*D$ which means that D is normal.

We prove the last statement and let $A = (a_{i,j})$. We also denote by $(\vec{r}_0, \ldots, \vec{r}_{k-1})$ the row vectors of A and by $(\vec{c}_0, \ldots, \vec{c}_{k-1})$ the column vectors of A. Then $A^*A = AA^*$ implies that for any $u \in \mathbb{Z}_k$ the entry of this matrix with row and column index u can be computed as $\vec{r}_u^* \vec{r}_u = \left\| \vec{r}_u \right\|^2$ and as $\vec{c}_u \vec{c}_u^* = \left\| \vec{c}_u \right\|$. So for $0 \le u < k$ we have

$$(2.4.58) \qquad \left\| \vec{r}_u \right\|^2 = \left\| \vec{c}_u \right\|^2.$$

We prove by induction on u that for $u = 0, 1, \ldots, k$ we have

$$(2.4.59) \qquad \vec{r}_i = \vec{c}_i = a_{i,i}\vec{e}_i, \quad 0 \le i < u.$$

For $u = k$ this implies that A is diagonal. For the base case $u = 0$ there is nothing to show. For the induction step, assume that $0 \le u < k$ and that (2.4.59) holds for u. Since A is upper triangular, it follows from (2.4.59) that

$$(2.4.60) \qquad \vec{c}_u = a_{u,u}\vec{e}_u$$

and

$$(2.4.61) \qquad \vec{r}_u = (0, \ldots, 0, a_{u,u}, a_{u,u+1}, \ldots, a_{u,k-1}).$$

So (2.4.58) implies $|a_{u,u}|^2 = \sum_{j=u}^{k-1} |a_{u,j}|^2$ and thus $a_{u,u+1} = \cdots = a_{u,k-1} = 0$ which shows that $\vec{r}_u = a_{u,u}\vec{e}_u$. $\qquad \square$

The next exercise shows that there are normal matrices that are neither unitary nor Hermitian.

Example 2.4.52. Show that the matrix

$$(2.4.62) \qquad \begin{pmatrix} 1 & 1+i & 1 \\ -1+i & 1 & 1 \\ -1 & -1 & 1 \end{pmatrix}$$

is normal but not Hermitian or unitary.

The next theorem states that normal matrices are diagonalizable.

Theorem 2.4.53. *Let $A \in \mathbb{C}^{(k,k)}$ be a normal matrix, let $l \in \mathbb{N}$, let $\lambda_0, \ldots, \lambda_{l-1}$ be the distinct eigenvalues of A, and let $m_0, \ldots, m_{l-1}$ be their algebraic multiplicities. Then there is a unitary matrix $U \in \mathbb{C}^{(k,k)}$ such that*

$$(2.4.63) \qquad U^*AU = \mathrm{diag}(\underbrace{\lambda_0, \ldots, \lambda_0}_{m_0}, \underbrace{\lambda_1, \ldots, \lambda_1}_{m_1}, \ldots, \underbrace{\lambda_{l-1}, \ldots, \lambda_{l-1}}_{m_{l-1}}).$$

Also, if we write the sequence of column vectors of U as

$$(2.4.64) \qquad U = (\underbrace{\vec{u}_{M_0}, \ldots, \vec{u}_{M_1-1}}_{U_0}, \underbrace{\vec{u}_{M_1}, \ldots, \vec{u}_{M_2-1}}_{U_1}, \ldots, \underbrace{\vec{u}_{M_{l-1}}, \ldots, \vec{u}_{M_l-1}}_{U_{l-1}})$$

with $M_j = \sum_{i=0}^{j-1} m_i$ for $0 \le j \le l$, then U_j is an orthonormal basis of the eigenspace associated with λ_j for all $j \in \mathbb{Z}_l$.

Proof. Let A be normal. By Theorem 2.4.46 there is a Schur decomposition $A = UTU^*$ where $U \in \mathbb{C}^{(k,k)}$ is a unitary matrix and T is an upper triangular matrix with diagonal

$$(2.4.65) \qquad (\underbrace{\lambda_0, \ldots, \lambda_0}_{m_0}, \underbrace{\lambda_1, \ldots, \lambda_1}_{m_1}, \ldots, \underbrace{\lambda_{l-1}, \ldots, \lambda_{l-1}}_{m_{l-1}}).$$

Since U is unitary, we can also write $T = U^*AU$. Now we have

$$
\begin{aligned}
(2.4.66) \qquad T^*T &= (U^*AU)^*(U^*AU) = U^*A^*UU^*AU \\
&= U^*A^*AU = U^*AA^*U = (U^*AU)(U^*A^*U) \\
&= (U^*AU)(U^*AU)^* = TT^*.
\end{aligned}
$$

Hence T is normal and upper triangular. So Proposition 2.4.51 implies that T is a diagonal matrix.

We prove the second assertion. Let $j \in \mathbb{Z}_l$ and let $\vec{u}$ be an element of U_j. Then $A\vec{u} = \lambda_j \vec{u}$. It follows that U_j is an orthonormal sequence of m_j elements of the eigenspace E_j of A associated with λ_j. By Theorem B.7.28, the dimension of E_j is m_j. So U_j is an orthonormal basis of E_j. $\square$

Note that in Theorem 2.4.53 we have $M_0 = 0$ and $M_l = k$.

Example 2.4.54. Consider the Pauli matrix

$$(2.4.67) \qquad X = \begin{pmatrix} 0 & 1 \\ 1 & 0 \end{pmatrix}$$

and set

$$(2.4.68) \qquad U = \frac{1}{\sqrt{2}} \begin{pmatrix} 1 & 1 \\ 1 & -1 \end{pmatrix}.$$

The we have $U^*XU = \mathrm{diag}(1, -1)$.

Exercise 2.4.55. Find the decomposition (2.4.63) for the Pauli matrices Y and Z and for the Hadamard matrix H.

From Theorem 2.4.53 we obtain the spectral theorem.

Theorem 2.4.56 (Spectral theorem). *Let $A \in \text{End}(\mathbb{H})$ be normal. Let Λ be the set of eigenvalues of A. For $\lambda \in \Lambda$ denote by P_λ the orthogonal projection onto the eigenspace E_λ corresponding to λ. Then the following are true.*

(1) *There are orthonormal bases B_λ of E_λ for all $\lambda \in \Lambda$ such that their concatenation is an orthonormal basis of $\mathbb{H}$.*

(2) *The eigenspaces E_λ are orthogonal to each other, and their sum is $\mathbb{H}$.*

(3) $P_\lambda = \sum_{|b\rangle \in B_\lambda} |b\rangle \langle b|$.

(4) $\sum_{\lambda \in \Lambda} P_\lambda = I_{\mathbb{H}}$.

(5) $A = \sum_{\lambda \in \Lambda} \lambda P_\lambda$. *This representation of A is called the* spectral decomposition *of A.*

Proof. Let $l = |\Lambda|$ and $\Lambda = \{\lambda_0, \ldots, \lambda_{l-1}\}$. Denote by A also the representation matrix of A with respect to an orthonormal basis C of $\mathbb{H}$. Use the notation of Theorem 2.4.53. Then

$$(2.4.69) \qquad B = CU = (|u_0\rangle, \ldots, |u_{k-1}\rangle)$$

is another orthonormal basis of $\mathbb{H}$. Let $j \in \mathbb{Z}_l$. It follows from the properties of U that $B_\lambda = (|u_{M_j}\rangle, \ldots, |u_{M_{j+1}-1}\rangle)$ is an orthonormal basis of E_λ. This proves the first assertion.

The second assertion follows immediately from the first. The second, third, and fourth assertions follow from Proposition 2.4.44. Using the fourth assertion we obtain

$$(2.4.70) \qquad A|\varphi\rangle = A \sum_{\lambda \in \Lambda} P_\lambda |\varphi\rangle = \sum_{\lambda \in \Lambda} A P_\lambda |\varphi\rangle = \sum_{\lambda \in \Lambda} \lambda P_\lambda |\varphi\rangle. \qquad \square$$

In the following, we adopt a simplified notation. When A is a normal operator in $\text{End}(\mathbb{H})$, we write its spectral as $A = \sum_{\lambda \in \Lambda} P_\lambda$. This notation assumes that Λ represents the set of eigenvalues of A, and for each $\lambda \in \Lambda$, P_λ represents the projection onto the eigenspace in $\mathbb{H}$ corresponding to λ without explicitly mentioning it.

The next example determines the spectral decomposition of the Pauli operators.

Example 2.4.57. Consider the following pairs of quantum states in $\mathbb{H}_1$:

$$(|x_+\rangle, |x_-\rangle) = \left(\frac{|0\rangle + |1\rangle}{\sqrt{2}}, \frac{|0\rangle - |1\rangle}{\sqrt{2}} \right),$$

$$(2.4.71) \qquad (|y_+\rangle, |y_-\rangle) = \left(\frac{|0\rangle + i|1\rangle}{\sqrt{2}}, \frac{|0\rangle - i|1\rangle}{\sqrt{2}} \right),$$

$$(|z_+\rangle, |z_-\rangle) = (|0\rangle, |1\rangle).$$

These pairs are orthonormal bases of $\mathbb{H}_1$. The eigenvalues of the Pauli operators X, Y, and Z are 1 and -1 and their spectral decomposition is

$$X = |x_+\rangle\langle x_+| - |x_-\rangle\langle x_-|,$$

$$(2.4.72) \qquad Y = |y_+\rangle\langle y_+| - |y_-\rangle\langle y_-|,$$

$$Z = |z_+\rangle\langle z_+| - |z_-\rangle\langle z_-|.$$

Exercise 2.4.58. Verify Example 2.4.57.

Here are some properties of the spectral decomposition of normal endomorphisms of $\mathbb{H}$.

Proposition 2.4.59. *Let $A \in \text{End}(\mathbb{H})$ be normal, let Λ be the set of eigenvalues of A, and let*

$$\tag{2.4.73} A = \sum_{\lambda \in \Lambda} \lambda P_\lambda$$

be the spectral decomposition of A. Also, let $m \in \mathbb{Z}_{\geq 0}$. Then we have

$$\tag{2.4.74} A^* = \sum_{\lambda \in \Lambda} \overline{\lambda} P_\lambda, \quad AA^* = \sum_{\lambda \in \Lambda} |\lambda|^2 P_\lambda, \quad A^m = \sum_{\lambda \in \Lambda} \lambda^m P_\lambda.$$

If A is invertible, then the last equation holds for all $m \in \mathbb{Z}$.

Proof. We have

$$
\begin{aligned}
A^* &= \left(\sum_{\lambda \in P} \lambda P_\lambda \right)^* && \text{by (2.4.73),} \\
&= \sum_{\lambda \in P} \overline{\lambda} P_\lambda^* && \text{by Proposition 2.3.9,} \\
&= \sum_{\lambda \in P} \overline{\lambda} P_\lambda && \text{by Proposition 2.4.41.}
\end{aligned}
$$

This proves the first assertion. The other two assertions can be verified using the fact that by Theorem 2.4.56 the eigenspaces $P_\lambda(\mathbb{H})$ are pairwise orthogonal. $\qquad\square$

We note that the second and third equations in (2.4.74) may show spectral decompositions, since the absolute values and powers of different eigenvalues of a normal operator may be the same. To obtain the spectral decompositions, we have to group the projections appropriately.

The spectral theorem allows us to characterize involutions, projections, and Hermitian and unitary operators by their eigenvalues.

Proposition 2.4.60. *Let $A \in \mathbb{C}^{(k,k)}$ or $A \in \text{End}(\mathbb{H})$ be normal and let Λ be the set of eigenvalues of A. Then the following hold.*

(1) *A is an involution if and only if $\Lambda \subset \{-1, 1\}$.*

(2) *A is a projection if and only if $\Lambda \subset \{0, 1\}$.*

(3) *A is Hermitian if and only if all its eigenvalues are real numbers.*

(4) *A is unitary if and only if all of its eigenvalues have absolute value 1.*

Proof. Let

$$\tag{2.4.75} A = \sum_{\lambda \in \Lambda} \lambda P_\lambda$$

be the spectral decomposition of A. Recall that by Proposition 2.4.44, the P_i are linearly independent. This will be used in all arguments.

Using Theorem 2.4.56 and Proposition 2.4.59 we see that A is an involution if and only if

$$(2.4.76) \qquad A^2 = \sum_{\lambda \in \Lambda} \lambda^2 P_\lambda = I_{\mathbb{H}} = \sum_\lambda P_\lambda.$$

Hence, A is an involution if and only if $\lambda^2 = 1$ for all $\lambda \in \Lambda$. This is true if and only if $\Lambda \subset \{1, -1\}$.

Next, A is a projection if and only if $A^2 = A$. By Theorem 2.4.56 and Proposition 2.4.59 this is true if and only if $\lambda^2 = \lambda$ for all $\lambda \in \Lambda$ which is equivalent to $\Lambda \subset \{0, 1\}$.

In addition, A is Hermitian if and only if $A^* = A$. By Theorem 2.4.56 and Proposition 2.4.59 this is equivalent to $\lambda = \bar{\lambda}$ and thus $\lambda \in \mathbb{R}$ for all $\lambda \in \Lambda$.

Finally, A is unitary if and only if A is invertible and $A^* = A^{-1}$. By Theorem 2.4.56 and Proposition 2.4.59 this is equivalent to $0 \notin \Lambda$ and $\bar{\lambda} = 1/\lambda$ for all $\lambda \in \Lambda$ which means that $|\lambda| = 1$ for all $\lambda \in \Lambda$. $\qquad \square$

As a consequence of Proposition 2.4.60, we obtain the following characterization of Hermitian matrices or operators.

Proposition 2.4.61. *A normal matrix $A \in \mathbb{C}^{(k,k)}$ or operator $A \in \mathrm{End}(\mathbb{H})$ is Hermitian if and only if $\langle \vec{u} | A | \vec{u} \rangle$ or $\langle \varphi | A | \varphi \rangle$ are real numbers for all $\vec{u} \in \mathbb{C}^k$ or $|\varphi\rangle \in \mathbb{H}$, respectively.*

Proof. It suffices to prove the assertion for $A \in \mathbb{C}^{(k,k)}$. First, assume that A is Hermitian and let $\vec{u} \in \mathbb{C}^k$. Then we have

$$(2.4.77) \qquad \overline{\langle \vec{u} | A\vec{u} \rangle} = \langle A\vec{u} | \vec{u} \rangle = \langle A^* \vec{u} | \vec{u} \rangle = \langle \vec{u} | A\vec{u} \rangle.$$

This shows that $\langle \vec{u} | A\vec{u} \rangle \in \mathbb{R}$.

Conversely, assume that $\langle \vec{u} | A\vec{u} \rangle \in \mathbb{R}$ for all $\vec{u} \in \mathbb{C}^k$. Let λ be an eigenvalue of A and let $\vec{u}$ be an eigenvector of A associated with λ of length 1. Then we have $\lambda = \langle \vec{u} | A | \vec{u} \rangle \in \mathbb{R}$. Therefore, Proposition 2.4.60 implies that A is Hermitian. $\qquad \square$

2.4.8. Definite operators and matrices. We define definite matrices and operators.

Definition 2.4.62. Let $A \in \mathrm{End}(\mathbb{H})$.

(1) A is called *positive definite* if $\langle \varphi | A | \varphi \rangle \in \mathbb{R}_{>0}$ for all nonzero $|\varphi\rangle \in \mathbb{H}$.

(2) A is called *positive semidefinite* if $\langle \varphi | A | \varphi \rangle \in \mathbb{R}_{\geq 0}$ for all $|\varphi\rangle \in \mathbb{H}$.

Definition 2.4.63. A matrix $A \in \mathbb{C}^{(k,k)}$ is called *positive definite* or *positive semidefinite* if the corresponding endomorphism of $\mathbb{C}^k$ has this property.

From Proposition 2.4.61 we obtain the following result.

Proposition 2.4.64. *All normal positive definite, positive semidefinite, negative definite, and negative semidefinite operators or matrices are Hermitian.*

We can also characterize positive definite operators by their eigenvalues.

Proposition 2.4.65. *Let* $A \in \mathbb{C}^{(k,k)}$ *or* $A \in \mathrm{End}(\mathbb{H})$.

(1) *If A is* positive definite, *then all eigenvalues of A are positive real numbers.*

(2) *If A is* positive semidefinite, *then all the eigenvalues of A are real numbers* ≥ 0.

Proof. Let λ be an eigenvalue of A and let $|\varphi\rangle \in \mathbb{H}$ be an eigenvector of A associated to λ of length 1. Then we have

$$(2.4.78) \qquad \langle\varphi|A|\varphi\rangle = \langle\varphi|\lambda\,|\varphi\rangle\rangle = \lambda.$$

This implies both assertions of the proposition. $\qquad\square$

We also present another characterization of normal positive semidefinite matrices and operators.

Proposition 2.4.66. *Let* $A \in \mathbb{C}^{(k,k)}$ *or* $A \in \mathrm{End}(\mathbb{H})$ *be normal. Then the following statements are equivalent.*

(1) *A is positive semidefinite.*

(2) $A = BB^*$ *for some normal* $B \in \mathbb{C}^{(k,k)}$ *or* $B \in \mathrm{End}(\mathbb{H})$, *respectively.*

(3) $A = B^2$ *for some Hermitian* $B \in \mathbb{C}^{(k,k)}$ *or* $B \in \mathrm{End}(\mathbb{H})$, *respectively.*

Proof. Let A be positive semidefinite. Then A is Hermitian by Proposition 2.4.64. Let

$$(2.4.79) \qquad A = \sum_{\lambda \in \Lambda} \lambda P_\lambda$$

be the spectral decomposition of A. Then by Proposition 2.4.65 we have $\lambda \in \mathbb{R}_{\geq 0}$ for all $\lambda \in \Lambda$. We set

$$(2.4.80) \qquad B = \sum_{\lambda \in \Lambda} \sqrt{\lambda} P_\lambda.$$

Then $B^* = B$ and Proposition 2.4.59 imply

$$(2.4.81) \qquad B^*B = BB^* = B^2 = \sum_{\lambda \in \Lambda} \lambda P_\lambda = A.$$

This shows that the first assertion implies the other two statements.

Now assume that there is a normal $B \in \mathrm{End}(\mathbb{H})$ such that $A = BB^*$. Let

$$(2.4.82) \qquad B = \sum_{\lambda \in \Lambda'} \lambda P_\lambda$$

be the spectral decomposition of B. Then we have

$$(2.4.83) \qquad A = \sum_{\lambda \in \Lambda'} |\lambda|^2 P_\lambda.$$

So A is positive semidefinite. Finally, assume that there is a Hermitian $B \in \mathrm{End}(\mathbb{H})$ such that $A = B^2$. Then we have $A = B^*B$ which implies that A is positive semidefinite. $\qquad\square$

2.4.9. Singular value decomposition. The next theorem is another consequence of the spectral theorem.

Theorem 2.4.67 (Singular value decomposition). *Let $k, l, r \in \mathbb{N}$ and let $A \in \mathbb{C}^{(k,l)}$ be
of rank r. Then there are unitary matrices $U \in \mathbb{C}^{(k,k)}$ and $V \in \mathbb{C}^{(l,l)}$ such that*

(2.4.84)
$$A = U \begin{pmatrix} \lambda_0 & \cdots & 0 & \vdots & & \vdots \\ \vdots & \ddots & \vdots & \cdots & 0 & \cdots \\ 0 & \cdots & \lambda_{r-1} & \vdots & & \vdots \\ \hline \vdots & & \vdots & \vdots & & \vdots \\ \cdots & 0 & \cdots & \cdots & 0 & \cdots \\ & \vdots & & & \vdots & \end{pmatrix} V^*$$

where $\lambda_0, \ldots, \lambda_{r-1}$ are positive real numbers. Such a representation is called a singular value decomposition *of the matrix A. In the decomposition, the diagonal entries
$\lambda_0, \ldots, \lambda_{r-1}$ are uniquely determined by A up to reordering. They are called the* singular values *of A.*

Proof. As shown in Exercise 2.4.68, the matrix A^*A is a positive semidefinite and Hermitian matrix in $\mathbb{C}^{(l,l)}$. It follows from Theorem 2.4.53 that there is a unitary matrix
$V \in \mathbb{C}^{(l,l)}$ such that

(2.4.85)
$$V^*A^*AV = D' = \begin{pmatrix} D & 0 \\ 0 & 0 \end{pmatrix}$$

where $D \in \mathbb{C}^{(m,m)}$ is a positive definite diagonal matrix, m is the number of nonzero
eigenvalues of A^*A, and these eigenvalues are positive real numbers and diagonal elements of D. By Theorem 2.4.53, the columns of V form an orthonormal basis of $\mathbb{C}^l$
consisting of eigenvectors of A^*A. The eigenvalue corresponding to the ith column of
V is the ith diagonal entry of D' for $0 \le i < k$. We write

(2.4.86)
$$V = \begin{pmatrix} V_1 & V_2 \end{pmatrix}$$

where $V_1 \in \mathbb{C}^{(l,m)}$ and $V_2 \in \mathbb{C}^{(l,l-m)}$. Then the columns of V_1 are linearly independent
eigenvectors corresponding to the nonzero eigenvalues of A^*A. Also, the columns of V_2
are eigenvectors corresponding to the eigenvalue 0 of A^*A. So (2.4.85) can be rewritten
as

(2.4.87)
$$\begin{pmatrix} V_1^* \\ V_2^* \end{pmatrix} A^*A \begin{pmatrix} V_1 & V_2 \end{pmatrix} = \begin{pmatrix} V_1^*A^*AV_1 & V_1^*A^*AV_2 \\ V_2^*A^*AV_1 & V_2^*A^*AV_2 \end{pmatrix} = \begin{pmatrix} D & 0 \\ 0 & 0 \end{pmatrix}.$$

This implies

(2.4.88)
$$(AV_1)^*AV_1 = V_1^*A^*AV_1 = D, \quad (AV_2)^*AV_2 = V_2^*A^*AV_2 = 0$$

which implies

(2.4.89)
$$\|AV_2\|^2 = \operatorname{tr}(V_2^*A^*AV_2) = 0.$$

Hence we have

(2.4.90)
$$AV_2 = 0.$$

The fact that V is unitary implies

(2.4.91)
$$V_1^*V_1 = I_l, \quad V_2^*V_2 = I_{l-m}, \quad V_1V_1^* + V_2V_2^* = I_l.$$

Now define

(2.4.92) $$U_1 = AV_1D^{-1/2}$$

where $D^{-1/2}$ is the diagonal matrix whose diagonal entries are the inverse square roots of the diagonal entries of D. Then the third equation in (2.4.91), (2.4.92), and (2.4.90) imply

(2.4.93)
$$U_1D^{1/2}V_1^* = AV_1D^{-1/2}D^{1/2}V_1^* = AV_1V_1^*$$
$$= A(I_k - V_2V_2^*) = A - (AV_2)V_2^* = A.$$

Also, (2.4.92) and (2.4.88) imply

(2.4.94) $$U_1^*U_1 = D^{-1/2}V_1^*A^*AV_1D^{-1/2} = D^{-1/2}DD^{-1/2} = I_m.$$

Hence, the columns of U_1 form an orthonormal sequence which by Theorem 2.2.33 can be extended to an orthonormal basis of $\mathbb{C}^l$. Therefore, we can choose $U_2 \in C^{(k,k-m)}$ so that

(2.4.95) $$U = \begin{pmatrix} U_1 & U_2 \end{pmatrix}$$

is a unitary matrix. Finally, we define the matrix $B \in \mathbb{C}^{(k,l)}$ as

(2.4.96) $$B = \begin{pmatrix} D^{1/2} & 0 \\ 0 & 0 \end{pmatrix}.$$

Then we have.

(2.4.97) $$UBV^* = \begin{pmatrix} U_1 & U_2 \end{pmatrix}\begin{pmatrix} D^{1/2} & 0 \\ 0 & 0 \end{pmatrix}\begin{pmatrix} V_1^* \\ V_2^* \end{pmatrix} = U_1D^{1/2}V_1^* = A. \qquad \square$$

Note that the singular values of a complex matrix A are the square roots of the nonzero eigenvalues of the normal square matrix A^*A.

Exercise 2.4.68. Let $A \in \mathbb{C}^{(k,k)}$ or $A \in \text{End}(\mathbb{H})$. Show that AA^* is positive semidefinite and Hermitian.

2.4.10. Functions of operators. In this section, we explain how functions of normal operators on $\mathbb{H}$ are defined. We start with a motivation. In many contexts, it is useful to write complex numbers γ with $|\gamma| = 1$ as

(2.4.98) $$\gamma = e^{i\beta} = \cos\beta + i\sin\beta$$

where i is the complex unit and $\beta \in \mathbb{R}$. The notions introduced in this section will allow us to write any unitary operator $U \in \text{End}(\mathbb{H})$ as $U = e^{iA}$ with a Hermitian operator $A \in \text{End}(\mathbb{H})$.

Definition 2.4.69. Let $f : \mathbb{C} \to \mathbb{C}$, let A be a normal linear operator on $\mathbb{H}$, let

(2.4.99) $$A = \sum_{\lambda \in \Lambda} \lambda P_\lambda$$

be the spectral decomposition of A. Then we define

(2.4.100) $$f(A) = \sum_{\lambda \in \Lambda} f(\lambda)P_\lambda.$$

It can be shown that in certain cases the function $f(A)$ can also be defined using power series, even if A is not normal. Also, we note that from (2.4.100), the spectral decomposition of $f(A)$ can be easily obtained.

Example 2.4.70. Consider the Pauli operator Z and the $\pi/8$ operator T which have the spectral decompositions

$$(2.4.101) \qquad Z = |0\rangle\langle 0| - |1\rangle\langle 1|, \quad T = |0\rangle\langle 0| + e^{i\pi/4}|1\rangle\langle 1|.$$

Also, let

$$(2.4.102) \qquad f : \mathbb{C} \to \mathbb{C}, \quad x \mapsto e^{-i\pi x/8}.$$

The eigenvalues of Z are 1 and -1. Therefore, we have

$$
\begin{aligned}
f(Z) &= e^{-i\pi Z/8} \\
&= e^{-i\pi/8}|0\rangle\langle 0| + e^{i\pi/8}|1\rangle\langle 1| \\
&= e^{-i\pi/8}(|0\rangle\langle 0| + e^{i\pi/4}|1\rangle\langle 1|) \\
&= e^{-i\pi/8}T.
\end{aligned}
$$
(2.4.103)

Exercise 2.4.71. Let A be a normal linear operator on $\mathbb{H}$ and let $\alpha, \beta \in \mathbb{R}$. Prove that $e^{iA(\alpha+\beta)} = e^{iA\alpha}e^{iA\beta}$.

The following theorem gives another characterization of unitary operators.

Theorem 2.4.72. *An operator $U \in \mathrm{End}(\mathbb{H})$ is unitary if and only if U can be written as $U = e^{iA}$ with a Hermitian operator $A \in \mathrm{End}(\mathbb{H})$.*

Proof. Let $A \in \mathrm{End}(\mathbb{H})$ be Hermitian and let

$$(2.4.104) \qquad A = \sum_{\lambda \in \Lambda} \lambda P_\lambda$$

be the spectral decomposition of A. Then we have

$$(2.4.105) \qquad e^{iA} = \sum_{\lambda \in \Lambda} e^{i\lambda} P_\lambda.$$

It follows that the eigenvalues of e^{iA} are $e^{i\lambda}$, $\lambda \in \Lambda$. Since A is Hermitian, Proposition 2.4.60 implies that $\Lambda \subset \mathbb{R}$. Hence, we have $|e^{i\lambda}| = 1$ for all $\lambda \in \Lambda$. Therefore, it follows from Proposition 2.4.60 that e^{iA} is unitary.

Now assume that $U \in \mathrm{End}(\mathbb{H})$ is unitary. Let Λ be the set of eigenvalues of U. Then it follows from Proposition 2.4.60 that $|\lambda| = 1$ for all $\lambda \in \Lambda$. Hence, for any $\lambda \in \Lambda$ there is $\alpha_\lambda \in \mathbb{R}$ such that $\lambda = e^{i\alpha_\lambda}$. Set

$$(2.4.106) \qquad A = \sum_{\lambda \in \Lambda} \alpha_\lambda P_\lambda.$$

Then A is a linear operator on $\mathbb{H}$ whose eigenvalues are all real numbers. Proposition 2.4.60 implies that A is Hermitian. Also, we see from (2.4.106) that $U = e^{iA}$. $\qquad\square$

From Theorem 2.4.72 we obtain the following characterization of $\mathrm{SU}(k)$.

Corollary 2.4.73. *An operator $U \in \mathrm{End}(\mathbb{H})$ is in $\mathrm{SU}(k)$ if and only if U can be written as $U = e^{iA}$. with a Hermitian operator $A \in \mathrm{End}(\mathbb{H})$ of trace 0.*

Proof. Let U be a unitary operator on $\mathbb{H}$. Then by Theorem 2.4.72 there is a Hermitian operator A on $\mathbb{H}$ with $U = e^{iA}$. Let Λ be the set of eigenvalues of A. For $\lambda \in \Lambda$ denote by a_λ the algebraic multiplicity of λ. Then we see from the proof of Theorem 2.4.72 and Proposition 2.4.1 that

$$(2.4.107) \qquad \det U = \prod_{\lambda \in \Lambda} e^{ia_\lambda \lambda} = e^{i \sum_{\lambda \in \Lambda} a_\lambda \lambda} = e^{i \operatorname{tr} A}.$$

If $\operatorname{tr} A = 0$, then $\det U = 1$. Conversely, if $\det U = 1$, then

$$(2.4.108) \qquad \operatorname{tr} A \equiv 0 \bmod 2\pi.$$

Change the eigenvalues of A modulo 2π so that $\operatorname{tr} A = 0$. Then we still have $U = e^{iA}$. $\qquad \square$

The following result will also be useful.

Proposition 2.4.74. *Let $A \in \operatorname{End}(\mathbb{H})$ be normal and an involution. Then we have $e^{ixA} = (\cos x)I_{\mathbb{H}} + i(\sin x)A$ for all $x \in \mathbb{R}$.*

Proof. Since A is an involution, it follows from Proposition 2.4.60 that the eigenvalues of A are in $\{1, -1\}$. If 1 is an eigenvalue of A, then we denote by P_1 the orthogonal projection onto the corresponding eigenspace. Otherwise, we set $P_1 = 0$. Likewise, if -1 is an eigenvalue of A, then we denote by P_{-1} the orthogonal projection onto the corresponding eigenspace. Otherwise, we set $P_1 = 0$. Then we have

$$(2.4.109) \qquad I_{\mathbb{H}} = P_1 + P_{-1}, \quad A = P_1 - P_{-1},$$

and therefore

$$(2.4.110) \qquad \begin{aligned} e^{iAx} &= e^{ix}P_1 + e^{-ix}P_{-1} \\ &= (\cos x + i\sin x)P_1 + (\cos(-x) + i\sin(-x))P_{-1} \\ &= (\cos x + i\sin x)P_1 + (\cos x - i\sin x)P_{-1} \\ &= \cos x(P_1 + P_{-1}) + i\sin x(P_1 - P_{-1}) \\ &= (\cos x)I_{\mathbb{H}} + i(\sin x)A. \end{aligned}$$

$\qquad \square$

2.5. Tensor products

In the previous sections, we have discussed finite-dimensional Hilbert spaces. In this section, we explain how to transfer many of the corresponding concepts and results to tensor products of such spaces. For an introduction to tensor products, refer to Section B.8.

2.5.1. Basics and notation. In this section, let $m \in \mathbb{N}$ and let $\mathbb{H}(0), \ldots, \mathbb{H}(m-1)$ be Hilbert spaces of finite dimension $k_0, \ldots, k_{m-1}$, respectively. The inner products on these Hilbert spaces are denoted by $\langle \cdot | \cdot \rangle$. We use the notation $\mathbb{H}(j)$ to distinguish this Hilbert space from the j-qubit state spaces $\mathbb{H}_j$. For each $j \in \mathbb{Z}_m$, let B_j be an orthonormal basis of $\mathbb{H}(j)$. We write these bases as

$$(2.5.1) \qquad B_j = (|b_{0,j}\rangle, \ldots, |b_{k_j,j}\rangle).$$

We study the tensor product

(2.5.2) $$\mathbb{H} = \mathbb{H}(0) \otimes \cdots \otimes \mathbb{H}(m-1).$$

Its dimension as a $\mathbb{C}$-vector space is

(2.5.3) $$k = \prod_{j=0}^{m-1} k_j.$$

Also, by Proposition B.9.11

(2.5.4) $$B = B_0 \otimes \cdots \otimes B_{m-1}$$

is a basis of $\mathbb{H}$.

In tensor products of kets, we sometimes omit the tensor symbols; i.e., if $|\varphi_j\rangle \in \mathbb{H}(j)$ for $0 \le j < m$, then we write

(2.5.5) $$|\varphi_0\rangle |\varphi_1\rangle \cdots |\varphi_{m-1}\rangle = |\varphi_0\rangle \otimes |\varphi_1\rangle \otimes \cdots \otimes |\varphi_{m-1}\rangle.$$

We present a few examples of tensor products of Hilbert spaces.

Example 2.5.1. Let $m = 2$, $\mathbb{H}(j) = \mathbb{H}_1$, and $B_j = (|0\rangle, |1\rangle)$ for $j = 0, 1$. Then we have

(2.5.6) $$\mathbb{H} = \mathbb{H}_1 \otimes \mathbb{H}_1$$

and

(2.5.7) $$B = (|0\rangle \otimes |0\rangle, |0\rangle \otimes |1\rangle, |1\rangle \otimes |0\rangle, |1\rangle \otimes |1\rangle).$$

By (2.5.5) we can write the basis B as

(2.5.8) $$B = (|0\rangle |0\rangle, |0\rangle |1\rangle, |1\rangle |0\rangle, |1\rangle |1\rangle).$$

If all Hilbert spaces $\mathbb{H}(j)$ are equal, then $\mathbb{H}$ is written as

(2.5.9) $$\mathbb{H}(0)^{\otimes m} = \underbrace{\mathbb{H}(0) \otimes \cdots \otimes \mathbb{H}(0)}_{m \text{ times}}.$$

Also, the elements of $\mathbb{H}(0)^m$ are written as

(2.5.10) $$|\varphi\rangle^{\otimes m} = \underbrace{|\varphi\rangle \cdots |\varphi\rangle}_{m \text{ times}}.$$

Example 2.5.2. We have

(2.5.11) $$\mathbb{H}_1^{\otimes 3} = \mathbb{H}_1 \otimes \mathbb{H}_1 \otimes \mathbb{H}_1$$

and

(2.5.12) $$|0\rangle^{\otimes 3} = |0\rangle |0\rangle |0\rangle.$$

We also show how to obtain the representation of a tensor product of elements of the Hilbert spaces $\mathbb{H}(j)$ with respect to the basis B from the representation of the components with respect to the bases B_j. We define

(2.5.13) $$\vec{k} = (k_0, \ldots, k_{m-1})$$

and

$$(2.5.14) \qquad \mathbb{Z}_{\vec{k}} = \prod_{j=0}^{m-1} \mathbb{Z}_{k_j}.$$

Also, for all $\vec{i} \in \mathbb{Z}_{\vec{k}}$, $\vec{i} = (i_0, \ldots, i_{m-1})$, we set

$$(2.5.15) \qquad |b_{\vec{i}}\rangle = \bigotimes_{j=0}^{m-1} |b_{i_j,j}\rangle = |b_{i_0,0}\rangle \cdots |b_{i_{m-1},m-1}\rangle.$$

Proposition 2.5.3. *For $0 \le j < m$ let $|\varphi_j\rangle \in \mathbb{H}(j)$ with*

$$(2.5.16) \qquad |\varphi_j\rangle = \sum_{i=0}^{k_j-1} \alpha_{i,j} |b_{u,j}\rangle$$

where $\alpha_{i,j} \in \mathbb{C}$ for all $i \in \mathbb{Z}_{k_j}$. Then

$$(2.5.17) \qquad \bigotimes_{j=0}^{m-1} |\varphi_j\rangle = \sum_{\vec{i} \in \mathbb{Z}_{\vec{k}}} \alpha_{\vec{i}} |b_{\vec{i}}\rangle$$

where for $\vec{i} = (i_0, \ldots, i_{k_j-1}) \in \mathbb{Z}_{\vec{k}}$ we have

$$(2.5.18) \qquad \alpha_{\vec{i}} = \prod_{j=0}^{m-1} \alpha_{i_j,j}.$$

Exercise 2.5.4. Prove Proposition 2.5.3.

2.5.2. Inner product. On the tensor product $\mathbb{H} = \mathbb{H}(0) \otimes \cdots \otimes \mathbb{H}(m-1)$ we use the *inner product induced by the inner products on the component spaces* which is defined as the Hermitian inner product with respect to the base $B = B_0 \otimes \cdots \otimes B_{m-1}$. So B is an orthonormal basis with respect to this inner product. Also, equipped with the inner product induced by the inner products of the component spaces, the tensor product $\mathbb{H}$ is a Hilbert space.

Proposition 2.5.5. *For $0 \le j < m$ let $|\varphi_j\rangle, |\psi_j\rangle \in \mathbb{H}(j)$. Then we have*

$$(2.5.19) \qquad \left\langle \bigotimes_{j=0}^{m-1} |\varphi_j\rangle \, \middle| \, \bigotimes_{j=0}^{m-1} |\psi_j\rangle \right\rangle = \prod_{j=0}^{m-1} \langle \varphi_j | \psi_j \rangle.$$

Proof. Write

$$(2.5.20) \qquad |\varphi\rangle = \bigotimes_{j=0}^{m-1} |\varphi_j\rangle \quad \text{and} \quad |\psi\rangle = \bigotimes_{j=0}^{m-1} |\psi_j\rangle.$$

For $0 \le j < m$ let

$$(2.5.21) \qquad |\varphi_j\rangle = \sum_{i=0}^{k_j-1} \alpha_{i,j} |b_{i,j}\rangle, \quad |\psi_j\rangle = \sum_{i=0}^{k_j-1} \beta_{i,j} |b_{i,j}\rangle$$

where $\alpha_{i,j}, \beta_{i,j} \in \mathbb{C}$ for all $i \in \mathbb{Z}_{k_j}$. Proposition 2.5.3 implies

$$(2.5.22) \qquad |\varphi\rangle = \sum_{\vec{i} \in \mathbb{Z}_{\vec{k}}} \alpha_{\vec{i}} |b_{\vec{i}}\rangle, \quad |\psi\rangle = \sum_{\vec{i} \in \mathbb{Z}_{\vec{k}}} \beta_{\vec{i}} |b_{\vec{i}}\rangle$$

where for all $\vec{i} = (i_0, \ldots, i_{m-1}) \in \mathbb{Z}_{\vec{k}}$ we have

$$(2.5.23) \qquad \alpha_{\vec{i}} = \prod_{j=0}^{m-1} \alpha_{i_j,j}, \quad \beta_{\vec{i}} = \prod_{j=0}^{m-1} \beta_{i_j,j}.$$

From the orthonormality of B, we obtain

$$(2.5.24) \qquad \langle\varphi|\psi\rangle = \left\langle \sum_{\vec{i} \in \mathbb{Z}_{\vec{k}}} \alpha_{\vec{i}} |b_{\vec{i}}\rangle \Big| \sum_{\vec{i} \in \mathbb{Z}_{\vec{k}}} \beta_{\vec{i}} |b_{\vec{i}}\rangle \right\rangle = \sum_{\vec{i} \in \mathbb{Z}_k} \overline{\alpha_{\vec{i}}} \beta_{\vec{i}}.$$

On the other hand, the orthonormality of the B_j implies

$$(2.5.25) \qquad \prod_{j=0}^{m-1} \langle\varphi_j|\psi_j\rangle = \prod_{j=0}^{m-1} \left\langle \sum_{i \in \mathbb{Z}_{k_j}} \alpha_{i,j} |b_{i,j}\rangle \Big| \sum_{i \in \mathbb{Z}_{k_j}} \beta_{i,j} |b_{i,j}\rangle \right\rangle = \sum_{\vec{i} \in \mathbb{Z}_k} \overline{\alpha_{\vec{i}}} \beta_{\vec{i}}. \qquad \square$$

Example 2.5.6. Let $m = 2$, $\mathbb{H}(j) = \mathbb{H}_1$, and $B_j = (|0\rangle, |1\rangle)$ for $0 \le j < 2$. For

$$(2.5.26) \qquad \begin{aligned} |\varphi\rangle_0 &= |0\rangle + i|1\rangle, \quad |\varphi\rangle_1 = |0\rangle - i|1\rangle \\ |\psi\rangle_0 &= |0\rangle + |1\rangle, \quad |\psi\rangle_1 = |0\rangle - |1\rangle \end{aligned}$$

we obtain

$$(2.5.27) \qquad \left\langle |\varphi_0\rangle|\varphi_1\rangle \,\Big|\, |\psi_0\rangle|\psi_1\rangle \right\rangle = \langle\varphi_0|\psi_0\rangle\langle\varphi_1|\psi_1\rangle = (1+i)(1-i) = 2.$$

2.5.3. State spaces as tensor products. We can use the construction in the previous section to identify the tensor product of state spaces with a larger state space. To explain this, let $m, n_0, \ldots, n_{m-1} \in \mathbb{N}$. Consider the tensor product

$$(2.5.28) \qquad \mathbb{H} = \mathbb{H}_{n_0} \otimes \cdots \otimes \mathbb{H}_{n_{m-1}}$$

of the n_j-qubit state spaces $\mathbb{H}_{n_j}$, $j \in \mathbb{Z}_m$.

Let $n = \sum_{j=0}^{m-1} n_j$. Denote by B the computational basis of $\mathbb{H}_n$. Then the linear map

$$(2.5.29) \qquad \mathbb{H} \to \mathbb{H}_n, \quad |\vec{b}_0\rangle|\vec{b}_1\rangle \cdots |\vec{b}_{m-1}\rangle \mapsto |\vec{b}_0\vec{b}_1 \cdots \vec{b}_{m-1}\rangle,$$

where $b_j \in \{0,1\}^{n_j}$ for $0 \le j < m$, is an isometry between $\mathbb{H}_{n_0} \otimes \cdots \otimes \mathbb{H}_{n_{m-1}}$ and $\mathbb{H}_n$. Using this isometry, we identify the elements of the tensor product $\mathbb{H}$ with the elements of $\mathbb{H}_n$.

Exercise 2.5.7. Show that the map (2.5.29) is an isometry.

Example 2.5.8. Let $m = 2$, $n_0, n_1 = 1$, and let

$$(2.5.30) \qquad |\varphi\rangle = |0\rangle + |1\rangle, \quad |\psi\rangle = |0\rangle - |1\rangle.$$

Then we have

$$
\begin{aligned}
|\varphi\rangle \otimes |\psi\rangle &= (|0\rangle + |1\rangle) \otimes (|0\rangle - |1\rangle) \\
&= |0\rangle |0\rangle - |0\rangle |1\rangle + |1\rangle |0\rangle - |1\rangle |1\rangle \\
&= |00\rangle - |01\rangle + |10\rangle - |11\rangle .
\end{aligned}
\tag{2.5.31}
$$

The isometry (2.5.29) can also be written as

$$
\mathbb{H}_{n_0} \oplus \cdots \oplus \mathbb{H}_{n_{m-1}} \to \mathbb{H}_n, \quad |b_0\rangle \cdots |b_{m-1}\rangle \mapsto \left| \sum_{j=0}^{m-1} b_j 2^{s_j} \right\rangle
\tag{2.5.32}
$$

where $b_j \in \mathbb{Z}_{2^{n_j}}$ and $s_j = \sum_{u=j}^{m-1} n_u$ for all $j \in \mathbb{Z}_m$.

2.5.4. Homomorphisms. Let $m \in \mathbb{N}$, and let $\mathbb{H}'(0), \dots, \mathbb{H}'(m-1)$ be finite-dimensional Hilbert spaces. Let

$$
\mathbb{H}' = \mathbb{H}(0)' \otimes \cdots \otimes \mathbb{H}(m-1)'.
\tag{2.5.33}
$$

Then, as shown in Section B.9.3, we identify $\bigotimes_{j=0}^{m-1} \operatorname{Hom}(\mathbb{H}(j), \mathbb{H}(j)')$ with $\operatorname{Hom}(\mathbb{H}, \mathbb{H}')$.

Proposition 2.5.9. *Let $(f_0, \dots, f_{m-1}) \in \prod_{j=0}^{m-1} \mathbb{H}(j)$. Then we have*

$$
f^* = f_0^* \otimes \cdots \otimes f_{m-1}^*
\tag{2.5.34}
$$

and

$$
f^k = f_0^k \otimes \cdots \otimes f_{m-1}^k
\tag{2.5.35}
$$

for all $k \in \mathbb{N}_0$ and for all $k \in \mathbb{Z}$ if f is invertible.

Exercise 2.5.10. Prove Proposition 2.5.9.

Example 2.5.11. Consider $H^{\otimes 2} = H \otimes H$ in $\operatorname{End}(\mathbb{H}_1 \otimes \mathbb{H}_1)$ where H is the Hadamard operator. Since we identify $\mathbb{H}_1 \otimes \mathbb{H}_1$ with $\mathbb{H}_2$, this map is in $\operatorname{End}(\mathbb{H}_2)$. From

$$
H |0\rangle = \frac{|0\rangle + |1\rangle}{\sqrt{2}}, \quad H |1\rangle = \frac{|0\rangle - |1\rangle}{\sqrt{2}}
\tag{2.5.36}
$$

it follows that

$$
\begin{aligned}
H^{\otimes 2} |0\rangle^{\otimes 2} &= H |0\rangle \otimes H |0\rangle \\
&= \frac{|0\rangle + |1\rangle}{\sqrt{2}} \otimes \frac{|0\rangle + |1\rangle}{\sqrt{2}} = \frac{1}{\sqrt{2}} \sum_{\vec{b} \in \{0,1\}^2} |\vec{b}\rangle .
\end{aligned}
\tag{2.5.37}
$$

Exercise 2.5.12. Show that for all $n \in \mathbb{N}$ we have

$$
H^{\otimes n} |0\rangle^{\otimes n} = \frac{1}{\sqrt{2^n}} \sum_{\vec{b} \in \{0,1\}^n} |\vec{b}\rangle .
\tag{2.5.38}
$$

We also note the following.

Proposition 2.5.13. *For* $0 \leq j < m$ *let* $|\varphi_j\rangle, |\psi_j\rangle \in \mathbb{H}(j)$. *Then we have*

$$\tag{2.5.39} \left\langle \bigotimes_{j=0}^{l-1} |\varphi_j\rangle \right| = \bigotimes_{j=0}^{l-1} \langle \varphi_j|$$

and

$$\tag{2.5.40} \left| \bigotimes_{j=0}^{l-1} |\varphi_j\rangle \right\rangle\!\!\left\langle \bigotimes_{j=0}^{l-1} |\psi_j\rangle \right| = \bigotimes_{j=0}^{l-1} |\varphi_j\rangle \langle \psi_j|.$$

Exercise 2.5.14. Prove Proposition 2.5.13.

2.5.5. Endomorphisms. We explain how the properties of the tensor product of endomorphisms are related to the properties of its components.

Proposition 2.5.15. *For* $0 \leq j < m$ *assume that* $A_j \in \mathrm{End}(\mathbb{H}(j))$ *and* Λ_j *is the set of eigenvalues of* A_j. *Also, for* $0 \leq j < m$ *and* $\lambda \in \Lambda_j$ *use the following notation:* $E_{\lambda,j}$ *denotes the eigenspace of* A_j *associated with the eigenvalue* λ, $B_{\lambda,j}$ *is an orthonormal basis of* $E_{\lambda,j}$, *and* $P_{\lambda,j}$ *is the orthogonal projection of* $\mathbb{H}(j)$ *onto* $E_{\lambda,j}$. *Finally, let*

$$\tag{2.5.41} A = A_0 \otimes \cdots \otimes A_{m-1}.$$

Then $A \in \mathrm{End}(\mathbb{H})$ *and the following hold.*

(1) *The set of eigenvalues of* A *is*

$$\tag{2.5.42} \Lambda = \left\{ \prod_{j=0}^{m-1} \lambda_j : \lambda_j \in \Lambda_j, \text{ for } 0 \leq j < m \right\}.$$

(2) *For all* $\lambda \in \Lambda$ *let*

$$\tag{2.5.43} L_\lambda = \left\{ (\lambda_0, \ldots, \lambda_{m-1}) \in \prod_{j=0}^{m-1} \Lambda_j : \lambda = \prod_{j=0}^{m-1} \lambda_j \right\}.$$

Then the eigenspace of A *associated with* λ *is*

$$\tag{2.5.44} E_\lambda = \sum_{(\lambda_0,\ldots,\lambda_{m-1}) \in L_\lambda} \bigotimes_{j=0}^{m-1} E_{\lambda_j, j}.$$

Also, the concatenation of all sequences $\bigotimes_{j=0}^{m-1} B_{\lambda_j, j}$, *where* $(\lambda_0, \ldots, \lambda_{m-1}) \in L_\lambda$, *is an orthonormal basis of* E_λ, *and the projection onto this eigenspace is*

$$\tag{2.5.45} P_\lambda = \sum_{(\lambda_0,\ldots,\lambda_{m-1}) \in L_\lambda} \bigotimes_{j=0}^{m-1} P_{\lambda_j, j}.$$

(3) *The operator* A *is a projection, an involution, normal, Hermitian, or unitary if and only if all of its components* A_j, $j \in \mathbb{Z}_m$, *have the respective properties.*

Exercise 2.5.16. Prove Proposition 2.5.15.

Example 2.5.17. Let $m = 2$, $\mathbb{H}(0) = \mathbb{H}(1) = \mathbb{H}_1$, and let A_0 and A_1 be the Pauli X operator introduced in Example 2.3.1 which sends $|0\rangle$ to $|1\rangle$ and vice versa. It has the eigenvalues 1 and -1. Also

$$(2.5.46) \qquad (|x_+\rangle) = \left(\frac{|0\rangle + |1\rangle}{\sqrt{2}}\right), \quad (|x_-\rangle) = \left(\frac{|0\rangle - |1\rangle}{\sqrt{2}}\right)$$

are orthonormal bases of the eigenspaces of X associated with the eigenvalues 1 and -1, respectively. The projections onto these eigenspaces are

$$(2.5.47) \qquad |x_+\rangle\langle x_+|, \quad |x_-\rangle\langle x_-|,$$

respectively.

We consider the tensor product

$$(2.5.48) \qquad A = A_0 \otimes A_1.$$

It is in $\mathrm{End}(\mathbb{H}_1 \otimes \mathbb{H}_1) = \mathrm{End}(\mathbb{H}_2)$. It follows from Proposition 2.5.15 that 1 and -1 are the eigenvalues of A, that the sequences

$$(2.5.49) \qquad \begin{aligned} B_1 &= (|x_+ x_+\rangle, |x_- x_-\rangle), \\ B_{-1} &= (|x_+ x_-\rangle, |x_- x_+\rangle) \end{aligned}$$

are orthonormal bases of the eigenspaces E_1 and E_{-1} associated with the eigenvalues 1 and -1, and that

$$(2.5.50) \qquad \begin{aligned} P_1 &= |x_+ x_+\rangle\langle x_+ x_+| + |x_- x_-\rangle\langle x_- x_-|, \\ P_{-1} &= |x_+ x_-\rangle\langle x_+ x_-| + |x_- x_+\rangle\langle x_- x_+| \end{aligned}$$

are the projections onto these eigenspaces.

Since X is a Hermitian unitary involution, the same is true for $X \otimes X$.

2.5.6. Schmidt decomposition theorem. In this section, we establish the renowned Schmidt decomposition theorem, credited to the mathematician Erhard Schmidt's discovery during the early 20th century. This theorem holds particular significance within the realm of quantum mechanics, playing a crucial role in the analysis of entanglement. The situation is the following.

It deals with the following problem. Let $\mathbb{H}(0)$ and $\mathbb{H}(1)$ be Hilbert spaces of dimensions k and l, respectively. Then $|\varphi_0\rangle \otimes |\varphi_1\rangle \in \mathbb{H}(0) \otimes \mathbb{H}(1)$ for all $|\varphi_0\rangle \in \mathbb{H}(0)$ and $|\varphi_1\rangle \in \mathbb{H}(1)$. But not all elements $|\varphi\rangle$ in this tensor product of Hilbert spaces can be written as a tensor product of elements in $\mathbb{H}(0)$ and $\mathbb{H}(1)$. The Schmidt decomposition theorem allows one to distinguish between these cases.

Theorem 2.5.18 (Schmidt decomposition theorem). *Let $m = \min\{k, l\}$ and let $|\varphi\rangle \in \mathbb{H}(0) \otimes \mathbb{H}(1)$. Then there are orthonormal sequences $(|u_0\rangle, \ldots, |u_{m-1}\rangle)$ in $\mathbb{H}(0)$ and $(|v_0\rangle, \ldots, |v_{m-1}\rangle)$ in $\mathbb{H}(1)$ and positive real numbers $r_0, \ldots, r_{m-1}$ such that*

$$(2.5.51) \qquad |\varphi\rangle = \sum_{i=0}^{m-1} r_i |u_i\rangle \otimes |v_i\rangle.$$

Up to reordering, the coefficients r_i are uniquely determined by $|\varphi\rangle$. The representation in (2.5.51) is called a Schmidt decomposition *of $|\varphi\rangle$.*

Proof. Let $B = (|b_0\rangle, \ldots, |b_k\rangle) \in \mathbb{H}(0)^k$ and $C = (|c_0\rangle, \ldots, |c_{l-1}\rangle) \in \mathbb{H}(1)^l$ be orthonormal bases of $\mathbb{H}(0)$ and $\mathbb{H}(1)$, respectively. Then we can write

$$(2.5.52) \qquad |\varphi\rangle = \sum_{i=0}^{k-1} \sum_{j=0}^{l-1} \alpha_{i,j} |b_i\rangle |c_j\rangle$$

with $\alpha_{i,j} \in \mathbb{C}$ for all $i \in \mathbb{Z}_k$ and $j \in \mathbb{Z}_l$. If we set $A = (\alpha_{i,j}) \in \mathbb{C}^{(k,l)}$, then (2.5.52) can also be written as

$$(2.5.53) \qquad |\varphi\rangle = BAC.$$

Without loss of generality, assume that $k \geq l$. Then by Theorem 2.4.67 there is a singular value decomposition

$$(2.5.54) \qquad A = U \begin{pmatrix} D \\ 0 \end{pmatrix} V^*$$

where $U \in \mathbb{C}^{(k,k)}$ and $V \in \mathbb{C}^{(l,l)}$ are unitary matrices and $D \in \mathbb{C}^{(l,l)}$ is a positive semidefinite diagonal matrix; that is,

$$(2.5.55) \qquad D = (r_0, \ldots, r_{l-1})$$

with $r_i \in \mathbb{R}_{\geq 0}$ for $0 \leq i < l$. Write

$$(2.5.56) \qquad U = (U_1\, U_2)$$

with $U_1 \in \mathbb{C}^{(k,l)}$ and $U_2 \in \mathbb{C}^{(k,k-l)}$. Then it follows from (2.5.54) that

$$(2.5.57) \qquad A = U_1 D V^*.$$

So (2.5.53) implies

$$(2.5.58) \qquad |\varphi\rangle = B U_1 D V^* C.$$

If we write

$$(2.5.59) \qquad (|u_0\rangle, \ldots, |u_{l-1}\rangle) = B U_1 \quad \text{and} \quad (|v_0\rangle, \ldots, |v_{l-1}\rangle) = V^* C,$$

we obtain

$$(2.5.60) \qquad |\varphi\rangle = \sum_{i=0}^{l-1} r_i |u_i\rangle \otimes |v_i\rangle.$$

Leaving out the summands with coefficients 0 we obtain a decomposition as in the theorem.

We show the uniqueness of the coefficients r_i. Suppose that we are given a Schmidt decomposition (2.5.51). From this, we can construct a singular value decomposition of the matrix A from (2.5.53) where the positive coefficients are the singular values of A. The details of the construction are worked out in Exercise 2.5.19. The singular values are uniquely determined by Theorem 2.4.67. This shows that these coefficients are uniquely determined up to reordering. $\square$

Exercise 2.5.19. Construct the singular value decomposition in the uniqueness proof of Theorem 2.5.18.

Definition 2.5.20. Let $\mathbb{H}(0)$ and $\mathbb{H}(1)$ be Hilbert spaces of dimension k and l, respectively, and let $m = \min\{k, l\}$. Let $|\varphi\rangle \in \mathbb{H}(0) \otimes \mathbb{H}(1)$. Let $r = (r_0, \ldots, r_{m-1})$ be the sequence of coefficients in a Schmidt decomposition of $|\varphi\rangle$. From Theorem 2.5.18 we know that the elements of r are nonnegative real numbers and that r is uniquely determined up to reordering.

(1) The elements of r are called the *Schmidt coefficients* of $|\varphi\rangle$.

(2) The number of elements in r counted with multiplicities is called the *Schmidt rank* or *Schmidt number* of $|\varphi\rangle$.

(3) The ket $|\varphi\rangle$ is called *separable* with respect to the decomposition $\mathbb{H} = \mathbb{H}(0) \otimes \mathbb{H}(1)$ if its Schmidt rank is 1, i.e., if it can be written as $|\varphi\rangle = |\psi\rangle \otimes |\xi\rangle$ with $|\psi\rangle \in \mathbb{H}(0)$ and $|\xi\rangle \in \mathbb{H}(1)$. Otherwise, $|\varphi\rangle$ is called *inseparable*.

We note that the notion of separability depends on the decomposition of a Hilbert space into the tensor product of two Hilbert spaces. The meaning of the term "entangled" will become clear in the next chapter. Entanglement is one of the most important concepts in quantum mechanics.

Example 2.5.21. Consider the decomposition $\mathbb{H}_2 = \mathbb{H}_1 \otimes \mathbb{H}_1$ and in $\mathbb{H}_2$ the element

$$(2.5.61) \qquad\qquad |\varphi\rangle = |0\rangle \frac{|0\rangle + |1\rangle}{\sqrt{2}} \in \mathbb{H}_1 \otimes \mathbb{H}_1.$$

This is a Schmidt decomposition of $|\varphi\rangle$, since both $|0\rangle$ and $(|0\rangle + |1\rangle)/\sqrt{2}$ have length 1. So 1 is the only Schmidt coefficient of $|\varphi\rangle$. It has multiplicity 1 in the Schmidt decomposition of $|\varphi\rangle$. Hence, the Schmidt rank or Schmidt number of $|\varphi\rangle$ (with respect to the chosen decomposition of $\mathbb{H}$) is 1, which means that $|\varphi\rangle$ is separable.

Example 2.5.22. Consider the so-called *Bell state*

$$(2.5.62) \qquad\qquad |\varphi\rangle = \frac{|0\rangle |0\rangle + |1\rangle |1\rangle}{\sqrt{2}} \in \mathbb{H}_1 \otimes \mathbb{H}_1.$$

The representation (2.5.62) is a Schmidt decomposition of $|\varphi\rangle$. So $1/\sqrt{2}$ is the only Schmidt coefficient of $|\varphi\rangle$. It has multiplicity 2 in the Schmidt decomposition of $|\varphi\rangle$. Hence, the Schmidt rank or Schmidt number of $|\varphi\rangle$ is 2 which means that $|\varphi\rangle$ is entangled.

Chapter 3

Quantum Mechanics

Quantum mechanics, discovered in the early 20th century, stands as one of the most revolutionary discoveries in physics. It was fundamentally shaped by the work of physicists like Max Planck, who received the Nobel Prize in 1918, and Albert Einstein, who received the Nobel Prize in 1921, although Einstein was later very critical of quantum mechanics. The theory was fully developed by remarkable scientists, including Niels Bohr, Nobel laureate of 1922, Werner Heisenberg, Nobel Prize 1932, Erwin Schrödinger, and Paul Dirac, Nobel Prize 1933, Wolfgang Pauli, Nobel Prize 1945, and Max Born, Nobel Prize 1954. In 1965, Richard P. Feynman, Julian Schwinger, and Sin-Itiro Tomonaga were awarded the Nobel Prize for their contributions to quantum electrodynamics. Recently, in 2022, Alain Aspect, John Clauser, and Anton Zeilinger received the Nobel Prize for experimentally verifying one of the most counterintuitive phenomena in quantum physics: entanglement.

One of the fundamental features of quantum mechanics is the concept that closed physical systems can exist in a superposition of many possible states. This intriguing property inspired the mathematician and physicist Yuri Manin [**Man80**] and the physicists Paul Benioff [**Ben80**] and Richard Feynman [**Fey82**] to conceive the idea of a quantum computer, where information is stored and processed in superposition. However, it soon became evident that designing practical and useful algorithms based on this concept is a challenging task, as described in the chapters following this one.

In order to grasp the functioning of these algorithms and their underlying principles, understanding quantum mechanics becomes essential. Therefore, the objective of this chapter is to introduce the reader to the quantum mechanical basis that underlies quantum computing.

Just like other fields of physics, quantum mechanics is built upon a set of postulates that establish correspondence between real-world objects and processes and their mathematical counterparts. This correspondence enables us to make predictions about quantum computing through rigorous mathematical reasoning. We commence by introducing these postulates and elucidating their significance within the realm of quantum computing. We illustrate how using these postulates, quantum bits and quantum registers are represented using Hilbert spaces, termed state spaces. Moreover, we explore how the evolution of such quantum systems is captured using unitary operators and how measurements unlock their content for subsequent classical computation. Our subsequent focus centers on visualizing quantum bits as points on the unit sphere with a radius of 1 within the three-dimensional real space, which is known as the Bloch sphere. Following this, we delve into an alternative description of quantum states using density operators. This approach provides a means of describing the state of components within composite quantum systems, further enhancing our understanding of the intricate quantum realm.

Throughout this chapter, we make the assumption that $\mathbb{H}$ is a Hilbert space of dimension $k \in \mathbb{N}$, equipped with the inner product $\langle \cdot | \cdot \rangle$.

3.1. State spaces

3.1.1. The State Space Postulate.
The first postulate that we discuss is the *State Space Postulate*. It specifies how the state of a closed or isolated physical system is modeled.

Postulate 3.1.1 (State Space Postulate). A closed physical system is associated with a Hilbert space, called the *state space* of the system. The system at a particular time is completely described by a unit vector in its state space, called the *state vector* or *state* of the physical system.

The term "closed" refers to the system not interacting with other systems, i.e., not exchanging energy or matter with them. In reality, the only closed system is the universe as a whole. However, in quantum computing, it is possible to construct quantum systems which can be described to a good approximation as being closed. The state vector of a quantum system is also called its *wave function*. The term "wave function" originates from the historical development of quantum mechanics, where the theory was initially formulated by analogy with classical wave phenomena.

3.1.2. Quantum bits.
In quantum computing, the basic unit of information is a *quantum bit* or *qubit* for short. A quantum bit is a physical system whose state space is the *single-qubit state space* $\mathbb{H}_1$ which we have specified in Definition 2.1.2. It is a two-dimensional complex Hilbert space with orthonormal basis $(|0\rangle, |1\rangle)$, which is called the *computational basis* of $\mathbb{H}_1$. By Postulate 3.1.1, an individual qubit at a particular time is completely described by its state vector

$$(3.1.1) \qquad |\varphi\rangle = \alpha_0 |0\rangle + \alpha_1 |1\rangle$$

where α_0 and α_1 are complex coefficients such that

$$(3.1.2) \qquad \|\varphi\|^2 = |\alpha_0|^2 + |\alpha_1|^2 = 1.$$

The linear combination in (3.1.1) is called a *superposition* of the basis states $|0\rangle$ and $|1\rangle$. So while the state of a classical bit is either 0 or 1, qubits are in a *superposition* of the basis states $|0\rangle$ and $|1\rangle$. The coefficients α_0 and α_1 are called the *amplitudes* of $|\varphi\rangle$ for the basis states $|0\rangle$ and $|1\rangle$, respectively. This is a special case of the following definition.

Definition 3.1.2. Let $l \in \mathbb{N}$, let $(|\varphi_0\rangle, \ldots, |\varphi_{l-1}\rangle) \in \mathbb{H}^l$ be linearly independent, and let $|\varphi\rangle \in \mathbb{H}$, $|\varphi\rangle = \alpha_0 |\varphi_0\rangle + \cdots + \alpha_{l-1} |\varphi_{l-1}\rangle$ with $\alpha_i \in \mathbb{C}$ for $i \in \mathbb{Z}_l$. Then, for each $i \in \mathbb{Z}_l$, the coefficient α_i is called the *amplitude* of $|\varphi\rangle$ for the state $|\varphi_i\rangle$.

In order to illustrate the physical meaning of the amplitudes of a single-qubit state, we give a preview of the concept of measurements that will be discussed in more detail in Section 3.4.1. A *measurement in the computational basis* of a qubit is an interaction of an observer with the qubit. If the qubit is in the state $|\varphi\rangle = \alpha_0 |0\rangle + \alpha_1 |1\rangle$, $\alpha_0, \alpha_1 \in \mathbb{C}$, $|\alpha_0|^2 + |\alpha_1|^2 = 1$, then the measurement gives 0 with probability $|\alpha_0|^2$ and 1 with probability $|\alpha_1|^2$. Note that $|\varphi\rangle$ must be unitary for the sum of the probabilities of the two measurement outcomes to be 1.

Example 3.1.3. Consider a physical system consisting of a single qubit. Then

$$(3.1.3) \qquad |\varphi\rangle = \frac{1}{\sqrt{2}} |0\rangle + \frac{i}{\sqrt{2}} |1\rangle$$

is a possible state of this qubit since

$$\||\varphi\|^2 = \left|\frac{1}{\sqrt{2}}\right|^2 + \left|\frac{i}{\sqrt{2}}\right|^2 = \frac{1}{2} + \frac{1}{2} = 1.$$

The amplitudes of $|\varphi\rangle$ for the states $|0\rangle$ and $|1\rangle$ are $\frac{1}{\sqrt{2}}$ and $\frac{i}{\sqrt{2}}$, respectively. If our qubit is in the state $|\varphi\rangle$ and an observer measures it, then she gets 0 or 1, each with probability 1/2. Also, immediately after the measurement, the qubit is in state $|0\rangle$ or $|1\rangle$, depending on the measurement outcome.

3.1.3. Spherical coordinates.
In many contexts, the geometric interpretation of single-qubit states as points on the so-called *Bloch sphere* is used. This interpretation requires spherical coordinates of vectors in the three-dimensional real space $\mathbb{R}^3$ which we explain in this section.

Let $\vec{p} = (x, y, z) \in \mathbb{R}^3$. The triplet of real numbers (x, y, z) is called the *Cartesian coordinate representation* of $\vec{p}$. The elements of this representation are called the *Cartesian coordinates* of $\vec{p}$. To represent $\vec{p}$, we also use the *Cartesian coordinate representation of $\vec{p}$ with respect to another basis B* of $\mathbb{R}^3$ by which we mean the Cartesian coordinate representation of $\vec{p}_B = B^{-1} \vec{p}$.

Example 3.1.4. Consider $\vec{p} = (1, 1, 1)$. Consider the alternative basis

$$(3.1.4) \qquad B = \begin{pmatrix} 1 & 0 & 0 \\ 0 & -1 & 0 \\ 0 & 0 & -1 \end{pmatrix}$$

of $\mathbb{R}^3$. The Cartesian coordinate representation of $\vec{p}$ with respect to B is $B^{-1}\vec{p} = (1, -1, -1)$.

As inner product on $\mathbb{R}^3$ we use the inner product on $\mathbb{C}^3$ restricted to $\mathbb{R}^3$. This is described in the next definition.

Definition 3.1.5. Let $\vec{p} = (p_x, p_y, p_z), \vec{q} = (q_x, q_y, q_z) \in \mathbb{R}^3$.

(1) The *inner product* of $\vec{p}$ and $\vec{q}$ is $\langle \vec{p}|\vec{q} \rangle = \vec{p} \cdot \vec{q} = p_x q_x + p_y q_y + p_z q_z s$.

(2) The *Euclidean norm* or *length* of $\vec{p}$ is $\|\vec{p}\| = \sqrt{\langle \vec{p}|\vec{p} \rangle}$.

(3) $\vec{p}$ is called a *unit vector* if its Euclidean length is 1.

(4) $\vec{p}$ and $\vec{q}$ are called *orthogonal* to each other if $\langle \vec{p}|\vec{q} \rangle = 0$.

(5) A basis of $\mathbb{R}^3$ is called *orthogonal* if its elements are pairwise orthogonal.

(6) A basis of $\mathbb{R}^3$ is called *orthonormal* if it is orthogonal and all its elements are unit vectors.

The next proposition provides properties of the inner product on $\mathbb{R}^3$ which are similar to those listed in Definition 2.2.1.

Proposition 3.1.6. *For all* $\vec{p}, \vec{q}, \vec{r} \in \mathbb{R}^3$ *and all* $\gamma \in \mathbb{R}$ *the following hold.*

(1) *Bilinearity:* $\langle \vec{p}+\vec{q}|\vec{r} \rangle = \langle \vec{p}|\vec{r} \rangle + \langle \vec{q}|\vec{r} \rangle$, $\langle \vec{p}|\vec{q}+\vec{r} \rangle = \langle \vec{p}|\vec{q} \rangle + \langle \vec{p}|\vec{r} \rangle$, *and* $\langle \gamma \vec{p}|\vec{q} \rangle = \langle \vec{p}|\gamma \vec{q} \rangle = \gamma \langle \vec{p}|\vec{q} \rangle$.

(2) *Positive definiteness:* $\langle \vec{p}|\vec{p} \rangle \geq 0$ *and* $\langle \vec{p}|\vec{p} \rangle = 0$ *if and only if* $\vec{p} = 0$. *This property is also called* positivity.

Exercise 3.1.7. Prove Proposition 3.1.6.

Example 3.1.8. The inner product of $(3, 2, 1)$ and $(-1, 1, 1)$ is $\langle (3, 2, 1)|(-1, 1, 1) \rangle = -3 + 2 + 1 = 0$. Therefore, these vectors are orthogonal to each other. The length of the first vector is $\|(3, 2, 1)\| = \sqrt{9 + 4 + 1} = \sqrt{14}$. So, $\frac{1}{\sqrt{14}}(3, 2, 1)$ is a unit vector.

In order to define spherical coordinates we need the following result.

Lemma 3.1.9. *Let* $x, y \in \mathbb{R}$ *with* $x^2 + y^2 = 1$. *Then there is a uniquely determined real number* γ *with* $0 \leq \gamma < 2\pi$ *such that*

$$(3.1.5) \qquad\qquad x = \cos \gamma \quad \text{and} \quad y = \sin \gamma.$$

Also, if $0 \leq x, y \leq 1$, *then* $\gamma = \arccos = \arcsin y$ *and* $0 \leq \gamma \leq \pi/2$.

Exercise 3.1.10. Prove Lemma 3.1.9.

Example 3.1.11. If $x = -\sqrt{2}/2$ and $y = \sqrt{2}/2$, then for $\gamma = 3/4\pi$ we have $(\cos \gamma, \sin \gamma) = (x, y)$.

Now we introduce spherical coordinates.

Proposition 3.1.12. *Let* $\vec{p} \in \mathbb{R}^3$, $\vec{p} \neq 0$. *Then there are uniquely determined real numbers* θ *and* ϕ *with*

$$(3.1.6) \qquad\qquad 0 \leq \theta \leq \pi \quad \text{and} \quad \begin{cases} \phi = 0 & \text{if } \theta \in \{0, \pi\}, \\ 0 < \phi < 2\pi & \text{otherwise} \end{cases}$$

such that

$$(3.1.7) \qquad\qquad \vec{p} = \|\vec{v}\|(\cos \phi \sin \theta, \sin \phi \sin \theta, \cos \theta).$$

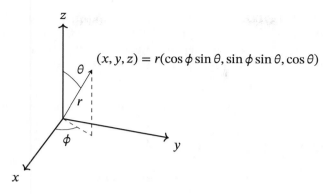

Figure 3.1.1. Spherical coordinates of (x, y, z).

The triplet $(\|\vec{p}\|, \theta, \phi)$ is called the spherical coordinate representation *of $\vec{p}$. Its elements are called the* spherical coordinates *of $\vec{p}$ and are referred to as $r(\vec{p}) = \|\vec{p}\|$, $\theta(\vec{p}) = \theta$, and $\phi(\vec{p}) = \phi$.*

Proof. Assume without loss of generality that $\|\vec{p}\| = 1$. Write $\vec{p} = (x, y, z)$. Then $|z| \leq 1$ and $\theta = \arccos z$ is the uniquely determined real number $\theta \in [0, \pi]$ that satisfies $\cos \theta = z$.

If $\vec{p} = (0, 0, 1)$, then $\theta = 0$. So, if we choose $\phi = 0$, then (3.1.7) is satisfied. Also, if $\vec{p} = (0, 0, -1)$, then $\theta = \pi$, and if we choose $\phi = 0$, then (3.1.7) is satisfied.

If $\vec{p} \neq (0, 0, \pm 1)$, then $0 < \theta < \pi$ and we have

$$(3.1.8) \qquad \left(\frac{x}{\sin \theta}\right)^2 + \left(\frac{y}{\sin \theta}\right)^2 h \frac{1 - z^2}{\sin^2 \theta} = \frac{1 - \cos^2 \theta}{\sin^2 \theta} = 1.$$

So it follows from Lemma 3.1.9 that there is a uniquely determined $\phi \in]0, 2\pi[$ such that (3.1.7) holds. $\qquad \square$

Figure 3.1.1 illustrates the spherical coordinates.

Example 3.1.13. We determine the spherical coordinate representation (r, θ, ϕ) of a vector in $\mathbb{R}^3$ with Cartesian coordinates $(1/2, 1/2, \sqrt{2}/2)$. We have $r^2 = 1/4 + 1/4 + 1/2 = 1$, $\theta = \arccos\left(\sqrt{2}/2\right) = \pi/4$. Also, we have $\cos \phi = \sqrt{2}/2$, $\sin \phi = \sqrt{2}/2$. So $\phi = \pi/4$.

3.1.4. The Bloch sphere. In this section, we present the Bloch sphere representation of single-qubit states. As we will see in Section 4.3, this representation allows for a geometric interpretation of the unitary operators in $\mathbb{H}_1$.

Definition 3.1.14. By the *Bloch sphere* we mean the set $\{\vec{p} \in \mathbb{R}^3 : \|\vec{p}\| = 1\}$ which is the surface of the sphere of radius 1 in $\mathbb{R}^3$. The elements of the Bloch sphere are referred to as *points on the Bloch sphere*.

In (3.1.1), elements $|\varphi\rangle$ of $\mathbb{H}_1$ are represented as superpositions of the computational basis elements $|0\rangle$ and $|1\rangle$ of $\mathbb{H}_1$. Since the two coefficients are complex numbers and $\mathbb{C}$ is a two-dimensional $\mathbb{R}$-vector space, $|\varphi\rangle$ can be described using four real numbers. But since single-qubit states have Euclidean length 1, these numbers are not independent of each other. We will now show that single-qubit states can be represented by three real numbers. For this, we need the following result which follows from Lemma 3.1.9.

Lemma 3.1.15. *If $\alpha \in \mathbb{C}$ with $|\alpha| = 1$, then there is a uniquely determined real number γ with $0 \le \gamma < 2\pi$ such that $\alpha = e^{i\gamma} = \cos\gamma + i\sin\gamma$.*

Exercise 3.1.16. Prove Lemma 3.1.15.

We give a few examples for the representation from Lemma 3.1.15.

Example 3.1.17. We have

$$1 = e^{i\cdot 0} = \cos 0 + i\sin 0,$$

$$i = e^{i\cdot\pi/2} = \cos\frac{\pi}{2} + i\sin\frac{\pi}{2},$$

(3.1.9)
$$\frac{1+i}{\sqrt{2}} = e^{i\cdot\pi/4} = \cos\frac{\pi}{4} + i\sin\frac{\pi}{4},$$

$$\frac{1-i}{\sqrt{2}} = e^{i\cdot 7\pi/4} = \cos\frac{7\pi}{4} + i\sin\frac{7\pi}{4}.$$

The next proposition presents the representation of single-qubit states by three real numbers.

Proposition 3.1.18. *Let $|\psi\rangle \in \mathbb{H}_1$ be a single-qubit state. Then there are uniquely determined real numbers γ, θ, and ϕ such that*

(3.1.10)
$$|\psi\rangle = e^{i\gamma}\left(\cos\left(\frac{\theta}{2}\right)|0\rangle + e^{i\phi}\sin\left(\frac{\theta}{2}\right)|1\rangle\right)$$

and

(3.1.11)
$$0 \le \theta \le \pi, \quad 0 \le \gamma, \phi < 2\pi, \quad \theta \in \{0, \pi\} \Rightarrow \phi = 0.$$

We write these numbers as $\gamma(\psi)$, $\theta(\psi)$, and $\phi(\psi)$.

Proof. Let $|\psi\rangle = \alpha_0|0\rangle + \alpha_1|1\rangle$ with $\alpha_0, \alpha_1 \in \mathbb{C}$. Since $|\psi\rangle$ is a single-qubit state, we have $|\alpha_0|^2 + |\alpha_1|^2 = 1$. Choose $\theta \in [0, \pi]$ such that

(3.1.12)
$$|\alpha_0| = \cos\frac{\theta}{2}, \quad |\alpha_1| = \sin\frac{\theta}{2}.$$

By Lemma 3.1.9, this is possible and θ is uniquely determined. To complete the proof, we distinguish three cases.

First, if $\alpha_0 = 0$, then $\theta = 0$, $|\alpha_1| = 1$, and by Lemma 3.1.15 we can write $|\alpha_1| = e^{i\gamma}$ with a uniquely determined $\gamma \in [0, 2\pi[$. If we set $\phi = 0$, then (γ, θ, ϕ) is the only triplet of real numbers that satifies (3.1.10) and (3.1.11).

Second, if $\alpha_1 = 0$, then $\theta = \pi$, $|\alpha_0| = 1$, and by Lemma 3.1.15 we can write $|\alpha_0| = e^{i\gamma}|0\rangle$ with a uniquely determined $\gamma \in [0, 2\pi[$. If we set $\phi = 0$, then (γ, θ, ϕ) is the only triplet of real numbers that satisfies (3.1.10) and (3.1.11).

Third, assume that $\alpha_0, \alpha_1 \neq 0$. Then it follows from Lemma 3.1.15 that there are uniquely determined real numbers $\gamma, \delta \in [0, 2\pi[$ such that

$$(3.1.13) \qquad \alpha_0 = e^{i\gamma} |\alpha_0| = e^{i\gamma} \cos \frac{\theta}{2}, \quad \alpha_1 = e^{i\delta} |\alpha_1| = e^{i\delta} \sin \frac{\theta}{2}.$$

Set $\phi = \delta - \gamma \bmod 2\pi$. Then we have

$$(3.1.14) \qquad |\varphi\rangle = e^{i\gamma} \left(\cos \frac{\theta}{2} |0\rangle + e^{i\phi} \sin \frac{\theta}{2} |1\rangle \right)$$

and (γ, θ, ϕ) is the uniquely determined triplet of real numbers that satisfies (3.1.10) and (3.1.11). $\qquad \square$

Building upon Proposition 3.1.18, we establish a correspondence between each single-qubit state and a location on the Bloch sphere and vice versa.

Definition 3.1.19. (1) To each single-qubit state $|\psi\rangle \in \mathbb{H}_1$ we assign the point $\vec{p}(\psi)$ on the Bloch sphere with spherical coordinates $(1, \theta(\psi), \phi(\psi))$ and Cartesian coordinates $(\sin \theta(\psi) \cos \phi(\psi), \sin \theta(\psi) \sin \phi(\psi), \cos \theta(\psi))$.

(2) To each point $\vec{p}$ on the Bloch sphere with spherical coordinates $(1, \theta, \phi)$ we assign the single-qubit state

$$(3.1.15) \qquad |\psi(\vec{p})\rangle = \cos\left(\frac{\theta}{2}\right) |0\rangle + e^{i\phi} \sin\left(\frac{\theta}{2}\right) |1\rangle.$$

The correspondence between single-qubit states and points on the Bloch sphere is illustrated in Example 3.1.20, Exercise 3.1.21, and Figure 3.1.2. There and in the remainder of this book, we write the unit vectors in the x-, y-, and z-directions in $\mathbb{R}^3$ as

$$(3.1.16) \qquad \hat{x} = (1,0,0), \quad \hat{y} = (0,1,0), \quad \hat{z} = (0,0,1).$$

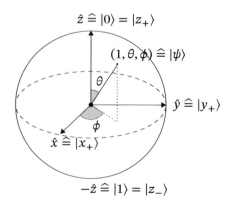

Figure 3.1.2. Points on the Bloch sphere corresponding to $|x_+\rangle$, $|y_+\rangle$, $|z_+\rangle = |0\rangle$, $|z_-\rangle = |1\rangle$, and a general single-qubit state $|\psi\rangle$.

We also recall that the orthonormal eigenbases of the Pauli operators X, Y, and Z on $\mathbb{H}_1$ (see Section 2.4.7) are

$$(|x_+\rangle, |x_-\rangle) = \left(\frac{|0\rangle + |1\rangle}{\sqrt{2}}, \frac{|0\rangle - |1\rangle}{\sqrt{2}} \right),$$

(3.1.17)
$$(|y_+\rangle, |y_-\rangle) = \left(\frac{|0\rangle + i|1\rangle}{\sqrt{2}}, \frac{|0\rangle - i|1\rangle}{\sqrt{2}} \right),$$

$$(|z_+\rangle, |z_-\rangle) = (|0\rangle, |1\rangle).$$

Example 3.1.20. The representation (3.1.10) of $|z_+\rangle = |0\rangle$ is

(3.1.18)
$$|z_+\rangle = |0\rangle = e^{i \cdot 0} \left(\cos \frac{0}{2} |0\rangle + e^{i \cdot 0} \sin \frac{0}{2} |1\rangle \right).$$

Hence, the spherical coordinate representation of the point on the Bloch sphere corresponding to this state is $(1, 0, 0)$ and its Cartesian coordinate representation is $(0, 0, 1) = \hat{z}$.

The representation (3.1.10) of $|z_-\rangle = |1\rangle$ is

(3.1.19)
$$|z_-\rangle = |1\rangle = e^{i \cdot 0} \left(\cos \frac{0}{2} |0\rangle + e^{i \cdot 0} \sin \frac{0}{2} |1\rangle \right).$$

Hence, the spherical coordinate representation of the point on the Bloch sphere corresponding to this state is $(1, \pi, 0)$ and its Cartesian coordinate representation $(0, 0, -1) = -\hat{z}$.

In the previous example, we have presented single-qubit states that correspond to the unit vectors on the Bloch sphere in the x-direction. The next exercise determines such states that correspond to the unit vectors in the y- and z-directions.

Exercise 3.1.21. Show that $(\vec{p}(x_+), \vec{p}(x_-)) = (\hat{x}, -\hat{x})$ and $(\vec{p}(y_+), \vec{p}(y_-)) = (\hat{y}, -\hat{y})$.

We now introduce global phase factors. For example, the term $e^{i\gamma}$ in the representation (3.1.10) is such a factor. The general definition is as follows.

Definition 3.1.22. Let $|\varphi\rangle, |\psi\rangle \in \mathbb{H}$ and let $\gamma \in \mathbb{R}$ be such that $|\psi\rangle = e^{i\gamma} |\varphi\rangle$. Then we say that $|\varphi\rangle$ and $|\psi\rangle$ *are equal up to the global phase factor* $e^{i\gamma}$ or that these states *differ by the global phase factor* $e^{i\gamma}$.

We note the following.

Proposition 3.1.23. *Let S be the set of all quantum states in the Hilbert space $\mathbb{H}$. Then the subset of S^2 of all pairs of quantum states that are equal up to a global phase factor is an equivalence relation on S. For $|\psi\rangle \in \mathbb{H}$, we denote the equivalence class of $|\psi\rangle$ with respect to this relation by $[\psi]$.*

Exercise 3.1.24. Prove Proposition 3.1.23

As we show in Theorem 3.4.4, global phase factors have no impact on measurement outcomes. So if $|\psi\rangle \in \mathbb{H}$ is a unit vector that describes the state of a closed physical system, then all elements of the equivalence class $[\psi]$, that is, all unit vectors in $\mathbb{H}$ that are equal to $|\psi\rangle$ up to a global phase factor, also completely describe the state of this system.

Next, we will show that there is a one-to-one correspondence between the points on the Bloch sphere and the equivalence classes $[\psi]$ of the quantum states $|\psi\rangle$ in $\mathbb{H}_1$. This means that the state of a single-qubit system is completely described by the corresponding point on the Bloch sphere.

Theorem 3.1.25. *Denote by S_1 the set of quantum states in $\mathbb{H}_1$ and by R_1 the equivalence relation on S_1 from Proposition 3.1.23. Then the map*

$$(3.1.20) \qquad S_1/R_1 \to \{\vec{p} \in \mathbb{R}^3 : \|\vec{p}\| = 1\}, \quad [\psi] \mapsto \vec{p}(\psi)$$

is a bijection. Its inverse is

$$(3.1.21) \qquad \{\vec{p} \in \mathbb{R}^3 : \|\vec{p}\| = 1\} \to S_1/R_1, \quad \vec{p} \mapsto [\psi(\vec{p})].$$

Proof. It follows from Proposition 3.1.12 that the map that sends the spherical coordinates of a point on the Bloch sphere to its Cartesian coordinates is a bijection. Therefore, it suffices to prove that the map

$$(3.1.22) \qquad S_1/R \to \{(0,0),(\pi,0)\} \cup (]0,\pi[\times[0,2\pi[), \quad [\psi] \mapsto (\theta(\psi),\phi(\psi))$$

is a bijection. Injectivity follows from Proposition 3.1.18. To see the surjectivity, we note that for a point $\vec{p}$ on the Bloch sphere with spherical coordinates (θ, ϕ), the equivalence class $[\psi(\vec{p})]$ is the inverse image of $\vec{p}$. $\qquad\square$

3.1.5. Quantum registers. To be able to perform complex computations, quantum systems are required that consist of more than one qubit. Let $n \in \mathbb{N}$. A quantum system comprising n qubits is called an *n-qubit quantum register*. The corresponding state space $\mathbb{H}_n$ and its computational basis have already been introduced in Definition 2.1.2. Recall that the computational basis of $\mathbb{H}_n$ is the lexicographically ordered sequence $B = \left(|\vec{b}\rangle\right)_{\vec{b}\in\{0,1\}^n}$ which can also be written as $\left(|b\rangle_n\right)_{b\in\mathbb{Z}_{2^n}}$ (see Example 2.1.5). Since the inner product on $\mathbb{H}_n$ is the Hermitian inner product with respect to B, this basis is orthonormal.

The state of an n-qubit quantum register can be written as

$$(3.1.23) \qquad |\varphi\rangle = \sum_{\vec{b}\in\{0,1\}^n} \alpha_{\vec{b}} |\vec{b}\rangle$$

with complex coefficients $\alpha_{\vec{b}}$ satisfying

$$(3.1.24) \qquad \sum_{\vec{b}\in\{0,1\}^n} |\alpha_{\vec{b}}|^2 = 1.$$

Such an element of $\mathbb{H}_n$ is called an *n-qubit state*. So, while the state of a classical n-bit register is an element $\vec{b} \in \{0,1\}^n$, the state of an n-qubit quantum register is a linear combination of the computational basis states $|\vec{b}\rangle$, $\vec{b} \in \{0,1\}^n$ which is also called a *superposition* of the basis elements.

Example 3.1.26. Consider the state space $\mathbb{H}_2$ of a 2-qubit system. The computational basis of $\mathbb{H}_2$ is $(|00\rangle, |01\rangle, |10\rangle, |11\rangle)$. It can also be written as $|0\rangle_2, |1\rangle_2, |2\rangle_2, |3\rangle_2$. For instance,

$$(3.1.25) \qquad |\varphi\rangle = \frac{1}{\sqrt{2}}|00\rangle - \frac{i}{\sqrt{2}}|11\rangle = \frac{1}{\sqrt{2}}|0\rangle_2 - \frac{i}{\sqrt{2}}|3\rangle_2$$

is a 2-qubit state. It is a superposition of the states $|00\rangle = |0\rangle_2$ and $|11\rangle = |3\rangle_2$.

Once again, we provide a preview of the concept of measurements, which will be discussed in more detail in Section 3.4.1. When measuring an n-qubit register in the computational basis of $\mathbb{H}_n$ that is in the state $\sum_{\vec{b}\in\{0,1\}^n} \alpha_{\vec{b}} |b\rangle$, where $\alpha_{\vec{b}} \in \mathbb{C}$, the probability of obtaining any specific $\vec{b} \in \{0,1\}^n$ is $|\alpha_{\vec{b}}|^2$. After a measurement with outcome $\vec{b} \in \{0,1\}^n$, the quantum register's state becomes $|\vec{b}\rangle$. If immediately after a measurement the register is measured again, then the result of the previous measurement is reproduced. Similar to single-qubits, when measuring an n-qubit register in two different states that are equal up to a global phase factor, the resulting probability distribution is the same.

3.2. State spaces of composite systems

In quantum computing, we need to be able to combine physical systems to larger physical systems and to operate on them. In this section, we explain how this is done.

3.2.1. The Composite Systems Postulate. The *Composite System Postulate* that we present now describes how to construct the state space of composite physical systems.

Postulate 3.2.1 (Composite Systems Postulate). The state space of the composition of finitely many physical systems is the tensor product of the state spaces of the component systems. Moreover, if we have systems numbered 0 through $m-1$ and if system i is in the state $|\psi_i\rangle$ for $0 \le i < m$, then the composite system is in state $|\psi_0\rangle \otimes \cdots \otimes |\psi_{m-1}\rangle$.

For quantum computing, compositions of the following type are frequently used:

$$(3.2.1) \qquad\qquad \mathbb{H} = \mathbb{H}_{n_0} \otimes \cdots \otimes \mathbb{H}_{n_{m-1}}$$

where $m \in \mathbb{N}$ and $n_i \in \mathbb{N}$ for all $i \in \mathbb{Z}_m$. As explained in Section 2.5.3, the composite space $\mathbb{H}$ is identified with the state space $\mathbb{H}_n$ where $n = \sum_{i=0}^{m-1} n_i$. The computational basis of $\mathbb{H} = \mathbb{H}_n$ is the tensor product of the computational bases of the Hilbert spaces $\mathbb{H}_{n_i}$.

3.2.2. Entangled states. An important reason for the superiority of quantum computing over classical computing is that quantum states can be entangled. Entanglement has already been introduced in Definition 2.5.20. The present section puts this concept into the context of quantum mechanics. We begin with an example.

Example 3.2.2. We consider the composition of two qubits with state space $\mathbb{H}_2 = \mathbb{H}_1 \otimes \mathbb{H}_1$. For all pairs $(|\varphi_0\rangle, |\varphi_1\rangle)$ of single-qubit states, $\mathbb{H}_2$ contains the composite state

$$(3.2.2) \qquad\qquad |\varphi\rangle = |\varphi_0\rangle \otimes |\varphi_1\rangle.$$

However, as we have seen in Example 2.5.22, the Bell state

$$(3.2.3) \qquad\qquad |\varphi\rangle = \frac{|00\rangle + |11\rangle}{\sqrt{2}}$$

cannot be written in this form, that is, as the tensor product of two single-qubit states. It is therefore called *entangled*.

The next definition generalizes Example 3.2.2.

Definition 3.2.3. A state of the composition of two physical systems is called *entangled* if it cannot be written as the tensor product of states of the component systems. Otherwise, this state is called *separable* or *nonentangled*.

Note that the concept of entanglement depends on the decomposition of a physical system into parts. Theorem 2.5.18 implies the following result.

Theorem 3.2.4. *The state of the composition of two quantum systems is separable if and only if its Schmidt rank is 1, and it is entangled if and only if its Schmidt rank is greater than 1.*

Exercise 3.2.5. Find an example of an entangled state in $\mathbb{H}_3$ and prove that it is entangled.

3.3. Time evolution

Quantum computers use a sequence of operations to transform the input state of a quantum register into its output state. In this section, we describe these operations.

3.3.1. Evolution Postulate.
How does the state of a quantum mechanical system change over time? This question is answered by the *Evolution Postulate*.

Postulate 3.3.1 (Evolution Postulate). The evolution of a closed quantum system is described by a unitary transformation. More precisely, if $t, t' \in \mathbb{R}$, $t < t'$, then the state $|\varphi'\rangle$ of the system at time t' is obtained from the state $|\varphi\rangle$ of the system at time t as $|\varphi'\rangle = U |\varphi\rangle$ where U is a unitary operator on the state space of the system that depends only on t and t'.

In general quantum mechanics where infinite-dimensional Hilbert spaces are used, the Schrödinger differential equation is used to formulate a more general Evolution Postulate. However, for our purposes the above formulation of the Evolution Postulate is appropriate. In the next section, we illustrate the Evolution Postulate in the context of quantum computing.

3.3.2. Quantum gates. Postulate 3.3.1 provides an understanding of how quantum computation works in principle. Such a computation uses a quantum register. Its initial state is the *input state* of the computation. A unitary transformation is applied to the input state, giving the *output state*. So, quantum computing relies on the implementation of unitary operators. For this, quantum circuits are used. They will be described in Section 3.3.4.

The building blocks of such circuits are quantum gates just as logic gates are the building blocks of Boolean circuits. Quantum gates implement simple unitary operators and are provided by the quantum computing platform that is used.

We now present two examples of quantum gates. Many more quantum gates are discussed in Chapter 4.

The first example is the *Hadamard gate* or *Hadamard operator*

$$(3.3.1) \qquad H : \mathbb{H}_1 \to \mathbb{H}_1, \quad |0\rangle \mapsto |x_+\rangle, \quad |1\rangle \mapsto |x_-\rangle$$

which was already introduced in Exercise 2.3.4. There, it is shown that its representation matrix with respect to the computational basis of $\mathbb{H}_1$ is

$$(3.3.2) \qquad H = \frac{1}{\sqrt{2}} \begin{pmatrix} 1 & 1 \\ 1 & -1 \end{pmatrix}.$$

This implies the following result.

Proposition 3.3.2. *The Hadamard operator is a Hermitian unitary involution; that is, we have $H^* = H = H^{-1}$ and $H^2 = I$.*

Exercise 3.3.3. Prove Proposition 3.3.2.

In quantum circuits, the Hadamard gate is represented by the symbol shown in Figure 3.3.1.

The second example of a quantum gate is the CNOT gate.

Definition 3.3.4. The *controlled*-NOT gate or CNOT gate for short is the linear operator

$$(3.3.3) \qquad \mathsf{CNOT} : \mathbb{H}_2 \to \mathbb{H}_2, \quad |c\rangle |t\rangle \mapsto |c\rangle X^c |t\rangle.$$

Definition 3.3.4 shows that the CNOT gate applies the Pauli X operator to a *target qubit* $|t\rangle$ if the *control qubit* $|c\rangle$ is $|1\rangle$. Otherwise, the target qubit remains unchanged. This means that the application of the Pauli X operator to the target qubit is controlled by the control qubit. Since the Pauli X operator is the quantum NOT operator, this explains the name "controlled-NOT gate". So, CNOT operates on the computational basis states of $\mathbb{H}_2$ in the following way:

$$(3.3.4) \qquad |00\rangle \mapsto |00\rangle, |01\rangle \mapsto |01\rangle, |10\rangle \mapsto |11\rangle, |11\rangle \mapsto |10\rangle,$$

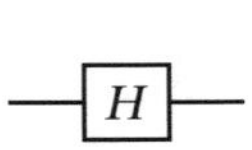

Figure 3.3.1. Symbol for the Hadamard gate in quantum circuits.

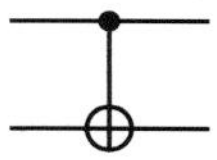

Figure 3.3.2. Symbol for the CNOT gate in quantum circuits.

which shows that the representation matrix of CNOT with respect to the computational basis of $\mathbb{H}_2$ is

$$(3.3.5) \qquad \text{CNOT} = \begin{pmatrix} 1 & 0 & 0 & 0 \\ 0 & 1 & 0 & 0 \\ 0 & 0 & 0 & 1 \\ 0 & 0 & 1 & 0 \end{pmatrix}.$$

This implies the following result.

Proposition 3.3.5. *The* CNOT *operator is a Hermitian unitary involution; that is, we have* $\text{CNOT}^* = \text{CNOT} = \text{CNOT}^{-1}$ *and* $\text{CNOT}^2 = I_2$.

Exercise 3.3.6. Prove Proposition 3.3.5.

In quantum circuits, the CNOT gate is represented by the symbol shown in Figure 3.3.2.

The definition of the CNOT gate might give the impression that this gate never changes the control qubit. However, as the next example shows, this impression is deceptive.

Example 3.3.7. We have

$$\text{CNOT} \,|x_+\rangle\,|x_-\rangle$$
$$= \frac{\text{CNOT}\,|0\rangle\,|0\rangle - \text{CNOT}\,|0\rangle\,|1\rangle + \text{CNOT}\,|1\rangle\,|0\rangle - \text{CNOT}\,|1\rangle\,|1\rangle}{2}$$
$$= \frac{|0\rangle\,|0\rangle - |0\rangle\,|1\rangle + |1\rangle\,|1\rangle - |1\rangle\,|0\rangle}{2}$$
$$= |x_-\rangle\,|x_-\rangle.$$

So applied to $|x_+\rangle\,|x_-\rangle$ the CNOT operator changes the control qubit but not the target qubit.

3.3.3. Composition of operators. Let $m \in \mathbb{N}$ and consider m quantum systems with the corresponding state spaces $\mathbb{H}(0),\dots,\mathbb{H}(m-1)$. According to Postulate 3.2.1, the state space of the composition of the m quantum systems is the tensor product

$$(3.3.6) \qquad \mathbb{H} = \mathbb{H}(0) \otimes \cdots \otimes \mathbb{H}(m-1).$$

Let $f_0,\dots,f_{m-1}$ be linear operators on $\mathbb{H}(0),\dots,\mathbb{H}(m-1)$, respectively. In Section 2.5.5 we have introduced the linear operator

$$(3.3.7) \qquad f = f_0 \otimes \cdots \otimes f_{m-1}$$

on $\mathbb{H}$ and presented its properties. In particular, we have seen that f is a projection, an involution, normal, Hermitian, or unitary if and only if all of its components have the corresponding property. This construction allows us to extend the action of an operator

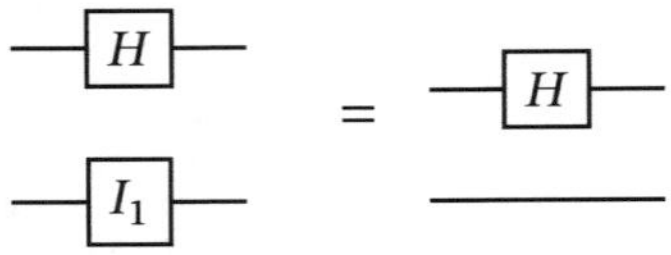

$$\begin{array}{c} -\boxed{H}- \\ -\boxed{I_1}- \end{array} \quad = \quad \begin{array}{c} -\boxed{H}- \\ ---- \end{array}$$

Figure 3.3.3. Extension of H acting on the first state space to an operator action on two qubits.

f_j on one of the component state spaces $\mathbb{H}(j)$, $j \in \mathbb{Z}_m$, to the composite state space by using the operator

(3.3.8) $$I_0 \otimes \cdots \otimes I_{j-1} \otimes f_j \otimes I_{j+1} \otimes \cdots \otimes I_{m-1}$$

where I_i is the identity operator on $\mathbb{H}(i)$ for $0 \leq i < m$.

Example 3.3.8. Consider the composition of two single-qubit systems with state spaces $\mathbb{H}_1$ each. The state space of the composite system is $\mathbb{H}_2$. The extension of the Hadamard operator acting on the first state space to an operator on the composite space is $H \otimes I$ where I is the identity operator on $\mathbb{H}_1$. In a quantum circuit, this extended operator is depicted on the right side of Figure 3.3.3. So, the identity operator is omitted. The two lines represent the two qubits. The box with H inside represents the Hadamard operator acting on the first qubit. The extended operator has the following effect on the computational basis elements of $\mathbb{H}_2$:

$$|00\rangle \mapsto \frac{1}{\sqrt{2}}(|00\rangle + |10\rangle), \quad |01\rangle \mapsto \frac{1}{\sqrt{2}}(|01\rangle + |11\rangle),$$

$$|10\rangle \mapsto \frac{1}{\sqrt{2}}(|00\rangle - |10\rangle), \quad |11\rangle \mapsto \frac{1}{\sqrt{2}}(|01\rangle - |11\rangle).$$

The matrix representation of the composite operator is

$$H \otimes I_2 = \frac{1}{\sqrt{2}}\begin{pmatrix} 1 & 1 \\ 1 & -1 \end{pmatrix} \otimes \begin{pmatrix} 1 & 0 \\ 0 & 1 \end{pmatrix} = \frac{1}{\sqrt{2}}\begin{pmatrix} 1 & 0 & 1 & 0 \\ 0 & 1 & 0 & 1 \\ 1 & 0 & -1 & 0 \\ 0 & 1 & 0 & -1 \end{pmatrix}.$$

3.3.4. Quantum circuits. Quantum circuits implement more complex unitary operators on a state space $\mathbb{H}_n$, $n \in \mathbb{N}$, by combining several quantum gates. We illustrate this concept by an example shown in Figure 3.3.4.

This quantum circuit has three wires. This indicates that the unitary operator U implemented by the quantum circuit acts on the state space $\mathbb{H}_3$ of 3-qubit registers.

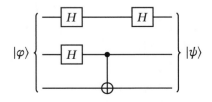

Figure 3.3.4. Quantum circuit that combines the Hadamard and CNOT gates.

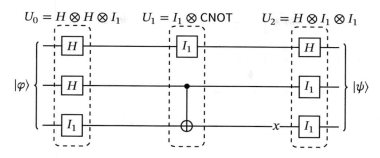

Figure 3.3.5. The quantum circuit from Figure 3.3.4 implements $U = U_0 \circ U_1 \circ U_2$.

The circuit transforms the 3-qubit input state $|\varphi\rangle$ into a 3-qubit output state $|\psi\rangle$. How is U constructed? As shown in Figure 3.3.5, U is the concatenation of three unitary operators; i.e.,

$$(3.3.9) \qquad U = U_2 \circ U_1 \circ U_0.$$

Each of these unitary operators is the tensor product of the unitary operators implemented by the gates that are one above the other. If there is no operator but only a wire, the identity operator is inserted. The composed operators are $U_0 = H \otimes H \otimes I_1$, $U_1 = I_1 \otimes \text{CNOT}$, and $U_2 = H \otimes I_1 \otimes I_1$.

We determine the state that is obtained when the input state of the quantum circuit in Figures 3.3.4 and 3.3.5 is $|000\rangle$:

$$
\begin{aligned}
|000\rangle &\underset{U_0}{\mapsto} \frac{1}{\sqrt{4}}(|0\rangle + |1\rangle)(|0\rangle + |1\rangle)\,|0\rangle \\
&= \frac{1}{\sqrt{4}}(|000\rangle + |010\rangle + |100\rangle + |110\rangle) \\
&\underset{U_1}{\mapsto} \frac{1}{\sqrt{4}}(|000\rangle + |011\rangle + |100\rangle + |111\rangle) \\
&= \frac{1}{\sqrt{4}}(|0\rangle + |1\rangle)(|00\rangle + |11\rangle) \\
&\underset{U_2}{\mapsto} \frac{1}{\sqrt{2}}\,|0\rangle\,(|00\rangle + |11\rangle) \\
&= \frac{1}{\sqrt{2}}(|000\rangle + |011\rangle).
\end{aligned}
$$

(3.3.10)

This computation is also illustrated in Figure 3.3.6.

Exercise 3.3.9. Determine the output state $U\,|\vec{b}\rangle$ of the circuit in Figure 3.3.4 for all $\vec{b} \in \{0,1\}^3$.

The concept of a quantum circuit can easily be generalized to circuits that operate on n-qubit registers for any $n \in \mathbb{N}$. This will be discussed in more detail in Section 4.7.

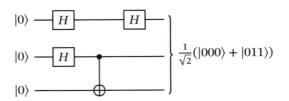

Figure 3.3.6. The quantum circuit from Figure 3.3.4 operating on $|000\rangle$.

3.4. Measurements

In the previous sections, we have seen the first steps of quantum computing. A state of an n-bit quantum register is prepared. This is the input state for a quantum circuit operating on n-qubit states. This quantum circuit implements a unitary operator on the state space of the quantum register. It is composed of several quantum gates and transforms the input state into an output state. This output state may either be used by a further quantum computation or the information may be extracted by a measurement. Such measurements are discussed in more detail in this section.

3.4.1. Measurement Postulate. The concept of measurements is controversial in quantum mechanics. Here, we present this concept according to the Copenhagen interpretation of quantum theory. This term was coined by Werner Heisenberg around 1955 for views of this theory developed in the second quarter of the 20th century. The idea is that a quantum mechanical system is initially closed. It then develops according to the state space and the Evolution Postulates. A measurement ends the seclusion of the quantum system. It is an interaction with a laboratory device. When this device makes a measurement, the state of the quantum system *collapses* irreversibly to an eigenstate of the *observable* implemented by the device. This observable is a Hermitian operator on the state space. The measurement makes a potentiality, the state of the quantum system in superposition of the eigenstates, become an actuality. Additionally, the device records the corresponding eigenvalue.

This view of quantum mechanical measurements is the content of the following *Measurement Postulate*. It describes projective measurements. In the sequel, we need only such measurements. The more general notion is that of a positive operator-valued measurement (POVM).

Postulate 3.4.1 (Measurement Postulate). A *projective measurement* is described by an *observable* O which is a Hermitian operator on the state space of the system being observed. Let $O = \sum_{\lambda \in \Lambda} \lambda P_\lambda$ be the spectral decomposition of O. The possible outcomes of the measurement are the eigenvalues λ of the observable. When measuring O while the quantum system is in the state $|\varphi\rangle$, the probability of getting the result λ is $\mathrm{Pr}_{O,\varphi}(\lambda) = \langle \varphi | P_\lambda | \varphi \rangle = \| P_\lambda | \varphi \rangle \|^2$. If this outcome occurs, the state of the quantum system immediately after the measurement is $\frac{P_\lambda | \varphi \rangle}{\| P_\lambda | \varphi \rangle \|}$.

The measurements in the postulate are called "projective" because they project the state of the quantum system onto one of the eigenspaces of the measured observable and normalize the length of this projection. We also note that measurement devices must use an appropriate encoding of the measurement outcomes, typically by finitely many bits. Examples of measurements are given in the following section.

Next, we discuss the expectation values of measurements. For every observable on the state space $\mathbb{H}$ and all $|\varphi\rangle \in \mathbb{H}$, Postulate 3.4.1 defines the discrete probability space $(\Lambda, \mathrm{Pr}_{O,\varphi})$. Since O is Hermitian, Proposition 2.4.61 implies that $\Lambda \subset \mathbb{R}$. So the identity map I_Λ on Λ is a random variable associated with the probability space. The next lemma determines its expectation value.

Lemma 3.4.2. *We have $E[I_\Lambda] = \langle \psi | O | \psi \rangle$.*

Proof. We note that

$$
\begin{aligned}
E[X] &= \sum_{\lambda \in \Lambda} \lambda \mathrm{Pr}(\lambda) && \text{definition of the expectation value,} \\
&= \sum_{\lambda \in \Lambda} \lambda \langle \psi | P_\lambda | \psi \rangle && \text{definition of Pr,} \\
&= \langle \psi | \sum_{\lambda \in \Lambda} \lambda P_\lambda | \psi \rangle && \text{linearity of the inner product,} \\
&= \langle \psi | O | \psi \rangle && \text{spectral theorem.}
\end{aligned}
$$

This completes the proof. $\qquad\square$

Lemma 3.4.2 motivates the following definition.

Definition 3.4.3. Let O be an observable of a quantum system with state space $\mathbb{H}$. Suppose that we measure this observable when the system is in the state $|\varphi\rangle \in \mathbb{H}$. Then the *expectation value of this measurement* is defined as $\langle \varphi | O | \varphi \rangle$.

In Section 3.1.4 we have introduced global phase factors of quantum states and have shown that equality up to a global phase factor is an equivalence relation. We have denoted the corresponding equivalence class of a quantum state $|\psi\rangle$ by $[\psi]$. We now show that global phase factors have no impact on quantum measurements.

Theorem 3.4.4. *Assume that we measure an observable O of a quantum system. Let λ be an eigenvalue of O and let $|\varphi\rangle$ and $|\psi\rangle$ be two states of the system that differ only by a global phase factor. Then the probability of measuring λ is the same regardless of whether the system is in the state $|\varphi\rangle$ or $|\psi\rangle$. Also, if the system is in one of the two states and the measurement outcome λ occurs, then the states immediately after the measurement are equal up to the same global phase factor.*

Proof. Let $\gamma \in \mathbb{R}$ such that $|\psi\rangle = e^{i\gamma} |\varphi\rangle$. Then we have

$$
(3.4.1) \qquad \|P_\lambda |\psi\rangle\| = \|e^{i\theta} P_\lambda |\varphi\rangle\| = |e^{i\theta}| \|P_\lambda |\varphi\rangle\| = \|P_\lambda |\varphi\rangle\|.
$$

Also, if the state of the system is $|\psi\rangle$, then the state immediately after the measurement is

$$
(3.4.2) \qquad \frac{P_\lambda |\psi\rangle}{\|P_\lambda |\varphi\rangle\|} = \frac{e^{i\gamma} P_\lambda |\varphi\rangle}{\|P_\lambda |\varphi\rangle\|}. \qquad\square
$$

Exercise 3.4.5. Let $\mathbb{H}$ be the state space of a quantum system. Show that $O = I_{\mathbb{H}}$ is an observable of the quantum system. Also, show that measuring O when the quantum system is in the state $|\varphi\rangle \in \mathbb{H}$ gives 1 with probability 1 and that immediately after the measurement, the quantum system is still in state $|\varphi\rangle$.

3.4.2. Measuring quantum systems in an orthonormal basis.

In quantum computing, it is common to measure n-qubit registers in the computational basis of their state space $\mathbb{H}_n$ for some $n \in \mathbb{N}$. We have already introduced these measurements in Sections 3.1.2 and 3.1.5. Now, we explain how to model them using the measurement postulate and start with an example.

Example 3.4.6. Let

$$(3.4.3) \qquad\qquad |\varphi\rangle = \alpha_0 |0\rangle + \alpha_1 |1\rangle$$

be a single-qubit state, where $\alpha_0, \alpha_1 \in \mathbb{C}$ with $|\alpha_0^2| + |\alpha_1|^2 = 1$. We would like the measurement to reflect this superposition; that is, we want the measurement outcome to be 0 with probability $|\alpha_0|^2$ and 1 with probability $|\alpha_1|^2$. Recall that, by Proposition 2.4.44, the projections onto the subspaces spanned by $|0\rangle$ or $|1\rangle$ are $P_0 = |0\rangle\langle 0|$ and $P_1 = |1\rangle\langle 1|$, respectively. Therefore, we use the observable O with the spectral decomposition

$$(3.4.4) \qquad\qquad O = 0 \cdot P_0 + 1 \cdot P_1 = |\varphi_1\rangle\langle\varphi_1|.$$

It is Hermitian, has the two eigenvalues $\lambda_0 = 0$ and $\lambda_1 = 1$, and the corresponding eigenspaces are $\mathbb{C}|0\rangle$ and $\mathbb{C}|1\rangle$. When the qubit is in the state $|\varphi\rangle$ from (3.4.3) and we measure O, the measurement produces $b \in \{0, 1\}$ with probability.

$$(3.4.5) \qquad\qquad \|P_b |\varphi\rangle\|^2 = |\alpha_b|^2.$$

Also, the state of the qubit immediately after the measurement is

$$(3.4.6) \qquad\qquad \frac{P_b |\varphi\rangle}{\|P_b |\varphi\rangle\|} = \frac{\alpha_b}{|\alpha_b|} |b\rangle.$$

This state is equal to $|b\rangle$ up to the global phase factor $\alpha_b/|\alpha_b|$. By (3.4.4), the expectation value of this measurement is

$$(3.4.7) \qquad\qquad \langle\varphi|O|\varphi\rangle = \langle\alpha_0 |0\rangle + \alpha_1 |1\rangle |\alpha_1 |1\rangle\rangle = |\alpha_1|^2.$$

For example, if the qubit is in the state $|0\rangle$, then the expectation value of this measurement is 0. Also, if the qubit is in the state $|1\rangle$, then the expectation value is 1. But if the qubit is in the superposition $|\psi\rangle = \frac{1}{\sqrt{2}} |0\rangle + \frac{1}{\sqrt{2}} |1\rangle$, then the expectation value of this measurement is $\frac{1}{2}$.

We generalize Example 3.4.6. Let $\mathbb{H}$ be the state space of a quantum system and let $B = (|b_0\rangle, \ldots, |b_{k-1}\rangle)$ be an orthonormal basis of $\mathbb{H}$. For example, if $\mathbb{H} = \mathbb{H}_n$, then $k = 2^n$ and we may use the computational basis $(|0\rangle_n, \ldots, |2^n - 1\rangle_n)$ of $\mathbb{H}_n$. For $j \in \mathbb{Z}_k$ the projection onto $\mathbb{C}|b_j\rangle$ is

$$(3.4.8) \qquad\qquad P_j = |b_j\rangle\langle b_j|.$$

We use the observable O with spectral decomposition

$$(3.4.9) \qquad O = \sum_{j=0}^{k-1} jP_j = \sum_{j=0}^{k-1} j\,|b_j\rangle\langle b_j|.$$

Its matrix representation with respect to the basis B is

$$(3.4.10) \qquad \begin{pmatrix} 0 & 0 & 0 & \cdots & 0 \\ 0 & 1 & 0 & \cdots & 0 \\ 0 & 0 & 2 & \cdots & 0 \\ \vdots & & & \ddots & \vdots \\ 0 & 0 & 0 & \cdots & k-1 \end{pmatrix}.$$

So O is Hermitian and has the k eigenvalues $\lambda_j = j$ with the corresponding eigenspaces $\mathbb{C}\,|b_j\rangle$, $j \in \mathbb{Z}_k$.

Let $|\varphi\rangle$ be a state of our quantum system; i.e.,

$$(3.4.11) \qquad |\varphi\rangle = \sum_{j=0}^{k-1} \alpha_j\,|b_j\rangle$$

with $\alpha_j \in \mathbb{C}$ for $0 \le j < k$ and $\sum_{i=0}^{k-1} |\alpha_i|^2 = 1$. According to Postulate 3.4.1, the possible outcomes when measuring the observable O are the eigenvalues $j \in \mathbb{Z}_k$ of O. Each integer $j \in \mathbb{Z}_k$ occurs with probability

$$(3.4.12) \qquad \|P_j\,|\varphi\rangle\|^2 = |\alpha_j|^2$$

and the state of the quantum system immediately after this outcome is

$$(3.4.13) \qquad \frac{P_j\,|\varphi\rangle}{\left\|P_j\,|\varphi\rangle\right\|} = \frac{\alpha_j}{|\alpha_j|}\,|b_j\rangle.$$

Up to a global phase factor, this state is equal to the basis state $|b_j\rangle$. The expectation value of the measurement is

$$(3.4.14) \qquad \langle\varphi|A|\varphi\rangle = \sum_{j=1}^{k-1} j|\alpha_j|^2.$$

This discussion motivates the following definition.

Definition 3.4.7. Let $\mathbb{H}$ be the state space of a quantum system and let $B = (|b_0\rangle, \dots, |b_{k-1}\rangle)$ be an orthonormal basis of $\mathbb{H}$. By measuring the quantum system in the basis B we mean measuring the observable

$$(3.4.15) \qquad O = \sum_{j=0}^{k-1} j\,|b_j\rangle\langle b_j|$$

of $\mathbb{H}$.

Example 3.4.8. Consider the circuit shown in Figure 3.3.6. As shown in the figure, the input state is $|000\rangle$ and the output state is $\frac{|000\rangle+|011\rangle}{\sqrt{2}}$. If we use integers in $\mathbb{Z}_3$ to denote the computational basis states, then the output state is $\frac{|0\rangle_3+|3\rangle_3}{\sqrt{2}}$. Measuring the 3-qubit register in the computational basis of $\mathbb{H}_3$ means measuring the observable $O = \sum_{j=1}^{7} j \, |j\rangle \langle j|$. The measurement outcome is one of the numbers 0 or 3, each with probability $\frac{1}{2}$. The expectation value of this measurement is $\frac{0+3}{2} = \frac{3}{2}$.

Exercise 3.4.9. Determine the measurement statistics and the expectation values for measuring the 3-qubit register of the circuit in Figure 3.3.4 in the computational basis for all input states $|\vec{b}\rangle$, $\vec{b} \in \{0, 1\}^3$.

3.4.3. Partial measurements.

In certain quantum computing scenarios, measurements are performed selectively on specific parts of a quantum system. We will now explain the modeling of such measurements in a particular situation. However, it allows for a straightforward generalization to other cases.

Assume that A and B are quantum systems with state spaces $\mathbb{H}_A$ and $\mathbb{H}_B$, respectively. Consider the composite quantum system AB with state space $\mathbb{H}_{AB} = \mathbb{H}_A \otimes \mathbb{H}_B$. Let O_A be an observable of system A with spectral decomposition

$$(3.4.16) \qquad O_A = \sum_{\lambda \in \Lambda} \lambda P_\lambda.$$

Denote by I_B the identity operator on $\mathbb{H}_B$. Then the composite operator

$$(3.4.17) \qquad O_{AB} = O_A \otimes I_B$$

is an observable of $\mathbb{H}_{AB}$. Its spectral decomposition is

$$(3.4.18) \qquad O_{AB} = \sum_{\lambda \in \Lambda} \lambda P_\lambda \otimes I_B.$$

Measuring this observable when the system AB is in the state $|\psi\rangle$, the outcome is $\lambda \in \Lambda$ with probability $\|P_\lambda \otimes I_B \, |\psi\rangle\|^2$ and the state after this outcome is $\frac{P_\lambda \otimes I_B |\psi\rangle}{\|P_\lambda \otimes I_B |\psi\rangle\|}$.

Example 3.4.10. Let A and B be single-qubit systems with state spaces $\mathbb{H}_A = \mathbb{H}_B = \mathbb{H}_1$. Our goal is to measure the first qubit of the Bell state

$$(3.4.19) \qquad |\psi\rangle = \frac{|00\rangle + |11\rangle}{\sqrt{2}}.$$

The corresponding observable is

$$(3.4.20) \qquad O = |1\rangle \langle 1| \otimes I_1.$$

If $b \in \{0, 1\}$, then we have

$$(3.4.21) \qquad P_b \otimes I_1 \, |\psi\rangle = \frac{P_b \otimes I_1 \, |0\rangle \, |0\rangle + P_b \otimes I_1 \, |1\rangle \, |1\rangle}{\sqrt{2}} = \frac{|b\rangle \, |b\rangle}{\sqrt{2}}.$$

So the probability that measuring O gives b is $\frac{1}{2}$ and the state after this measurement outcome is $|b\rangle \, |b\rangle$.

Example 3.4.10 shows that measuring the first qubit of the entangled Bell state and leaving the second qubit alone changes both qubits. This observation is central to the famous EPR thought experiment. It is named after its inventors Albert Einstein, Boris Podolsky, and Nathan Rosen. They published it in 1935 to demonstrate that quantum mechanics is incomplete. We present a simplified description of their idea. Prepare two qubits in the entangled state (3.4.19). Give one to Alice and the other to Bob. Then Alice travels a long way, taking her qubit with her. Upon arriving, she measures her qubit. As we have seen in Example 3.4.10, this measurement will with probability 1/2 put both qubits into the state $|0\rangle$ or $|1\rangle$. After this measurement, Alice knows with certainty the state of Bob's qubit. Einstein, Podolsky, and Rosen claimed that this instantaneous change of Bob's qubit contradicts relativity theory which says that the maximum possible speed is the speed of light. They concluded that the quantum mechanics explanation must be incomplete. However, much later experiments confirmed the prediction of quantum mechanics and thus showed that the arguments of Einstein, Podolsky, and Rosen were not correct. We note that from the perspective of information theory, Alice and Bob do not exchange information. They only obtain a uniformly distributed random bit, which they can also produce by tossing a coin.

As we will see now, the situation is much easier if the composite system AB is in a separable state

$$(3.4.22) \qquad\qquad |\psi\rangle = |\varphi\rangle\,|\xi\rangle$$

with $|\varphi\rangle, |\xi\rangle \in \mathbb{H}_1$. Then for $\lambda \in \Lambda$ we have

$$(3.4.23) \qquad P_\lambda \otimes I_B \psi = P_\lambda \otimes I_B\,|\varphi\rangle\,|\xi\rangle = P_\lambda\,|\varphi\rangle \otimes |\xi\rangle.$$

Hence, the probability of measuring λ is

$$(3.4.24) \qquad \|P_\lambda\,|\varphi\rangle \otimes |\xi\rangle\| = \|P_\lambda\,|\varphi\rangle\|\,\|\xi\| = \|P_\lambda\,|\varphi\rangle\|.$$

This is the probability for obtaining λ when only system A is measured. Also, the state after measuring λ is

$$(3.4.25) \qquad\qquad \frac{P_\lambda\,|\varphi\rangle}{\|P_\lambda\,|\varphi\rangle\|} \otimes |\xi\rangle.$$

This state is the tensor product of the state after measuring system A with the state $|\xi\rangle$ of system B before the measurement.

Example 3.4.11. Consider the separable quantum state

$$(3.4.26) \qquad\qquad |\psi\rangle = |x_-\rangle\,|x_-\rangle.$$

Measuring only the first qubit in the computational basis of $\mathbb{H}_1$ gives 0 or 1, each with probability $\frac{1}{2}$. If the measurement outcome $b \in \{0,1\}$ occurs, then the state immediately after the measurement is $|b\rangle\,|x_-\rangle$.

Exercise 3.4.12. Write down the observable that only measures the first and last qubit of a 3-qubit quantum register. Determine the measurement statistics for the quantum state $|\varphi\rangle = \frac{|000\rangle + i|111\rangle}{\sqrt{2}}$.

3.5. Density operators

In this section, we introduce density operators on state spaces of quantum systems. We show how they can be used instead of state vectors to describe the states of quantum systems and generalizations of such states, the so-called mixed states. Let Q be a quantum system with state space $\mathbb{H}$.

3.5.1. Definition.

Definition 3.5.1. A *density operator* on $\mathbb{H}$ is a linear operator on $\mathbb{H}$ that satisfies the following conditions.

(1) *Trace condition:* $\operatorname{tr} \rho = 1$,

(2) *Positivity condition:* ρ is positive semidefinite.

Example 3.5.2. If $|\varphi\rangle$ is a state of Q, then

$$\rho = |\varphi\rangle\langle\varphi| \tag{3.5.1}$$

is a density operator on $\mathbb{H}$. In fact, by Proposition 2.4.27 and since quantum states have norm 1, we have

$$\operatorname{tr}\rho = \operatorname{tr}|\varphi\rangle\langle\varphi| = \langle\varphi|\varphi\rangle = 1. \tag{3.5.2}$$

This proves the trace condition. Also, it follows from Proposition 2.4.27 that for all $|\psi\rangle \in \mathbb{H}$ we have

$$\langle\psi|\rho|\psi\rangle = \langle\psi|\varphi\rangle\langle\varphi|\psi\rangle = |\langle\varphi|\psi\rangle|^2 \geq 0. \tag{3.5.3}$$

This proves the positivity condition.

We note that density operators on $\mathbb{H}$ are Hermitian since by Proposition 2.4.64 positive semidefinite operators are Hermitian. Next, we introduce mixed states of quantum systems. They allow us to describe the probabilistic behavior of quantum systems in situations where we don't have complete information about the system.

Definition 3.5.3. (1) A *mixed state* of the quantum system Q is a sequence

$$((p_0, |\psi_0\rangle), \dots (p_{k-1}, |\psi_{l-1}\rangle)) \tag{3.5.4}$$

where $l \in \mathbb{N}$, the $|\psi_i\rangle$ are quantum states in $\mathbb{H}$ for $0 \leq i < l$, and $p_i \in \mathbb{R}_{\geq 0}$ for $0 \leq i < l$ such that $\sum_{i=0}^{l-1} p_i = 1$.

(2) A *pure state* of the quantum system Q is a quantum state in its state space $\mathbb{H}$.

We note that there is a one-to-one correspondence between the pure states $|\psi\rangle$ and the mixed states $(1, |\psi\rangle)$ of Q. Using this correspondence, we identify pure states and these mixed states. A mixed state of Q with more than 1 component describes the situation where the exact state of the quantum system is inaccessible. For instance, such mixed states arise as the states of parts of composite quantum systems that are in an entangled state. This will be discussed in Section 3.7. The next theorem associates each mixed state with a density operator.

Proposition 3.5.4. *Let* $((p_0, |\psi_0\rangle), \dots (p_{l-1}, |\psi_{l-1}\rangle))$ *be a mixed state of the quantum system Q. Then*

$$\rho = \sum_{i=0}^{l-1} p_i |\psi_i\rangle\langle\psi_i| \tag{3.5.5}$$

is a density operator on the state space $\mathbb{H}$ of Q.

Proof. We know from Proposition B.5.25 that the trace is $\mathbb{C}$-linear. Therefore, we have

$$\text{tr}(\rho) = \sum_{i=0}^{l-1} p_i \, \text{tr} \, |\psi_i\rangle\langle\psi_i| \qquad \text{linearity of the trace,}$$

$$= \sum_{i=0}^{l-1} p_i\langle\psi_i|\psi_i\rangle \qquad \text{Proposition 2.4.27(6),}$$

$$= \sum_{i=0}^{l-1} p_i = 1 \qquad \text{Definition 3.5.3.}$$

This proves the trace condition. To show the positivity condition, we note that for all $|\xi\rangle \in \mathbb{H}$ we have

$$\langle\xi|\rho|\xi\rangle = \left\langle \xi \left| \sum_{i=0}^{l-1} p_i \, |\psi_i\rangle\langle\psi_i|\xi \right\rangle \right. \qquad \text{definition of } \rho,$$

$$= \sum_{i=0}^{l-1} p_i\langle\xi|\psi_i\rangle\langle\psi_i|\xi\rangle \qquad \text{linearity of the inner product,}$$

$$= \sum_{i=0}^{l-1} p_i|\langle\psi_i|\xi\rangle|^2 \geq 0 \qquad \text{conjugate symmetry of the inner product.}$$

This concludes the proof of the proposition. $\qquad\qquad\qquad\qquad\qquad\qquad$ $\square$

Proposition 3.5.4 justifies the following definition.

Definition 3.5.5. (1) The density operator of a mixed state

(3.5.6) $$S = (p_0, |\psi_0\rangle), \ldots, (p_{l-1}, |\psi_{l-1}\rangle)$$

of Q is defined as

(3.5.7) $$\rho_S = \sum_{i=0}^{l-1} p_i \, |\psi_i\rangle\langle\psi_i|.$$

(2) The density operator of a pure state $|\psi\rangle \in \mathbb{H}$ is defined as

(3.5.8) $$\rho_\psi = |\psi\rangle\langle\psi|.$$

Example 3.5.6. Consider the mixed state

(3.5.9) $$S = \left(\left(\frac{1}{2}, |0\rangle\right), \left(\frac{1}{2}, |1\rangle\right)\right).$$

The corresponding density operator is

(3.5.10) $$\rho_S = \frac{1}{2}\left(|0\rangle\langle0| + |1\rangle\langle1|\right).$$

We show that there is no pure state $|\psi\rangle$ such that $\rho = \rho_\psi$. Let $|\psi\rangle = \alpha_0 |0\rangle + \alpha_1 |1\rangle$ be a pure state where $\alpha_0, \alpha_1 \in \mathbb{C}$ such that $|\alpha_0|^2 + |\alpha_1|^2 = 1$. Its density operator is

(3.5.11) $$\rho_\psi = |\alpha_0|^2 |0\rangle\langle0| + \alpha_0\overline{\alpha_1} |0\rangle\langle1| + \alpha_1\overline{\alpha_0} |1\rangle\langle0| + |\alpha_1|^2 |1\rangle\langle1|.$$

Now $\rho = \rho_\psi$ implies $|\alpha_0|^2 = |\alpha_1|^2 = \frac{1}{2}$ and $\alpha_0 = 0$ or $\alpha_1 = 0$. But this cannot be true.

Exercise 3.5.7. Let $B = (|b_0\rangle, \ldots, |b_{k-1}\rangle)$ be an orthonormal basis of $\mathbb{H}$. Furthermore, consider the quantum state

$$(3.5.12) \qquad\qquad |\varphi\rangle = \sum_{i=0}^{k-1} \alpha_i |b_i\rangle$$

where $\alpha_i \in \mathbb{C}$ for all $i \in \mathbb{Z}_k$ such that $\sum_{i=0}^{k-1} |\alpha_i|^2 = 1$. Show that the density operators of the pure state $|\varphi\rangle$ and the mixed state $((|\alpha_0|^2, |b_0\rangle), \ldots, (|\alpha_{k-1}|^2, |b_{k-1}\rangle))$ are the same.

3.5.2. Correspondence between mixed states and density operators. This section explains the correspondence between mixed states and density operators. To begin, we present the following observation.

Proposition 3.5.8. *Every density operator on $\mathbb{H}$ is the density operator of some mixed state of the quantum system Q.*

Proof. Let ρ be a density operator on $\mathbb{H}$. Then ρ is Hermitian. It follows from Theorem 2.4.56 that we can write

$$(3.5.13) \qquad\qquad \rho = \sum_{i=0}^{k-1} \lambda_i |b_i\rangle\langle b_i|$$

where $B = (|b_0\rangle, \ldots, |b_{k-1}\rangle)$ is an orthonormal basis of eigenvectors of ρ and λ_i is the eigenvalue associated with the eigenvector $|b_i\rangle$ for all $i \in \mathbb{Z}_k$. It follows from the trace condition and Proposition 2.4.1 that

$$(3.5.14) \qquad\qquad 1 = \mathrm{tr}(\rho) = \sum_{i=0}^{k-1} \lambda_i.$$

The positivity condition and Proposition 2.4.65 imply that $\lambda_i \geq 0$ for $0 \leq i < k$. Hence,

$$((\lambda_0, |b_0\rangle), \ldots, (\lambda_{k-1}, |b_{k-1}\rangle))$$

is a mixed state of the quantum system with density operator ρ. $\qquad\square$

The proof of Proposition 3.5.8 contains a method for constructing a mixed state that corresponds to a given density operator. This is illustrated in the next example.

Example 3.5.9. Consider the operator

$$(3.5.15) \qquad\qquad \rho = \frac{1}{2} |0\rangle\langle 0| + \frac{1}{2} |1\rangle\langle 1|.$$

This is the representation of ρ as in (3.5.13). Also, ρ is positive semidefinite and has trace 1. Therefore, ρ is a density operator. The construction in the proof of Proposition 3.5.8 gives the mixed state

$$(3.5.16) \qquad\qquad \left(\left(\frac{1}{2}, |0\rangle \right), \left(\frac{1}{2}, |1\rangle \right) \right)$$

that we already know from Example 3.5.6. Its density operator is ρ.

Proposition 3.5.8 shows that the map that sends a mixed state to its density operator is surjective. However, in general, this map is not injective. The next proposition allows us to determine the mixed states that are associated with the same density operator.

Proposition 3.5.10. *Let $l \in \mathbb{N}$ and let*

(3.5.17) $$S = (|\varphi_0\rangle, \ldots, |\varphi_{l-1}\rangle), \quad T = (|\psi_0\rangle, \ldots, |\psi_{l-1}\rangle) \in \mathbb{H}^l.$$

Then we have

(3.5.18) $$\sum_{i=0}^{l-1} |\varphi_i\rangle \langle \varphi_i| = \sum_{i=0}^{l-1} |\psi_i\rangle \langle \psi_i|$$

if and only if there is a unitary matrix $U \in \mathbb{C}^{(l,l)}$ such that

(3.5.19) $$T = SU.$$

Proof. Let $U \in \mathbb{C}^{(l,l)}$ be a unitary matrix such that

(3.5.20) $$T = SU.$$

We write $U = (u_{i,j})$ and $U^* = (u^*_{i,j})$ with $u_{i,j}, u^*_{i,j} \in \mathbb{C}$ for $0 \leq i, j < l$. Then we have

(3.5.21) $$u^*_{i,j} = \overline{u_{j,i}}, \quad 0 \leq i, j < l.$$

Since U is unitary, we have $U^*U = I_l$. Hence, for $0 \leq m, n < l$ we have

(3.5.22) $$\sum_{i=0}^{l-1} \overline{u_{i,m}} u_{i,n} = \sum_{i=0}^{l-1} u^*_{m,i} u_{i,n} = \delta_{m,n}.$$

This identity and the rules in Proposition 2.4.27 imply

$$\sum_{i=0}^{l-1} |\psi_i\rangle \langle \psi_i|$$

$$= \sum_{i=0}^{l-1} \left| \sum_{m=0}^{l-1} u_{i,m} |\varphi_m\rangle \right\rangle \left\langle \sum_{n=0}^{l-1} u_{i,n} |\varphi_n\rangle \right|$$

$$= \sum_{i,m,n=0}^{l-1} \overline{u_{i,m}} u_{i,n} |\varphi_m\rangle \langle \varphi_n|$$

$$= \sum_{m,n=0}^{l-1} \left(\sum_{i=0}^{l-1} u^*_{m,i} u_{i,n} \right) |\varphi_m\rangle \langle \varphi_n|$$

$$= \sum_{m,n=0}^{l-1} \delta_{m,n} |\varphi_m\rangle \langle \varphi_n|$$

$$= \sum_{m=0}^{l-1} |\varphi_m\rangle \langle \varphi_m|.$$

Now, assume that the two operators are equal. Call them ρ. Since ρ is Hermitian it follows from the Spectral Theorem 2.4.56 that there is a decomposition

(3.5.23) $$\rho = \sum_{i=0}^{m-1} \lambda_i |b_i\rangle \langle b_i|$$

of ρ where $m \in \mathbb{N}$, $(|b_0\rangle, \ldots, |b_{m-1}\rangle)$ is an orthonormal sequence of eigenvectors of ρ, and $\lambda_i \in \mathbb{R}$ is a nonzero eigenvalue associated with $|b_i\rangle$ for all $i \in \mathbb{Z}_m$. Since ρ

is a density operator, these eigenvalues are positive. We show that the vector space $V = \mathrm{Span}(|b_0\rangle, \ldots, |b_{m-1}\rangle)$ is equal to the vector space $V' = \mathrm{Span}(|\varphi_0\rangle, \ldots, |\varphi_{l-1}\rangle)$. Since for all $j \in \mathbb{Z}_m$ we have

$$(3.5.24) \qquad \lambda_j |b_j\rangle = \rho |b_j\rangle = \sum_{i=0}^{l-1} \langle \varphi_i | b_j \rangle |\varphi_i\rangle$$

and $\lambda_j \neq 0$ it follows that

$$(3.5.25) \qquad V \subset V'.$$

Next, set $|\xi_i\rangle = \sqrt{\lambda_i} |b_i\rangle$ for all $i \in \mathbb{Z}_m$. Then we have

$$(3.5.26) \qquad \rho = \sum_{i=0}^{m-1} |\xi_i\rangle \langle \xi_i|$$

and $(|\xi_0\rangle, \ldots, |\xi_{m-1}\rangle)$ is an orthogonal basis of V. Let $|\psi\rangle \in V^\perp$. Then

$$(3.5.27) \qquad \langle \psi | \rho | \psi \rangle = \left\langle \psi \left| \sum_{i=0}^{m-1} \langle \xi_i | \psi \rangle |\xi_i\rangle \right. \right\rangle = \langle \psi | 0 \rangle = 0.$$

This implies

$$(3.5.28) \qquad \begin{aligned} 0 = \langle \psi | \rho | \psi \rangle &= \left\langle \psi \left| \sum_{i=0}^{l-1} \langle \varphi_i | \psi \rangle |\varphi_i\rangle \right. \right\rangle \\ &= \sum_{i=0}^{l-1} \langle \psi | \varphi_i \rangle \langle \varphi_i | \psi \rangle = \sum_{i=0}^{l-1} |\langle \varphi_i | \psi \rangle|^2. \end{aligned}$$

Hence, $|\langle \varphi_i | \psi \rangle|^2 = 0$ for $0 \leq i < l$ and all $|\psi\rangle \in V^\perp$. Therefore $|\varphi_i\rangle \in (V^\perp)^\perp = V$ for $0 \leq i < l$ which together with (3.5.25) implies

$$(3.5.29) \qquad V = V' \quad \text{and} \quad m \leq l.$$

So we can write

$$(3.5.30) \qquad |\varphi_j\rangle = \sum_{i=0}^{m-1} u_{i,j} |\xi_i\rangle$$

for all $j \in \mathbb{Z}_l$ with complex coefficients $u_{i,j}$. Denote the matrix $(u_{i,j}) \in \mathbb{C}^{(m,l)}$ by U. Also, denote the entries of the adjoint U^* of U by $u_{i,j}^*$. Now we have

$$(3.5.31) \qquad \begin{aligned} \sum_{p=0}^{m-1} |\xi_p\rangle \langle \xi_p| &= \sum_{p=0}^{l-1} |\varphi_p\rangle \langle \varphi_p| \\ &= \sum_{p=0}^{l-1} \left| \sum_{i=0}^{m-1} u_{i,p} |\xi_i\rangle \right\rangle \left\langle \sum_{j=0}^{m-1} u_{j,p} |\xi_j\rangle \right| \\ &= \sum_{i,j=0}^{m-1} \left(\sum_{p=0}^{l-1} u_{i,p} u_{p,j}^* \right) |\xi_i\rangle \langle \xi_j|. \end{aligned}$$

It follows from Proposition 2.4.31 that the sequence $(|\xi_p\rangle \langle \xi_p|)$ is linearly independent. Hence, (3.5.31) implies

$$(3.5.32) \qquad \sum_{p=0}^{l-1} u_{i,p} u_{p,j}^* = \delta_{i,j}$$

for $0 \le i, j < m$. So the sequence of row vectors of U is orthonormal and it follows from Theorem 2.2.33 that we can add $l - m$ rows to U such that the new U is a unitary matrix. Recall that by (3.5.29) we have $m \le l$. We set $|\xi_m\rangle, \ldots, |\xi_{l-1}\rangle = 0$. Then we have

$$(3.5.33) \qquad (|\varphi_0\rangle, \ldots, |\varphi_{l-1}\rangle) = (|\xi_0\rangle, \ldots, |\xi_{l-1}\rangle)U.$$

In the same way, we can show that there is a unitary matrix $U' \in \mathbb{C}^{(l,l)}$ such that

$$(3.5.34) \qquad (|\psi_0\rangle, \ldots, |\psi_{l-1}\rangle) = (|\xi_0\rangle, \ldots, |\xi_{l-1}\rangle)U'.$$

So we obtain

$$(3.5.35) \qquad (|\psi_0\rangle, \ldots, |\psi_{l-1}\rangle) = (|\varphi_0\rangle, \ldots, |\varphi_{l-1}\rangle)U^*U'.$$

Since U^*U' is a unitary matrix, this completes the proof. $\qquad \square$

From Proposition 3.5.10 we obtain the following theorem.

Theorem 3.5.11. (1) *The density operators of two pure states of Q are the same if and only if these states are equal up to a global phase factor.*

(2) *Let $l \in \mathbb{N}$. The density operators of two mixed states*

$$(3.5.36) \qquad ((p_0, |\varphi_0\rangle), \ldots, (p_{l-1}, |\varphi_{l-1}\rangle)), \quad ((q_0, |\psi_0\rangle), \ldots, (q_{l-1}, |\psi_{l-1}\rangle))$$

of Q are the same if and only if there is a unitary matrix $U \in \mathbb{C}^{(l,l)}$ such that

$$(3.5.37) \qquad (\sqrt{p_0}\,|\varphi_0\rangle), \ldots, \sqrt{p_{l-1}}\,|\varphi_{l-1}\rangle) = (\sqrt{q_0}\,|\psi_0\rangle), \ldots, \sqrt{q_{l-1}}\,|\psi_{l-1}\rangle)U.$$

Proof. To prove the first assertion, we let $|\varphi\rangle$ and $|\psi\rangle$ be pure states of Q. The density operators of these states are the density operators of the mixed states $((1, |\varphi\rangle))$ and $((1, |\psi\rangle))$, respectively. Therefore, it follows from Proposition 3.5.10 that the density operators of these states are equal if and only if there is a complex number u of norm 1 such that $|\psi\rangle = u\,|\varphi\rangle$. This proves the first assertion. The second assertion follows immediately from Proposition 3.5.10. $\qquad \square$

Theorem 3.5.11 can also be used to characterize the mixed states of different lengths that correspond to the same density operator. As is shown in Exercise 3.5.12 we can extend any mixed state of length l to a mixed state of length $k > l$ with the same density operator by appending $k - l$ pairs $(0, 0)$ to it.

Exercise 3.5.12. Show that appending pairs $(0, 0)$ to a mixed state gives a mixed state with the same density operator.

We generalize the equivalence relation introduced in Proposition 3.1.23.

Theorem 3.5.13. *The set R of all pairs of mixed states of Q with the same density operator is an equivalence relation on the set of all mixed states of Q.*

Exercise 3.5.14. Prove Theorem 3.5.13.

The next theorem gives a criterion that allows one to distinguish between density operators of pure states and mixed states.

Theorem 3.5.15. *Let ρ be a density operator on $\mathbb{H}$. Then the following statements hold.*

(1) *ρ is the density operator of a pure state if and only if $\rho^2 = \rho$, which is true if and only if $\operatorname{tr} \rho^2 = 1$.*

(2) *ρ is not the density operator of a pure state if and only if $\rho^2 \neq \rho$, which is true if and only if $\operatorname{tr} \rho^2 < 1$.*

Proof. First, we show that $\operatorname{tr} \rho^2 \leq 1$. Since ρ is positive semidefinite, Proposition 2.4.64 tells us that ρ is Hermitian. By Theorem 2.4.56 we can write

$$(3.5.38) \qquad \rho = \sum_{i=0}^{k-1} \lambda_i |b_i\rangle \langle b_i|$$

where $(|b_0\rangle, \ldots, |b_{k-1}\rangle)$ is an orthonormal basis of eigenvectors of ρ and λ_i is the eigenvalue associated with $|b_i\rangle$ which is a real number for all $i \in \mathbb{Z}_k$. Proposition 2.4.59 implies

$$(3.5.39) \qquad \rho^2 = \sum_{i=0}^{k-1} \lambda_i^2 |b_i\rangle \langle b_i|.$$

The trace and positivity conditions imply

$$(3.5.40) \qquad \operatorname{tr} \rho^2 = \sum_{i=0}^{k-1} \lambda_i^2 \leq \left(\sum_{i=0}^{k-1} \lambda_i \right)^2 = (\operatorname{tr} \rho)^2 = 1.$$

We now prove the first assertion of the theorem. Let $\rho = |\varphi\rangle \langle \varphi|$ with a quantum state $|\varphi\rangle \in \mathbb{H}$. Since $\langle \varphi | \varphi \rangle = 1$, it follows that ρ is a projection; that is, $\rho^2 = \rho$. Now assume that $\rho^2 = \rho$. Then the trace condition implies $\operatorname{tr} \rho^2 = \operatorname{tr} \rho = 1$. Finally, let $\operatorname{tr} \rho^2 = 1$. Then it follows from (3.5.40) that there is $l \in \mathbb{Z}_k$ with $|\lambda_l| = 1$ and $\lambda_i = 0$ for $i \neq l$. Therefore, $\rho = |b_l\rangle \langle b_l|$. The second assertion is proved in Exercise 3.5.16. $\qquad \square$

Exercise 3.5.16. Prove the second assertion of Theorem 3.5.15.

Example 3.5.17. Consider the density operator ρ from Example 3.5.9. Since

$$(3.5.41) \qquad \rho^2 = \frac{1}{4}(|0\rangle \langle 0| + |1\rangle \langle 1|) \neq \rho$$

it follows from Theorem 3.5.15 that ρ is not the density operator of a pure state. This we have already directly verified in Example 3.5.6.

Note that for density operators ρ with $\rho^2 = \rho$ or $\operatorname{tr} \rho^2 = 1$, the proof of Theorem 3.5.15 contains a method for determining a pure state $|\varphi\rangle$ such that $\rho = |\varphi\rangle \langle \varphi|$.

3.6. The quantum postulates for mixed states

The set of mixed states of a quantum system Q can be viewed as a superset of the set of all pure states of Q if we identify a pure state $|\varphi\rangle$ with the mixed state $(1, |\varphi\rangle)$. In this section, we generalize the postulates of quantum mechanics to mixed states.

3.6.1. State Space Postulate.
We begin with the generalized State Space Postulate.

Postulate 3.6.1 (State Space Postulate — density operator version). Associated with any physical system is a Hilbert space, called the state space of the system. The system is completely described by a density operator on the state space.

By Theorem 3.5.11, modeling pure quantum states using density operators is coarser than modeling them using state vectors. This theorem tells us that the density operators of two state vectors are equal as long as they are equal up to a global phase factor. However, since by Theorem 3.4.4 all state vectors that are equal up to a global phase factor behave the same in measurements, this is of no importance.

The description of quantum systems using density operators includes mixed states. This becomes essential in scenarios like when a component of a composite system is discarded, leaving only the remaining system to be described.

The Composite Systems Postulate for density operators is the following.

Postulate 3.6.2 (Composite Systems Postulate — density operator version). The state space of the composition of finitely many physical systems is the tensor product of the state spaces of the component physical systems. Moreover, if we have systems numbered 0 through $m-1$ and if system i is in the state ρ_i where ρ_i is a density operator on the state space of the ith component system for $0 \leq i < k$, then the composite system is in the state $\rho_0 \otimes \cdots \otimes \rho_{m-1}$.

3.6.2. Evolution Postulate.
We also present a density operator analog of the Evolution Postulate 3.3.1. The postulate for pure states tells us that the evolution of a closed quantum system is described by unitary transformations. Let U be a unitary transformation on a state space $\mathbb{H}$ that describes an evolution of the quantum system. Then after this evolution, a state vector $|\psi\rangle$ becomes $U|\psi\rangle$. The corresponding density operator is

$$(3.6.1) \qquad |U|\psi\rangle\rangle\langle U|\psi\rangle| = U|\psi\rangle\langle\psi|U^*.$$

This motivates the following modified Evolution Postulate.

Postulate 3.6.3 (Evolution Postulate — density operator version). The evolution of a quantum system with state space $\mathbb{H}$ is described by a unitary transformation on $\mathbb{H}$. More precisely, let $t, t' \in \mathbb{R}$, $t < t'$. Assume that the state of the system at time t is described by the density operator ρ on $\mathbb{H}$. Then the state of the system at time t' is obtained from ρ as $\rho' = U\rho U^*$ where U is a unitary operator on $\mathbb{H}$ that only depends on t and t'.

Analogous to what we have described in Section 3.3.3, we can use composite unitary operators to describe the time evolution of composite quantum systems that are in a mixed state.

Exercise 3.6.4. Suppose that at time t a 2-qubit quantum register is in the state $\rho = |00\rangle\langle00|$ and that the state of the system at time $t' > t$ is obtained from ρ by applying the CNOT operator. Determine the density operator that describes the state of the system at time t'.

3.6.3. Measurement Postulate.
Next, we adapt the Measurement Postulate 3.4.1 to density operators. First, we motivate this modification.

Let O be an observable of a quantum system and let $O = \sum_\lambda \lambda P_\lambda$ be its spectral decomposition. Assume that the quantum system is in the state $|\varphi\rangle$. Then the corresponding density operator is

$$(3.6.2) \qquad\qquad \rho = |\varphi\rangle\langle\varphi|.$$

The Measurement Postulate tells us that measuring the observable O when the quantum system is in the state $|\varphi\rangle$ gives one of the eigenvalues λ of O with probability $\Pr(\lambda) = \langle\varphi|P_\lambda|\varphi\rangle$. From Proposition 2.4.29 we obtain

$$(3.6.3) \qquad\qquad \Pr(\lambda) = \langle\varphi|P_\lambda|\varphi\rangle = \operatorname{tr}(P_\lambda |\varphi\rangle\langle\varphi|).$$

Furthermore, if the result of the measurement is λ, then the state of the system immediately after the measurement is

$$(3.6.4) \qquad\qquad \frac{P_\lambda |\psi\rangle}{\sqrt{\Pr(\lambda)}}.$$

By Proposition 2.4.29 the density operator of this state is

$$(3.6.5) \qquad\qquad \frac{P_\lambda |\varphi\rangle\langle\varphi| P_\lambda}{\Pr(\lambda)} = \frac{P_\lambda\rho P_\lambda}{\Pr(\lambda)}.$$

This motivates the following modified Measurement Postulate.

Postulate 3.6.5 (Measurement Postulate — density operator version). A projective measurement is described by an observable O that is a Hermitian operator on the state space of the system being observed. Let $O = \sum_{\lambda \in \Lambda} \lambda P_\lambda$ be the spectral decomposition of O. The possible outcomes of the measurement are the eigenvalues of the observable. When measuring the state ρ the probability of getting the result corresponding to λ is $\Pr(\lambda) = \operatorname{tr}(P_\lambda\rho)$. If this outcome occurs, the state immediately after the measurement is $\frac{P_\lambda\rho P_\lambda}{\Pr(\lambda)}$.

In the situation of the Measurement Postulate 3.6.5, the *expectation value* of the random variable that sends a measurement outcome to the corresponding eigenvalue is

$$(3.6.6) \qquad \sum_{\lambda \in \Lambda} \lambda\Pr(\lambda) = \sum_{\lambda \in \Lambda} \lambda \operatorname{tr}(P_\lambda\rho) = \operatorname{tr}\left\langle \left(\sum_{\lambda \in \Lambda} \lambda P_\lambda\right)\rho\right| = \operatorname{tr}(O\rho).$$

This motivates the following definition.

Definition 3.6.6. Let O be an observable of a quantum system with state space $\mathbb{H}$. Suppose that we measure this observable when the system is in a state described by the density operator ρ. Then the *expectation value of this measurement* is defined as $\text{tr}(O\rho)$.

The following proposition applies the Measurement Postulate for density operators to explain what happens when mixed states are measured in an orthonormal basis.

Proposition 3.6.7. *Suppose that we measure the quantum system in the orthonormal basis $B = (|b_0\rangle, \ldots, |b_{k-1}\rangle)$ when it is in the mixed state. Then measuring the observable $\sum_{\lambda=0}^{k-1} \lambda |\lambda_j\rangle \langle \lambda_j|$ gives $\lambda \in \mathbb{Z}_k$ with probability $\Pr(\lambda) = \sum_{i=0}^{l-1} p_i|\langle b_\lambda|\varphi_i\rangle|^2$. Immediately after this measurement, the quantum system is in the state $|b_\lambda\rangle \langle b_\lambda|$.*

Proof. The density operator corresponding to the mixed state of the quantum system is

$$(3.6.7) \qquad \rho = \sum_{i=0}^{l-1} p_i |\varphi_i\rangle \langle \varphi_i|.$$

For all $\lambda \in \mathbb{Z}_k$ we have

$$(3.6.8) \qquad P_\lambda \rho = P_\lambda \sum_{i=0}^{l-1} p_i |\varphi_i\rangle \langle \varphi_i| = \sum_{i=0}^{l-1} p_i \langle b_\lambda|\varphi_i\rangle |b_\lambda\rangle \langle \varphi_i|.$$

So it follows from Proposition 2.4.27 that measuring O gives $\lambda \in \mathbb{Z}_k$ with probability

$$(3.6.9) \qquad \Pr(\lambda) = \text{tr}(P_\lambda \rho) = \sum_{i=0}^{l-1} p_i|\langle b_\lambda|\varphi_i\rangle|^2.$$

When the measurement outcome is λ, then immediately after the measurement the quantum system is in the state

$$(3.6.10) \qquad \frac{P_\lambda \rho P_\lambda}{\Pr(\lambda)} = \frac{\sum_{i=0}^{l-1} p_i |b_\lambda\rangle \langle b_\lambda|\varphi_i\rangle \langle \varphi_i|b_\lambda\rangle \langle b_\lambda|}{\Pr(\lambda)}$$
$$= \frac{P_\lambda \sum_{i=0}^{l-1} p_i|\langle b_\lambda|\varphi_i\rangle|^2}{\Pr(\lambda)} = P_\lambda. \qquad \square$$

The concepts and results regarding partial measurements in Section 3.4.3 carry over to the mixed state situation. We only need to replace the formulas for the measurement probabilities and for the states immediately after measurement with the formulas that hold for mixed states. In Exercise 3.6.8 this is done for quantum systems that are the composition of two quantum systems.

Exercise 3.6.8. Assume that A and B are quantum systems with state spaces $\mathbb{H}_A$ and $\mathbb{H}_B$. Consider the composite quantum system A with state space $\mathbb{H}_{AB} = \mathbb{H}_A \otimes \mathbb{H}_B$. Let O_A be an observable of system A with spectral decomposition $O_A = \sum_{\lambda \in \Lambda} \lambda P_\lambda$. Also, let ρ_A, ρ_B be states of the systems A and B, respectively. Prove that the following hold.

(1) The composed operator $O_{AB} = O_A \otimes I_B$ is an observable of the composite system AB with spectral decomposition

$$(3.6.11) \qquad O_{AB} = \sum_{\lambda \in \Lambda} \lambda (P_\lambda \otimes I_B).$$

(2) Measuring O_{AB} when AB is in the nonentangled state $\rho_A \otimes \rho_B$ we obtain the eigenvalue λ with probability $\Pr(\lambda) = \mathrm{tr}(O_A \rho_A)$. If this outcome occurs, the state of system AB immediately after the measurement is $\frac{(P_\lambda \rho_A P_\lambda) \otimes \rho_B}{\mathrm{tr}(P_\lambda \rho_A)}$. Also, the expectation of O_{AB} is $\mathrm{tr}(O_A \rho_A)$.

3.6.4. The descriptions by state vectors and density operators are equivalent.
We have introduced the description of the states of quantum systems using state vectors and density operators. Modeling quantum systems using density operators is more general since it also covers the situation where a quantum system is described by a mixed state. However, when we restrict our attention to pure states, the two descriptions are equivalent. This means the following. Consider a quantum system with state space $\mathbb{H}$. Suppose that it evolves in k steps. Initially, it is in the pure state s_0, then in the pure state s_1, etc., until it is finally in the pure state s_l. These states can be described by state vectors or density operators. In both versions of the State Space Postulate, each transition is associated with a unitary operator on $\mathbb{H}$. So let $U_0, U_1, \ldots, U_{l-1}$ be a sequence of unitary operators on $\mathbb{H}$ such that state s_{i+1} is obtained from state s_i using U_i for $0 \le i < l$. Assume that the states s_i are represented by state vectors $|\varphi_i\rangle \in \mathbb{H}$ for $0 \le i < l$. Then we have

$$(3.6.12) \qquad |\varphi_k\rangle = U |\varphi_0\rangle$$

with

$$(3.6.13) \qquad U = U_{l-1} \cdots U_0.$$

Next, assume that the state s_0 is represented by the density operator

$$(3.6.14) \qquad \rho_0 = |\varphi_0\rangle\langle\varphi_0|.$$

Also, for $0 \le i < l$ set

$$(3.6.15) \qquad \rho_{i+1} = U_i \rho_i U_i^*.$$

Then from (3.6.1) we obtain

$$(3.6.16) \qquad \rho_l = U \rho_0 U^* = |\varphi_l\rangle\langle\varphi_l|.$$

As shown in Section 3.6.3 the measurement statistics and the quantum state immediately after the measurement are the same for the state described by $|\varphi_l\rangle$ and ρ_l. This shows that from the perspective of quantum mechanics, the two descriptions of the state evolution are equivalent.

3.7. Partial trace and reduced density operators

In Section B.9.4 we have introduced the partial trace. In this section, we use it to model states of subsystems of composite physical systems.

We consider two quantum systems A and B with state spaces $\mathbb{H}_A$ and $\mathbb{H}_B$ of dimensions M and N, respectively. Without loss of generality, we assume that $M \le N$.

Let $(|a_0\rangle, \ldots, |a_{M-1}\rangle)$ be an orthonormal basis of $\mathbb{H}_A$ and let $(|b_0\rangle, \ldots, |b_{N-1}\rangle)$ be an orthonormal basis of $\mathbb{H}_B$. We also denote the composition of A and B by AB. Its state space is $\mathbb{H}_{AB} = \mathbb{H}_A \otimes \mathbb{H}_B$ and $(|a_i\rangle |b_j\rangle)$ is an orthonormal basis of $\mathbb{H}$. Furthermore, $(|a_i\rangle \langle a_j|)$ and $(|b_k\rangle \langle b_l|)$ are orthonormal bases of $\mathrm{End}(\mathbb{H}_A)$ and $\mathrm{End}(\mathbb{H}_B)$. Hence $(|a_i\rangle \langle a_j| \otimes |b_k\rangle \langle b_l|)$ is an orthonormal basis of $\mathrm{End}(\mathbb{H}_{AB})$.

3.7.1. Partial trace on $\mathbb{H}_{AB}$. In this section, we discuss the partial trace on $\mathbb{H}_{AB}$ over $\mathbb{H}_B$. We refer to it as the *partial trace over B* and denote it by tr_B. The definition of the partial trace tr_A over A and the corresponding terminology are analogous.

From the definition of the partial trace in Section B.9.4 we obtain the following formula.

Proposition 3.7.1. *Let* $U \in \mathrm{End}(\mathbb{H}_{AB})$,

$$(3.7.1) \qquad U = \sum_{i,j \in \mathbb{Z}_M, k,l \in \mathbb{Z}_N} u_{i,j,k,l} |a_i\rangle \langle a_j| \otimes |b_k\rangle \langle b_l|$$

with complex coefficients $u_{i,j,k,l}$. *Then*

$$(3.7.2) \qquad \mathrm{tr}_B U = \sum_{k \in \mathbb{Z}_N} U_k$$

where

$$(3.7.3) \qquad U_k = \sum_{i,j \in \mathbb{Z}_M} u_{i,j,k,k} |a_i\rangle \langle a_j|.$$

Exercise 3.7.2. Prove Proposition 3.7.1

Example 3.7.3. Consider

$$(3.7.4) \qquad U = |x_+\rangle \langle x_+| \otimes |0\rangle \langle 0|.$$

It follows from Proposition 3.7.1 that

$$(3.7.5) \qquad \mathrm{tr}_B U = |x_+\rangle \langle x_+|.$$

The next proposition shows how to determine the trace over B of a projection $|\varphi\rangle \langle \varphi|$ for $|\varphi\rangle \in \mathbb{H}_{AB}$ from the Schmidt decomposition of $|\varphi\rangle$.

Proposition 3.7.4. *Let* $\varphi \in \mathbb{H}_{AB}$ *and let*

$$(3.7.6) \qquad |\varphi\rangle = \sum_{i=0}^{l-1} r_i |\psi_i\rangle |\xi_i\rangle$$

be a Schmidt decomposition of φ *as described in Theorem 2.5.18. Then we have*

$$(3.7.7) \qquad \mathrm{tr}_B(|\varphi\rangle \langle \varphi|) = \sum_{i=0}^{l-1} r_i^2 |\psi_i\rangle \langle \psi_i|.$$

Proof. We have

$$\mathrm{tr}_B\,|\varphi\rangle\langle\varphi| = \mathrm{tr}_B\left(\sum_{i,j=0}^{l-1} r_i^2\,|\psi_i\rangle\,|\xi_i\rangle\langle\psi_j|\,\langle\xi_j|\right)$$

$$= \sum_{i,j=0}^{l-1} r_i^2\,\mathrm{tr}_B(|\psi_i\rangle\langle\psi_j|\otimes|\xi_i\rangle\langle\xi_j|)$$

(3.7.8)
$$= \sum_{i,j=0}^{l-1} r_i^2\,|\psi_i\rangle\langle\psi_j|\,\mathrm{tr}(|\xi_i\rangle\langle\xi_j|)$$

$$= \sum_{i,j=0}^{l-1} r_i^2\,|\psi_i\rangle\langle\psi_j|\,\delta_{i,j}$$

$$= \sum_{i=0}^{l-1} r_i^2\,|\psi_i\rangle\langle\psi_i|. \qquad\qquad\square$$

Next, we prove that the partial trace preserves positive semidefiniteness.

Proposition 3.7.5. *If $U \in \mathrm{End}(\mathbb{H}_{AB})$ is positive semidefinite, then $\mathrm{tr}_B(U)$ is positive semidefinite.*

Proof. Let $U \in \mathrm{End}(\mathbb{H}_{AB})$ be positive semidefinite. Use the representation (3.7.1) of U. For $k, l \in \mathbb{Z}_N$ let

(3.7.9)
$$U_{k,l} = \sum_{i,j\in\mathbb{Z}_M} u_{i,j,k,l}\,|a_i\rangle\langle a_j|.$$

Then we have

(3.7.10)
$$U = \sum_{k,l\in\mathbb{Z}_N} U_{k,l}\otimes|b_k\rangle\langle b_l|.$$

Let $|\varphi\rangle \in \mathbb{H}_A$. For $x \in \mathbb{Z}_N$ set

(3.7.11)
$$|\varphi_x\rangle = |\varphi\rangle\otimes|b_x\rangle.$$

Then (3.7.11), U being unitary, and Proposition 3.7.1 imply

(3.7.12)
$$0 \le \sum_{x\in\mathbb{Z}_N}\langle\varphi_x|U|\varphi_x\rangle = \sum_{x\in\mathbb{Z}_N}\sum_{k,l\in\mathbb{Z}_N}\langle\varphi|U_{k,l}|\varphi\rangle\langle b_x|b_k\rangle\langle b_l|b_x\rangle$$

$$= \sum_{x\in\mathbb{Z}_N}\langle\varphi|U_{x,x}|\varphi\rangle = \langle\varphi|\,\mathrm{tr}_B(U)|\varphi\rangle. \qquad\square$$

3.7.2. Tracing out subsystems. In this section, we study the following question. Suppose that the state of a composite quantum system AB at time t is ρ and we discard subsystem B at this time. Discarding component B refers to intentionally disregarding the quantum state of this part of the composite system while focusing solely on the remaining part A. What is the state of A after time t? We will show that it must be the partial trace $\mathrm{tr}_B \rho$ of ρ over B. In the whole section, states of quantum systems are described by density operators.

We start with the following observation.

Proposition 3.7.6. *Let ρ be a density operator on $\mathbb{H}_{AB}$. Then $\mathrm{tr}_B(\rho)$ is a density operator on $\mathbb{H}_A$.*

Proof. We must show that $\mathrm{tr}_B(\rho)$ satisfies the trace condition and the positivity condition. As a density operator, ρ satisfies these conditions. Since by Proposition B.9.25 the partial trace is trace-preserving, we have $\mathrm{tr}(\mathrm{tr}_B(\rho)) = \mathrm{tr}(\rho) = 1$. Therefore, $\mathrm{tr}_B(\rho)$ satisfies the trace condition. Also, $\mathrm{tr}_B(\rho)$ satisfies the positivity condition by Proposition 3.7.5. $\qquad\square$

The previous proposition justifies the next definition.

Definition 3.7.7. If ρ is a density operator on $\mathbb{H}_{AB}$, then $\mathrm{tr}_B(\rho)$ is called the *reduced density operator* of ρ on the subsystem A. This operator is denoted by ρ^A.

Our next objective is to demonstrate that by discarding subsystem B subsystem A assumes the state ρ^A. The argument proceeds as follows: Let O_A be an observable of subsystem A, and let us assume that the composite system is in the state ρ. There are two approaches to measuring O_A. We can either measure the composite observable $O_{AB} = O_A \otimes I_B$ for system AB, where I_B denotes the identity operator on the state space of subsystem B, or we can discard subsystem B and measure O_A only for the subsystem A. The measurement statistics in both scenarios must be identical. Specifically, the expectation values of both measurements must be equal. The following theorem states that the equality of these expectation values implies $\rho_A = \rho^A$. Consequently, after discarding system B, subsystem A assumes the state ρ^A.

Theorem 3.7.8. (1) *Let O_A be an observable of system A, let $O_{AB} = O_A \otimes I_B$, and assume that the state of the quantum system is ρ. Then the expectation value of O_{AB} is the same as the expectation value of O_A when system A is in the reduced state ρ^A; i.e.,*

$$\text{(3.7.13)} \qquad\qquad \mathrm{tr}(O_{AB}\rho) = \mathrm{tr}(O_A \rho^A).$$

(2) *The function*

$$\text{(3.7.14)} \qquad\qquad \mathrm{End}(\mathbb{H}_{AB}) \to \mathrm{End}(\mathbb{H}_A), \quad \rho \mapsto \rho^A = \mathrm{tr}_B(\rho)$$

is the only linear map that satisfies (3.7.13) for all observables O_A of A and all states ρ of AB.

Proof. Let (S_i) and (T_j) be bases of $\mathrm{End}(\mathbb{H}_A)$ and $\mathrm{End}(\mathbb{H}_B)$, respectively. Then $(S_i \otimes T_j)$ is a basis of $\mathrm{End}(\mathbb{H}_{AB})$. Since the trace is a linear map, it suffices to prove (3.7.13) for the basis elements $S_i \otimes T_j$. So, let $i \in M^2$, $j \in N^2$, and $\rho = S_i \otimes T_j$. We use the fact that the partial trace is trace-preserving (see Proposition B.9.25) and obtain

$$
\begin{aligned}
\mathrm{tr}(O_{AB}\rho) &= \mathrm{tr}\big((O_A \otimes I_B)(S_i \otimes T_j)\big) = \mathrm{tr}(O_A S_i \otimes T_j) = \mathrm{tr}(\mathrm{tr}_B(O_A S_i \otimes T_j)) \\
&= \mathrm{tr}(O_A S_i \, \mathrm{tr}\, T_j) \\
&= \mathrm{tr}(O_A \rho^A).
\end{aligned}
$$

To prove the second assertion, consider a linear map $f : \text{End}(\mathbb{H}_{AB}) \to \text{End}(\mathbb{H}_A)$ satisfying

$$(3.7.15) \qquad\qquad \text{tr}(O_{AB}\rho) = \text{tr}(O_A f(\rho))$$

for all observables O_A of A, $O_{AB} = O_A \otimes I_B$, and all states ρ of system AB. We show that this map is the partial trace. Denote by K the dimension of the linear space of all Hermitian operators on $\mathbb{H}_A$, let $(S_i)_{0 \le i < K}$ be an orthonormal basis of this space with respect to the Hilbert-Schmidt inner product, and let $\rho \in \text{End}(\mathbb{H}_{AB})$. Then expanding $f(\rho)$ in this basis and noting that the basis elements S_i are observables of system A, we obtain from (3.7.15)

$$(3.7.16) \qquad\qquad f(\rho) = \sum_{i=0}^{K-1} \text{tr}(S_i^* f(\rho)) S_i = \sum_{i=0}^{K-1} \text{tr}(S_i^* \otimes I_B \rho) S_i.$$

The expression on the right side of (3.7.16) is independent of f. Hence, f must be equal to tr_B since by the first assertion $f = \text{tr}_B$ satisfies (3.7.15). $\qquad\square$

As explained above, Theorem 3.7.8 tells us the following. Suppose that AB is a quantum system that is composed of the two quantum systems A and B. Assume that AB is in the state ρ. We discard system B and only keep system A. Then A is in the state $\rho^A = \text{tr}_B \rho$. Therefore, we refer to the process of discarding system B as *tracing out system B*.

Example 3.7.9. We determine the state of the first qubit of the Bell state

$$(3.7.17) \qquad\qquad |\psi\rangle = \frac{1}{\sqrt{2}}(|00\rangle + |11\rangle)$$

when the second qubit is discarded. The density operator of $|\psi\rangle$ is

$$\rho = |\psi\rangle\langle\psi|$$

$$(3.7.18) \qquad\qquad = \frac{1}{2}(|00\rangle + |11\rangle)(\langle 00| + \langle 11|)$$

$$= \frac{1}{2}(|00\rangle\langle 00| + |00\rangle\langle 11| + |11\rangle\langle 00| + |11\rangle\langle 11|).$$

We note that

$$(3.7.19) \qquad\qquad \text{tr}_B(|ik\rangle\langle jl|) = |i\rangle\langle j| \, \delta_{kl}.$$

We trace out the second qubit to obtain the reduced density operator of the first qubit. From (3.7.19) we obtain

$$\rho^A = \text{tr}_B(\rho)$$

$$(3.7.20) \qquad = \frac{1}{2}(\text{tr}_B(|00\rangle\langle 00|) + \text{tr}_B(|00\rangle\langle 11|) + \text{tr}_B(|11\rangle\langle 00|) + \text{tr}_B(|11\rangle\langle 11|))$$

$$= \frac{1}{2}(|0\rangle\langle 0| + |1\rangle\langle 1|).$$

This is the density operator of the mixed state $((\frac{1}{2}, |0\rangle), (\frac{1}{2}, |1\rangle))$.

It follows from Theorem 3.5.15 that the density operator in (3.7.20) is not the density operator of a pure state because the trace of its square is $1/2$.

The next proposition generalizes Example 3.7.9.

Proposition 3.7.10. *Let $l \in \mathbb{N}$ and for $0 \le i < l$ let $|\varphi_i\rangle$ and $|\psi_i\rangle$ be quantum states in $\mathbb{H}_A$ and $\mathbb{H}_B$, respectively, such that the states $|\psi_i\rangle$ are orthogonal to each other. Also, let ρ be the density operator of the state*

$$(3.7.21) \qquad |\xi\rangle = \frac{1}{\sqrt{l}} \sum_{i=0}^{l-1} |\varphi_i\rangle |\psi_i\rangle.$$

Then ρ^A is the density operator of the mixed state

$$(3.7.22) \qquad \left(\left(\frac{1}{l}, |\varphi_0\rangle \right), \ldots, \left(\frac{1}{l}, |\varphi_{l-1}\rangle \right) \right).$$

In other words, if the composite system AB is in the state $\rho = |\xi\rangle\langle\xi|$, then the state of system A after tracing out system B can be described by the mixed state (3.7.22).

Exercise 3.7.11. Prove Proposition 3.7.10.

From Proposition 3.7.10 we obtain the following consequence.

Corollary 3.7.12. *Assume that the composite system AB is in the state $\rho = |\xi\rangle\langle\xi|$ where $|\xi\rangle = |\varphi\rangle|\psi\rangle$ with $|\varphi\rangle \in \mathbb{H}_A$ and $|\psi\rangle \in \mathbb{H}_B$. Then $\rho^A = |\varphi\rangle\langle\varphi|$. This means that the state of system A after tracing out system B can be described by the state vector $|\varphi\rangle$.*

Now we characterize the states of composite systems whose partial trace is not a pure state.

Theorem 3.7.13. *Let $|\varphi\rangle$ be the state of the composite system AB and let $\rho = |\varphi\rangle\langle\varphi|$ be its density operator. Then $|\varphi\rangle$ is entangled with respect to the decomposition of AB into the subsystems A and B if and only if the reduced density operator ρ^A is not the density operator of a pure state.*

Proof. Let

$$(3.7.23) \qquad |\varphi\rangle = \sum_{i=0}^{s-1} r_i |\psi_i\rangle |\xi_i\rangle$$

be a Schmidt decomposition of $|\varphi\rangle$. Then, as shown in Exercise 3.7.14, we have

$$(3.7.24) \qquad \rho = \sum_{i=0}^{s-1} r_i^2 |\psi_i\rangle |\xi_i\rangle \langle\psi_i| \langle\xi_i|.$$

Hence, we have

$$(3.7.25) \qquad \operatorname{tr} \rho = \sum_{i=0}^{s-1} r_i^2.$$

By Proposition 3.7.4 we have

$$(3.7.26) \qquad \rho^A = \sum_{i=0}^{s-1} r_i^2 |\psi_i\rangle\langle\psi_i|.$$

So it follows from Proposition 2.4.59 that

$$(3.7.27) \qquad (\rho^A)^2 = \sum_{i=0}^{s-1} r_i^4 |\psi_i\rangle\langle\psi_i|.$$

Assume that $|\varphi\rangle$ is entangled. Then by Theorem 3.2.4 we have $s > 1$. Since the Schmidt coefficients r_i are positive real numbers, the trace condition for ρ and (3.7.25) imply that $0 < r_i < 1$ for $0 \leq i < s$. So, by (3.7.26) and (3.7.27) we have $\rho^A \neq (\rho^A)^2$ and it follows from Theorem 3.5.15 that ρ^A is not the density operator of a pure state.

Conversely, suppose that $|\varphi\rangle$ is separable. Then Theorem 3.2.4 implies that $s = 1$. It follows from the trace condition and (3.7.25) that $r_0 = 1$. So (3.7.26) and (3.7.27) imply that $\rho^A = (\rho^A)^2$. Theorem 3.5.15 shows that ρ^A is the density operator of a pure state. $\qquad\square$

Exercise 3.7.14. Verify (3.7.24).

Chapter 4

The Theory of Quantum Algorithms

In this chapter, we introduce fundamental facets of quantum algorithms. Building upon the concepts previously introduced in Chapter 1, our journey commences by shedding light on the pivotal building blocks that constitute quantum circuits: single-qubit operators. Initially, the chapter introduces the Pauli and Hadamard gates, unraveling their properties. It then delves into the theory of rotations in real three-space, revealing that all single-qubit operators can essentially be viewed as such rotations. This insight leads to the derivation of decomposition theorems for single-qubit operators, drawing parallels from the decomposition of three-dimensional rotations.

However, to construct general quantum circuits, single-qubit operators alone are insufficient. Therefore, the chapter explores multiple-qubit operators, specifically controlled operators. It covers a spectrum ranging from simple controlled-NOT operators to the intricacies of general controlled operators. The chapter also acquaints readers with ancillary and erasure gates, instrumental in adding and erasing quantum bits. By utilizing these components, the groundwork for defining general quantum circuits is laid. Furthermore, drawing upon analogous results in classical reversible circuits, the chapter demonstrates that every Boolean function can be implemented by a quantum circuit.

An intriguing question arises concerning the necessary gates for implementing any quantum circuit. While the NAND gate suffices in the classical case, no finite set of quantum gates can adequately fulfill this purpose. Instead, finite sets of quantum gates are presented, enabling the approximation of operators implemented by arbitrary quantum circuits. These sets leverage the decomposition theorems for single-qubit operators as their foundation.

Following the exploration of the computing power of quantum gates and circuits, the chapter ventures into quantum complexity theory. Here, quantum algorithms are defined as probabilistic algorithms capable of utilizing elements from uniform families of quantum circuits as subroutines. This approach facilitates the transfer of complexity analysis from probabilistic algorithms to quantum algorithms and culminates in the introduction of the complexity class BQP-representing bounded-error quantum polynomial time.

In the whole chapter, we always denote by k, l, m, n positive integers.

4.1. Simple single-qubit operators

This section discusses important single-qubit operators, that is, unitary operators on $\mathbb{H}_1$. Since they are typically used as building blocks of quantum circuits, we also refer to them as *single-qubit gates*. We identify these operators with their representation matrices with respect to the computational basis $(|0\rangle, |1\rangle)$ of $\mathbb{H}_1$. To begin, we summarize the properties of the identity, Pauli, and Hadamard gates that were already introduced in Chapters 2 and 3. Subsequently, we explain rotations in $\mathbb{R}^3$ and rotation gates. We will prove that the set of all such gates is the special unitary group SU(2). Finally, we introduce phase shift gates, which are special rotation operators.

4.1.1. The identity gate. The *identity gate*

$$(4.1.1) \qquad\qquad I = \begin{pmatrix} 1 & 0 \\ 0 & 1 \end{pmatrix}$$

is the most elementary single-qubit gate. It is a Hermitian and unitary involution, its only eigenvalue is 1, and its spectral decomposition is

$$(4.1.2) \qquad\qquad I = |0\rangle\langle 0| + |1\rangle\langle 1|.$$

4.1.2. The Pauli gates. The Pauli gates have already been introduced in Section 2.3.1. They are named after the physicist Wolfgang Pauli (1900–1958) and are of great importance for the construction of quantum circuits and the implementation of quantum algorithms. We recall their definition and discuss their properties.

Definition 4.1.1. The *Pauli gates* or *Pauli operators* are

$$(4.1.3) \qquad\qquad X = \begin{pmatrix} 0 & 1 \\ 1 & 0 \end{pmatrix}, \quad Y = \begin{pmatrix} 0 & -i \\ i & 0 \end{pmatrix}, \quad Z = \begin{pmatrix} 1 & 0 \\ 0 & -1 \end{pmatrix}.$$

Sometimes, the Pauli gates are also denoted by $\sigma_1, \sigma_2, \sigma_3$ or by $\sigma_x, \sigma_y, \sigma_z$. We can also write them as

$$(4.1.4) \qquad X = |0\rangle\langle 1| + |1\rangle\langle 0|, \quad Y = i(|1\rangle\langle 0| - |0\rangle\langle 1|), \quad Z = |0\rangle\langle 0| - |1\rangle\langle 1|.$$

So the effect of the Pauli gates on the computational basis vectors of $\mathbb{H}_1$ is as follows:

$$(4.1.5) \qquad \begin{aligned} X|0\rangle &= |1\rangle, & X|1\rangle &= |0\rangle, \\ Y|0\rangle &= i|1\rangle, & Y|1\rangle &= -i|0\rangle, \\ Z|0\rangle &= |0\rangle, & Z|1\rangle &= -|1\rangle. \end{aligned}$$

This shows that the Pauli X gate can be considered as the quantum equivalent of the classical NOT gate since it sends $|b\rangle$ to $|\neg b\rangle$ for all $b \in \{0, 1\}$. It is also called the *quantum* NOT *gate* or the *bit-flip gate*. Also, the Pauli Z gate is sometimes called the *phase-flip* gate since it flips the phase of $|1\rangle$ from 1 to -1.

In Example 2.4.57 we have determined the spectral decomposition of the Pauli operators. They are

(4.1.6)
$$X = |x_+\rangle\langle x_+| - |x_-\rangle\langle x_-|,$$
$$Y = |y_+\rangle\langle y_+| - |y_-\rangle\langle y_-|,$$
$$Z = |z_+\rangle\langle z_+| - |z_-\rangle\langle z_-|$$

where

(4.1.7)
$$(|x_+\rangle, |x_-\rangle) = \left(\frac{|0\rangle + |1\rangle}{\sqrt{2}}, \frac{|0\rangle - |1\rangle}{\sqrt{2}}\right),$$
$$(|y_+\rangle, |y_-\rangle) = \left(\frac{|0\rangle + i|1\rangle}{\sqrt{2}}, \frac{|0\rangle - i|1\rangle}{\sqrt{2}}\right),$$
$$(|z_+\rangle, |z_-\rangle) = (|0\rangle, |1\rangle).$$

Now we present important properties of the Pauli gates.

Theorem 4.1.2. *The Pauli gates are Hermitian and unitary involutions*

(4.1.8) $$XY = iZ = -YX, \quad ZX = iY = -XZ, \quad YZ = iX = -ZY,$$

and

(4.1.9) $$-iXYZ = I.$$

Exercise 4.1.3. Prove Theorem 4.1.2.

The following proposition will be very useful when we discuss rotation gates.

Proposition 4.1.4. *The sequence (I, X, Y, Z) is a $\mathbb{C}$-basis of $\mathrm{End}(\mathbb{H}_1)$ which is orthogonal with respect to the Hilbert-Schmidt inner product.*

Proof. Let $\alpha, \beta, \gamma, \delta \in \mathbb{C}$. Then we have

$$\alpha I + \beta X + \gamma Y + \delta Z = \begin{pmatrix} \alpha + \delta & \beta - i\gamma \\ \beta + i\gamma & \alpha - \delta \end{pmatrix}.$$

So, $\alpha I + \beta X + \gamma Y + \delta Z = 0$ implies

(4.1.10) $$0 = \alpha + \delta = \alpha - \delta = \beta + i\gamma = \beta - i\gamma.$$

This implies $\alpha = -\delta$ and $\alpha = \delta$ and therefore $\alpha = \delta = 0$. This also implies $\beta = -i\gamma$ and $\beta = i\gamma$ and therefore $\beta = \gamma = 0$. Hence, the sequence (I, X, Y, Z) is linearly independent. Since the dimension of $\mathrm{End}(\mathbb{H}_1)$ as a complex vector space is 4, the sequence is a basis of $\mathrm{End}(\mathbb{H}_1)$. The orthogonality can be verified by matrix multiplication which is done in Exercise 4.1.5. $\square$

Exercise 4.1.5. Verify that (I, X, Y, Z) is orthogonal with respect to the Hilbert-Schmidt inner product.

The symbols representing the identity and the Pauli gates in quantum circuits are shown in Figure 4.1.1.

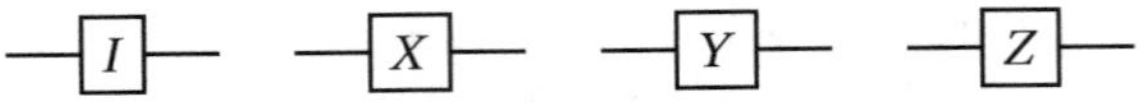

$$—\boxed{I}—\qquad—\boxed{X}—\qquad—\boxed{Y}—\qquad—\boxed{Z}—$$

Figure 4.1.1. Symbols for the identity and the Pauli gates in quantum circuits.

4.1.3. The Hadamard gate. Another important single-qubit gate, the *Hadamard gate* or *Hadamard operator*, has already been introduced in Section 3.3.2. We recall that this gate is

$$(4.1.11) \qquad\qquad H = \frac{1}{\sqrt{2}}\begin{pmatrix} 1 & 1 \\ 1 & -1 \end{pmatrix}.$$

The Hadamard operator is a unitary and Hermitian involution and we have shown in Exercise 2.3.4 that

$$(4.1.12) \qquad\qquad HXH = Z, \quad HYH = -Y, \quad HZH = X.$$

We also note that

$$(4.1.13) \qquad\qquad H = \frac{1}{\sqrt{2}}(X + Z).$$

The symbol representing the Hadamard gate is shown in Figure 4.1.2.

$$—\boxed{H}—$$

Figure 4.1.2. Symbol for the Hadamard gate in quantum circuits.

4.2. More geometry in $\mathbb{R}^3$

In Section 3.1.4, we have shown that any single-qubit quantum state corresponds to a point on the Bloch sphere. In Section 4.2.2, we will prove the important fact that any unitary operator on $\mathbb{H}_1$ can be seen as a rotation of the Bloch sphere. This proof requires more concepts and results from the geometry of $\mathbb{R}^3$, which are presented in this section. Specifically, we will discuss rotations in $\mathbb{R}^3$.

We will identify triplets $(\vec{a}, \vec{b}, \vec{c})$ of vectors in $\mathbb{R}^3$ with the matrices that have $\vec{a}, \vec{b}, \vec{c}$ as column vectors; e.g., the unit matrix

$$(4.2.1) \qquad\qquad I_3 = \begin{pmatrix} 1 & 0 & 0 \\ 0 & 1 & 0 \\ 0 & 0 & 1 \end{pmatrix}$$

is identified with the basis

$$(4.2.2) \qquad\qquad (\hat{x}, \hat{y}, \hat{z}) = ((1,0,0),(0,1,0),(0,0,1)).$$

We also equate the endomorphisms of $\mathbb{R}^3$ with their representation matrices corresponding to the basis I_3.

4.2.1. General spherical coordinates. We discuss spherical coordinates with respect to any pair of orthogonal axes. We start by defining the angle between two nonzero vectors $\vec{a}$ and $\vec{b}$ in $\mathbb{R}^3$. For this, we note that the Cauchy-Schwarz inequality (see Proposition 2.2.25) implies

$$\tag{4.2.3} \frac{\langle \vec{a}|\vec{b} \rangle}{\left\| \vec{a} \right\| \left\| \vec{b} \right\|} \le 1$$

for any two nonzero vectors $\vec{a}, \vec{b} \in \mathbb{R}^3$. So the following definition makes sense.

Definition 4.2.1. Let $\vec{a}, \vec{b} \in \mathbb{R}^3$ be nonzero. Then the *angle between $\vec{a}$ and $\vec{b}$* is defined as

$$\tag{4.2.4} \angle(\vec{a}, \vec{b}) = \arccos \frac{\langle \vec{a}|\vec{b} \rangle}{\left\| \vec{a} \right\| \left\| \vec{b} \right\|}.$$

The angle between two vectors in $\mathbb{R}^3$ is illustrated in Figure 4.2.1. As shown in Proposition 4.2.3, such an angle is always between 0 and π.

Example 4.2.2. The angle between $\hat{x}$ and $\hat{y}$ is

$$\tag{4.2.5} \angle(\hat{x}, \hat{y}) = \arccos \frac{\langle \hat{x}|\hat{y} \rangle}{\left\| \hat{x} \right\| \left\| \hat{y} \right\|} = \arccos(0) = \frac{\pi}{2}.$$

The angle between $\hat{x}$ and $-\hat{x}$ is

$$\tag{4.2.6} \angle(\hat{x}, -\hat{x}) = \arccos \frac{\langle \hat{x}| - \hat{x} \rangle}{\left\| \hat{x} \right\| \left\| -\hat{x} \right\|} = \arccos(-1) = \pi.$$

Let

$$\tag{4.2.7} \hat{a} = \frac{\hat{x} + \hat{y}}{\sqrt{2}}.$$

Then the angle between $\hat{x}$ and $\hat{a}$ is

$$\tag{4.2.8} \angle(\hat{x}, \hat{a}) = \arccos \frac{\langle \hat{x}|\hat{a} \rangle}{\left\| \hat{x} \right\| \left\| \hat{a} \right\|} = \arccos \frac{1}{\sqrt{2}} = \frac{\pi}{4}.$$

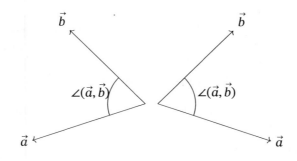

Figure 4.2.1. Angle between two vectors.

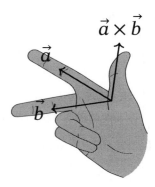

$$\vec{a} \times \vec{b}$$

Figure 4.2.2. Visualization of the cross product by the right-hand rule.

The angle between two vectors in $\mathbb{R}^3$ has the following properties.

Proposition 4.2.3. *Let $\vec{a}, \vec{b} \in \mathbb{R}^3$ be nonzero vectors. Then we have the following.*

(1) $0 \leq \angle(\vec{a}, \vec{b}) = \angle(\vec{b}, \vec{a}) \leq \pi.$

(2) $\angle(\vec{a}, \vec{b}) = 0$ *if and only if $\vec{b} = r\vec{a}$ with $r \in \mathbb{R}_{>0}$.*

(3) $\angle(\vec{a}, \vec{b}) = \pi/2$ *if and only if $\langle \vec{a}|\vec{b}\rangle = 0$; that is, $\vec{a}$ and $\vec{b}$ are orthogonal to each other.*

(4) $\angle(\vec{a}, \vec{b}) = \pi$ *if and only if $\vec{b} = r\vec{a}$ with $r \in \mathbb{R}_{<0}$.*

Exercise 4.2.4. Prove Proposition 4.2.3.

Next, we define the cross products or outer product of two vectors in $\mathbb{R}^3$.

Definition 4.2.5. Let $\vec{a} = (a_x, a_y, a_z), \vec{b} = (b_x, b_y, b_z) \in \mathbb{R}^3$. Then the *cross product* or *outer product* of $\vec{a}$ and $\vec{b}$ is

$$(4.2.9) \qquad \vec{a} \times \vec{b} = (a_y b_z - a_z b_y, a_z b_x - a_x b_z, a_x b_y - a_y b_x).$$

Example 4.2.6. Let $\vec{a} = \hat{x} = (1, 0, 0), \vec{b} = \hat{y} = (0, 1, 0)$. Then $\vec{a} \times \vec{b} = \hat{z} = (0, 0, 1)$.

If $\vec{a}, \vec{b} \in \mathbb{R}^3$ are linearly independent, then the cross product has the following geometric interpretation by the *right-hand rule* which is illustrated in Figure 4.2.2. If $\vec{a}$ points in the direction of the index finger of the right hand and $\vec{b}$ points in the direction of the middle finger, then $\vec{a} \times \vec{b}$ is a vector orthogonal to the plane spanned by $\vec{a}$ and $\vec{b}$ that points in the direction of the thumb.

Here are some important properties of the outer product.

Proposition 4.2.7. *Let $\vec{a}, \vec{b} \in \mathbb{R}^3$ and let θ be the angle between $\vec{a}$ and $\vec{b}$. Then the following hold.*

(1) $\left\| \vec{a} \times \vec{b} \right\| = \left\| \vec{a} \right\| \left\| \vec{b} \right\| \sin \theta.$

(2) $\det(\vec{a}, \vec{b}, \vec{c}) = \langle \vec{a} \times \vec{b}|\vec{c}\rangle$ *for all $\vec{c} \in \mathbb{R}^3$.*

(3) $\vec{a} \times \vec{b}$ *is orthogonal to* $\vec{a}$ *and* $\vec{b}$.

(4) $\vec{a} \times \vec{b} = 0$ *if and only if* $\vec{a}$ *and* $\vec{b}$ *are linearly dependent.*

Proof. It follows from Definition 4.2.1 that

$$\left\| \vec{a} \right\|^2 \|\vec{b}\|^2 \sin^2 \theta = \left\| \vec{a} \right\|^2 \|\vec{b}\|^2 - \langle \vec{a}|\vec{b} \rangle^2. \tag{4.2.10}$$

Now we have

$$\left\| \vec{a} \right\|^2 \|\vec{b}\|^2 = (a_x^2 + a_y^2 + a_z^2)(b_x^2 + b_y^2 + b_z^2) \tag{4.2.11}$$
$$= a_x^2 b_x^2 + a_x^2 b_y^2 + a_x^2 b_z^2 + a_y^2 b_x^2 + a_y^2 b_y^2 + a_y^2 b_z^2 + a_z^2 b_x^2 + a_z^2 b_y^2 + a_z^2 b_z^2$$

and

$$\langle \vec{a}|\vec{b} \rangle^2 = (a_x b_x + a_y b_y + a_z b_z)^2 \tag{4.2.12}$$
$$= a_x^2 b_x^2 + a_y^2 b_y^2 + a_z^2 b_z^2 + 2(a_x b_x a_y b_y + a_x b_x a_z b_z + a_y b_y a_z b_z).$$

So

$$\left\| \vec{a} \right\|^2 \|\vec{b}\|^2 - \langle \vec{a}|\vec{b} \rangle^2 \tag{4.2.13}$$
$$= a_x^2 b_y^2 + a_x^2 b_z^2 + a_y^2 b_x^2 + a_y^2 b_z^2 + a_z^2 b_x^2 + a_z^2 b_y^2$$
$$- 2(a_x b_x a_y b_y + a_x b_x a_z b_z + a_y b_y a_z b_z).$$

On the other hand, we have

$$\left\| \vec{a} \times \vec{b} \right\|^2 = (a_y b_z - a_z b_y)^2 + (a_z b_x - a_x b_z)^2 + (a_y b_z - a_z b_y)^2 \tag{4.2.14}$$
$$= a_y^2 b_z^2 + a_z^2 b_y^2 + a_z^2 b_x^2 + a_x^2 b_z^2 + a_y^2 b_z^2 + a_z^2 b_y^2$$
$$- 2(a_y b_z a_z b_y + a_z b_x a_x b_z + a_y b_z a_z b_y).$$

So the first assertion follows from (4.2.13) and (4.2.14). Also, Theorem B.5.16, the Laplace expansion formula for determinants, implies the second assertion. The second assertion and the alternating property of the determinant imply the third assertion. Finally, the first assertion implies the fourth assertion. $\qquad\square$

From Proposition 4.2.7 we obtain the following important result.

Theorem 4.2.8. *Let* $\hat{a}, \hat{b} \in \mathbb{R}^3$ *be unit vectors that are orthogonal to each other. Then* $\hat{p} = \hat{r} = \hat{a} \times \hat{b}$ *and* $\hat{q} = \hat{b} \times \hat{a}$ *are the uniquely determined vectors in* $\mathbb{R}^3$ *such that* $(\hat{p}, \hat{a}, \hat{b})$, $(\hat{a}, \hat{q}, \hat{b})$, *and* $(\hat{a}, \hat{b}, \hat{r})$ *are orthonormal bases of* $\mathbb{R}^3$ *with determinant* 1.

Proof. Since $\hat{a}$ and $\hat{b}$ are unit vectors and since they are orthogonal to each other, it follows from the first assertion in Proposition 4.2.7 that $\hat{p}$, $\hat{q}$, and $\hat{r}$ are unit vectors. Also, the second and third assertion of Proposition 4.2.7 imply that $(\hat{a}, \hat{b}, \hat{r})$ is an orthonormal basis of $\mathbb{R}^3$ with determinant 1.

Assume that $(\hat{a}, \hat{b}, \hat{r}')$ is another orthonormal basis of $\mathbb{R}^3$ with determinant 1. Then there are $\alpha, \beta, \gamma \in \mathbb{R}$ such that

$$\hat{r}' = \alpha \hat{a} + \beta \hat{b} + \gamma \hat{r}. \tag{4.2.15}$$

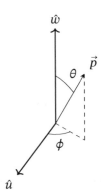

Figure 4.2.3. The spherical coordinate representation of $\vec{p}$ with respect to $(\hat{u}, \hat{v})$ is $(\|\vec{p}\|, \theta, \phi)$.

Since $\hat{a}$ is a unit vector and since it is orthogonal to $\hat{b}$, $\hat{r}$, and $\hat{r}'$, we have $\alpha = \langle \hat{a}|\hat{r}'\rangle = 0$. In the same way, we see that $\beta = 0$. Since $\hat{r}$ and $\hat{r}'$ are unit vectors and $\det(\hat{a}, \hat{b}, \hat{r}) = \det(\hat{a}, \hat{b}, \hat{r}') = 1$, it follows that $\gamma = 1$. So, we have $\hat{r} = \hat{r}'$, as asserted.

The assertions for $\hat{p}$ and $\hat{q}$ follow by swapping the columns of $(\hat{a}, \hat{b}, \hat{r})$ and applying Proposition B.5.11.   □

Exercise 4.2.9. Let $\hat{a} = \left(\frac{1}{\sqrt{2}}, \frac{1}{\sqrt{2}}, 0\right)$ and $\hat{b} = \left(\frac{1}{\sqrt{2}}, -\frac{1}{\sqrt{2}}, 0\right)$. Find $\hat{c} \in \mathbb{R}^3$ so that $(\hat{a}, \hat{b}, \hat{c})$ is an orthogonal matrix.

We now introduce general spherical coordinates.

Definition 4.2.10. Let $\hat{u}, \hat{w}$ be unit vectors that are orthogonal to each other. Let $B = (\hat{u}, \hat{v}, \hat{w})$ be an orthonormal basis of $\mathbb{R}^3$ with determinant 1 which according to Theorem 4.2.8 exists and is uniquely determined. Also, let $\vec{p} \in \mathbb{R}^3$. Then the *spherical coordinate representation of $\vec{p}$ with respect to the azimuth reference $\hat{u}$ and the zenith $\hat{w}$ or with respect to $(\hat{u}, \hat{w})$ for short* is defined as the spherical coordinate representation of $B^{-1}\vec{p}$.

Figure 4.2.3 illustrates the generalized spherical coordinate representation.

Example 4.2.11. We determine the spherical coordinate representation (r, θ, ϕ) of $\vec{p} = (\frac{1}{2}, \frac{1}{2}, \sqrt{2}/2)$ with respect to the azimuth reference $\hat{v} = (1, 0, 0)$ and the zenith $\hat{w} = (0, 0, -1)$. First,

(4.2.16)
$$B = (\hat{u}, \hat{v}, \hat{w}) = \begin{pmatrix} 1 & 0 & 0 \\ 0 & -1 & 0 \\ 0 & 0 & -1 \end{pmatrix}$$

is an orthonormal basis of $\mathbb{R}^3$ with determinant 1. Also, we have $B^{-1} = B$ and $B^{-1}\vec{p} = (\frac{1}{2}, -\frac{1}{2}, -\sqrt{2}/2)$. So using arguments analogous to those in Example 3.1.13 we obtain $r = 1$, $\theta = 3\pi/4$, and $\phi = 5\pi/4$.

Next, we show how the spherical coordinate representation changes if the azimuth reference is changed.

Proposition 4.2.12. *Let* $\hat{u}, \hat{u}', \hat{w} \in \mathbb{R}^3$ *be unit vectors and assume that both* $\hat{u}$ *and* $\hat{u}'$ *are orthogonal to* $\hat{w}$. *Then the following hold.*

(1) *The spherical coordinate representation of* $\hat{u}'$ *with respect to* $(\hat{u}, \hat{w})$ *is* $(1, \pi/2, \delta)$ *where* $\cos\delta = \langle \hat{u} | \hat{u}' \rangle$ *and* $\sin\delta = \langle \hat{w} \times \hat{u} | \hat{u}' \rangle$.

(2) *Let* $\vec{p} \in \mathbb{R}^3$ *and let* (r, θ, ϕ) *and* (r', θ', ϕ') *be the spherical coordinate representations of* $\vec{p}$ *with respect to* $(\hat{u}, \hat{w})$ *and* $(\hat{u}', \hat{w})$, *respectively. Then we have* $r' = r$, $\theta' = \theta$, *and*

$$(4.2.17) \qquad \phi' = \begin{cases} 0 & \text{if } \phi = 0, \\ \phi - \delta \bmod 2\pi & \text{otherwise.} \end{cases}$$

Proof. Set $\hat{v} = \hat{w} \times \hat{u}$ and $\hat{v}' = \hat{w}' \times \hat{u}'$. Then it follows from Theorem 4.2.8 that $B = (\hat{u}, \hat{v}, \hat{w})$ and $B' = (\hat{u}', \hat{v}', \hat{w})$ are the uniquely determined orthonormal bases of $\mathbb{R}^3$ with determinant 1 and first and last column $\hat{u}$, $\hat{w}$ and $\hat{u}'$, $\hat{w}'$, respectively. Let $(1, \varepsilon, \delta)$ be the spherical coordinate representation of $\hat{u}'$ with respect to $(\hat{u}, \hat{w})$. Then by Proposition 3.1.12 we have

$$(4.2.18) \qquad \hat{u}' = \cos\delta \sin\varepsilon\, \hat{u} + \sin\delta \sin\varepsilon\, \hat{v} + \cos\varepsilon\, \hat{w}.$$

Since $\hat{w}$ is a unit vector and since it is orthogonal to $\hat{u}'$, $\hat{u}$, and $\hat{v}$, it follows that $\cos\varepsilon = 0$ and thus $\varepsilon = \pi/2$. Since $\hat{u}$ and $\hat{v}$ are unit vectors and since they are orthogonal to each other, it follows that $\cos\delta = \langle \hat{u} | \hat{u}' \rangle$ and $\sin\delta = \langle \hat{v} | \hat{u}' \rangle$.

Now we turn to the second assertion. Set

$$(4.2.19) \qquad M = \begin{pmatrix} \cos\delta & -\sin\delta & 0 \\ \sin\delta & \cos\delta & 0 \\ 0 & 0 & 1 \end{pmatrix}.$$

Then BM is an orthonormal basis of $\mathbb{R}^3$ with determinant 1 with first vector $\hat{u}'$ and last vector $\hat{w}$. Since by Theorem 4.2.8 there is only one such basis, it follows that $\hat{v}' = -\sin\delta\, \hat{u} + \cos\delta\, \hat{v}$. Let $\vec{p} \in \mathbb{R}^3$ with spherical coordinate representations (r, θ, ϕ) and (r', θ', ϕ') with respect to $(\hat{u}, \hat{w})$ and $(\hat{u}', \hat{w})$, respectively. Then $r = \|\vec{p}\| = r'$. As shown in Exercise 4.2.13 we have

$$(4.2.20) \qquad \vec{q}' = M^{-1}\vec{q} = (\cos(\phi - \delta)\sin\theta, \sin(\phi - \delta)\sin\theta, \cos\theta).$$

So Definition 4.2.10 and Proposition 3.1.12 imply the assertion. $\qquad\square$

Exercise 4.2.13. Verify (4.2.20) in the proof of Proposition 4.2.12 using the trigonometric identities (A.5.3) and (A.5.6).

Example 4.2.14. Let $\hat{u} = \hat{x} = (1, 0, 0)$, $\hat{u}' = \hat{y} = (0, 1, 0)$, and $\hat{w} = \hat{z} = (0, 0, 1)$. Then $\hat{v} = \hat{w} \times \hat{u} = (0, 1, 0) = \hat{y}$, $\langle \hat{u} | \hat{u}' \rangle = 0$, $\langle \hat{w} \times \hat{u} | \hat{u}' \rangle = 1$. Hence, the spherical coordinate representation of $\hat{u}'$ with respect to $(\hat{u}, \hat{w})$ is $(1, \pi/2, \pi/2)$. Also, let $\vec{p} \in \mathbb{R}^3$ with Cartesian coordinates $(\sqrt{2}, \sqrt{2}, 0)$. Then the spherical coordinate representation of $\vec{p}$ is $(2, \pi/2, \pi/4)$. So, by Proposition 4.2.12, the sperical coordinate representation of $\vec{p}$ with respect to $(\hat{y}, \hat{z})$ is $(2, \pi/2, 7\pi/4)$.

4.2.2. Rotations. In this section, we explain the geometry of rotations in $\mathbb{R}^3$. First, we define orthogonal matrices.

Definition 4.2.15. (1) A matrix $O \in \mathbb{R}^{(3,3)}$ is called *orthogonal* if O is invertible and $O^{-1} = O^{\mathrm{T}}$.

(2) The set of all orthogonal matrices is denoted by $O(3)$.

Exercise 4.2.16. Show that the determinant of orthogonal matrices is in $\{\pm 1\}$.

We note that orthogonal matrices are unitary matrices in $\mathbb{C}^{(3,3)}$ with real entries. Therefore, Proposition 2.4.18 implies the following characterization of orthogonal matrices.

Proposition 4.2.17. *Let $O \in \mathbb{R}^{(3,3)}$. Then the following statements are equivalent.*

(1) *$O \in O(3)$.*

(2) *The columns of O form an orthonormal basis of $\mathbb{R}^3$.*

(3) *The rows of O form an orthonormal basis of $\mathbb{R}^3$.*

(4) *$\langle O\hat{v}|O\hat{w}\rangle = \langle \hat{v}|\hat{w}\rangle$ for all $\hat{v}, \hat{w} \in \mathbb{R}^3$.*

(5) *$\|O\hat{v}\| = \|\hat{v}\|$ for all $\hat{v} \in \mathbb{R}^3$.*

Exercise 4.2.18. Prove Proposition 4.2.17.

It follows from the equivalence of the first two statements in Proposition 4.2.17 that there is a one-to-one correspondence between orthogonal matrices and orthonormal bases of $\mathbb{R}^3$.

Now we introduce the orthogonal and the special orthogonal group.

Theorem 4.2.19. (1) *The set $O(3)$ of all orthogonal matrices is a group with respect to matrix multiplication. It is called the* orthogonal group *of rank* 3.

(2) *The set of all orthogonal matrices with determinant 1 is a subgroup of $O(3)$. It is denoted by $SO(3)$ and is called the* special orthogonal group *of rank* 3.

Exercise 4.2.20. Prove Theorem 4.2.19.

The next theorem introduces rotations in $\mathbb{R}^3$. They are illustrated in Figure 4.2.4.

Theorem 4.2.21. *Let $\hat{u}, \hat{w} \in \mathbb{R}^3$ be unit vectors and let them be orthogonal to each other, and let $\gamma \in \mathbb{R}$. Consider the map $\mathbb{R}^3 \to \mathbb{R}^3$ that sends $\vec{p} \in \mathbb{R}^3$ with spherical coordinates (r, θ, ϕ) with respect to $(\hat{u}, \hat{w})$ to the vector in R^3 with the following spherical coordinate representation with respect to $(\hat{u}, \hat{w})$:*

(4.2.21)
$$\begin{cases} (r, \theta, \phi) & \text{if } \theta \in \{0, \pi\}, \\ (r, \theta, (\phi + \gamma) \bmod 2\pi) & \text{otherwise.} \end{cases}$$

Then this map depends only on $\hat{w}$ and γ and is independent of $\hat{u}$. It is denoted by $\mathrm{Rot}_{\hat{w}}(\gamma)$ and is called the rotation about $\hat{w}$ through the angle γ. *Also, $\hat{w}$ and γ are called the* axis *and the* angle *of this rotation, respectively.*

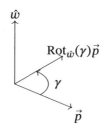

Figure 4.2.4. Rotation of $\vec{p}$ about $\hat{w}$ through the angle γ.

Proof. We show that the map defined in the theorem is independent of $\vec{u}$. Let $\hat{u}'$ be another unit vector in $\mathbb{R}^3$ that is orthogonal to $\hat{w}$. Denote the map in the theorem by Rot. We show that Rot is the same map regardless of whether we use $\hat{u}$ or $\hat{u}'$ for its definition. By Proposition 4.2.12, the spherical coordinate representation of $\hat{u}'$ with respect to $(\hat{u}, \hat{w})$ is $(1, \pi/2, \delta)$ with $\delta \in [0, 2\pi[$. Let $\vec{p} \in \mathbb{R}^3$ with spherical coordinate representation (r, θ, ϕ) with respect to $(\hat{u}, \hat{w})$. If $\theta \in \{0, \pi\}$, then by Proposition 4.2.12 the spherical coordinate representation of $\vec{p}$ with respect to $(\hat{u}', \hat{w})$ is also (r, θ, ϕ). So Rot$(\vec{p}) = \vec{p}$, regardless of whether we choose $\hat{u}$ or $\hat{u}'$ for its definition. If $\theta \neq 0, \pi$, then by Proposition 4.2.12 the spherical coordinate representation of $\vec{p}$ with respect to $(\hat{u}', \hat{w})$ is $(r, \theta, (\phi - \delta) \bmod 2\pi)$. So if we use $\hat{u}'$ to define Rot, we obtain $(r, \theta, (\phi - \delta + \gamma) \bmod 2\pi)$ as the spherical coordinate representation of Rot$(\vec{p})$ with respect to $(\hat{u}', \hat{w})$. Proposition 4.2.12 shows that the spherical coordinate representation of this vector with respect to $(\hat{u}, \hat{w})$ is $(r, \theta, (\phi + \gamma) \bmod 2\pi)$. But this is the spherical coordinate representation of Rot$(\vec{p})$ if we use $\hat{u}$ to define Rot. $\qquad \square$

Figure 4.2.4 shows that applying Rot$_{\hat{w}}(\gamma)$ to $\vec{p} \in \mathbb{R}^3$ rotates this vector about the axis $\hat{w}$ counterclockwise through an angle γ.

In the remainder of this section, we will prove the following theorem.

Theorem 4.2.22. *The set of rotations in $\mathbb{R}^3$ is $x\mathrm{SO}(3)$.*

We first determine the rotations about the axes $\hat{x} \doteq (1, 0, 0)$, $\hat{y} = (0, 1, 0)$, and $\hat{z} = (0, 0, 1)$ explicitly.

Proposition 4.2.23. *Let $\gamma \in \mathbb{R}$. Then we have*

$$(4.2.22) \qquad \mathrm{Rot}_{\hat{x}}(\gamma) = \begin{pmatrix} 1 & 0 & 0 \\ 0 & \cos\gamma & -\sin\gamma \\ 0 & \sin\gamma & \cos\gamma \end{pmatrix},$$

$$(4.2.23) \qquad \mathrm{Rot}_{\hat{y}}(\gamma) = \begin{pmatrix} \cos\gamma & 0 & -\sin\gamma \\ 0 & 1 & 0 \\ \sin\gamma & 0 & \cos\gamma \end{pmatrix},$$

$$(4.2.24) \qquad \mathrm{Rot}_{\hat{z}}(\gamma) = \begin{pmatrix} \cos\gamma & -\sin\gamma & 0 \\ \sin\gamma & \cos\gamma & 0 \\ 0 & 0 & 1 \end{pmatrix}.$$

Exercise 4.2.24. Prove Proposition 4.2.23.

Note that by Proposition 4.2.23, the rotations about the x-, y-, and z-axes are in $SO(3)$. The next proposition provides explicit formulas for all rotations in $\mathbb{R}^3$ and shows that all rotations are in $SO(3)$.

Proposition 4.2.25. *Let $B = (\hat{u}, \hat{v}, \hat{w}) \in SO(3)$ and let $\gamma \in \mathbb{R}$. Then we have*

$$(4.2.25) \qquad \mathrm{Rot}_{\hat{u}}(\gamma) = B \, \mathrm{Rot}_{\hat{x}}(\gamma) B^{-1},$$

$$(4.2.26) \qquad \mathrm{Rot}_{\hat{v}}(\gamma) = B \, \mathrm{Rot}_{\hat{y}}(\gamma) B^{-1},$$

$$(4.2.27) \qquad \mathrm{Rot}_{\hat{w}}(\gamma) = B \, \mathrm{Rot}_{\hat{z}}(\gamma) B^{-1}$$

and these rotation operators are in $SO(3)$.

Proof. We first prove (4.2.27). Let $\vec{p} \in \mathbb{R}^3$ with spherical coordinates (r, θ, ϕ) with respect to $(\hat{u}, \hat{w})$. Then we have

$$(4.2.28) \qquad \vec{p} = rB(\cos\phi\sin\theta, \sin\phi\sin\theta, \cos\theta).$$

Applying (4.2.24) and the trigonometric identities (A.5.2) and (A.5.5) we obtain

$$(4.2.29) \qquad \begin{aligned} B\,\mathrm{Rot}_{\hat{z}}(\gamma)B^{-1}\vec{p} &= B\,\mathrm{Rot}_{\hat{z}}(\gamma)(\cos\phi\sin\theta, \sin\phi\sin\theta, \cos\theta) \\ &= rB(\cos(\phi+\gamma)\sin\theta, \sin(\phi+\gamma)\sin\theta, \cos\theta). \end{aligned}$$

On the other hand, by (3.1.7) we have

$$(4.2.30) \qquad \mathrm{Rot}_{\hat{w}}(\gamma)\vec{p} = rB(\cos(\phi+\gamma)\sin\theta, \sin(\phi+\gamma)\sin\theta, \cos\theta).$$

So (4.2.30) and (4.2.29) imply (4.2.27). Next, we prove (4.2.25). With the permutation matrix

$$(4.2.31) \qquad P = \begin{pmatrix} 0 & 0 & 1 \\ 0 & 1 & 0 \\ 1 & 0 & 0 \end{pmatrix}$$

we have

$$(4.2.32) \qquad BP = (\hat{w}, \hat{v}, \hat{u})$$

and

$$(4.2.33) \qquad P\,\mathrm{Rot}_{\hat{z}}(\gamma)P = \mathrm{Rot}_{\hat{z}}(\gamma).$$

So it follows from (4.2.27), (4.2.32), and (4.2.33) that

$$(4.2.34) \qquad \begin{aligned} \mathrm{Rot}_{\hat{u}}(\gamma) &= BP\,\mathrm{Rot}_{\hat{z}}(\gamma)(BP)^{-1} \\ &= BP\,\mathrm{Rot}_{\hat{z}}(\gamma)PB^{-1} = B\,\mathrm{Rot}_{\hat{x}}(\gamma)B^{-1}. \end{aligned}$$

The identity (4.2.26) can be proved analogously. $\qquad\qquad\square$

The following exercise is an application of Proposition 4.2.25.

Exercise 4.2.26. Let $\hat{w} \in \mathbb{R}^3$ be a unit vector and let $\gamma \in \mathbb{R}$. Prove that

$$(4.2.35) \qquad \mathrm{Rot}_{\hat{w}}(-\gamma) = \mathrm{Rot}_{-\hat{w}}(\gamma).$$

We now prove that every operator in $SO(3)$ is a rotation and we explain how rotations can be represented.

Proposition 4.2.27. *Let $O \in SO(3)$. Then the following hold.*

(1) *We have $O = I_3$ if and only if $O = \text{Rot}_{\hat{w}}(\gamma)$ with an arbitrary unit vector $\hat{w} \in \mathbb{R}^3$ and $\gamma \in \mathbb{R}$ such that $\gamma \equiv 0 \mod 2\pi$.*

(2) *If $O \neq I_3$, then there is a unit vector $\hat{w} \in \mathbb{R}^3$ such that $\pm\hat{w}$ are the only unit eigenvectors of O associated with the eigenvalue 1 and there is a modulo 2π uniquely determined $\gamma \in \mathbb{R}$ such that $O = \text{Rot}_{\hat{w}}(\gamma) = \text{Rot}_{-\hat{w}}(-\gamma)$.*

Proof. Let $\hat{w} \in \mathbb{R}^3$ be a unit vector and let $\gamma \in \mathbb{R}$.

If $\gamma \equiv 0 \mod 2\pi$, then $\text{Rot}_{\hat{w}}(\gamma) = I_3$ by Theorem 4.2.21. Conversely, let $\text{Rot}_{\hat{w}}(\gamma) = I_3$. Let $\hat{u} \in \mathbb{R}^3$ be a unit vector that is orthogonal to $\hat{w}$. Then the spherical coordinate representation of $\hat{u}$ with respect to $(\hat{u}, \hat{w})$ is $(1, \pi/2, 0)$. So by Theorem 4.2.21 the spherical coordinate representation of $\text{Rot}_{\hat{w}}(\gamma)\hat{u}$ with respect to $(\hat{u}, \hat{w})$ is $(1, \pi/2, \gamma \mod 2\pi)$. But since $\text{Rot}_{\hat{w}}(\gamma) = I_3$, it follows that $\text{Rot}_{\hat{w}}(\gamma)\hat{u} = \hat{u}$. So $\gamma \equiv 0 \mod 2\pi$.

To prove the second assertion, assume that $O \neq I_3$. The characteristic polynomial of O has degree 3 and real coefficients. So it follows from Proposition B.7.21 and Exercise A.4.54 that O has only real eigenvalues or one real eigenvalue and a pair of complex conjugate eigenvalues.

Since by Proposition 4.2.17 the orthogonal operator O is length preserving, the absolute value of these eigenvalues is 1. But because $O \neq I_3$, $\det O = 1$, and $\det O$ is the product of the three eigenvalues by Proposition 2.4.1, it follows that exactly one of the eigenvalues O is 1. This implies that the eigenspace associated with the eigenvalue 1 has dimension 1. Let $\hat{w}$ be a unit eigenvector of U associated with the eigenvalue 1. So $-\hat{w}$ is the only other unit eigenvector of O associated with the eigenvalue 1.

Let $B = (\hat{u}, \hat{v}, \hat{w}) \in SO(3)$ which exists by Theorem 4.2.8. Set

(4.2.36)
$$R = B^{-1}OB.$$

Now we show that $R = \text{Rot}_{\hat{z}}(\gamma)$ with $\gamma \in \mathbb{R}$. Then $O = \text{Rot}_{\hat{w}}(\gamma)$ by Proposition 4.2.25.

It follows from (4.2.36) that $\hat{w} = O\hat{w}$ is the last column of BR. This implies that

(4.2.37)
$$R = \begin{pmatrix} a & b & 0 \\ c & d & 0 \\ e & f & 1 \end{pmatrix}$$

with real entries a, b, c, d, e, f. Since R is orthogonal, Proposition 4.2.17 implies that the columns of R are orthogonal to each other. Hence, we have $e = f = 0$. Also, since R is orthogonal we have

(4.2.38)
$$R = \begin{pmatrix} a & -b & 0 \\ b & a & 0 \\ 0 & 0 & 1 \end{pmatrix}$$

and

(4.2.39)
$$a^2 + b^2 = 1.$$

Therefore, Lemma 3.1.9 implies that $R = \text{Rot}_{\hat{z}}(\gamma)$ with $\gamma \in \mathbb{R}$.

We conclude the proof by showing the uniqueness statements for $\hat{w}$ and γ. So let $\hat{w}' \in \mathbb{R}^3$ be a unit vector, and let $\gamma' \in \mathbb{R}$ be such that $O = \text{Rot}_{\hat{w}'}(\gamma')$. Then we have $O\hat{w}' = \mathrm{Rot}_{\hat{w}'}(\gamma')\hat{w}' = \hat{w}'$. So $\hat{w}'$ is a unit eigenvector of O associated with the eigenvalue 1. As seen above, this implies $\hat{w}' = \hat{w}$ or $\hat{w}' = -\hat{w}$. The uniqueness modulo 2π is proved in Exercise 4.2.28. $\square$

Exercise 4.2.28. Prove the uniqueness of γ modulo 2π in Proposition 4.2.27.

Now we can prove Theorem 4.2.22. It follows from Proposition 4.2.25 that all rotations of $\mathbb{R}^3$ operators are in SO(3). Proposition 4.2.27 shows that all elements of SO(3) are rotations of $\mathbb{R}^3$. Therefore, the set of all rotations of $\mathbb{R}^3$ is indeed SO(3).

4.2.3. Decomposition of rotations. Next, we give two decomposition theorems for rotations in $\mathbb{R}^3$. For this, we need the following lemma.

Lemma 4.2.29. *Let $\hat{u}, \hat{u}', \hat{w} \in \mathbb{R}^3$ be unit vectors such that $\hat{u}$ and $\hat{u}'$ are orthogonal to $\hat{w}$. Then there is a modulo 2π uniquely determined $\delta \in \mathbb{R}$ such that $\mathrm{Rot}_{\hat{w}}(\delta)\hat{u} = \hat{u}'$.*

Proof. It follows from Proposition 4.2.12 that the spherical coordinate representation of $\hat{u}'$ with respect to $(\hat{u}, \hat{w})$ is $(1, \pi/2, \delta)$ with a modulo 2π uniquely determined $\delta \in \mathbb{R}$. Therefore, the definition of rotations in Theorem 4.2.21 implies the assertion. $\square$

Here is our first decomposition theorem.

Theorem 4.2.30. *For every $O \in$ SO(3) there are $\alpha, \beta, \gamma \in \mathbb{R}$ such that*

$$(4.2.40) \qquad\qquad O = \mathrm{Rot}_{\hat{z}}(\alpha)\,\mathrm{Rot}_{\hat{y}}(\beta)\,\mathrm{Rot}_{\hat{z}}(\gamma).$$

The real numbers α, β, γ are called the Euler angles *of O.*

Proof. Let $O \in$ SO(3). Denote by $\hat{x}', \hat{y}', \hat{z}'$ the column vectors of O. If $\hat{z}' = \hat{z}$ or $\hat{z}' = -\hat{z}$, then, as shown in the proof of Proposition 4.2.27, we have $O = \mathrm{Rot}_{\hat{z}}(\gamma)$ with $\gamma \in \mathbb{R}$. So, if we set $\alpha = \beta = 0$, then (4.2.40) holds.

Assume that $\hat{z}' \neq \pm\hat{z}$. The proof for this case is illustrated in Figure 4.2.5. The intersection of the plane spanned by $\hat{x}$ and $\hat{y}$ and the plane spanned by $\hat{x}$ and $\hat{y}'$ is a line, that is, a one-dimensional subspace of $\mathbb{R}^3$. Denote by $\hat{v}$ a unit vector that generates this space. Both $\hat{y}$ and $\hat{v}$ are orthogonal to $\hat{z}$. By Lemma 4.2.29 we can choose $\alpha \in \mathbb{R}$ such that $\mathrm{Rot}_{\hat{z}}(\alpha)\hat{y} = \hat{v}$. This rotation does not change $\hat{z}$ and maps $\hat{x}$ to some unit vector $\hat{x}_1 \in \mathbb{R}^3$. If we apply this rotation to the standard basis I_3 of $\mathbb{R}^3$, we obtain the orthonormal basis

$$(4.2.41) \qquad\qquad B_1 = \mathrm{Rot}_{\hat{z}}(\alpha)I_3 = (\hat{x}_1, \hat{v}, \hat{z}) \in \mathrm{SO}(3).$$

Next, we observe that $\hat{z}$ and $\hat{z}'$ are orthogonal to $\hat{v}$. By Lemma 4.2.29 we can choose $\beta \in \mathbb{R}$ so that $\mathrm{Rot}_{\hat{v}}(\beta)\hat{z} = \hat{z}'$. This rotation does not change $\hat{v}$ and maps $\hat{x}_1$ to some unit vector $\hat{x}_2 \in \mathbb{R}^3$. By Proposition 4.2.25 and (4.2.41) we can write this rotation as

$$(4.2.42) \qquad\qquad \mathrm{Rot}_{\hat{v}}(\beta) = B_1\,\mathrm{Rot}_{\hat{y}}(\beta)B_1^{-1} = \mathrm{Rot}_{\hat{z}}(\alpha)\,\mathrm{Rot}_{\hat{y}}(\beta)B_1^{-1}.$$

Applying this rotation to B_1 we obtain the basis

$$(4.2.43) \qquad\qquad B_2 = \mathrm{Rot}_{\hat{v}}(\beta)B_1 = \mathrm{Rot}_{\hat{z}}(\alpha)\,\mathrm{Rot}_{\hat{y}}(\beta)I_3 = (\hat{x}_2, \hat{v}, \hat{z}') \in \mathrm{SO}(3).$$

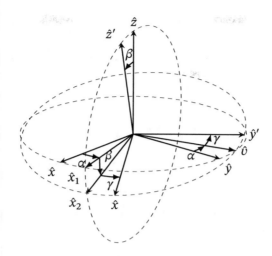

Figure 4.2.5. Euler angles.

Finally, we note that $\hat{v}$ and $\hat{y}'$ are both orthogonal to $\hat{z}'$. By Lemma 4.2.29 we can choose $\gamma \in \mathbb{R}^3$ such that $\mathrm{Rot}_{\hat{z}'}(\gamma)\hat{v} = \hat{y}'$. This rotation does not change $\hat{z}'$ and maps $\hat{x}_2$ to some unit vector $\hat{x}_3$. By Proposition 4.2.25 and (4.2.43) this rotation is

$$(4.2.44) \qquad \mathrm{Rot}_{\hat{z}'}(\gamma) = B_2 \, \mathrm{Rot}_{\hat{z}}(\alpha)B_2^{-1} = \mathrm{Rot}_{\hat{z}}(\alpha) \, \mathrm{Rot}_{\hat{y}}(\beta) \, \mathrm{Rot}_{\hat{z}}(\gamma)B_2^{-1}.$$

Applying this rotation to B_2 we obtain the basis

$$(4.2.45) \qquad B_3 = \mathrm{Rot}_{\hat{z}}(\alpha) \, \mathrm{Rot}_{\hat{y}}(\beta) \, \mathrm{Rot}_{\hat{z}}(\gamma) = (\hat{x}_3, \hat{y}', \hat{z}') \in \mathrm{SO}(3).$$

Since $O = (\hat{x}, \hat{y}', \hat{z}') \in \mathrm{SO}(3)$, it follows from Proposition 4.2.7 that $\hat{x}_3 = \hat{x}$ and $O = B_3$. This concludes the proof. $\qquad\square$

The next exercise generalizes Theorem 4.2.30.

Exercise 4.2.31. Let $\hat{v}, \hat{w} \in \mathbb{R}^3$ be unit vectors that are orthogonal to each other. Show that for all $O \in \mathrm{SO}(3)$ there are $\alpha, \beta, \gamma \in \mathbb{R}$ such that

$$(4.2.46) \qquad O = \mathrm{Rot}_{\hat{w}}(\alpha) \, \mathrm{Rot}_{\hat{v}}(\beta) \, \mathrm{Rot}_{\hat{w}}(\gamma).$$

If in Exercise 4.2.31 the rotation axes $\hat{v}$ and $\hat{w}$ are not orthogonal, then, in general, a decomposition (4.2.46) does not exist. However, we can prove a weaker result, for which we need the following lemma.

Lemma 4.2.32. *Let* $\hat{w}, \hat{w}' \in \mathbb{R}^3$ *be unit vectors, and let* $\gamma \in \mathbb{R}$. *Also, let* $O \in \mathrm{SO}(3)$ *with* $O\hat{w} = \hat{w}'$. *Then*

$$(4.2.47) \qquad \mathrm{Rot}_{\hat{w}'}(\gamma) = O \, \mathrm{Rot}_{\hat{w}}(\gamma)O^{-1}.$$

Proof. Let $B = (\hat{u}, \hat{v}, \hat{w}) \in \mathrm{SO}(3)$. This matrix exists by Theorem 4.2.8. Set $B' = OB = (\hat{u}', \hat{v}', \hat{w}')$. Then it follows from Proposition 4.2.25 that $\mathrm{Rot}_{\hat{w}'}(\gamma) = B' \, \mathrm{Rot}_{\hat{z}}(\gamma)(B')^{-1} = OB \, \mathrm{Rot}_{\hat{z}}(\gamma)B^{-1}O^{-1} = OR_{\hat{w}}(\gamma)O^{-1}$. $\qquad\square$

Here is our second decomposition theorem with respect to nonparallel rotation axes that are not required to be orthogonal to each other.

Theorem 4.2.33. *Let $\hat{a}, \hat{b} \in \mathbb{R}^3$ be nonparallel unit vectors. Denote by φ the angle between $\hat{a}$ and $\hat{b}$. Then for all $O \in SO(3)$ there are $k \in \mathbb{N}$ and $\alpha_1, \ldots, \alpha_k, \beta_1, \ldots, \beta_k \in \mathbb{R}$ such that $k = O(1/\varphi)$ and*

$$(4.2.48) \qquad O = \prod_{i=1}^{k} \mathrm{Rot}_{\hat{a}}(\alpha_i) \, \mathrm{Rot}_{\hat{b}}(\beta_i).$$

Proof. Let $O \in SO(3)$. Then by Proposition 4.2.27 we can write

$$(4.2.49) \qquad\qquad O = \mathrm{Rot}_{\hat{w}}(\gamma)$$

with a unit vector $\hat{w} \in \mathbb{R}^3$ and $\gamma \in \mathbb{R}$. We will show that there are $l \in \mathbb{N}$ and real numbers $\alpha_1, \ldots, \alpha_l, \beta_1, \ldots, \beta_l$ such that for

$$(4.2.50) \qquad\qquad O_1 = \prod_{i=1}^{l} \mathrm{Rot}_{\hat{a}}(\alpha_i) \, \mathrm{Rot}_{\hat{b}}(\beta_l)$$

we have $\hat{b} = O_1 \hat{w}$. It then follows from Lemma 4.2.32 that

$$(4.2.51) \qquad\qquad O = O_1^{-1} \, \mathrm{Rot}_{\hat{b}}(\gamma) O_1$$

which is a decomposition as in (4.2.48). We will also show that $l = O(1/\varphi)$. Then the theorem is proved. In order to keep the proof as simple as possible, we will give geometric arguments. They can be verified algebraically using the terminology introduced so far.

First, we observe that a rotation about $\hat{a}$ brings $\hat{w}$ into the plane P spanned by $\hat{a}$ and $\hat{b}$. So, we assume that $\hat{w}$ is in this plane. Next, we may assume that the initial positions of the three vectors $\hat{a}$, $\hat{b}$, and $\hat{w}$ are as shown in Figure 4.2.6. Therefore, they are in the half-plane above the dashed line orthogonal to $\hat{a}$ and also in the half-plane to the right of $\hat{a}$. This can be achieved as follows. Vectors that are below the dashed line are multiplied by -1. This is justified by Exercise 4.2.26. If $\hat{b}$ is on the wrong side of $\hat{a}$, then we exchange $\hat{a}$ and $\hat{b}$. Also, if $\hat{w}$ is on the wrong side of $\hat{a}$, then we apply a rotation about $\hat{a}$ through an angle of π to $\hat{w}$.

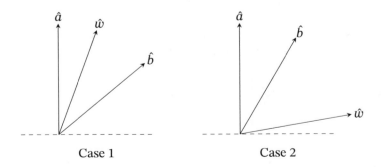

Case 1   Case 2

Figure 4.2.6. The two cases in the proof of Theorem 4.2.33 for possible initial positions of the vectors $\hat{a}$, $\hat{b}$, and $\hat{w}$.

In the situation shown in Figure 4.2.6 there are two possible cases. In Case 1, the vector $\hat{w}$ is on the left of $\hat{b}$, and in Case 2, it is on the right of $\hat{b}$. We prove the assertion in both cases.

The proof in the first case is illustrated in Figure 4.2.7. In this case, $\hat{w}$ is in the half-plane to the left of $\hat{b}$ and the angle θ between $\hat{a}$ and $\hat{w}$ is at most as big as the angle φ between $\hat{a}$ and $\hat{b}$. If $\theta = \varphi$, then we are done. Therefore, we assume that $\theta < \varphi$. Suppose that we rotate $\hat{w}$ about $\hat{b}$ through an angle of $\beta \in [0, \pi]$. Denote the rotated vector by $\hat{w}(\beta)$ and the angle between $\hat{w}(\beta)$ and $\hat{a}$ by $\theta(\beta)$. Then $\theta(0) = \theta < \varphi$ and $\theta(\pi) > \varphi$. Since the function $[0, \pi] \to \mathbb{R}$, $\beta \mapsto \theta(\beta)$ is continuous, it follows from the intermediate value theorem that there is $\beta \in [0, \pi]$ such that $\theta(\beta) = \varphi$. We apply the rotation $\text{Rot}_{\hat{b}}(\beta)$ to $\hat{w}$ and obtain $\hat{w}'$ such that the angle between $\hat{a}$ and $\hat{w}'$ is equal to the angle between $\hat{a}$ and $\hat{b}$. A rotation of $\hat{w}'$ about $\hat{a}$ through some angle $\alpha \in \mathbb{R}$ sends $\hat{w}'$ to $\hat{b}$.

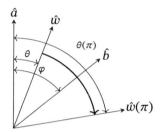

Figure 4.2.7. Illustration of the proof of Theorem 4.2.33 in Case 1.

Now we turn to the second case where $\hat{w}$ is in the half-plane on the right side of $\hat{b}$. We show how to use rotations of $\hat{w}$ about $\hat{a}$ and $\hat{b}$ to obtain Case 1. This construction is illustrated in Figure 4.2.8. We set $\hat{w}_0 = \hat{w}$ and construct a finite sequence $\hat{w}_1, \ldots, \hat{w}_m$, $m \in \mathbb{N}$, such that $\hat{w}_i$ is obtained from $\hat{w}_{i-1}$ by rotations about $\hat{a}$ and $\hat{b}$ and $\hat{w}_m$ is for the first time between $\hat{a}$ and $\hat{b}$. For $i \in \{0, \ldots, m\}$ we denote by α_i the angle between $\hat{a}$ and $\hat{w}_i$ and by β_i the angle between $\hat{w}_i$ and $\hat{b}$. Furthermore, we denote by φ the angle between $\hat{a}$ and $\hat{b}$. To construct $\hat{w}_1$ from $\hat{w}_0$, we rotate $\hat{w}_0$ about $\hat{b}$ through an angle π. If $\beta_0 \leq \phi$, then $\hat{w}_1$ is between $\hat{a}$ and $\hat{b}$ and we are in Case 1. In Figure 4.2.8 this is not the case. If $\beta_0 > \varphi$, then we have $\alpha_1 = \beta_1 - \varphi = \beta_0 - \varphi$. Since $\beta_0 < \alpha_0$, it follows that $\alpha_1 < \alpha_0 - \varphi$. Next, we construct $\hat{w}_2$ by a rotation of $\hat{w}_1$ about $\hat{a}$ through

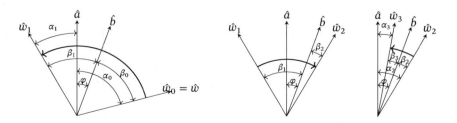

Figure 4.2.8. Illustration of the proof of Theorem 4.2.33 in Case 2.

an angle π. If $\alpha_1 \leq \varphi$, then $\hat{w}_2$ is between $\hat{a}$ and $\hat{b}$ and we are in Case 1. Otherwise, we have $\beta_2 = \alpha_2 - \varphi = \alpha_1 - \varphi < \alpha_0 - 2\varphi$. If we continue this construction, we obtain $\alpha_{2i+1} < \alpha_0 - (2i+1)\varphi$ as long as $\beta_{2i} > \varphi$. Also, we obtain $\beta_{2i+2} < \alpha_0 - (2i+2)\varphi$ as long as $\alpha_{2i+1} > \varphi$. Since $0 \leq \alpha_0 \leq \pi/2$, this implies that $m = O(1/\varphi)$. $\qquad\square$

4.3. Rotation operators

In this section, we introduce rotation operators. We will show that applying such operators to a quantum state $|\psi\rangle$ in $\mathbb{H}_1$ means applying a rotation to the corresponding point $\vec{p}(\psi)$ on the Bloch sphere. We will prove that the set of all these operators is the special unitary group SU(2), i.e., the group of all unitary operators on $\mathbb{H}_1$ with determinant 1, which was introduced in Theorem 2.4.20. We will also construct an isomorphism between SU(2)/$\{\pm I\}$ and the group SO(3) of rotations in $\mathbb{R}^3$. This allows us to obtain decomposition theorems for rotation operators from Theorems 4.2.30 and 4.2.33.

4.3.1. Basics. This section introduces rotation operators.

Let

(4.3.1) $\sigma = (X, Y, Z)$

be the triplet of Pauli operators. For all $\vec{p} = (p_x, p_y, p_z) \in \mathbb{R}^3$ set

(4.3.2) $\vec{p} \cdot \sigma = p_x X + p_y Y + p_z Z.$

Example 4.3.1. We have $\hat{x} \cdot \sigma = (1, 0, 0) \cdot \sigma = X$, $\hat{y} \cdot \sigma = (0, 1, 0) \cdot \sigma = Y$, and $\hat{z} \cdot \sigma = (0, 0, 1) \cdot \sigma = Z$.

For the definition of rotation operators, we need the following proposition.

Proposition 4.3.2. *Let $\hat{p} \in \mathbb{R}^3$ be a unit vector. Then $\hat{p} \cdot \sigma$ is a Hermitian unitary involution with trace 0 and eigenvalues ± 1.*

Proof. We use Theorem 4.1.2 and obtain the following. Since the Pauli operators are Hermitian operators with trace 0, the operator $\hat{p} \cdot \sigma$ is also Hermitian and has trace 0. Let $\hat{p} = (p_x, p_y, p_z)$. Due to $\|\hat{p}\| = 1$ we have

$$\begin{aligned}
(\hat{p} \cdot \sigma)^2 &= (p_x X + p_y Y + p_z Z)^2 \\
&= p_x^2 X^2 + p_y^2 Y^2 + p_z^2 Z^2 + p_x p_y (XY + YX) \\
&\quad + p_x p_z (XZ + YZ) + p_y p_z (YZ + ZY) \\
&= (p_x^2 + p_y^2 + p_z^2) I = I.
\end{aligned}$$

So $\hat{p} \cdot \sigma$ is an involution. But Hermitian involutions are unitary. Also, since $\hat{p} \cdot \sigma$ is a Hermitian involution of trace 0, this operator is diagonalizable by Theorem 2.4.53 and its eigenvalues are in $\{\pm 1\}$ by Proposition 2.4.60. But since (I, X, Y, Z) is a $\mathbb{C}$-basis of End$(\mathbb{H}_1)$ by Proposition 4.1.4, it follows that $\hat{p} \cdot \sigma \neq I$. So, the set of eigenvalues of $\hat{p} \cdot \sigma$ is $\{\pm 1\}$. $\qquad\square$

From Proposition 4.3.2, Corollary 2.4.73, and Proposition 2.4.74 we obtain the following result.

Proposition 4.3.3. *For all unit vectors $\hat{w} \in \mathbb{R}^3$ and all $\gamma \in \mathbb{R}$*

$$(4.3.3) \qquad e^{-i\gamma\, \hat{w}\cdot\sigma/2} = \cos\frac{\gamma}{2} I - i\sin\frac{\gamma}{2} \hat{w}\cdot\sigma$$

is a unitary operator on $\mathbb{H}_1$ with determinant 1, i.e., in SU(2).

Proposition 4.3.3 justifies the following definition.

Definition 4.3.4. A *rotation gate* or *rotation operator* is an operator

$$(4.3.4) \qquad R_{\hat{w}}(\gamma) = e^{-i\gamma\, \hat{w}\cdot\sigma/2} = \cos\frac{\gamma}{2} I - i\sin\frac{\gamma}{2} \hat{w}\cdot\sigma$$

on $\mathbb{H}_1$ where $\hat{w} \in \mathbb{R}^3$ is a unit vector and $\gamma \in \mathbb{R}$.

The name "rotation operator" comes from the fact that applying the operator from (4.3.4) to a quantum state in $\mathbb{H}_1$ means applying the rotation $\text{Rot}_{\hat{w}}(\gamma)$ to the corresponding point on the Bloch sphere. This will be shown in Theorem 4.3.20. The next exercise verifies this for the special case where $\hat{w} = \hat{z} = (0, 0, 1)$.

Exercise 4.3.5. Show that for every $\gamma \in \mathbb{R}$ and every quantum state $|\psi\rangle \in \mathbb{H}_1$ we have

$$(4.3.5) \qquad \hat{p}\left(R_{\hat{z}}(\gamma)\,|\psi\rangle\right) = \text{Rot}_{\hat{z}}(\gamma)\hat{p}(\psi).$$

From Exercise 2.4.71 we obtain the following result.

Proposition 4.3.6. *For all unit vectors $\hat{w} \in \mathbb{R}^3$ and all $\beta, \gamma \in \mathbb{R}$ we have*

$$(4.3.6) \qquad R_{\hat{w}}(\beta)R_{\hat{w}}(\gamma) = R_{\hat{w}}(\beta + \gamma).$$

We define special rotations.

Definition 4.3.7. Let $\gamma \in \mathbb{R}$. The *rotation operators about the x-, y-, and z-axes through the angle γ* are defined as

$$(4.3.7) \qquad R_{\hat{x}}(\gamma) = e^{-i\gamma X/2}, \quad R_{\hat{y}}(\gamma) = e^{-i\gamma Y/2}, \quad R_{\hat{z}}(\gamma) = e^{-i\gamma Z/2},$$

respectively.

Here is another representation of the rotation operators that we have just introduced.

Proposition 4.3.8. *Let $\gamma \in \mathbb{R}$. Then we have*

$$(4.3.8) \qquad R_{\hat{x}}(\gamma) = \begin{pmatrix} \cos\frac{\gamma}{2} & -i\sin\frac{\gamma}{2} \\ -i\sin\frac{\gamma}{2} & \cos\frac{\gamma}{2} \end{pmatrix},$$

$$(4.3.9) \qquad R_{\hat{y}}(\gamma) = \begin{pmatrix} \cos\frac{\gamma}{2} & -\sin\frac{\gamma}{2} \\ \sin\frac{\gamma}{2} & \cos\frac{\gamma}{2} \end{pmatrix},$$

$$(4.3.10) \qquad R_{\hat{z}}(\gamma) = \begin{pmatrix} e^{-i\frac{\gamma}{2}} & 0 \\ 0 & e^{i\frac{\gamma}{2}} \end{pmatrix}.$$

Exercise 4.3.9. Prove Proposition 4.3.8

The next exercise shows that iX, iY, iZ, and XH are rotation operators.

Exercise 4.3.10. Show that $R_{\hat{x}}(\pi) = iX$, $R_{\hat{y}}(\pi) = iY$, $R_{\hat{z}}(\pi) = iZ$, and $H = XR_{\hat{y}}\left(\frac{\pi}{2}\right)$.

4.3.2. The group of rotation operators. Our next goal is to show that the set of rotation operators on $\mathbb{H}_1$ is SU(2). For this, we use the following characterization of SU(2) which follows from Corollary 2.4.73:

(4.3.11) $\text{SU}(2) = \{e^{iA} : A \text{ Hermitian and } \operatorname{tr} A = 0\}.$

The following definition will simplify our discussion.

Definition 4.3.11. The set of all Hermitian operators on $\mathbb{H}_1$ with trace 0 is denoted by su(2).

We note that we slightly deviate from the standard notation in mathematics where su(2) typically denotes the Lie algebra that consists of all 2×2 skew-Hermitian matrices with trace 0.

The elements of su(2) can be characterized as follows.

Lemma 4.3.12. *Let* $A \in \mathbb{C}^{(2,2)}$. *Then* $A \in \text{su}(2)$ *if and only if there are* $a \in \mathbb{R}$ *and* $b \in \mathbb{C}$ *such that*

(4.3.12) $$A = \begin{pmatrix} a & \bar{b} \\ b & -a \end{pmatrix}.$$

Exercise 4.3.13. Prove Lemma 4.3.12.

The next proposition uses Lemma 4.3.12 to describe the structure of su(2).

Proposition 4.3.14. *The set* su(2) *is a real three-dimensional vector space. The triplet* $\sigma = (X, Y, Z)$ *of the three Pauli operators is an* $\mathbb{R}$-*basis of* su(2) *that is orthogonal with respect to the Hilbert-Schmidt inner product.*

Proof. Let $A \in \text{su}(2)$. By Lemma 4.3.12 we can write

(4.3.13) $A = (\mathfrak{R}b)X + (\mathfrak{I}b)Y + aZ$

where $a \in \mathbb{R}$ and $b \in \mathbb{C}$. This implies that (X, Y, Z) is a generating system of su(2). But it follows from Proposition 4.1.4 that the triplet (X, Y, Z) is linearly independent over $\mathbb{R}$ and orthogonal with respect to the Hilbert-Schmidt inner product. Therefore, su(2) is a real three-dimensional vector space. $\square$

We now show that the operators in SU(2) are rotation operators.

Theorem 4.3.15. *The set of rotation operators on* $\mathbb{H}_1$ *is* SU(2). *Moreover, if* $U \in \text{SU}(2)$, *then the following hold.*

(1) $U = I$ *if and only if* $U = R_{\hat{w}}(\gamma)$ *with a unit vector* $\hat{w} \in \mathbb{R}^3$ *and* $\gamma/2 \equiv 0 \mod 2\pi$.

(2) $U = -I$ *if and only if* $U = R_{\hat{w}}(\gamma)$ *with a unit vector* $\hat{w} \in \mathbb{R}^3$ *and* $\gamma/2 \equiv \pi \mod 2\pi$.

(3) *Let* $U \neq \pm I$.
   (a) *There are a unit vector* $\hat{w} \in \mathbb{R}^3$ *and* $\gamma \in \mathbb{R}$ *such that* $U = R_{\hat{w}}(\gamma)$.
   (b) *If* $\hat{w}' \in \mathbb{R}^3$ *is a unit vector and* $\gamma' \in \mathbb{R}$, *then* $U = R_{\hat{w}'}(\gamma')$ *if and only if* $\hat{w} = \hat{w}'$ *and* $\gamma/2 \equiv \gamma'/2 \mod 2\pi$ *or* $\hat{w} = -\hat{w}'$ *and* $\gamma/2 \equiv -\gamma'/2 \mod 2\pi$.

Proof. Let $\hat{w} \in \mathbb{R}^3$ be a unit vector. It can be easily verified that $R_{\hat{w}}(\gamma) = I$ if $\gamma/2 \equiv 0 \bmod 2\pi$. Also, by Proposition 4.1.4 the coefficients of a representation of I as a linear combination of I, X, Y, Z are uniquely determined. So if $I = R_{\hat{w}}(\gamma)$ with $\gamma \in \mathbb{R}$, then it follows from (4.3.3) that we have $\cos \gamma/2 = 1$ and $\sin \gamma/2 = 0$. This implies $\gamma/2 \equiv 0 \bmod 2\pi$. The second assertion can be proved analogously.

We prove the third statement. It follows from (4.3.11) that $U = e^{iA}$ with $A \in \mathfrak{su}(2)$. By Proposition 4.3.14 there is a uniquely determined $\vec{p} \in \mathbb{R}^3$ such that $A = \vec{p} \cdot \sigma$. Since $U \neq \pm I$ it follows that $\vec{p}$ is nonzero. Set $\gamma = 2 \|\vec{p}\| \bmod 4\pi$ and $\hat{w} = -\vec{p}/\|\vec{p}\|$. Then $\hat{w}$ is a unit vector, and we have $U = e^{-i\gamma \hat{w} \cdot \sigma/2}$.

Next, let $\hat{w}' \in \mathbb{R}^3$ be a unit vector and let $\gamma' \in \mathbb{R}$ be such that $U = R_{\hat{w}'}(\gamma')$. Then $\cos \gamma/2 \neq \pm 1$ and $\sin \gamma/2 \neq 0$. The uniqueness of the coefficient of I in (4.3.3) implies $\gamma/2 \equiv \pm \gamma'/2 \bmod 2\pi$. If $\gamma/2 \equiv \gamma'/2 \bmod 2\pi$, then $\sin \gamma/2 = \sin \gamma'/2$ and due to the uniqueness of the coefficients of X, Y, and Z in (4.3.3) we have $\hat{w}' = \hat{w}$. If $\gamma/2 \equiv -\gamma'/2 \bmod 2\pi$, then $\sin \gamma/2 = -\sin \gamma'/2$ and because of the uniqueness of the coefficients of X, Y, and Z in (4.3.3) we have $\hat{w}' = -\hat{w}$. $\square$

Theorem 4.3.15 implies that, up to a global phase factor, every unitary operator on $\mathbb{H}_1$ is a rotation operator. This is what the next corollary says.

Corollary 4.3.16. *Let $U \in U(2)$. Then there is $\delta \in \mathbb{R}$ such that $e^{-i\delta} U$ is a rotation operator on $\mathbb{H}_1$.*

Proof. Since $|\det U| = 1$ we can choose $\delta \in \mathbb{R}$ such that $\det U = e^{i2\delta}$. So $\det(e^{-i\delta} U) = 1$ which implies that $e^{-i\delta} U \in SU(2)$. So by Theorem 4.3.15, $e^{-i\delta} U$ is a rotation operator. $\square$

Theorem 4.3.15 also implies the next corollary which, in turn, allows us to assign a uniquely determined rotation of $\mathbb{R}^3$ to each rotation operator on $\mathbb{H}_1$ in the most obvious way.

Corollary 4.3.17. *Let $U \in SU(2)$. Then for all unit vectors $\hat{w} \in \mathbb{R}$ and $\gamma \in \mathbb{R}$ such that $U = R_{\hat{w}}(\gamma)$ the rotation $\text{Rot}_{\hat{w}}(\gamma)$ is the same.*

Exercise 4.3.18. Use Theorem 4.3.15 and Proposition 4.2.27 to prove Corollary 4.3.17.

Corollary 4.3.17 justifies the following definition.

Definition 4.3.19. Let $U \in SU(2)$ and let $U = R_{\hat{w}}(\gamma)$ with a unit vector $\hat{w} \in \mathbb{R}^3$ and $\gamma \in \mathbb{R}$. Then we set $\text{Rot}(U) = \text{Rot}_{\hat{w}}(\gamma)$.

4.3.3. Rotation operators and rotations on the Bloch sphere.

After the preparations of the preceding section, we are now ready to prove the following important theorem.

Theorem 4.3.20. *The map*

$$(4.3.14) \qquad \text{Rot} : SU(2) \to SO(3), \quad U \mapsto \text{Rot}(U)$$

is a surjective group homomorphism with kernel $\pm I$. Furthermore, for all $U \in SU(2)$ and all quantum states $|\psi\rangle$ in $\mathbb{H}_1$ the point on the Bloch sphere corresponding to $|\psi\rangle$ is

$$(4.3.15) \qquad \vec{p}(U |\psi\rangle) = \text{Rot}(U) \vec{p}(\psi).$$

We start by verifying the surjectivity of Rot. Let $O \in \mathrm{SO}(3)$. By Proposition 4.2.27 we have $O = \mathrm{Rot}_{\hat{w}}(\gamma)$ with a unit vector $\hat{w} \in \mathbb{R}^3$ and $\gamma \in \mathbb{R}$. Set $U = R_{\hat{w}}(\gamma)$. Then $U \in \mathrm{SU}(2)$ by Proposition 4.3.3 and $O = \mathrm{Rot}(U)$ by Definition 4.3.19.

In order to prove the other properties of Rot, we need some notation and several auxiliary results, which we present now.

Definition 4.3.21. Let $\tau = (\tau_u, \tau_v, \tau_w) \in \mathrm{su}(2)^3$.

(1) For all $\vec{p} = (p_u, p_v, p_w) \in \mathbb{R}^3$ we set

$$(4.3.16) \qquad \vec{p} \cdot \tau = p_u \tau_u + p_v \tau_v + p_w \tau_w.$$

(2) If $B = (\hat{u}, \hat{v}, \hat{w}) \in \mathbb{R}^{(3,3)}$, then we define

$$(4.3.17) \qquad B \cdot \tau = (\hat{u} \cdot \tau, \hat{v} \cdot \tau, \hat{w} \cdot \tau).$$

Lemma 4.3.22. *Let* $\tau \in \mathrm{su}(2)^3$, $B \in \mathbb{R}^{(3,3)}$, *and* $\vec{p} \in \mathbb{R}^3$. *Then we have*

$$(4.3.18) \qquad (B\vec{p}) \cdot \tau = \vec{p} \cdot (B \cdot \tau).$$

Exercise 4.3.23. Prove Lemma 4.3.22.

Lemma 4.3.24. *Let* $\vec{p}, \vec{q} \in \mathbb{R}^3$. *Then we have*

$$(4.3.19) \qquad (\vec{p} \cdot \sigma)(\vec{q} \cdot \sigma) = \langle \vec{p} | \vec{q} \rangle I + i\, \vec{p} \times \vec{q} \cdot \sigma.$$

Proof. Let $\vec{p} = (p_x, p_y, p_z), \vec{q} = (q_x, q_y, q_z) \in \mathbb{R}^3$. Then we obtain from Theorem 4.1.2

$$\begin{aligned}
(\vec{p} \cdot \sigma)(\vec{q} \cdot \sigma) &= (p_x X + p_y Y + p_z Z)(q_x X + q_y Y + q_z Z) \\
&= p_x q_x X^2 + p_y q_y Y^2 + p_z q_z Z^2 \\
&\quad + p_x q_y XY + p_y q_x YX + p_x q_z XZ + p_z q_x ZX + p_y q_z YZ + p_z q_y ZY \\
&= \langle \vec{p} | \vec{q} \rangle I + i(Z(p_x q_y - p_y q_x) + Y(p_z q_x - p_x q_z) + Z(p_x q_y - p_y q_z)) \\
&= \langle \vec{p} | \vec{q} \rangle I + i\, \vec{p} \times \vec{q} \cdot \sigma.
\end{aligned}$$

This proves the assertion. $\qquad \square$

Proposition 4.3.25. *Let* $B \in \mathrm{SO}(3)$ *and write* $\tau = (\tau_u, \tau_v, \tau_w) = B \cdot \sigma$. *Then we have*

$$(4.3.20) \qquad \tau_u^2 = \tau_v^2 = \tau_w^2 = I,$$

$$(4.3.21) \qquad \tau_u \tau_v = i\tau_w = -\tau_v \tau_u, \quad \tau_w \tau_u = i\tau_v = -\tau_u \tau_w, \quad \tau_v \tau_w = i\tau_u = -\tau_w \tau_v,$$

and

$$(4.3.22) \qquad -i\tau_u \tau_v \tau_w = I.$$

Proof. First, (4.3.20) follows from Proposition 4.3.2. Let $B = (\hat{u}, \hat{v}, \hat{w})$ with column vectors $\hat{u}, \hat{v}, \hat{w} \in \mathbb{R}^3$. From Lemma 4.3.24 and Theorem 4.2.8 we obtain

$$(4.3.23) \qquad \tau_u \tau_v = (\hat{u} \cdot \sigma)(\hat{v} \cdot \sigma) = \langle \hat{u} | \hat{v} \rangle I + i(\hat{u} \times \hat{v}) \cdot \sigma = i\, \hat{w} \cdot \sigma = i\tau_w.$$

The other identities in (4.3.21) can be proved analogously. Finally, from (4.3.20) and (4.3.21) we obtain

$$(4.3.24) \qquad -i\tau_u \tau_v \tau_w = (-i)i\tau_w \tau_w = I. \qquad \square$$

Lemma 4.3.26. *Let $U \in \mathrm{SU}(2)$ and let $\vec{p} \in \mathbb{R}^3$. Then we have*

(4.3.25)
$$(\mathrm{Rot}(U)\vec{p}) \cdot \sigma = U(\vec{p} \cdot \sigma)U^{-1}.$$

Proof. Let $U = R_{\hat{w}}(\gamma)$ with a unit vector $\hat{w} \in \mathbb{R}^3$ and $\gamma \in \mathbb{R}$ which exist by Theorem 4.3.15. Let $B = (\hat{u}, \hat{v}, \hat{w}) \in \mathrm{SO}(3)$ which exists by Theorem 4.2.8. Set

(4.3.26)
$$\tau = (\tau_u, \tau_v, \tau_w) = B \cdot \sigma.$$

Also let $\vec{q} = B^{-1}\vec{p}$. Then Proposition 4.2.25 and Lemma 4.3.22 imply

(4.3.27)
$$(\mathrm{Rot}(U)\vec{p}) \cdot \sigma = (\mathrm{Rot}_{\hat{w}}(\gamma)\vec{p}) \cdot \sigma = (B\,\mathrm{Rot}_{\hat{z}}(\gamma)B^{-1}\vec{p}) \cdot \sigma$$
$$= (B\,\mathrm{Rot}_{\hat{z}}(\gamma)\vec{q}) \cdot \sigma = (\mathrm{Rot}_{\hat{z}}(\gamma)\vec{q}) \cdot \tau$$

and

(4.3.28)
$$U(\vec{p} \cdot \sigma)U^{-1} = U(B\vec{q} \cdot \sigma)U^{-1} = U(\vec{q} \cdot \tau)U^{-1}.$$

So it suffices to show that

(4.3.29)
$$(\mathrm{Rot}_{\hat{z}}(\gamma)\vec{q}) \cdot \tau = U(\vec{q} \cdot \tau)U^{-1}.$$

Since the expressions on the left and right side of (4.3.29) are linear in $\vec{q}$, it suffices to prove this identity for $\vec{q} \in \{\hat{x}, \hat{y}, \hat{z}\}$. This is done in Exercise 4.3.27. $\square$

Exercise 4.3.27. Verify (4.3.29) in the proof of Lemma 4.3.26 for $\vec{q} = \hat{x} = (1, 0, 0)$, $\vec{q} = \hat{y} = (0, 1, 0)$, and $\vec{q} = \hat{z} = (0, 0, 1)$ using Proposition 4.3.25 and the trigonometric identities in Section A.5.

The linearity of the map Rot can now be seen as follows. Lemma 4.3.26 implies that for all $U_1, U_2 \in \mathrm{SU}(2)$ and all $\vec{p} \in \mathbb{R}^3$ we have

(4.3.30)
$$\left(\mathrm{Rot}(U_1 U_2)\vec{p}\right) \cdot \sigma = U_1 U_2(\vec{p} \cdot \sigma)U_2^{-1}U_1^{-1}$$
$$= U_1\left((\mathrm{Rot}(U_2)\vec{p}) \cdot \sigma\right)U_1^{-1} = \left(\mathrm{Rot}(U_1)\,\mathrm{Rot}(U_2)\vec{p}\right) \cdot \sigma.$$

We determine the kernel of Rot. Let $U \in \mathrm{SU}(2)$. Write $U = R_{\hat{w}}(\gamma)$ as in Theorem 4.3.15. Then it follows from the definition of rotations in Theorem 4.2.21 that $\mathrm{Rot}_{\hat{w}}(\gamma) = I_3$ if and only if $\gamma \equiv 0 \bmod 2\pi$. This is true if and only if $\gamma/2 \equiv 0 \bmod \pi$. Therefore, Theorem 4.3.15 implies that $\mathrm{Rot}(U) = I_3$ if and only if $U = \pm I$.

Next, we prove the second assertion of Theorem 4.3.20. For this, we need further auxiliary results.

Lemma 4.3.28. *Let $\vec{p} \in \mathbb{R}^3$ with spherical coordinate representation $(1, \theta, \phi)$. Then we have*

(4.3.31)
$$\vec{p} \cdot \sigma = \begin{pmatrix} \cos\theta & e^{-i\phi}\sin\theta \\ e^{i\phi}\sin\theta & -\cos\theta \end{pmatrix}.$$

Exercise 4.3.29. Prove Lemma 4.3.28.

From Lemma 4.3.28 we obtain the following representation of the density operators corresponding to quantum states in $\mathbb{H}_1$.

Proposition 4.3.30. *Let $|\psi\rangle$ be a quantum state in $\mathbb{H}_1$. Then we have*

(4.3.32)
$$|\psi\rangle\langle\psi| = \frac{1}{2}(I + \vec{p}(\psi) \cdot \sigma).$$

Proof. Let $(1, \theta, \phi)$ be the spherical coordinate representation of $\vec{p}(\psi)$. Without loss of generality let

$$(4.3.33) \qquad |\psi\rangle = \cos\frac{\theta}{2}|0\rangle + e^{i\phi}\sin\frac{\theta}{2}|1\rangle.$$

Using the trigonometric identities (A.5.4) and (A.5.7) we obtain

$$(4.3.34) \quad \begin{aligned} (2|\psi\rangle\langle\psi| - I)|0\rangle &= 2|\psi\rangle\langle\psi|0\rangle - |0\rangle = 2\cos\frac{\theta}{2}|\psi\rangle - |0\rangle \\ &= \left(2\cos^2\frac{\theta}{2} - 1\right)|0\rangle + 2e^{i\phi}\cos\frac{\theta}{2}\sin\frac{\theta}{2}|1\rangle = \cos\theta|0\rangle + e^{i\phi}\sin\theta|1\rangle \end{aligned}$$

and

$$(4.3.35) \quad \begin{aligned} (2|\psi\rangle\langle\psi| - I)|1\rangle &= 2|\psi\rangle\langle\psi|1\rangle - |1\rangle = 2e^{-i\phi}\sin\frac{\theta}{2}|\psi\rangle - |1\rangle \\ &= 2e^{-i\phi}\sin\frac{\theta}{2}\cos\frac{\theta}{2}|0\rangle + \left(2\sin^2\frac{\theta}{2} - 1\right)|1\rangle \\ &= e^{-i\phi}\sin\theta|0\rangle - \cos\theta|1\rangle. \end{aligned}$$

So the assertion follows from Lemma 4.3.28. $\qquad\square$

With these results, we can prove the second assertion of Theorem 4.3.20. For this, let $U \in \mathrm{SU}(2)$ and let $|\psi\rangle$ be a quantum state in $\mathbb{H}_1$. Set $\vec{p} = \vec{p}(\psi)$ and $\vec{q} = \vec{p}(U|\psi\rangle)$. Then it follows from Proposition 4.3.30 that

$$(4.3.36) \qquad |U|\psi\rangle\rangle\langle U|\psi\rangle| = \frac{1}{2}(I + \vec{q}\cdot\sigma)$$

and

$$(4.3.37) \qquad U|\psi\rangle\langle\psi|U^{-1} = \frac{1}{2}(I + U\vec{p}\cdot\sigma U^{-1}).$$

But Lemma 4.3.26 gives

$$(4.3.38) \qquad U\vec{p}\cdot\sigma U^{-1} = (\mathrm{Rot}(U)\vec{p})\cdot\sigma.$$

So (4.3.36), (4.3.37), and (4.3.38) imply the assertion.

4.3.4. Decomposition of rotation operators.

From Corollary 4.3.16, Theorem 4.2.33, and Theorem 4.3.20 we obtain the following decomposition result for rotation gates.

Theorem 4.3.31. *For every $U \in \mathrm{U}(2)$ there are $\alpha, \beta, \gamma, \delta \in \mathbb{R}$ such that*

$$(4.3.39) \qquad U = e^{i\delta}R_{\hat{z}}(\alpha)R_{\hat{y}}(\beta)R_{\hat{z}}(\gamma).$$

If $U \in \mathrm{SU}(2)$, then there is such a representation with $\delta = 0$.

We will now use Theorem 4.3.31 to give another representation of unitary single-qubit operators, since Section 4.4 allows us to implement controlled operators. For this, we need the following lemma.

Lemma 4.3.32. *For all $\gamma \in \mathbb{R}$ we have $XR_{\hat{y}}(\gamma)X = R_{\hat{y}}(-\gamma)$ and $XR_{\hat{z}}(\gamma)X = R_{\hat{z}}(-\gamma)$.*

Proof. By Theorem 4.1.2 we have $X^2 = I$ and $XYX = -XXY = -Y$. So from (4.3.3) we obtain

$$XR_{\hat{y}}(\gamma)X = X\left(\cos\frac{\gamma}{2}I - i\sin\frac{\gamma}{2}Y\right)X$$
$$= \cos\frac{\gamma}{2}X^2 - i\sin\frac{\gamma}{2}XYX$$
$$= \cos\frac{-\gamma}{2}I - i\sin\frac{-\gamma}{2}Y$$
$$= R_{\hat{y}}(-\gamma).$$

The second assertion can be proved analogously. $\qquad\square$

Theorem 4.3.33. *Let U be a unitary operator on $\mathbb{H}_1$. Let $\alpha, \beta, \gamma, \delta \in \mathbb{R}$ such that*

$$(4.3.40) \qquad\qquad U = e^{i\delta}R_{\hat{z}}(\alpha)R_{\hat{y}}(\beta)R_{\hat{z}}(\gamma).$$

Such a representation exists by by Theorem 4.3.31. Set

$$(4.3.41) \qquad\qquad \begin{aligned} A &= R_{\hat{z}}(\alpha)R_{\hat{y}}\left(\frac{\beta}{2}\right), \\ B &= R_{\hat{y}}\left(-\frac{\beta}{2}\right)R_{\hat{z}}\left(-\frac{\alpha+\gamma}{2}\right), \\ C &= R_{\hat{z}}\left(-\frac{\alpha-\gamma}{2}\right). \end{aligned}$$

Then we have

$$(4.3.42) \qquad\qquad ABC = I \quad\text{and}\quad U = e^{i\delta}AXBXC.$$

Proof. Proposition 4.3.6 implies

$$(4.3.43) \qquad ABC = R_{\hat{z}}(\alpha)R_{\hat{y}}\left(\frac{\beta}{2}\right)R_{\hat{y}}\left(-\frac{\beta}{2}\right)R_{\hat{z}}\left(-\frac{\alpha+\gamma}{2}\right)R_{\hat{z}}\left(-\frac{\alpha-\gamma}{2}\right) = I.$$

Using $X^2 = I$ and Lemma 4.3.32 we find that

$$(4.3.44) \qquad \begin{aligned} XBX &= XR_{\hat{y}}\left(-\frac{\beta}{2}\right)XXR_{\hat{z}}\left(-\frac{\alpha+\gamma}{2}\right)X \\ &= R_{\hat{y}}\left(\frac{\beta}{2}\right)R_{\hat{z}}\left(\frac{\alpha+\gamma}{2}\right). \end{aligned}$$

Applying Proposition 4.3.6 again, we obtain

$$(4.3.45) \qquad \begin{aligned} AXBXC &= R_{\hat{z}}(\alpha)R_{\hat{y}}\left(\frac{\beta}{2}\right)R_{\hat{y}}\left(\frac{\beta}{2}\right)R_{\hat{z}}\left(\frac{\alpha+\gamma}{2}\right)R_{\hat{z}}\left(-\frac{\alpha-\gamma}{2}\right) \\ &= R_{\hat{z}}(\alpha)R_{\hat{y}}(\gamma)R_{\hat{z}}(\gamma). \end{aligned}$$

Hence(4.3.40) implies $U = e^{i\gamma}AXBXC$. Together with (4.3.43) this concludes the proof. $\qquad\square$

The next exercise gives a representation as in Theorem 4.3.33 for rotation operators.

Exercise 4.3.34. Let $\hat{w} \in \mathbb{R}^3$ be a unit vector and let $\gamma \in \mathbb{R}$. Set $A = R_{\hat{w}}(\gamma/2)$, $B = R_{\hat{w}}(-\gamma/2)$, and $C = I_2$. Show that $R_{\hat{w}}(\gamma) = AXBXC$ and that $ABC = I_2$.

From Corollary 4.3.16, Proposition 4.3.14, and Theorem 4.3.20 we obtain the following decomposition result.

Theorem 4.3.35. *Let $\vec{a}, \vec{b} \in \mathbb{R}^3$ be nonparallel unit vectors. Denote by φ the angle between $\vec{a}$ and $\vec{b}$. Then for all unitary operators U on $\mathbb{H}_1$ there are $k \in \mathbb{N}$ and real numbers $\alpha_1, \ldots, \alpha_k, \beta_1, \ldots, \beta_k, \delta$ such that $k = O(1/\varphi)$ and*

$$(4.3.46) \qquad U = e^{i\delta} \prod_{i=1}^{k} R_{\vec{a}}(\alpha_i) R_{\vec{b}}(\beta_i).$$

If $U \in SU(2)$, then there is such a representation with $\delta = 0$.

4.3.5. Phase shift gates. In this section, we introduce the following special class of rotation operators.

Definition 4.3.36. For $\gamma \in \mathbb{R}$ the *phase shift gate* $P(\gamma)$ is defined as

$$(4.3.47) \qquad P(\gamma) = \begin{pmatrix} 1 & 0 \\ 0 & e^{i\gamma} \end{pmatrix}.$$

It shifts the phase of the amplitude of $|1\rangle$ by an angle γ while it does not change the amplitude of $|0\rangle$.

The phase shift gate can also be written as

$$(4.3.48) \qquad P(\gamma) = e^{i\frac{\gamma}{2}} R_{\hat{z}}(\gamma).$$

The inverse and adjoint of $P(\gamma)$ is $P(-\gamma)$.

Next, we introduce special phase shift gates. For $k \in \mathbb{N}$ we set

$$(4.3.49) \qquad R_k = \begin{pmatrix} 1 & 0 \\ 0 & e^{\frac{2\pi i}{2^k}} \end{pmatrix} = e^{\frac{2\pi i}{2^{k-1}}} R_{\hat{z}}\left(\frac{2\pi}{2^{k-1}}\right).$$

For $k = 2$ we obtain the *phase gate*

$$(4.3.50) \qquad S = R_2 = P\left(\frac{\pi}{2}\right) = \begin{pmatrix} 1 & 0 \\ 0 & i \end{pmatrix}.$$

Also, for $k = 3$ this gives the $\pi/8$ *gate*

$$(4.3.51) \qquad T = R_3 = P\left(\frac{\pi}{4}\right) = \begin{pmatrix} 1 & 0 \\ 0 & e^{i\pi/4} \end{pmatrix}.$$

It is called the "$\pi/8$ gate" since it can be written as

$$(4.3.52) \qquad T = e^{\frac{i\pi}{8}} R_{\hat{z}}\left(\frac{\pi}{8}\right).$$

We note that

$$(4.3.53) \qquad T^2 = S.$$

Figure 4.3.1. Symbols for the phase and $\pi/8$ gates in quantum circuits.

The symbols that represent the phase gate and the $\pi/8$ gate are shown in Figure 4.3.1.

4.4. Controlled operators

So far, we have discussed single-qubit operators. However, they are not sufficient to construct quantum circuits that implement unitary operators on $\mathbb{H}_n$ for $n > 1$. For this purpose, multiple-qubit operators must be used. An important class of multiple-qubit operators is controlled operators, which are discussed in this section. They apply a unitary operator to some qubits, called the *target qubits*, conditioned on some other qubits, called the *control qubits*, being in certain quantum states. We start with the description of controlled-NOT gates. Then general controlled operators and several special cases, such as quantum Toffoli operators, are introduced. In our discussion, we let n be a positive integer and identify linear operators on the state space $\mathbb{H}_n$ with their representation matrices with respect to the computational basis of $\mathbb{H}_n$. For several gates that we introduce here, there are classical equivalents, which are presented in Section 1.7. The classical and the corresponding quantum gates are referred to by the same names. Also, some of these gates are already explained in Chapter 3 .

4.4.1. Controlled-NOT gates. We begin with controlled-NOT gates, also known as CNOT gates, which were introduced in Section 3.3.2. In Section 1.7.1, their classical equivalent CNOT was presented. Figure 4.4.1 illustrates four different types of such gates. All apply the Pauli X gate, i.e., the quantum NOT gate, to a target qubit based on the state of a control qubit. The control qubit can be either the first or the second qubit, and correspondingly, the target qubit can be either the second or the first qubit. Furthermore, the Pauli X gate may be applied to the target qubit depending on whether the control qubit is $|1\rangle$ or $|0\rangle$. The upper-left CNOT gate in Figure 4.4.1 is called the *standard* CNOT *gate*.

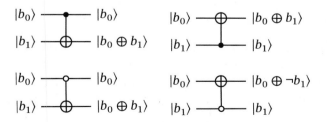

Figure 4.4.1. The four CNOT gates. The upper-left CNOT gate is called the *standard* CNOT *gate*.

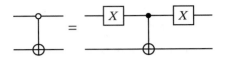

Figure 4.4.2. Implementation of the lower-left CNOT gate from Figure 4.4.1.

As seen in (3.3.5), the representation matrix of the standard CNOT gate with respect to the computational basis of $\mathbb{H}_2$ is

(4.4.1)
$$\text{CNOT} = \begin{pmatrix} 1 & 0 & 0 & 0 \\ 0 & 1 & 0 & 0 \\ 0 & 0 & 0 & 1 \\ 0 & 0 & 1 & 0 \end{pmatrix}.$$

Exercise 4.4.1. Determine matrix representations of all CNOT gates in Figure 4.4.1 with respect to the computational basis of $\mathbb{H}_2$.

We note that the lower-left CNOT gate in Figure 4.4.1 can be implemented using the standard CNOT gate and the Pauli X gate as shown in Figure 4.4.2.

Exercise 4.4.2. (1) Show that the implementation of the CNOT gate in Figure 4.4.2 is correct.

(2) Give an implementation of the lower-right CNOT gate in Figure 4.4.1 using the upper-right CNOT gate and the Pauli X gate.

Next, we note that the roles of qubits as control and target qubits in the CNOT gates depend on the choice of the basis of $\mathbb{H}_2$. To see this, consider the orthonormal basis

(4.4.2)
$$(|x_+\rangle, |x_-\rangle) = (H|0\rangle, H|1\rangle) = \left(\frac{|0\rangle + |1\rangle}{\sqrt{2}}, \frac{|0\rangle - |1\rangle}{\sqrt{2}} \right)$$

of $\mathbb{H}_1$ where H is the Hadamard operator. We have seen in Section 4.1.2 that it is an eigenbasis of the Pauli X operator. Also,

(4.4.3)
$$(|x_+ x_+\rangle, |x_- x_+\rangle, |x_+ x_-\rangle, |x_- x_-\rangle)$$

is an orthonormal basis of $\mathbb{H}_2$. As shown in Exercise 4.4.3, applying CNOT to the elements of this basis has the following effect:

(4.4.4)
$$\text{CNOT}|x_+ x_+\rangle = |x_+ x_+\rangle, \quad \text{CNOT}|x_- x_+\rangle = |x_- x_+\rangle,$$
$$\text{CNOT}|x_+ x_-\rangle = |x_- x_-\rangle, \quad \text{CNOT}|x_- x_-\rangle = |x_+ x_-\rangle.$$

Exercise 4.4.3. Prove (4.4.4).

We see that in (4.4.4) the CNOT operator exchanges the basis states $|x_+\rangle$ and $|x_-\rangle$ of the first qubit conditioned on the second qubit being in state $|x_-\rangle$. So in this representation of CNOT, the first qubit is the target, while the second qubit is the control. As shown in Figure 4.4.3, this observation can be used to implement the upper-right CNOT gate in Figure 4.4.1 using the standard CNOT gate and the Hadamard gate.

Exercise 4.4.4. Show that all CNOT gates in Figure 4.4.1 can be implemented using the standard CNOT gate, the Pauli X gate, and the Hadamard gate H.

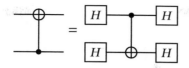

Figure 4.4.3. Implementation of the upper-right CNOT gate in Figure 4.4.1 using the Hadamard gate and the standard CNOT gate.

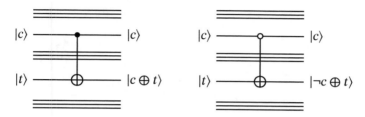

Figure 4.4.4. CNOT operators acting on quantum registers of length > 2.

The CNOT gates from Figure 4.4.1 can also be applied to quantum registers of length > 2. This is shown in Figure 4.4.4. Such CNOT operators are specified by their action on the computational basis vectors $|b_0 \cdots b_{n-1}\rangle$, $(b_0 \cdots b_{n-1}) \in \{0, 1\}^n$, as follows. There is a control qubit $|c\rangle$ and a target qubit $|t\rangle$, $c, t \in \mathbb{Z}_n, c \neq t$. The target qubit $|t\rangle$ is mapped to $X^c |t\rangle$. All other qubits remain unchanged.

4.4.2. Controlled-U operators. We generalize the construction of the CNOT operators by replacing the Pauli X gate in Figure 4.4.1 by an arbitrary unitary single-qubit operator U. The resulting operators are shown in Figure 4.4.5. They apply U to the target qubit depending on the control qubit being $|0\rangle$ or $|1\rangle$, respectively, and are called *controlled-U operators*.

For any unitary single-qubit operator U, the controlled-U operators in Figure 4.4.5 can be implemented using a decomposition $U = e^{i\delta} AXBXC$ with $ABC = I$ which exists by Theorem 4.3.33. For the upper-left operator in Figure 4.4.5 this is shown in Figure 4.4.7. The correctness of this construction is stated in the next theorem.

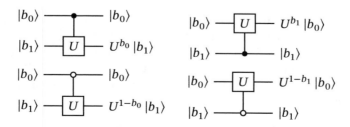

Figure 4.4.5. Controlled-U operators.

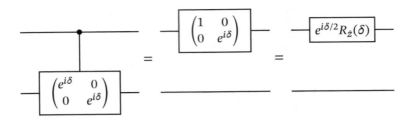

Figure 4.4.6. Two circuits implementing the same operator.

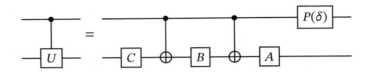

Figure 4.4.7. Implementation of the controlled-U gate with the first qubit as control using the decomposition $U = e^{i\delta}AXBXC$.

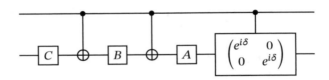

Figure 4.4.8. Circuit that implements the same operator as the circuit in Figure 4.4.7.

Theorem 4.4.5. *Let U be a unitary operator on $\mathbb{H}_1$ and let $U = e^{i\delta}AXBXC$ where $\delta \in \mathbb{R}$ and A, B, C are unitary single-qubit operators with $ABC = I$. Then the upper-left controlled-U operator in Figure 4.4.5 can be implemented as shown in Figure 4.4.7 using A, B, C, and $P(\delta)$ and two* CNOT *gates .*

Proof. We show that the implementation in Figure 4.4.7 is correct. We first note that the two circuits in Figure 4.4.6 implement the same operator since when applied to the computational basis states of $\mathbb{H}_2$, both have the following effect:

$$(4.4.5) \qquad |00\rangle \mapsto |00\rangle, \quad |01\rangle \mapsto |01\rangle, \quad |10\rangle \mapsto e^{i\delta}|10\rangle, \quad |11\rangle \mapsto e^{i\delta}|11\rangle.$$

This means that the circuit in Figure 4.4.8 implements the same operator as the circuit in Figure 4.4.7. We show that the circuit in Figure 4.4.8 implements the controlled-U operator with the first qubit as control. If the first qubit is $|1\rangle$, then the circuit applies $U = e^{i\delta}AXBXC$ to the second qubit. If the first qubit is $|0\rangle$, then the circuit applies ABC to the second qubit. Since $ABC = I$, the circuit does not change the second qubit. This proves the claim. $\square$

Exercise 4.4.6. Show how Theorem 4.4.5 can be used to implement the controlled-Y, Z, S, and T operators.

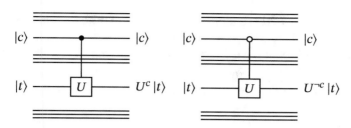

Figure 4.4.9. Controlled-U operators acting on quantum states with more than two qubits.

Like the CNOT gates, the controlled-U gates can be applied to quantum registers of length > 2. This is shown in Figure 4.4.9. Theorem 4.4.5 implies that these generalized controlled-U operators can be implemented using four unitary single-qubit operators and two CNOT gates.

4.4.3. General controlled operators. Now we present the most general controlled operators. An example of such an operator is shown in Figure 4.4.10. This operator applies the unitary operator U on $\mathbb{H}_2$ to the qubits $|b_4 b_5\rangle$ conditioned on the qubit $|b_1\rangle$ being $|0\rangle$ and the qubit $|b_2\rangle$ being $|1\rangle$. The other qubits are not changed. So, it acts on the computational basis states of $\mathbb{H}_7$ as follows:

$$(4.4.6) \qquad |b_0 \ldots b_6\rangle \mapsto |b_0 \ldots b_3\rangle\, U^{(1-b_1)b_2}\, |b_4 b_5\rangle\, |b_6\rangle .$$

We describe general controlled operators on $\mathbb{H}_n$ formally.

Definition 4.4.7. Let C_0, C_1, and T be pairwise disjoint subsets of the index set $\mathbb{Z}_n$. Let $m = |T| > 0$, and let $T = \{t, t+1, \ldots, t+m-1\}$ with $t \in \mathbb{Z}_n$. So T is a set of m consecutive integers in the index set $\mathbb{Z}_n$. Also, let U be a unitary operator on $\mathbb{H}_m$. Then

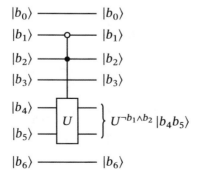

Figure 4.4.10. Example for a general controlled operator.

the linear operator $C^{C_0,C_1,T}(U)$ is defined by its action on the computational basis states $|b_0 \cdots b_{n-1}\rangle$ of $\mathbb{H}_n$ as follows. It applies U to the *target qubits* $|b_t \cdots b_{t+m-1}\rangle$ conditioned on the *control qubits* $|b_i\rangle$ with $i \in C_0$ being $|0\rangle$ and the *control qubits* $|b_i\rangle$ with $i \in C_1$ being $|1\rangle$; i.e.,

$$
(4.4.7) \quad
\begin{aligned}
& C^{C_0,C_1,T}(U)\,|b_0 \cdots b_{n-1}\rangle \\
& = |b_0 \cdots b_{t-1}\rangle\, U^c\, |b_t \cdots b_{t+m-1}\rangle\, |b_m \cdots b_{n-1}\rangle
\end{aligned}
$$

where

$$
(4.4.8) \qquad c = \prod_{i \in C_0}(1 - b_i) \prod_{i \in C_1} b_i.
$$

If any of the index sets C_0, C_1, or T has only one element, then we replace the set in the superscript by this element.

In this definition, we could drop the requirement that the set T of target qubits be a set of consecutive numbers. However, this would complicate the definition and we can achieve the same effect by using SWAP gates, described in Section 4.5. As shown in Exercise 4.4.8, general multiply controlled operators are unitary.

Exercise 4.4.8. Prove that every multiply controlled operator as specified in Definition 4.4.7 is unitary.

Example 4.4.9. Using the notation from Definition 4.4.7 the operator in Figure 4.4.10 can be written as

$$
(4.4.9) \qquad C^{1,2,\{4,5\}}(U).
$$

Example 4.4.10. Using the notation from Definition 4.4.7 we see that for the left operator in Figure 4.4.4 we have $C_0 = \emptyset$, $C_1 = \{c\}$, $T = \{t\}$, and $U = X$. So, the operator is $C^{\emptyset,c,t}(X)$. Also, the right operator is $C^{c,\emptyset,t}(X)$.

The implementation of general controlled operators is discussed in Section 4.9.2. Also, the next sections present further important instances of general controlled operators.

4.4.4. The quantum Toffoli gate.
Figure 4.4.11 shows the *quantum Toffoli gate*, also called CCNOT. Its classical counterpart has been introduced in Section 1.7.1.

The CCNOT gate applies the Pauli X gate to the target qubit $|t\rangle$ conditioned on the control qubits $|c_0\rangle$ and $|c_1\rangle$ both being $|1\rangle$; that is, it acts on the computational basis vectors of $\mathbb{H}_3$ as follows:

$$
(4.4.10) \qquad |c_0 c_1 t\rangle \mapsto |c_0 c_1\rangle X^{c_0 c_1}\,|t\rangle.
$$

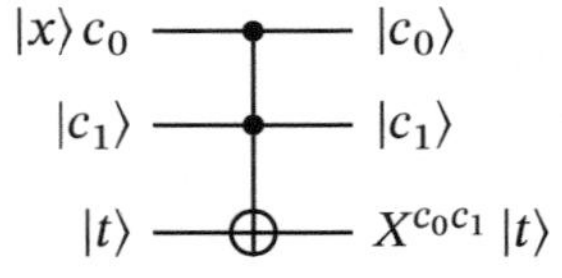

Figure 4.4.11. The quantum Toffoli or CCNOT gate.

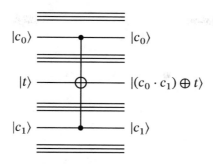

Figure 4.4.12. Generalized CCNOT operator.

Using the terminology from Definition 4.4.7 we can write

(4.4.11) $$\text{CCNOT} = C^{\emptyset,\{0,1\},2}(X).$$

We may also apply CCNOT to states that have more than three qubits or change the order of the control and target qubits. This is shown in Figure 4.4.12. The implementation of the Toffoli gate is discussed in Section 4.9.1.

4.4.5. The $C^k(U)$ operators. We introduce the controlled operators $C^k(U)$. Such an operator is shown in Figure 4.4.13. To specify it, we use $k, m \in \mathbb{N}$ with $n = k + m$ and a unitary operator U on $\mathbb{H}_m$. We write the computational basis vectors of $\mathbb{H}_n$ as $|c_0 \cdots c_{k-1} t_0 \cdots t_{m-1}\rangle$ instead of $|b_0 \cdots b_{n-1}\rangle$ to distinguish between control and target qubits. Then we have

(4.4.12) $$C^k(U)|c_0 \cdots c_{k-1} t_0 \cdots t_{m-1}\rangle = |c_0 \cdots c_{k-1}\rangle U^{\prod_{i=0}^{k-1}|t_0 \cdots t_{m-1}\rangle}$$

or using the notation of Definition 4.4.7

(4.4.13) $$C^k(U) = C^{\emptyset,\{0,\ldots,k-1\},\{k,\ldots,k+m-1\}}(U).$$

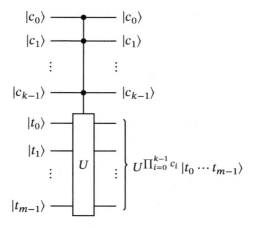

Figure 4.4.13. The operator $C^k(U)$.

For example, we can write

(4.4.14) $$\text{CCNOT} = C^2(X).$$

4.4.6. Transposition operators. As a last example for controlled operators, we present transposition operators.

Definition 4.4.11. Let $t \in \mathbb{Z}_n$ and let

(4.4.15) $$\vec{c} = c_0 \cdots c_{t-1} * c_{t+1} \cdots c_{n-1}$$

with $c_i \in \{0, 1\}$ for $i \in \mathbb{Z}_n, i \neq t$. This is a vector of length n with entries from $\{0, 1\}$ except that the entry with index t is "$*$". The *transposition operator* $\text{TRANS}^{\vec{c}}$ exchanges the computational basis elements

(4.4.16) $$|c_0 \cdots c_{t-1}\rangle |0\rangle |c_{t+1} \cdots c_{n-1}\rangle$$

and

(4.4.17) $$|c_0 \cdots c_{t-1}\rangle |1\rangle |c_{t+1} \cdots c_{n-1}\rangle$$

and leaves all other computational basis elements unchanged.

Using the notation from Definition 4.4.7 we can write

(4.4.18) $$\text{TRANS}^{\vec{c}} = C^{C_0, C_1, t}(X)$$

where

(4.4.19) $$C_0 = \{i \in \mathbb{Z}_n : c_i = 0\}, \quad C_1 = \{i \in \mathbb{Z}_n : c_i = 1\}.$$

An example of such a transposition operator can be seen in Figure 4.4.14. In this example, we have $\vec{c} = 01 * 0$.

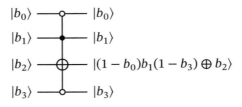

Figure 4.4.14. The operator $\text{TRANS}^{(01*0)}$ which exchanges $|0100\rangle$ and $|0110\rangle$.

4.5. Swap and permutation operators

Another important multiple-qubit operator is the quantum SWAP operator which corresponds to the classical SWAP gate introduced in Section 1.7.1. Applied to a computational basis element $|b_0 b_1\rangle$ of $\mathbb{H}_2$ it swaps b_0 and b_1; i.e., we have

(4.5.1) $$\text{SWAP}\,|b_0 b_1\rangle = |b_1 b_0\rangle.$$

This gate is shown in Figure 4.5.1 together with an implementation that only uses CNOT gates.

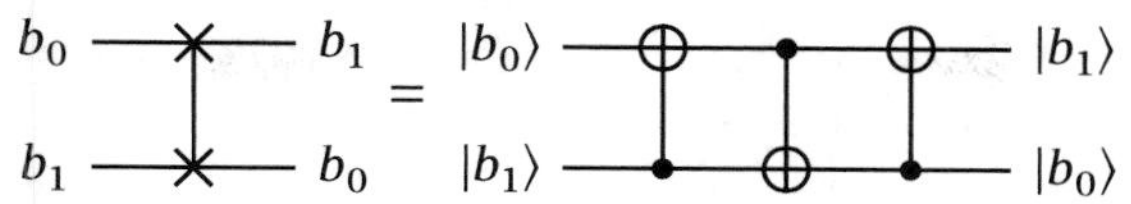

Figure 4.5.1. The quantum SWAP gate and its implementation using CNOT gates.

Exercise 4.5.1. Verify that the implementation of the SWAP gate in Figure 4.5.1 is correct.

We generalize the quantum SWAP gates. Let $n \in \mathbb{N}$, $i, j \in \mathbb{Z}_n$ with $i \leq j$. Then the *quantum swap gate* $\text{SWAP}_n(i, j)$ is the unitary operator on $\mathbb{H}_n$ that is defined by its effect on the computational basis states $|b_0 \cdots b_{n-1}\rangle$ of $\mathbb{H}_n$ as follows:

$$(4.5.2) \qquad \text{SWAP}_n(i, j) |b_0 \cdots b_i \cdots b_j \cdots b_{n-1}\rangle = |b_0 \cdots b_j \cdots b_i \cdots b_{n-1}\rangle.$$

Generalizing the implementation of the simple SWAP gate in Figure 4.5.1, we see that every quantum SWAP gate can be implemented using 3 CNOT gates.

We note that SWAP gates apply a transposition to the index sequence of the computational basis states of $\mathbb{H}_n$. This suggests the generalization of quantum swap gates as follows. For $\pi \in S_n$ the *quantum permutation operator* U_π is defined by its action on the computational basis states $|b_0 \cdots b_{n-1}\rangle$ as follows:

$$(4.5.3) \qquad U_\pi |b_0 \cdots b_{n-1}\rangle = |b_{\pi(0)} \cdots b_{\pi(n-1)}\rangle.$$

Using Proposition 1.7.2 the following result can be proved.

Proposition 4.5.2. *Let $n \in \mathbb{N}$ and let $\pi \in S_n$. Then the permutation operator U_π can be implemented by a quantum circuit that uses at most $n - 1$ SWAP gates or at most $3n - 3$ CNOT gates, respectively.*

Exercise 4.5.3. Prove Proposition 4.5.2.

4.6. Ancillary and erasure gates

In the previous sections and in Section 3.3.2, we have introduced quantum gates that implement unitary operators on some state space $\mathbb{H}_n$ and which can be used as building blocks of more complex quantum circuits. Such quantum gates are called *unitary gates*. Now it turns out that the construction of many quantum circuits also requires the use of two types of quantum gates that do not implement unitary operators: *ancillary gates* and *erasure gates*. This section introduces these gates. The quantum circuit in Figure 4.6.1 illustrates why these gates are useful. It implements the operator $C^3(U)$ for some single-qubit operator U. The evolution of the basis states of the quantum register in this circuit is shown in Table 4.6.1. The circuit uses two ancilla qubits $|a_0\rangle$ and $|a_1\rangle$. They are inserted between the control and target qubits and are initialized to $|0\rangle$ using two ancillary gates. The first CCNOT gate changes the state of $|a_0\rangle$ to $|c_0 \cdot c_1\rangle$. The second CCNOT gate changes the state of $|a_1\rangle$ to $|c_0 \cdot c_1 \cdot c_2\rangle$. The ancillary qubit $|a_1\rangle$ controls the application of U to the target qubit $|t\rangle$. So after the application of this controlled operation, the target qubit is in the state $U^{c_0 \cdot c_1 \cdot c_2} |t\rangle$. The two further CCNOT gates change the ancilla bits back to $|a_0\rangle = |a_1\rangle = |0\rangle$. The two erasure gates trace out $|a_0\rangle$

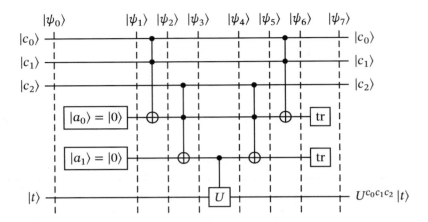

Figure 4.6.1. Implementation of $C^3(U)$ using ancillary and erasure gates.

Table 4.6.1. Evolution of the states in the circuit from Figure 4.6.1.

i	$|\psi_i\rangle$
0	$|c_0\rangle|c_1\rangle|c_2\rangle|t\rangle$
1	$|c_0\rangle|c_1\rangle|c_2\rangle|0\rangle|0\rangle|t\rangle$
2	$|c_0\rangle|c_1\rangle|c_2\rangle|c_0 \cdot c_1\rangle|0\rangle|t\rangle$
3	$|c_0\rangle|c_1\rangle|c_2\rangle|c_0 \cdot c_1\rangle|c_0 \cdot c_1 \cdot c_2\rangle|t\rangle$
4	$|c_0\rangle|c_1\rangle|c_2\rangle|c_0 \cdot c_1\rangle|c_0 \cdot c_1 \cdot c_2\rangle U^{c_0 \cdot c_1 \cdot c_1}|t\rangle$
5	$|c_0\rangle|c_1\rangle|c_2\rangle|c_0 \cdot c_1\rangle|0\rangle U^{c_0 \cdot c_1 \cdot c_1}|t\rangle$
6	$|c_0\rangle|c_1\rangle|c_2\rangle|0\rangle|0\rangle U^{c_0 \cdot c_1 \cdot c_1}|t\rangle$
7	$|c_0\rangle|c_1\rangle|c_2\rangle U^{c_0 \cdot c_1 \cdot c_1}|t\rangle = C^3(U)|\psi\rangle_0$

and $|a_1\rangle$. Since the state of the quantum register used in the circuit is separable with respect to the decomposition into ancillary and nonancillary qubits, it follows from Corollary 3.7.12 that this does not change the other qubits and the resulting state is $C^3(U)|\psi\rangle$.

Exercise 4.6.1. Verify Table 4.6.1.

4.7. Quantum circuits revisited

We have already introduced quantum circuits in Section 3.3.4. In this section, we expand the definition of quantum circuits to include ancillary and erasure gates and provide a more formal description of quantum circuits and their computations.

Definition 4.7.1. A *quantum circuit* Q is specified by two positive integers n and k and a finite sequence $(q_0, \ldots, q_{k-1})$. Here, n is the number of input qubits and for all $i \in \mathbb{Z}_k$ the component q_i contains the following:

(1) a tuple of quantum gates that are either all ancillary gates, all unitary gates, or all erasure gates and

(2) the information about how the ancilla qubits are initialized and where they are inserted or to which qubits the unitary or erasure gates are applied, respectively. At most one gate is applied to each qubit.

We illustrate Definition 4.7.1.

Example 4.7.2. Consider the quantum circuit shown in Figure 4.6.1. Its representation as explained in Definition 4.7.1 is illustrated in Figure 4.7.1. In this circuit, we have $n = 4$ since there are the input qubits $|c_0\rangle$, $|c_1\rangle$, $|c_2\rangle$, and $|t\rangle$. Also, we have $k = 7$. The component q_0 contains two ancillary gates that insert two ancillary qubits and initialize them to $|0\rangle$. Also, q_0 contains the information that the two ancilla qubits are inserted behind the control qubits. The sequence elements q_1 and q_2 contain one CCNOT gate each and the information about to which of the six qubits these gates are applied. The element q_3 contains a $C^1(U)$ gate and q_4 and q_5 each contain two CCNOT gates. Additionally, these gates include information about the qubits to which they are applied. Finally, q_6 contains two erasure gates and the information that they trace out the two ancillary qubits. The quantum operator implemented by this circuit is $C^3(U)$.

Definition 4.7.3. The size of a quantum circuit is the number of quantum gates it contains plus the number of input qubits it operates on.

The computation of a quantum circuit Q as in Definition 4.7.1 can be described as an evolution

(4.7.1) $$|\psi_0\rangle, |\psi_1\rangle, \ldots, |\psi_k\rangle$$

of quantum states that are defined as follows.

(1) The initial state $|\psi_0\rangle \in \mathbb{H}_n$ is the input of the computation.

(2) For $i \in \mathbb{Z}_k$ the state $|\psi_{i+1}\rangle$ is obtained by applying the quantum gates in q_i to $|\psi_i\rangle$ as specified in q_i.

(3) The final state is $|\psi_k\rangle = |c_0 \cdots c_{m-1}\rangle \in \mathbb{H}_m$ with $m = n + n_a - n_e$ where n_a is the number of ancillary gates and n_e is the number of erasure gates used in the quantum circuit.

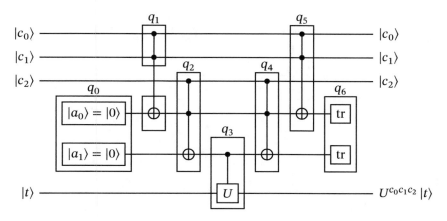

Figure 4.7.1. Illustration of Definition 4.7.1.

The *quantum operator implemented by Q* is

(4.7.2) $$\mathbb{H}_n \to \mathbb{H}_m, \quad |\psi_0\rangle \mapsto |\psi_k\rangle.$$

Example 4.7.4. As shown in Section 4.6, the quantum operator shown in Figures 4.6.1 and 4.7.1 implements $C^3(U)$.

We make some remarks about this construction. Since tracing out quantum bits is allowed in this general definition of quantum circuits, the resulting quantum state may be a mixed state. But frequently the situation is simpler. Denote by A the quantum system of all output qubits that are not traced out and by B the quantum system of the qubits that are traced out and assume that the state of system AB is separable. Then by Corollary 3.7.12, tracing out system B means omitting the corresponding quantum bits. This is the case in Figure 4.6.1.

Next, we call the quantum circuit Q *unitary* if $m = n$ and the quantum operator implemented by it is unitary. If Q uses no ancillary and erasure gates, then it is always unitary. If Q uses ancillary or erasure gates, it may or may not be unitary. For example, if $m \neq n$, then Q is not unitary.

But every quantum circuit can be transformed into a unitary quantum circuit. To see how this works, let Q be a quantum circuit. The transformed quantum circuit R is obtained by removing the ancillary gates and adding the corresponding ancillary qubits as new input qubits. Also, we remove the erasure gates and add the corresponding qubits as new output qubits, and the new quantum circuit R is called the *purification* of the quantum circuit Q. This purification is unitary. Furthermore, this process is referred to as *purification* of Q. Figure 4.7.2 shows the purification R of the quantum circuit Q from Figure 4.6.1. However, note that the quantum circuit Q in Figure 4.6.1 is already unitary. So we see that purification may not be required to make a quantum circuit unitary.

It is also possible to represent quantum circuits as algorithms that are specified using pseudocode. An example is the algorithmic representation of the quantum circuit in Figure 4.6.1 which is shown in Algorithm 4.7.5.

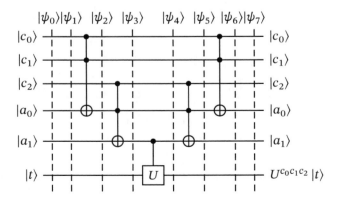

Figure 4.7.2. Purification of the quantum circuit in Figure 4.6.1.

Algorithm 4.7.5. Implementation of $C^3(U)$ using only CCNOT and $C^1(U)$

Input: $|c_0\rangle |c_1\rangle |c_2\rangle |t\rangle$
Output: $|c_0\rangle |c_1\rangle |c_2\rangle U^{c_0 \cdot c_1 \cdot c_2} |t\rangle$

1:   $C^3(U)$
2:   Insert ancilla qubits $|a_0\rangle, |a_1\rangle$ after $|c_3\rangle$ and initialize them to $|0\rangle$
3:   $|a_0\rangle \leftarrow X^{c_0 \cdot c_1} |a_0\rangle$
4:   $|a_1\rangle \leftarrow X^{c_2 \cdot a_0} |a_1\rangle$
5:   $|t\rangle \leftarrow U^{a_1} |t\rangle$
6:   $|a_1\rangle \leftarrow X^{c_2 \cdot a_0} |a_1\rangle$
7:   $|a_0\rangle \leftarrow X^{c_0 \cdot c_1} |a_0\rangle$
8:   Trace out $|a_0\rangle$ and $|a_1\rangle$
9:   The final state is $|c_0\rangle |c_1\rangle |c_2\rangle |t\rangle$
10: **end**

Exercise 4.7.6. Write an algorithm that represents the quantum circuit in Figure 4.7.2.

From Theorem 1.7.12 we can deduce the following result.

Theorem 4.7.7. *Let $n, m \in \mathbb{N}$ and let $f : \{0,1\}^n \to \{0,1\}^m$. Then there is a quantum circuit Q of size $O(|f|_F)$ that only uses quantum Toffoli, ancillary, and erasure gates and implements the quantum operator*

$$(4.7.3) \qquad U : \mathbb{H}_{n+m} \to \mathbb{H}_{n+m}, \quad |\vec{x}\rangle |\vec{y}\rangle \mapsto |\vec{x}\rangle |\vec{y} \oplus f(\vec{x})\rangle.$$

Proof. Consider the quantum circuit Q_r which is obtained by replacing the classical Toffoli gates in the circuit D_r from Theorem 1.7.12 with quantum Toffoli gates. It implements a unitary operator

$$(4.7.4) \qquad U_r : \mathbb{H}_n \otimes \mathbb{H}_{n+p} \otimes \mathbb{H}_m \to \mathbb{H}_n \otimes \mathbb{H}_{n+p} \otimes \mathbb{H}_m$$

where $p \in \mathbb{N}$, $p \le 2|f|_F$, such that for all $\vec{x} \in \{0,1\}^n$ and all $\vec{y} \in \{0,1\}^m$ we have

$$(4.7.5) \qquad U_r |\vec{x}\rangle |\vec{0}\rangle |\vec{y}\rangle = |\vec{x}\rangle |\vec{0}\rangle |\vec{y} \oplus f(\vec{x})\rangle.$$

This quantum circuit is the purification of a quantum circuit Q that is constructed as follows. The input state is $|\vec{x}\rangle |\vec{y}\rangle$. First, Q inserts $n + p$ ancilla qubits behind $|x\rangle$ that are all initialized to $|0\rangle$. Then it applies Q_r. The result is the separable state in (4.7.5). Finally, Q traces out the $n + p$ ancilla qubits. By Corollary 3.7.12, this gives the asserted quantum state. $\qquad \square$

4.8. Universal sets of quantum gates

In Definition 1.5.6, a universal set S of logic gates is defined by the property that every function $f : \{0,1\}^n \to \{0,1\}^m$ can be implemented by a Boolean circuit that uses only the gates from S. For example, by Theorems 1.5.7 and 1.5.9, the sets {AND, OR, NOT} and {NAND} are universal for classical computation. In this section, we introduce and discuss universal sets of quantum gates.

Our first theorem shows that the notion of a universal set of quantum gates cannot be obtained as a straightforward generalization of the corresponding definition in classical computing.

Theorem 4.8.1. *Let S be a set of quantum gates such that for every $n \in \mathbb{N}$ and every unitary operator U on $\mathbb{H}_n$ there is a quantum circuit that implements U and uses only gates from S. Then S is uncountable.*

Proof. Let $n \in \mathbb{N}$. By Theorem 4.3.15 the rotation gates $R_{\hat{x}}(\theta)$ with $\theta/2 \in [0, 2\pi[$ are pairwise different. Since the set $[0, 2\pi[$ is uncountable, it follows that the set of unitary operators on $\mathbb{H}_n$ is uncountable. But if S is a countable set of quantum gates, then the set of all quantum circuits that can be constructed using the gates in S is also countable.   □

Theorem 4.8.1 implies that there are no finite or even countable universal sets of quantum gates in the classical sense. We will therefore call a set of quantum gates universal if it can be used to approximate every unitary operator to an arbitrary precision and we will show that finite sets of quantum gates with this property exist. The notions of universality can be generalized to general quantum operators. But in this book, we restrict ourselves to unitary quantum operators. For the discussion of universality, we need the following definition. It uses the *supremum* sup S of a set S of real numbers that is bounded from above. It is the least upper bound of S in $\mathbb{R}$ and it can be shown that it always exists and is uniquely determined.

Definition 4.8.2. Let U and V be two unitary operators on $\mathbb{H}_n$. Then we define the *error when V is implemented instead of U* as

(4.8.1) $E(U, V) = \sup\{\|(U - V)|\varphi\rangle\| : |\varphi\rangle \in \mathbb{H}_n, \langle \psi|\psi \rangle = 1\}.$

We also call this error the *distance* between U and V.

The next proposition uses the distance between two unitary operators U and V on $\mathbb{H}_n$ to estimate the difference between the probabilities of measuring a certain outcome when U and V are applied to a quantum state.

Proposition 4.8.3. *Let U and V be unitary operators on $\mathbb{H}_n$, let O be an observable on $\mathbb{H}_n$, let $O = \sum_{\lambda \in \Lambda} \lambda P_\lambda$ be the spectral decomposition of O. Then for all $\lambda \in \Lambda$ and all quantum states $|\psi\rangle \in \mathbb{H}_n$ we have*

(4.8.2) $\left| \left\langle U|\psi\rangle \big| P_\lambda \big| U|\psi\rangle \right\rangle - \left\langle V|\psi\rangle \big| P_\lambda \big| V|\psi\rangle \right\rangle \right| \leq 2E(U, V).$

Proof. Let $\lambda \in \Lambda$ and let $|\psi\rangle \in \mathbb{H}_n$ be a quantum state. Since P_λ is an orthogonal projection, it is Hermitian by Proposition 2.4.41. Also, U is unitary and $|\psi\rangle$ has Euclidean length 1. From Exercise 2.4.38 we obtain

(4.8.3) $\left\| P_\lambda^* U|\psi\rangle \right\| = \left\| P_\lambda U|\psi\rangle \right\| \leq \left\| U|\psi\rangle \right\| = \|\psi\| = 1.$

This inequality, the linearity of the inner product on $\mathbb{H}_n$, the Cauchy-Schwarz inequality, and Exercise 2.4.38 imply

$$
\begin{aligned}
&\left| \left\langle U \left| \psi \right\rangle \left| P_\lambda \right| U \left| \psi \right\rangle \right\rangle - \left\langle V \left| \psi \right\rangle \left| P_\lambda \right| V \left| \psi \right\rangle \right\rangle \right| \\
&= \left| \left\langle U \left| \psi \right\rangle \left| P_\lambda \right| (U - V) \left| \psi \right\rangle \right\rangle + \left\langle (U - V) \left| \psi \right\rangle \left| P_\lambda \right| V \left| \psi \right\rangle \right\rangle \right| \\
&= \left| \left\langle P_\lambda^* U \left| \psi \right\rangle \left| (U - V) \left| \psi \right\rangle \right\rangle + \left\langle (U - V) \left| \psi \right\rangle \left| P_\lambda V \left| \psi \right\rangle \right\rangle \right| \\
&\leq \left\| P_\lambda^* U \left| \psi \right\rangle \right\| \left\| (U - V) \left| \psi \right\rangle \right\| + \left\| (U - V) \left| \psi \right\rangle \right\| \left\| P_\lambda V \left| \psi \right\rangle \right\| \\
&\leq 2E(U, V). \qquad\qquad \square
\end{aligned}
$$

(4.8.4)

Now we can define universal sets of quantum gates. In this definition, we use the term *unitary quantum gate*. By this we mean a quantum gate that implements a unitary operator; i.e., this quantum gate is neither an ancillary nor an erasure gate.

Definition 4.8.4. Let S be a set of unitary quantum gates.

(1) We say that S is *universal* for a set T of unitary quantum operators if for every $\varepsilon \in \mathbb{R}_{>0}$ and every $U \in T$ there is a unitary operator V which can — up to a global phase factor — be implemented by a quantum circuit that uses only gates from S, ancillary gates, and erasure gates such that $E(U, V) < \varepsilon$.

(2) We say that S is *universal for quantum computation* if S is universal for the set of all unitary quantum operators.

We also define a direct analog of the classical notion of universality.

Definition 4.8.5. We call a set S of unitary quantum gates *perfectly universal for quantum computation* or *perfectly universal* for short if all unitary operators U on $\mathbb{H}_n$ can, up to a global phase factor, be implemented by a quantum circuit that uses only gates from S, ancillary gates, and erasure gates.

4.9. Implementation of controlled operators

This section discusses the implementation of general controlled unitary operators that were defined in Section 4.4.3. They are required in many contexts, for example in the proofs that certain sets of gates are universal or perfectly universal.

We first note the following. In Section 4.4.2, we demonstrated how to implement $C^1(U)$ for single-qubit operators U when a certain representation of U is known. However, in general, implementing $C^1(U)$ for unitary operators on $\mathbb{H}_n$ when $n > 1$ is more involved. To create implementations of such controlled operators, additional information about U is required, as will be discussed in Section 4.12.3. Therefore, this section exclusively discusses implementations of controlled-U operators that may use $C^1(V)$ gates for certain operators V.

4.9.1. Implementations of $C^2(U)$ **operators.** We start by presenting implementations of $C^2(U)$ for unitary operators U for which a square root is given. This will allow the implementation of the quantum Toffoli gate. We first show that any unitary operator possesses a square root.

Proposition 4.9.1. *Let U be a unitary operator on $\mathbb{H}_n$. Then there is a unitary operator V on $\mathbb{H}_n$ with $V^2 = U$.*

Proof. Let

$$(4.9.1) \qquad\qquad U = \sum_{\lambda \in \Lambda} \lambda P_\lambda$$

be the spectral decomposition of U. Set

$$(4.9.2) \qquad\qquad V = \sum_{\lambda \in \Lambda} \sqrt{\lambda} P_\lambda$$

where $\sqrt{\lambda}$ is a square root of λ in $\mathbb{C}$. This is the spectral decomposition of V and we have $V^2 = U$. Also, by Proposition 2.4.60 we have $|\lambda| = 1$ for all $\lambda \in \Lambda$. This implies that $|\sqrt{\lambda}| = 1$ for all $\lambda \in \Lambda$. Therefore, Proposition 2.4.60 implies that V is unitary. $\square$

Using the method from the proof of Proposition 4.9.1, we can determine square roots of the Pauli operators.

Proposition 4.9.2. *The operator $V = (1 + i)(I - iX)/2$ is unitary, and we have $V^2 = X$.*

Proof. By Example 2.4.57 the spectral decomposition of the Pauli X operator is $X = |x_+\rangle\langle x_+| - |x_-\rangle\langle x_-|$ with $|x_+\rangle, |x_-\rangle$ from (4.1.7). Hence, if we set $V = |x_+\rangle\langle x_+| + i|x_-\rangle\langle x_-|$, then V is a unitary single-qubit operator, we have $V^2 = X$, and $V = (1 + i)(I - iX)$ as shown in Exercise 4.9.3. $\square$

Exercise 4.9.3. Verify that $|x_+\rangle\langle x_+| + i|x_-\rangle\langle x_-| = (1 + i)(I - iX)/2$.

Exercise 4.9.4. Determine square roots of the Pauli operators Y and Z and the representations of these square roots as a linear combination of the basis elements I, X, Y, Z of $\text{End}(\mathbb{H}_1)$.

Figure 4.9.1 shows an implementation of $C^2(U)$ that uses a square root V of U. Its correctness is stated in Proposition 4.9.5.

Proposition 4.9.5. *Let U, V be unitary operators such that $V^2 = U$. Then the circuit on the right side of Figure 4.9.1 implements $C^2(U)$. It uses two CNOT gates, two $C^1(V)$ gates, and one $C^1(V^*)$ gates.*

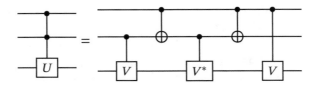

Figure 4.9.1. Implementation of $C^2(U)$ if $V^2 = U$.

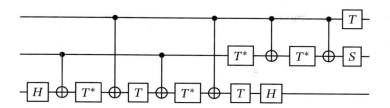

Figure 4.9.2. An implementation of the Toffoli gate that uses the Hadamard, phase, CNOT, and $\pi/8$ gates.

Exercise 4.9.6. Prove Proposition 4.9.5.

It follows from Proposition 4.9.2 that the circuit in Figure 4.9.1 can be used to implement the quantum Toffoli gate. Figure 4.9.2 shows another implementation of this gate. It uses the phase, $\pi/8$, and CNOT gates and its correctness is stated in Proposition 4.9.7.

Proposition 4.9.7. *The circuit in Figure 4.9.2 implements the Toffoli gate. It uses two Hadamard, one phase, seven (inverse) $\pi/8$, and six CNOT gates.*

Exercise 4.9.8. Prove Proposition 4.9.7.

4.9.2. Implementation of general controlled operators. We now present implementations of general controlled operators as described in Definition 4.4.7, using only Toffoli gates and controlled gates with a single control qubit. We begin by providing an explanation of the implementations of all operators $C^k(U)$ introduced in Section 4.4.5, for every unitary operator U on $\mathbb{H}_m$, where k and m are both natural numbers and $k \geq 2$. The circuits use $k - 1$ ancilla qubits and operate on quantum registers of length $2k + m - 1$. The corresponding circuit implementation of $C^3(U)$ has already been presented in Section 4.6 and Figure 4.6.1 to motivate the use of ancillary qubits. The circuit for $k = 5$ and $m = 1$ is shown in Figure 4.9.3. The general implementation of $C^k(U)$ is presented in Algorithm 4.9.9.

The idea of this construction is the following. Fix k, m, U. After inserting the ancilla qubits, the circuit operates on states of the form

$$(4.9.3) \qquad |c_0 \cdots c_{k-1}\rangle |a_0 \cdots a_{k-2}\rangle |t_0 \cdots t_{m-1}\rangle$$

where $|c_0 \cdots c_{k-1}\rangle$ are the k control qubits, $|a_0 \cdots a_{k-2}\rangle$ are the $k-1$ ancilla qubits, and $|t_0 \cdots t_{m-1}\rangle$ are the m target qubits. The first $k-1$ steps of the computation change the ancilla qubits to

$$(4.9.4) \qquad |a_j\rangle = \left| \prod_{i=0}^{j+1} c_i \right\rangle, \quad 0 \leq j \leq k - 2,$$

and have no effect on the other qubits. In step k the circuit changes the target qubits $|t_0 \cdots t_{m-1}\rangle$ to $U^{a_{k-2}} |t_0 \cdots t_{m-1}\rangle$ which by (4.9.4) is

$$(4.9.5) \qquad U^{\prod_{i=0}^{k-1} c_i} |t_0 \cdots t_{m-1}\rangle.$$

Algorithm 4.9.9. Implementation of $C^k(U)$ using only CCNOT and $C^1(U)$

Input: $|c_0 \cdots c_{k-1}\rangle |t_0 \cdots t_{m-1}\rangle$

Output: $C^k(U)|c_0 \cdots c_{k-1}\rangle |t_0 \cdots t_{m-1}\rangle$

1:   $C^k(U)$

2:     Insert $k-1$ ancilla qubits $|a_0\rangle, \ldots, |a_{k-2}\rangle$ after the control qubit $|c_{k-1}\rangle$ and initialize them to $|0\rangle$

3:     $|a_0\rangle \leftarrow X^{c_0 \cdot c_1} |a_0\rangle$

4:     **for** $j = 1 \cdots k-2$ **do**

5:        $|a_j\rangle \leftarrow X^{c_{j+1} \cdot a_{j-1}} |0\rangle$

6:     **end for**

7:     $|t_0 \cdots t_{m-1}\rangle \leftarrow U^{a_{k-2}} |t_0 \cdots t_{m-1}\rangle$

8:     **for** $j = k-2, \ldots, 1$ **do**

9:        $|a_j\rangle \leftarrow X^{c_{j+1} \cdot a_{j-1}} |a_j\rangle$

10:    **end for**

11:    $|a_0\rangle \leftarrow X^{c_0 \cdot c_1} |a_0\rangle$

12:    Trace out $|a_0 \cdots a_{k-2}\rangle$

13:    The final state is $|c_0 \cdots c_{k-1}\rangle |t_0 \cdots t_{m-1}\rangle$

14: **end**

This step has no effect on the other qubits. The next $k-1$ steps of the circuit change the ancilla qubits back to $|0\rangle$ without an effect on the other qubits. Now the state of

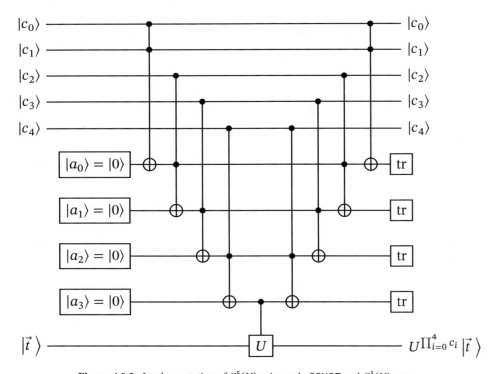

Figure 4.9.3. Implementation of $C^5(U)$ using only CCNOT and $C^1(U)$ gates.

the composition AB, where A consists of the control and target qubits and system B consists of the ancilla qubits, is separable. Hence, it follows from Corollary 3.7.12 that tracing out the ancilla qubits does not change the control or target qubits and yields

$$(4.9.6) \qquad C^k(U) |c_0 \cdots c_{k-1}\rangle |t_0 \cdots t_{m-1}\rangle = |c_0 \cdots c_{k-1}\rangle U^{\prod_{i=0}^{k-1} c_i} |t_0 \cdots t_{m-1}\rangle.$$

The next proposition states the correctness of the construction.

Proposition 4.9.10. *Let U be a unitary operator on $\mathbb{H}_m$ and let $k \in \mathbb{N}$, $k \geq 2$. Then Algorithm 4.9.9 implements $C^k(U)$. It uses $2k - 2$ CCNOT gates, one $C^1(U)$ gate, and $k - 1$ ancillary and erasure gates.*

Proof. We show by induction that for $j = 0, \ldots, k - 2$, the first steps $j + 1$ carried out by the algorithm starting in line 3 yield the ancilla qubits

$$(4.9.7) \qquad |a_j\rangle = \left| \prod_{i=0}^{j+1} c_i \right\rangle$$

and do not change the other qubits. We start with $j = 0$, that is, with the statement in line 3 of the algorithm. It produces the state

$$(4.9.8) \qquad |a_0\rangle = |c_0 \cdot c_1\rangle.$$

All other qubits remain unchanged.

Now let $j \in \mathbb{N}$, $1 \leq j < k - 2$, and assume that after j steps we have

$$(4.9.9) \qquad |a_{j-1}\rangle = \left| \prod_{i=0}^{j} c_j \right\rangle.$$

Step $j + 1$ is the iteration with loop index j of the **for** loop starting in line 4. After this iteration, we have

$$(4.9.10) \qquad |a_j\rangle = X^{a_{j-1} \cdot c_{j+1}} |0\rangle = |a_{j-1} \cdot c_{j+1}\rangle.$$

So (4.9.9) and (4.9.10) imply (4.9.7). Next, it follows from (4.9.7) with $j = k - 2$ that the statement in line 7 has the following effect:

$$(4.9.11) \qquad |t\rangle \leftarrow C^{a_{k-2}} |t_0 \cdots t_{m-1}\rangle = C^{\prod_{i=0}^{k-1} c_i} |t_0 \cdots t_{m-1}\rangle.$$

Finally, the **for** loop starting in line 8 carries out $k - 2$ iterations with loop indices $j = k - 2, \ldots, 1$. Iteration with loop index $j \in \{1, \ldots, k - 2\}$ changes the qubit $|a_j\rangle$ and no other qubit. Also, after the jth iteration we have

$$(4.9.12) \qquad |a_j\rangle = |0\rangle.$$

This can also be seen by induction on j. The statement in line 11 yields $|a_0\rangle = 0$. So (4.9.12) holds for $j = 0, \ldots, k - 2$. In addition, the state of the composition AB, where A consists of the control and target qubits and system B consists of the ancillary qubits, is separable. So, by Corollary 3.7.12, tracing out the qubits $|a_0\rangle, \ldots, |a_{k-2}\rangle$ has no effect on the other qubits. This concludes the proof. $\qquad \square$

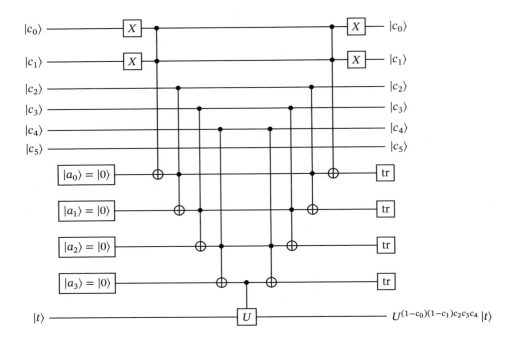

Figure 4.9.4. Implementation of $C^{\{0,1\},\{2,3,4\},6}(U)$ using only X, CCNOT, and $C^1(U)$ gates.

From Propositions 4.9.10 and 4.9.7 we obtain the following result.

Proposition 4.9.11. *Let U be a unitary operator on $\mathbb{H}_m$ and let $k \in \mathbb{N}$, $k \geq 2$. Then Algorithm 4.9.9 implements $C^k(U)$ using $O(k)$ Hadamard, phase, $\pi/8$, inverse $\pi/8$, CNOT, ancillary, and erasure gates, and one $C^1(U)$ gate.*

Exercise 4.9.12. Prove Proposition 4.9.11.

Based on the construction of the $C^k(U)$ quantum circuit, we can implement general $C^{C_0,C_1,T}$ operators as in Definition 4.4.7. As an example, Figure 4.9.4 shows a quantum circuit that implements such an operator with $n = 7$, $C_0 = \{0,1\}$, $C_1 = \{2,3,4\}$, and $T = \{6\}$. In this circuit, the Pauli X operator is applied to the qubits $|b_i\rangle$ with $i \in C_0$ at the beginning and the end of the computation. This idea is the basis for the following theorem.

Theorem 4.9.13. *Let C_0, C_1, and T be pairwise disjoint subsets of $\mathbb{Z}_n$, let $m = |T| > 0$, and assume that $T = \{i, i+1, \ldots, i+m-1\}$. Also, let U be a unitary operator on $\mathbb{H}_m$. Set $k_0 = |C_0|$, $k_1 = |C_1|$, $k = k_0 + k_1$. Then the unitary operator $C^{C_0,C_1,T}(U)$ can be implemented by a quantum circuit that uses $2k_0$ Pauli X gates, $2k - 2$ CCNOT gates, one $C^1(U)$ gate, and $k - 1$ ancillary and erasure gates.*

For the special case where U is a single-qubit operator, Theorem 4.9.13 implies the following result.

Theorem 4.9.14. *Let U be a unitary single-qubit operator, let C_0, C_1 be disjoint subsets of $\mathbb{Z}_n$, let $k_0 = |C_0|$, $k_1 = |C_1|$, $k = k_0 + k_1 < n$, and $t \in \mathbb{Z}_n \setminus (C_0 \cup C_1)$. Then the* *unitary operator $C^{C_0,C_1,t}(U)$ can be implemented by a quantum circuit that uses $O(k)$ Pauli X, Hadamard, $\pi/8$, inverse $\pi/8$, CNOT, ancillary, and erasure gates and four other single-qubit gates.*

Proof. It follows from Theorem 4.9.13 that $C^{C_0,C_1,t}(U)$ can be implemented using $2k_0$ Pauli X gates, $2k - 2$ CCNOT gates, one $C^1(U)$ operator, and $k - 1$ ancillary and erasure gates. By Proposition 4.9.7 each CCNOT gate can be implemented using 2 Hadamard gates H, one phase gate S, 7 (inverse) $\pi/8$ gates T, and 6 standard CNOT gates. Also, by Theorem 4.4.5 the operator $C^1(U)$ can be implemented using 2 standard CNOT and 4 single-qubit gates A, B, C, $e^{i\delta/2}R_{\hat{z}}(\delta)$ such that $U = e^{i\delta}AXBXC$. Since by (4.3.53) we have $S = T^2$, it follows that $C^{C_0,C_1,t}(U)$ can be implemented using $O(k)$ Pauli X, Hadamard, (inverse) $\pi/8$, and standard CNOT gates and four other single-qubit gates. $\qquad\square$

From Theorem 4.9.13 we obtain the following result.

Theorem 4.9.15. *Every transposition operator can be implemented by a quantum circuit that uses $O(n)$ Pauli X, Hadamard, (inverse) $\pi/8$, standard CNOT, ancillary and erasure gates.*

Proof. The proof is analogous to the proof of Theorem 4.9.14 and uses the fact that for a transposition operator, we have $U = X$. $\qquad\square$

4.10. Perfectly universal sets of quantum gates

In this section, we construct two perfectly universal sets of quantum gates.

4.10.1. Two-level operators.
For the first construction, we need the following definition. We recall that we identify linear operators on $\mathbb{H}_n$ with their representation matrices with respect to the computational basis of $\mathbb{H}_n$.

Definition 4.10.1. Let $A \in \mathbb{C}^{(k,k)}$. Then A is called a *two-level matrix, two-level operator*, or *two-level gate* if there are $i, j \in \mathbb{Z}_k$ such that for every $\hat{v} \in \mathbb{C}^k$ all entries of the vectors $\hat{v}$ and $A\hat{v}$ with indices different from i and j are equal.

We note that in this definition we do not require i and j to be different. This implies that all matrices in $\mathbb{C}^{(1,1)}$ are two-level matrices that will simplify our reasoning. We also note that all matrices in $\mathbb{C}^{(2,2)}$ are two-level matrices.

Example 4.10.2. Consider the matrix
$$A = \begin{pmatrix} 1 & 1 & 0 \\ 1 & -1 & 0 \\ 0 & 0 & 1 \end{pmatrix}.$$
For every $\hat{v} = (v_0, v_1, v_2) \in \mathbb{C}^3$ we have
$$A\hat{v} = \begin{pmatrix} v_0 + v_1 \\ v_0 - v_1 \\ v_2 \end{pmatrix}.$$
Hence, A is a two-level matrix.

Example 4.10.3. Consider the matrix

$$A = \begin{pmatrix} 1 & 1 & 0 \\ 1 & -1 & 0 \\ 0 & 1 & 1 \end{pmatrix}.$$

For every $\hat{v} = (v_0, v_1, v_2) \in \mathbb{C}^3$ we have

$$A\hat{v} = \begin{pmatrix} v_0 + v_1 \\ v_0 - v_1 \\ v_1 + v_2 \end{pmatrix}.$$

Hence, for $\hat{v} = (1, 1, 1)$ we have $A\hat{v} = (2, 0, 2)$. So A is not a two-level matrix.

The main theorem of this section is the following.

Theorem 4.10.4. *Let $U \in \mathbb{C}^{(k,k)}$ be unitary. Then U can be written as a product of $k(k-1)/2$ unitary two-level matrices.*

Proof. We prove the assertion by induction on k. For $k = 1$ and $k = 2$ the assertion holds. Now assume that $k > 2$ and that the assertion is proved for $k - 1$. For the inductive step, let $U \in \mathbb{C}^{(k,k)}$ be a unitary matrix. We first show that there are unitary two-level matrices $U_1, \ldots, U_{k-1} \in \mathbb{C}^{(k,k)}$ such that the product

$$(4.10.1) \qquad\qquad V = U_{k-1} \cdots U_1 U$$

is of the form

$$(4.10.2) \qquad\qquad V = \begin{pmatrix} 1 & 0 & \cdots & 0 \\ 0 & * & \cdots & * \\ \vdots & \vdots & & \vdots \\ 0 & * & \cdots & * \end{pmatrix}.$$

To prove this claim, we use the following construction. Let $W = (w_{i,j}) \in \mathbb{C}^{(k,k)}$ be a unitary matrix. For $i = 1, \ldots, k - 1$ define the matrix $T_i(W) = (t_{p,q}) \in \mathbb{C}^{(k,k)}$ as follows. If $w_{i,0} = 0$, we set $T_i(W) = I_k$. If $w_{i,0} \neq 0$, we initialize $T_i(W)$ to I_k and then change four of the entries of $T_i(W)$ in the following way. Set

$$(4.10.3) \qquad\qquad c = \frac{1}{\sqrt{|w_{0,0}|^2 + |w_{i,0}^2|}}$$

and

$$(4.10.4) \qquad t_{0,0} = c\overline{w_{0,0}}, \quad t_{0,i} = c\overline{w_{i,0}}, \quad t_{i,0} = cw_{i,0}, \quad t_{i,i} = -cw_{0,0}.$$

Then $T_i(W)$ is a two-level matrix. Also, this matrix is unitary. To see this, we observe that the columns with indices $j \neq 0, i$ are equal to the unit vectors $\vec{e}_j$. Therefore, they have length 1 and are pairwise orthogonal. Furthermore, the length of the first column is $c^2(|w_{0,0}|^2 + |w_{i,0}|^2) = 1$, the length of the ith column is $c^2(|w_{i,0}|^2 + |w_{0,0}|^2) = 1$, the inner product of the first and the ith columns is $c^2(w_{0,0}\overline{w_{i,0}} - \overline{w_{i,0}}w_{0,0}) = 0$, and the inner product of the first and second columns with the other columns is 0.

We determine the entries in the first column of the product $T_i(W)W$. The entries with indices different from 0 and i are the same as the corresponding entries in the first column of W. The entry with index 0 is

(4.10.5)
$$t_{0,0}w_{0,0} + t_{0,i}w_{i,0} = c(|w_{0,0}|^2 + |w_{i,0}|^2) = \frac{1}{c}.$$

The entry with index i is

(4.10.6)
$$t_{i,0}w_{0,0} + t_{i,i}w_{i,0} = c(w_{i,0}w_{0,0} - w_{0,0}w_{i,0}) = 0.$$

So if we set $U_0 = U$ and $U_j = T_j(U_{j-1} \cdots U_2 U)$ for $j = 1, \ldots, k - 1$, then $V = U_{k-1} \cdots U_1 U$ has the form

(4.10.7)
$$V = \begin{pmatrix} x & * & \cdots & * \\ 0 & * & \cdots & * \\ \vdots & \vdots & & \vdots \\ 0 & * & \cdots & * \end{pmatrix}$$

with a positive real number x. As a product of unitary matrices, this matrix is unitary. So it follows from Proposition 2.4.18 that the length of the first row and column is 1. This implies $x = 1$ and that the entries in the first row with indices $j \in \{1, \ldots, k - 1\}$ are 0. Hence, V is of the form

(4.10.8)
$$V = \begin{pmatrix} 1 & \vec{0} \\ \vec{0} & V' \end{pmatrix}$$

where $\vec{0}$ denotes the row and column vectors in $\mathbb{C}^{k-1}$ which have only zero entries, respectively, and $V' \in \mathbb{C}^{(k-1,k-1)}$ is the minor obtained from V by deleting the first row and column of this matrix. Since V is unitary, Proposition 2.4.18 implies that V' is unitary.

According to the induction hypothesis, there are $m \in \mathbb{N}$ and unitary two-level matrices $V_1', \ldots, V_m' \in \mathbb{C}^{(k-1,k-1)}$ such that $m \leq (k-1)(k-2)/2$ and $V' = V_0' \cdots V_{m-1}'$. For $0 \leq i < m$ we set

(4.10.9)
$$V_i = \begin{pmatrix} 1 & \vec{0} \\ \vec{0} & V_i' \end{pmatrix}.$$

So V_i is obtained from V_i' by prepending the unit vector $(1, 0, \ldots, 0) \in \mathbb{C}^k$ as first row and column. Then the V_i are unitary two-level matrices and we have $V = V_0 \cdots V_{m-1}$. From (4.10.1) we obtain

(4.10.10)
$$U = U_1^* \cdots U_{k-1}^* V_0 \cdots V_{m-1}.$$

This is a decomposition of U into a product of two-level unitary matrices. The number of factors is $m + k - 1 \leq (k-1)(k-2)/2 + k - 1 = (k^2 - 3k + 2 + 2k - 2)/2 = (k^2 - k)/2 = k(k-1)/2$. $\qquad\square$

The proof of Theorem 4.10.4 also contains a method to construct the decomposition of a unitary matrix into a product of two-level matrices and thus a method to construct a circuit that implements a given unitary operator and uses only two-level gates. Therefore, Theorem 4.10.4 implies the following corollary.

Corollary 4.10.5. *The set of all two-level unitary operators is perfectly universal for quantum computation.*

4.10.2. Another perfectly universal set of quantum gates. In this section, we prove the following theorem.

Theorem 4.10.6. *The set that contains all rotation gates and the standard CNOT gate is perfectly universal for quantum computing.*

To prove Theorem 4.10.6 we will show that every unitary two-level operator can be implemented by a quantum circuit that uses only single-qubit gates and the standard CNOT gate. Then Theorem 4.10.6 follows from this statement and Corollaries 4.10.5 and 4.3.16. For the proof of Theorem 4.10.6 we also need the following definition and result.

Definition 4.10.7. Let $\vec{s}, \vec{t} \in \{0,1\}^n$. A *Gray code connecting $\vec{s}$ and $\vec{t}$* is a sequence $G = (\vec{g}_0, \ldots, \vec{g}_m)$ of pairwise distinct vectors in $\{0,1\}^n$ such that $\vec{g}_0 = \vec{s}$, $\vec{g}_m = \vec{t}$, and the successive elements of G differ by exactly one bit.

Example 4.10.8. Let $\vec{s} = (0,0,0)$ and $\vec{t} = (1,1,1)$. Then

(4.10.11) $$G = ((0,0,0), (1,0,0), (1,1,0), (1,1,1))$$

is a Gray code that connects $\vec{s}$ and $\vec{t}$.

Exercise 4.10.9. Find the shortest Gray code that connects $(1,1,0)$ and $(0,1,1)$.

The next proposition is also required for the proof of Theorem 4.10.6.

Proposition 4.10.10. *Let $\vec{s}, \vec{t} \in \{0,1\}^n$. Then there is a Gray code of length $\leq n + 1$ that connects $\vec{s}$ and $\vec{t}$.*

Proof. We prove the theorem by induction on n. Let $n = 1$. Then $\vec{s}, \vec{t} \in \{0,1\}$. If $\vec{s} = \vec{t}$, then $(\vec{s})$ is a gray code of length $1 \leq 2 = n + 1$ that connects $\vec{s}$ and $\vec{t}$. If $\vec{s} \neq \vec{t}$, then $(\vec{s}, \vec{t})$ is a Gray code of length $2 = n + 1$ connecting $\vec{s}$ and $\vec{t}$. This proves the base case.

For the induction step, assume that the assertion is true for $n-1$. Denote by $\vec{s'}$ and $\vec{t'}$ the vectors that are obtained from $\vec{s}$ and $\vec{t}$, respectively, by deleting the last entry. Then it follows from the induction hypothesis that there is a Gray code $G' = (\vec{g}_0', \ldots, \vec{g}_{m-1}')$ of length $m \leq n$ that connects $\vec{s'}$ and $\vec{t'}$. Let b be the last entry of $\vec{s}$. Append b to all elements of G' as the new last entry. Denote the resulting sequence by $G = (\vec{g}_0, \ldots, \vec{g}_{m-1})$. Then G is a sequence of length m in $\{0,1\}^n$ and the successive elements of G differ by exactly one bit. Also, we have $\vec{g}_0 = \vec{s}$ and the vectors $\vec{g}_{m-1}$ and $\vec{t}$ are either equal or differ exactly in the last bit. In the first case, G is a Gray code connecting $\vec{s}$ and $\vec{t}$. In the second case, $(\vec{g}_0, \ldots, \vec{g}_{m-1}, \vec{t})$ is such a Gray code. $\square$

Note that the proof of Proposition 4.10.10 contains a method for constructing a connecting gray code.

Exercise 4.10.11. Find an algorithm that on input of $\vec{s}, \vec{t} \in \{0,1\}^n$ computes a Gray code of length at most $n + 1$ that connects $\vec{s}$ and $\vec{t}$ and analyze its complexity.

Next, we will prove a statement that together with Corollary 4.10.5 implies Theorem 4.10.6.

Theorem 4.10.12. *For every two-level unitary operator U on $\mathbb{H}_n$ there is a unitary single-qubit operator V such that U can be implemented by a quantum circuit that uses V and $O(n^2)$ Pauli X, Hadamard, (inverse) $\pi/8$, standard CNOT, ancillary, and erasure gates and four other single-qubit gates.*

Proof. Let U be a two-level unitary operator on $\mathbb{H}_n$. We prove the theorem by constructing V and a circuit with the required properties.

We start by constructing V. Since U is two-level, we can choose $\vec{s}, \vec{t} \in \{0, 1\}^n$ and $\alpha, \beta, \gamma, \delta \in \mathbb{C}$ such that

$$(4.10.12) \qquad U|\vec{s}\rangle = \alpha|\vec{s}\rangle + \beta|\vec{t}\rangle, \quad U|\vec{t}\rangle = \gamma|\vec{s}\rangle + \delta|\vec{t}\rangle,$$

and U leaves all other computational basis states of $\mathbb{H}_n$ unchanged. We define the single-qubit operator V by its action on the computational basis elements of $\mathbb{H}_1$ as follows:

$$(4.10.13) \qquad V|0\rangle = \alpha|0\rangle + \beta|1\rangle, \quad V|1\rangle = \gamma|0\rangle + \delta|1\rangle.$$

We also write

$$(4.10.14) \qquad |\vec{t}\rangle = |t_0 t_1 \cdots t_{n-1}\rangle.$$

The operator V is unitary because U is unitary. This is verified in Exercise 4.10.13.

Next, we construct the quantum circuit that has the asserted properties. This construction has three steps.

(1) We show how to find $i \in \mathbb{Z}_n$ and a unitary operator P that can be implemented using at most n transposition operators such that

$$(4.10.15) \qquad \begin{aligned} P|\vec{s}\rangle &= |t_0 \cdots t_{i-1}\rangle|0\rangle|t_{i+1} \cdots t_{n-1}\rangle, \\ P|\vec{t}\rangle &= |t_0 \cdots t_{i-1}\rangle|1\rangle|t_{i+1} \cdots t_{n-1}\rangle, \end{aligned}$$

and P leaves the other computational basis vectors of $\mathbb{H}_n$ unchanged.

(2) We let

$$(4.10.16) \qquad C_0 = \{j \in \mathbb{Z}_n \setminus \{i\} : t_j = 0\}, \quad C_1 = \{j \in \mathbb{Z}_n \setminus \{i\} : t_j = 1\}$$

and show that

$$(4.10.17) \qquad U = P^* C^{C_0, C_1, i}(V) P.$$

(3) We construct the quantum circuit using (1) and (2) and apply Theorems 4.9.14 and 4.9.15 to estimate the number of required gates.

We start with (1) and show how to construct the unitary operator P. Let $G = (\vec{g}_0, \ldots, \vec{g}_m)$ be a Gray code with $m \le n$ that connects $\vec{s}$ and $\vec{t}$. It exists by Proposition 4.10.10. Let $j \in \{1, \ldots, m\}$. Then $\vec{g}_{j-1}$ and $\vec{g}_j$ differ in exactly one bit. Denote by T_j the transposition operator from Definition 4.4.11 that satisfies

$$(4.10.18) \qquad |\vec{g}_j\rangle = T_j|\vec{g}_{j-1}\rangle \quad \text{and} \quad |\vec{g}_{j-1}\rangle = T_j|\vec{g}_j\rangle.$$

It does not change the other computational basis states of $\mathbb{H}_n$. Set

(4.10.19) $$P = T_{m-1} \cdots T_1.$$

Then

(4.10.20) $$P|\vec{s}\rangle = |\vec{g}_{m-1}\rangle \quad \text{and} \quad T_m|g_{m-1}\rangle = |\vec{g}_m\rangle = |\vec{t}\rangle.$$

Since the elements of the Gray code G are pairwise different, it follows that $\vec{t}$ is different from the first $m-1$ elements in the Gray code which implies $P|\vec{t}\rangle = |\vec{t}\rangle$. Also, $P|\vec{s}\rangle$ and $|\vec{t}\rangle$ differ in exactly one qubit. Let i be its index. If $t_i = 1$, then (4.10.15) holds. If $t_i = 0$, then we replace P by TP where T is $T = \mathsf{TRANS}^{\vec{c}}$ with $\vec{c} = (t_0, \ldots, t_{i-1}, *, t_{i+1}, \ldots, t_{n-1})$. Note that P is the product of at most n transposition operators.

Next, assertion (2) is verified in Exercise 4.10.14.

Finally, we deduce the assertion of the theorem from (1) and (2) and Theorems 4.9.14 and 4.9.15. Since P is the product of $O(n)$ transposition operators, it follows from Theorem 4.9.15 that P and P^* can be implemented by a quantum circuit that contains $O(n^2)$ Pauli X, Hadamard, (inverse) $\pi/8$, standard CNOT, ancillary, and erasure gates. Also, it follows from Theorem 4.9.14 that $C^{C_0,C_1,i}(V)$ can be implemented by a quantum circuit that uses $O(n)$ Pauli X, Hadamard, (inverse) $\pi/8$, standard CNOT, ancillary, and erasure gates and four other single-qubit gates. This concludes the proof of the theorem. □

Exercise 4.10.13. Show that the operator V from the proof of Theorem 4.10.12 is unitary.

Exercise 4.10.14. Show that in the proof of Theorem 4.10.12 assertion (2) is correct.

4.11. A universal set of quantum gates

The goal of this section is to prove the following theorem.

Theorem 4.11.1. *The set containing the Hadamard, $\pi/8$, and standard CNOT gates is universal for quantum computation.*

The main work in proving this theorem is to show the following theorem.

Theorem 4.11.2. *The set containing the Hadamard and $\pi/8$ gates is universal for the set of all unitary single-qubit operators.*

Then Theorem 4.11.1 follows from Theorems 4.10.6 and 4.11.2.

To prove Theorem 4.11.2, we first estimate the distance between two products of quantum operators in terms of the distance of the factors.

Proposition 4.11.3. *Let $k \in \mathbb{N}$ and let U_i, V_i be unitary operators on $\mathbb{H}_n$ for $1 \le i \le k$. Then we have*

(4.11.1) $$E\left(\prod_{i=1}^{k} U_i, \prod_{i=1}^{k} V_i\right) \le \sum_{i=1}^{k} E(U_i, V_i).$$

Proof. We prove the assertion by induction on k. For the base case $k = 1$ the assertion is obviously true. For the inductive step, let $k > 1$,

$$(4.11.2) \qquad U = \prod_{i=1}^{k-1} U_i, \quad V = \prod_{i=1}^{k-1} V_i,$$

and assume that

$$(4.11.3) \qquad E(U, V) \le \sum_{i=1}^{k-1} E(U_i, V_i).$$

Also, let $|\psi\rangle \in \mathbb{H}_n$ be a quantum state. Then we have

$$\left\| (U_k U - V_k V) |\psi\rangle \right\|$$
$$= \left\| U_k U |\psi\rangle - V_k V |\psi\rangle \right\|$$
$$= \left\| U_k (U - V) |\psi\rangle + (U_k - V_k) V |\psi\rangle \right\|$$
$$\underset{(1)}{\le} \left\| U_k (U - V) |\psi\rangle \right\| + \left\| (U_k - V_k) V |\psi\rangle \right\|$$
$$\underset{(2)}{=} \left\| (U - V) |\psi\rangle \right\| + \left\| (U_k - V_k) V |\psi\rangle \right\|$$
$$\underset{(3)}{\le} E(U, V) + E(U_k, V_k)$$
$$\underset{(4)}{\le} \sum_{i=1}^{k} E(U_i, V_i).$$

These equations and inequalities are valid for the following reasons: inequality (1) uses the triangle inequality which holds by Proposition 2.2.25, equation (2) holds because U_k is unitary, inequality (3) uses Definition 4.8.2, and inequality (4) follows from an application of the induction hypothesis (4.11.3). $\square$

Next, we prove Theorem 4.11.2 using Theorem 4.3.35. Define

$$(4.11.4) \qquad \vec{n} = \left(\cos \frac{\pi}{8}, \sin \frac{\pi}{8}, \cos \frac{\pi}{8} \right).$$

Then we have

$$(4.11.5) \qquad \|\vec{n}\|^2 = 2 \cos^2 \frac{\pi}{8} + \sin^2 \frac{\pi}{8} = \cos^2 \frac{\pi}{8} + 1.$$

We normalize $\vec{n}$ and obtain the unit vector

$$(4.11.6) \qquad \hat{n} = \frac{\vec{n}}{\|\vec{n}\|}.$$

We also set

$$(4.11.7) \qquad \vec{m} = \left(\cos \frac{\pi}{8}, -\sin \frac{\pi}{8}, \cos \frac{\pi}{8} \right).$$

Then $\|\vec{m}\| = \|\vec{n}\| = \cos^2 \frac{\pi}{8} + 1$. Normalizing $\vec{m}$ we obtain

$$(4.11.8) \qquad \hat{m} = \frac{\vec{m}}{\|\vec{m}\|}.$$

We write

(4.11.9) $\hat{n} = (n_x, n_y, n_z), \quad \hat{m} = (m_x, m_y, n_z)$

and use the following observation.

Lemma 4.11.4. *For all $\gamma \in \mathbb{R}$ we have*

(4.11.10) $R_{\hat{m}}(\gamma) = H R_{\hat{n}}(\gamma) H.$

Proof. Let $\gamma \in \mathbb{R}$. It follows from Proposition 4.3.3 that

(4.11.11) $R_{\hat{n}}(\gamma) = \cos \frac{\gamma}{2} I - i \sin \frac{\gamma}{2}(n_x X + n_y Y + n_z Z).$

From (4.1.12) and

(4.11.12) $n_x = n_z = m_x = m_z, \quad n_y = -m_y$

we obtain from (2.3.17)

$$HR_{\hat{n}}(\gamma)H = \cos \frac{\gamma}{2} I - i \sin \frac{\gamma}{2}(n_x HXH + n_y HYH + n_z HZH)$$

$$= \cos \frac{\gamma}{2} I - i \sin \frac{\gamma}{2}(n_x X - n_y Y + n_z Z)$$

(4.11.13) $= \cos \frac{\gamma}{2} I - i \sin \frac{\gamma}{2}(m_x X + m_y Y + n_z Z)$

$$= R_{\hat{m}}(\gamma). \qquad \qquad \square$$

Now let

(4.11.14) $\theta = 2 \arccos\left(\cos^2 \frac{\pi}{8}\right).$

The next lemma shows that, up to a global phase factor, we can implement the rotation operator $R_{\hat{n}}(\theta)$ by a circuit that uses only Hadamard and $\pi/8$ gates.

Lemma 4.11.5. *We have $R_{\hat{n}}(\theta) = e^{-i\frac{\pi}{4}} THTH.$*

Proof. It follows from (4.3.52) and (4.3.10) that

(4.11.15) $e^{-i\frac{\pi}{8}} T = R_{\hat{z}}\left(\frac{\pi}{8}\right) = e^{-i\frac{\pi}{8}} |0\rangle\langle 0| + e^{i\frac{\pi}{8}} |1\rangle\langle 1|.$

Next, recall that by (2.4.72),

(4.11.16) $X = H |0\rangle\langle 0| H - H |1\rangle\langle 1| H$

is the spectral decomposition of the Pauli X gate. So it follows from (4.11.15) and Definition 2.4.69 that

$$e^{-i\frac{\pi}{8}} HTH = e^{-i\frac{\pi}{8}} H |0\rangle\langle 0| H + e^{i\frac{\pi}{8}} H |1\rangle\langle 1| H$$

(4.11.17) $= e^{-i\frac{\pi}{8}} |x_+\rangle\langle x_+| + e^{i\frac{\pi}{8}} |x_-\rangle\langle x_-|$

$$= e^{-i\pi X/8}.$$

Now we observe that by (4.11.5) and (4.11.14) we have

(4.11.18) $\sin^2 \frac{\pi}{8} = 1 - \cos^2 \frac{\pi}{8} = \frac{1 - \cos^4 \frac{\pi}{8}}{1 + \cos^2 \frac{\pi}{8}} = \frac{\sin^2 \frac{\theta}{2}}{\|\vec{n}\|^2}.$

This implies

$$(4.11.19) \qquad \sin\frac{\pi}{8} = \frac{\sin\frac{\theta}{2}}{\|\vec{n}\|}.$$

So we obtain

$$e^{-i\pi/4}THTH$$

$$\underset{(1)}{=} e^{-i\pi Z/8}e^{-i\pi X/8}$$

$$\underset{(2)}{=} \left(\cos\frac{\pi}{8}I - i\sin\frac{\pi}{8}Z\right)\left(\cos\frac{\pi}{8}I - i\sin\frac{\pi}{8}X\right)$$

$$= \cos^2\frac{\pi}{8}I - i\sin\frac{\pi}{8}\cos\frac{\pi}{8}(X+Z) - i\sin^2\frac{\pi}{8}ZX$$

$$\underset{(3)}{=} \cos\frac{\theta}{2}I - i\sin\frac{\theta}{2}\frac{1}{\|\vec{n}\|}\left(\cos\frac{\pi}{8}X + \sin\frac{\pi}{8}Y + \cos\frac{\pi}{8}Z\right)$$

$$\underset{(4)}{=} R_{\hat{n}}(\theta).$$

In these equations we use the following arguments: equation (1) follows from (4.11.15) and (4.11.17), equaltion (2) is obtained from (4.3.3), equation (3) holds because of (4.11.14), $ZX = -Y$ (see Theorem 4.1.2), (4.11.19), and equation (4) is true because of (4.3.3). □

To show that $\{H, T\}$ is universal for the set of all unitary single-qubit operators we also need the following auxiliary results. Their proofs require some algebraic number theory which is beyond the scope of this book. We refer to [IR10] which is an excellent introduction to the subject.

Lemma 4.11.6. *If $v, u \in \mathbb{Z}$ with $v > 0$, then $2\cos\left(\frac{u\pi}{v}\right)$ is an algebraic integer.*

Proof. We will show that for all $v \in \mathbb{N}$ and all $y \in \mathbb{R}$, there exists a polynomial $P_v \in \mathbb{Z}[x]$ that is monic, has degree v, and satisfies

$$(4.11.20) \qquad P_v(2\cos y) = 2\cos vy.$$

So we have

$$(4.11.21) \qquad P_v\left(2\cos\frac{u\pi}{v}\right) = 2\cos u\pi$$

which implies the assertion. The polynomials P_v are constructed inductively. We set $P_0(x) = 2$, $P_1(x) = x$. Then (4.11.20) holds for $v = 0, 1$. Also, for $v \geq 1$ we set

$$(4.11.22) \qquad P_{v+1}(x) = xP_v(x) - P_{v-1}(x).$$

Assume that (4.11.20) holds for all $v' \leq v$. Then (A.5.8), (4.11.22), and the induction hypothesis imply

$$(4.11.23) \qquad P_{v+1}(2\cos y) = 2\cos y P_v(2\cos y) - P_{v-1}(2\cos y)$$

$$= 4\cos y\cos vy - 2\cos(v-1)y = 2\cos(v+1)y. \qquad □$$

Lemma 4.11.7. *The fraction $\frac{\theta}{\pi}$ is irrational.*

Proof. Using (A.5.7) we obtain

$$(4.11.24) \qquad \cos\frac{\theta}{2} = \cos^2\frac{\pi}{8} = \frac{1}{2}\left(\cos\frac{\pi}{4} + 1\right) = \frac{1}{2} + \frac{\sqrt{2}}{4}.$$

Now assume that $\frac{\theta}{2\pi} = \frac{u}{v}$ with $u, v \in \mathbb{Z}, v > 0$. Then it follows from Lemma 4.11.6 that

$$(4.11.25) \qquad 2\cos\frac{\theta}{2} = 2\cos\frac{u\pi}{v}$$

is an algebraic integer. From (4.11.24) we see that $2\cos\frac{\theta}{2}$ is a quadratic irrationality of norm

$$(4.11.26) \qquad 4\left(\frac{1}{2} + \frac{\sqrt{2}}{4}\right)\left(\frac{1}{2} - \frac{\sqrt{2}}{4}\right) = 1 - \frac{1}{2} = \frac{1}{2}.$$

But this is not the norm of an algebraic integer, a contradiction. $\qquad\square$

In the next lemma, we need the following notion.

Definition 4.11.8. Let S and T be sets of real numbers. Then we say that T is *dense* in S if for every $\varepsilon > 0$ and every $s \in S$ there is $t \in T$ such that $|s - t| < \varepsilon$.

Lemma 4.11.9. *Let $\alpha \in \mathbb{R}$ be an irrational number. Then the set $\{u\alpha \bmod 1 : u \in \mathbb{N}\}$ is dense in $[0, 1[$.*

Proof. For $u \in \mathbb{Z}$ set

$$(4.11.27) \qquad \alpha_u = (u\alpha) \bmod 1.$$

Let $x \in [0, 1[$ and $\varepsilon \in \mathbb{R}_{>0}$. We must show that there is $u \in \mathbb{N}$ such that

$$(4.11.28) \qquad |\alpha_u - x| < \varepsilon.$$

To construct this u, we select $N \in \mathbb{N}$ such that

$$(4.11.29) \qquad \frac{1}{N} < \varepsilon.$$

Using the Pigeonhole Principle and the irrationality of α we see that there are $k, l, k > l$, such that

$$(4.11.30) \qquad 0 < \alpha_k - \alpha_l < \frac{1}{N}$$

or

$$(4.11.31) \qquad -\frac{1}{N} < \alpha_k - \alpha_l < 0.$$

First, assume that (4.11.30) is true. This inequality implies that there is $v \in \mathbb{N}$ such that

$$(4.11.32) \qquad v(\alpha_k - \alpha_l) \in [0, 1[$$

and

$$(4.11.33) \qquad |v(\alpha_k - \alpha_l) - x| < \frac{1}{N} < \varepsilon.$$

It follows from (4.11.32) that

(4.11.34) $$\alpha_{v(k-l)} = v(\alpha_k - \alpha_l).$$

So if we set $u = v(k - l)$, then (4.11.28) holds.

Next, assume that (4.11.31) is true. Then we can select $v \in \mathbb{N}$ such that

(4.11.35) $$v(\alpha_k - \alpha_l) \in]-1, 0]$$

and

(4.11.36) $$|v(\alpha_k - \alpha_l) - (x - 1)| < \frac{1}{N} < \varepsilon.$$

It follows from (4.11.34) that

(4.11.37) $$\alpha_{v(k-l)} = v(\alpha_k - \alpha_l) + 1.$$

Using this equality in (4.11.36) we see that (4.11.28) holds for $u = v(k - l)$. $\qquad \square$

The next proposition shows that the rotation operator $R_{\hat{n}}(\theta)$ can be used to approximate every other rotation about the $\hat{n}$-axis with arbitrary precision.

Proposition 4.11.10. *For all $\varepsilon \in \mathbb{R}_{>0}$ and all $\gamma \in \mathbb{R}$ there is $k \in \mathbb{N}$ such that*

(4.11.38) $$E(R_{\hat{n}}(\gamma), R_{\hat{n}}(\theta)^k) < \varepsilon.$$

Proof. Let $\varepsilon \in \mathbb{R}_{>0}$ and $\gamma \in \mathbb{R}$. We approximate $R_{\hat{n}}(\gamma)$ to precision ε. By Theorem 4.3.15, we may choose $\gamma \in [0, 2\pi[$. For $k \in \mathbb{N}$ set

(4.11.39) $$\theta_k = (k\theta) \bmod 2\pi.$$

It follows from Lemma 4.11.7 and Lemma 4.11.9 that there is $k \in \mathbb{N}$ such that

(4.11.40) $$|\gamma - \theta_k| < \frac{\varepsilon}{2}.$$

Let $|\psi\rangle$ be a quantum state in $\mathbb{H}_1$. Using the triangle inequality, the fact that $\hat{n} \cdot \sigma$ is unitary (Proposition 4.3.2), Lemma A.5.2, and (4.11.40) we obtain

(4.11.41)
$$\begin{aligned}
&\|(R_{\hat{n}}(\gamma) - R_{\hat{n}}(\theta)^k)|\psi\rangle\| \\
&= \|(\cos\gamma - \cos\theta_k)|\psi\rangle - i(\sin\gamma - \sin\theta_k)(\hat{n} \cdot \sigma)|\psi\rangle\| \\
&\leq |\cos\gamma - \cos\theta_k| + |\sin\gamma - \sin\theta_k)| \\
&\leq 2|\gamma - \theta_k| \\
&< \varepsilon.
\end{aligned}$$
$\qquad \square$

Now we can prove Theorem 4.11.2. Let U be a unitary single-qubit operator. It follows from Theorem 4.3.35 that up to a global phase factor the operator U can be written as

(4.11.42) $$U = \prod_{i=0}^{k-1} R_{\hat{n}}(\alpha_i) R_{\hat{m}}(\beta_i)$$

where $k \in \mathbb{N}$, $k = O(1)$, $\alpha_i, \beta_i \in \mathbb{R}$ for $i \in \mathbb{Z}_k$. By Lemma 4.11.4 this implies

$$(4.11.43) \qquad U = \prod_{i=0}^{k-1} R_{\hat{n}}(\alpha_i) H R_{\hat{n}}(\beta_i) H.$$

According to Proposition 4.11.10, we can select positive integers $a_0, \ldots, a_{k-1}$ and $b_0, \ldots, b_{k-1}$ such that

$$(4.11.44) \qquad E(R_{\hat{n}}(\alpha_i), R_{\hat{n}}^{a_i}(\theta)) < \varepsilon/2k, \quad E(R_{\hat{n}}(\beta_i), R_{\hat{n}}^{b_i}(\theta)) < \varepsilon/2k$$

for all $i \in \mathbb{Z}_k$. Now we set

$$(4.11.45) \qquad V = \prod_{i=0}^{k-1} R_{\hat{n}}(\theta)^{a_i} H R_{\hat{n}}(\theta)^{b_i} H.$$

From Proposition 4.11.5 we know that $R_{\hat{n}}(\theta) = e^{-i\frac{\pi}{4}} THTH$. Hence, there is $\gamma \in \mathbb{C}$ such that

$$(4.11.46) \qquad V = e^{i\gamma} \prod_{i=0}^{k-1} (THTH)^{a_i} H (THTH)^{b_i} H.$$

From Proposition 4.11.3 we obtain

$$
\begin{aligned}
E(U, V) &= E\left(\prod_{i=0}^{k-1} R_{\hat{n}}(\alpha_i) H R_{\hat{n}}(\beta_i) H, \prod_{i=0}^{k-1} R_{\hat{n}}(\theta)^{a_i} H R_{\hat{n}}(\theta)^{b_i} H \right) \\
&\leq \sum_{i=1}^{k-1} E(R_{\hat{n}}(\alpha_i), R_{\hat{n}}(\theta)^{a_i}) + \sum_{i=1}^{k-1} E\left(R_{\hat{n}}(\beta_i), R_{\hat{n}}(\theta)^{b_i} \right) \\
&< 2k\varepsilon/2k = \varepsilon.
\end{aligned}
$$
$(4.11.47)$

Theorem 4.11.2 now follows from (4.11.46) and (4.11.47).

Now we prove Theorem 4.11.1. Let U be a unitary operator on $\mathbb{H}_n$. By Theorem 4.10.6, there are $k \in \mathbb{N}$ and unitary single-qubit operators $U_0, \ldots, U_{k-1}$, such that — up to a global phase factor — the operator U is the product of the operators U_i (applied to certain qubits) and some CNOT gates (applied to certain pairs of qubits). Let $\varepsilon \in \mathbb{R}_{>0}$. It follows from Theorem 4.11.2 that for all $i \in \mathbb{Z}_k$ there are unitary single-qubit operators V_i that — up global phase factors — can be written as products of Hadamard and $\pi/8$ gates and satisfy

$$(4.11.48) \qquad E(U_i, V_i) < \frac{\varepsilon}{k}.$$

Let V be the unitary operator that is obtained as follows. In the representation of U as the product of the single-qubit operators U_i and certain CNOT gates replace all U_i by V_i. Then Exercise 4.11.11 and Proposition 4.11.3 imply

$$(4.11.49) \qquad E(U, V) < \varepsilon.$$

This concludes the proof of the theorem.

Exercise 4.11.11. Let U and V be single-qubit operators, and let $i \in \mathbb{Z}_n$. Denote by $U(i)$ and $V(i)$ the unitary operator on $\mathbb{H}_n$ that applies U and V to the ith qubit of a quantum register of length n. Show that $E(U, V) = E(U(i), V(i))$.

4.11.1. Efficiency of approximation. We have observed in Theorem 4.10.6 that the set comprising all rotation gates and the standard CNOT gate is perfectly universal. This implies that, disregarding a global phase factor, it is possible to implement all unitary operators on any state space $\mathbb{H}_n$, where $n \in \mathbb{N}$, using gates from this set. Consequently, when discussing the complexity of quantum computing, it is highly convenient to assume that the available quantum computing platform offers all rotation gates and the standard CNOT gate. This is our approach in Section 4.12.2.

However, quantum computing platforms may also provide only a finite number of single-qubit gates, which, in conjunction with the CNOT gate, form a universal set of quantum gates. According to Theorem 4.11.1, the set containing the Hadamard, $\pi/8$, and standard CNOT gates possesses this property. As stated in Theorem 4.11.2, all single-qubit gates can be approximated with arbitrary precision through compositions of the Hadamard and $\pi/8$ gates. When transitioning from a platform that offers all rotation gates to one that provides specific rotation gates capable of approximating all rotation gates, the impact on complexity results depends on the efficiency of this approximation. Unfortunately, Theorem 4.11.2 does not address the issue of approximation efficiency. This is where the Solovay-Kitaev Theorem proves invaluable. Initially announced by Robert M. Solovay in 1995 and independently proven by Alexei Kitaev in 1997, it establishes the existence of highly efficient approximations.

Theorem 4.11.12. *Let G be a finite set of rotation gates containing its own inverses which is universal for the set of all rotation operators. Then for all* $\varepsilon \in \mathbb{R}_{>0}$ *and all rotation operators U there is* $l \in \mathbb{N}$ *and a sequence* $V_0, \ldots, V_{l-1}$ *such that* $l = \mathrm{O}(\log^c 1/\varepsilon)$ *and* $E\left(U, \prod_{i=0}^{l-1} V_i\right) < \varepsilon$.

For the proof of Theorem 4.11.12 we refer the reader to [**NC16**, Appendix 3].

4.12. Quantum algorithms and quantum complexity

So far, we have focused on quantum circuits designed for fixed-length inputs. This section introduces quantum algorithms capable of handling inputs of any length through the utilization of quantum circuit families. Following this, we delve into quantum complexity theory, which builds upon classical complexity theory as elucidated in Chapter 1. This theory empowers the assessment of quantum algorithm efficiency.

4.12.1. Quantum algorithms. To be able to define quantum algorithms, we need families of quantum circuits, which are defined now. The classical analog is presented in Section 1.6.3.

Definition 4.12.1. A *family of quantum circuits* is a sequence $(Q_n)_{n \in \mathbb{N}}$ of quantum circuits Q_n such that Q_n operates on n-qubit input registers for all $n \in \mathbb{N}$.

In the theory of Boolean circuits, classical algorithms correspond to uniform families of such circuits. This has been explained in Section 1.6.3. Analogously, for the construction of quantum algorithms, we use uniform families of quantum circuits. To define such families, quantum circuits are encoded by finite bit strings. Definition 4.7.1 shows how such an encoding can be constructed. We assume that any such encoding has the following properties, which are mentioned in [**Wat09**] and which we have already used in Section 1.6.3.

(1) The encoding is *sensible*: every quantum circuit is encoded by at least one bit string and every bit string encodes at most one quantum circuit.

(2) The encoding is *efficient*: there is $c \in \mathbb{N}$ such that every quantum circuit Q of size N has an encoding of length at least size Q and at most N^c. Information about the structure of a circuit must be computable in polynomial time from an encoding of the circuit.

(3) The length of every encoding of a quantum circuit is at least the size of the circuit.

The term "structure information" may, for example, refer to information regarding the input qubits and the quantum gates used in quantum circuits, including the qubits on which these gates operate.

Now, uniform quantum circuit families can be defined analogously to classical uniform circuit families in Definition 1.6.7. Since, by Theorem 4.7.7, the computing power of quantum circuits is the same as the computing power of classical circuits, it follows that quantum computing is Turing complete.

Next, similar to classical circuit complexity theory, quantum complexity theory requires P-uniform quantum circuit families. Let us provide a formal definition of them.

Definition 4.12.2. A quantum circuit family (Q_n) is called P-*uniform* if there is a deterministic polynomial time algorithm that on input of I^n, $n \in \mathbb{N}$, outputs an encoding of Q_n.

Our next goal is to define quantum algorithms. A simple example of such an algorithm is the quantum implementation of coinToss presented in Algorithm 4.12.3. This algorithm uses the quantum circuit QcoinToss shown in Figure 4.12.1, where the Hadamard operator is applied to $|0\rangle$, followed by measuring the resulting state. The algorithm provides the measurement result, which can be 0 or 1, both occurring with equal probabilities of $\frac{1}{2}$.

Algorithm 4.12.3. Quantum coin toss

Input: $\emptyset$
Output: 0 or 1
  1: coinToss
  2:     $|\psi\rangle \leftarrow |0\rangle$
  3:     $b \leftarrow$ QcoinToss $|\psi\rangle$
  4:     **return** b
  5: **end**

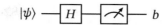

Figure 4.12.1. The quantum circuit QcoinToss where $|\psi\rangle$ is a single-qubit quantum state and $b \in \{0, 1\}$.

Generalizing this example, we define general quantum algorithms as follows.

Definition 4.12.4. A *quantum algorithm* is a probabilistic algorithm with the following additional features.

(1) The algorithm may invoke elements from a P-uniform quantum circuit family. To do so, it prepares an input state for the quantum circuit unless this state is already part of this circuit. The return value is the outcome of the final measurement performed in the quantum circuit.

(2) The algorithm may also invoke other quantum algorithms as subroutines if they terminate on any input.

Example 4.12.5. Consider Algorithm 4.12.6 that implements the probabilistic operation randomString from Section 1.2.1. It uses the family of quantum circuits (QrandomString_n) whose elements are shown in Figure 4.12.2. On input of a string length $n \in \mathbb{N}$, it prepares the input state $|0\rangle^{\otimes n}$ and applies the quantum circuit QrandomString_n to this state. This circuit applies the Hadamard operator to all input qubits and measures the resulting state in the computational basis of $\mathbb{H}_n$. The return value is one of the vectors $\vec{x} \in \{0, 1\}^n$, each with probability $\frac{1}{2}^n$.

Algorithm 4.12.6. Quantum random string selection

Input: $n \in \mathbb{N}$
Output: $\vec{b} \in \{0, 1\}^n$
1:   randomString(n)
2:     $|\psi\rangle \leftarrow |0\rangle^{\otimes n}$
3:     $\vec{x} \leftarrow \text{QrandomString}_n |\psi\rangle$
4:     **return** $\vec{x}$
5: **end**

This definition allows for the smooth transition of concepts and results from probabilistic algorithms to quantum algorithms. For example, quantum Monte Carlo algorithms and quantum Las Vegas algorithms can be defined in a straightforward manner. Also, quantum Monte Carlo algorithms can be either error-free or not. Additionally, there are quantum Bernoulli algorithms that correspond to error-free quantum Monte

Figure 4.12.2. The quantum circuit QrandomString_n where $n \in \mathbb{N}$, $|\psi\rangle$ is a quantum state in $\mathbb{H}_n$, and $\vec{x} \in \{0, 1\}^n$.

Carlo algorithms. Lastly, quantum decision algorithms and their properties are analogous to probabilistic decision algorithms discussed in Section 1.2.2.

4.12.2. The quantum computing platform. In the forthcoming exposition of quantum complexity theory and in the complexity analysis of the quantum algorithms presented in the subsequent chapters, we assume the availability of a quantum computing platform.

According to Theorem 4.10.6, the set that includes all rotation gates and the standard CNOT gate is perfectly universal. This means that it is possible to implement all unitary operators on any state space using the gates from this set, up to a global phase factor. Also, all physical realizations of quantum computers allow for the implementation of rotation gates. Since we are only interested in implementing quantum operators up to a global phase factor, we, therefore, assume that our quantum computing platform makes all rotation gates, the CNOT gate, the ancillary and the erasure gates available. To use the rotation gates in the platform, the rotation axis and the rotation angle must be known. For the convenience of the exposition, we also assume that the platform includes the Pauli, Hadamard, and phase shift gates R_k for all $k \in \mathbb{N}$ since for these gates, we know representations as rotation gates, up to global phase factors. For the Pauli and Hadamard gates, this is shown in Exercise 4.3.10 and for the phase shift gates R_k in (4.3.49). Note that the platform also includes the phase gate $S = R_2$ from (4.3.50) and the $\pi/8$ gate $T = R_3$ from (4.3.51). Furthermore, the platform includes the Toffoli gate CCNOT. As we have seen in Figure 4.9.2, it can be implemented using O(1) of the previous gates. We refer to the gates listed so far as *elementary gates*.

In addition, specific quantum circuits may make use of certain operators linked to the particular computational problem that they solve. A notable example is the Deutsch algorithm, which will be explored in the next chapter. This algorithm uses a black-box that implements an operator U_f, tailored for a function $f : 0, 1 \to 0, 1$ where by a black-box we mean a system or device that can only be observed in terms of its input and output, without revealing any knowledge of its inner workings. The purpose of the algorithm is to determine $f(0) \oplus f(1)$. This incorporation of problem-specific operators adds a layer of versatility to quantum circuits, allowing them to encapsulate the intricacies of the problems they seek to solve.

As already discussed in Section 4.11.1, it is also possible that quantum computing platforms only provide a finite set of single-qubit gates which, in conjunction with the CNOT gate, form a universal set of quantum gates. Theorem 4.11.12 reveals that these gates can be used to efficiently approximate all single-qubit operators. Consequently, the complexity results obtained with our larger platform undergo minimal changes if such reduced platforms are utilized. However, discussing this is beyond the scope of this book.

4.12.3. Implementing $C^1(U)$. In several quantum circuits constructed in the following, controlled-U operators will be required for certain unitary operators U. We have seen in Theorem 4.9.13 that general controlled operators with k control bits, where $k \geq 2$, can be implemented using O(k) elementary gates and one $C^1(U)$ gate. This raises the question of how to implement the $C^1(U)$ gate. Unfortunately, there is no generic construction of $C^1(U)$ when U is given as a black-box. However, as the next theorem shows, the situation is different if a quantum circuit implementation of U is provided, using only elementary gates.

Theorem 4.12.7. *Let $n \in \mathbb{N}$ and let U be a unitary operator on $\mathbb{H}_n$. Assume that there is a quantum circuit Q that implements U and uses $k \in \mathbb{N}$ elementary and no other quantum gates. Then there is a quantum circuit Q' that implements the controlled operator $C^1(U)$ and uses $O(k)$ elementary gates and no other gates.*

Proof. The quantum circuit Q' is constructed from the quantum circuit Q by replacing all unitary elementary gates V with their controlled counterparts $C^1(V)$. The elementary single-qubit gates are rotation gates, up to global phase factors. Therefore, it suffices to consider rotation gates. Let $\hat{w} \in \mathbb{R}^3$ be a unit vector, $\gamma \in \mathbb{R}$, and consider the rotation operator $V = R_{\hat{w}}(\gamma)$. As demonstrated in Exercise 4.3.34, we can express V as $V = AXBXC$, where $A = R_{\hat{w}}(\gamma/2)$, $B = R_{\hat{w}}(-\gamma/2)$, and $C = I_2$. Consequently, it follows from Theorem 4.4.5 that $C^1(V)$ can be implemented by a quantum circuit that uses the three rotation gates A, B, and C with known axis and angle of rotation, along with two CNOT gates. There are two more unitary elementary gates: the CNOT gate and the CCNOT gate. It follows from Proposition 4.9.10 that the corresponding controlled versions can be implemented using $O(1)$ elementary gates. These arguments imply the assertion of the theorem. $\qquad\square$

From Theorem 4.12.7 we obtain the following corollary.

Corollary 4.12.8. *For any unitary elementary gate U the controlled-U operator can be implemented using a quantum circuit that uses $O(1)$ elementary quantum gates.*

Exercise 4.12.9. Prove Corollary 4.12.8.

4.12.4. Time and space complexity. The goal of this section is to introduce the time and space complexity of quantum algorithms which are defined as probabilistic algorithms that may invoke quantum subroutines. Therefore, we must define the complexity of such subroutines. For this, we first explain the running time and space requirements of quantum circuits.

Definition 4.12.10. Let Q be a quantum circuit.

(1) The *running time* or *time complexity* of Q is its *size*, i.e., the number of input qubits plus the number of gates used by the circuit.

(2) The *space complexity* of Q is the number of input qubits plus the number of ancilla qubits used by Q.

The complexity of quantum algorithms associated with P-uniform quantum circuit families is defined next.

Definition 4.12.11. Let (Q_n) be a P-uniform family of quantum circuits and let A be the quantum algorithm corresponding to it.

(1) The *time complexity* or *running time* of A is the function qTime $: \mathbb{N} \to \mathbb{N}$ that sends an input length $n \in \mathbb{N}$ to the maximum time complexity of a quantum circuit used in the execution of A with an input of length n.

(2) The *space complexity* of A is the function qSpace $: \mathbb{N} \to \mathbb{N}$ that sends an input length $n \in \mathbb{N}$ to the maximum space complexity of a quantum circuit used in the execution of A with an input of length n.

The complexity of quantum algorithms is now defined using the corresponding concepts for probabilistic algorithms found in Definitions 1.1.25 and 1.1.26, while also accounting for the complexity of quantum subroutines. Furthermore, the names of the asymptotic time and space complexities in Table 1.1.7 apply directly to quantum algorithms. Additionally, the concepts of expected running time discussed in Section 1.3.3, success probability outlined in Section 1.3.2, and its amplification as described in Section 1.3.4 all seamlessly carry over to quantum algorithms.

Example 4.12.12. The time and space complexity of Algorithm 4.12.6 is exponential since the input is the number of bits of the integer returned by the algorithm.

4.12.5. Quantum complexity classes. The complexity theory for probabilistic algorithms also carries over to quantum algorithms.

To say that a quantum Monte Carlo or Las Vegas algorithm solves a computational problem is analogous to stating that a classical Monte Carlo or Las Vegas algorithm solves such a problem (see Definition 1.4.4). Accordingly, if $f : \mathbb{N} \to \mathbb{R}_{>0}$ is a function, we say that an algorithmic problem can be solved in quantum time $O(f)$ if there is a quantum Monte Carlo algorithm that solves this problem with success probability $\geq \frac{2}{3}$ and has running time $O(f)$. We also say that a computational problem is solvable in quantum linear, quasilinear, quadratic, cubic, polynomial, subexponential, and exponential time if there is a quantum Monte Carlo algorithm with this running time that solves this problem with success probability $\geq \frac{2}{3}$ (see Definition 1.4.11).

Finally, we define the complexity class BQP in analogy to BPP (see Definition 1.4.18).

Definition 4.12.13. The *complexity class* BQP is the set of all languages L for which there is a quantum polynomial time Monte Carlo algorithm A that decides L and satisfies $\Pr(A(\vec{s}) = 1) \geq \frac{2}{3}$ for all $\vec{s} \in L$ and $\Pr(A(\vec{s}) = 0) \geq \frac{2}{3}$ for all $\vec{s} \in \{0,1\}^* \setminus L$.

It is important to observe that by Exercise 1.4.12 the value $\frac{2}{3}$ in the definition of the quantum time complexities and the quantum complexity class BQP may be replaced by any real number in $]\frac{1}{2}, 1]$. Also, we note that the following inclusions are satisfied.

Theorem 4.12.14. *We have* $P \subset BPP \subset BQP \subset PSPACE$.

Exercise 4.12.15. Sketch the proof of Theorem 4.12.14.

Chapter 5

The Algorithms of Deutsch and Simon

After the initial concepts of quantum computers emerged, a fundamental question arose: Can these new computers speed up the process of solving complex problems? This chapter delves into the world of early quantum algorithms that firmly answer this question in the affirmative.

Before we provide an overview of the content of this chapter, it is important to note the following. None of the algorithms presented here are primarily intended for practical applications. Instead, their primary purpose is to illustrate the superiority of quantum computing over classical computing in terms of complexity. Additionally, these algorithms demonstrate fundamental components and techniques of quantum algorithms and have been a source of inspiration for researchers in the development of more practical quantum algorithms.

One of the pioneering researchers who offered a positive response was David Deutsch in 1985 [**Deu85**]. We kick off this chapter by introducing his clever and concise quantum algorithm, which has the capability of calculating the value $f(0) \oplus f(1)$ for a function $f : \{0, 1\} \to \{0, 1\}$. Notably, this quantum algorithm achieves its result with just one evaluation of the quantum counterpart of the function $f : \{0, 1\} \to \{0, 1\}$, whereas in the classical context, achieving the same result would require two evaluations of f. This algorithm already unveils key techniques that underpin a majority of quantum algorithms: quantum parallelism, the phase-kickback trick, and quantum interference.

Subsequently, we discuss the generalization of the Deutsch algorithm, jointly conceived by David Deutsch and Richard Jozsa [**DJ92**] in 1992 and subsequently refined by Richard Cleve, Artur Ekert, Leah Henderson, Chaira Macchiavello, and Michele Mosca [**CEH**$^+$**98**]. This algorithm determines whether a function $f : \{0, 1\}^n \to \{0, 1\}$ consistently yields the same output or provides an equal distribution of 0 and 1. While the classical deterministic alternative requires $2^{n-1} + 1$ evaluations of f, the Deutsch-Jozsa algorithm achieves its result with just one evaluation of a quantum operator corresponding to f. This significant efficiency boost lends the algorithm its renown, although its advantage disappears when compared to a relatively straightforward probabilistic algorithm.

However, the story does not end here. The next groundbreaking step was the development of quantum algorithms that offer exponential speedup compared to all efficient classical algorithms for solving the same computational problem that they address. So following the Deutsch and Deutsch-Jozsa algorithms, this chapter proceeds to present the first such algorithm introduced in 1994 by Daniel R. Simon [**Sim94**]. It represents yet another quantum computing breakthrough. In fact, it laid the groundwork for Peter Shor's renowned polynomial time quantum algorithms, which are described in the following chapter and target problems like integer factorization and discrete logarithm computation.

So, the algorithms presented in this chapter serve as pivotal steps towards more sophisticated algorithms, for instance, those explained in the following chapters, which provide much more efficient solutions to algorithmic problems in many application domains than their classical counterparts.

5.1. The Deutsch algorithm

The first quantum algorithm that we discuss is the Deutsch algorithm. It was first proposed by David Deutsch in 1985 [**Deu85**] and in 1998 improved by Cleve et al. [**CEH**$^+$**98**].

5.1.1. The classical version. Consider a function

(5.1.1) $$f : \{0, 1\} \to \{0, 1\}.$$

Assume that this function is provided by a black-box. By "black-box", we mean a system or device that can only be observed in terms of its input and output, without revealing any knowledge of its inner workings. This concept was previously introduced in Section 4.12.2. The black-box implementing f can be queried with an input value $b \in \{0, 1\}$, returns $f(b)$, and this constitutes the only information obtained from this query. If we are given a black-box that implements a function f, we also refer to it as having "black-box access" to f.

Next, we call f from (5.1.1) *constant* if $f(0) = f(1)$, that is, $f(0) \oplus f(1) = 0$, and we call f *balanced* if $f(0) \neq f(1)$, that is, $f(0) \oplus f(1) = 1$.

The classical Deutsch problem is to find out if a function $f : \{0, 1\} \to \{0, 1\}$ is constant or balanced given black-box access to this function.

More formally, this can be described as follows.

Problem 5.1.1 (Deutsch problem — classical version).

Input: A black-box implementing a function $f : \{0, 1\} \to \{0, 1\}$.

Output: $f(0) \oplus f(1)$.

At first glance, this problem may appear trivial, as with just two queries to the black-box function f, we can determine the answer. However, it's essential to make a significant observation: there is no way to solve the Deutsch problem with fewer queries. To understand this, let's assume that the first query yields $f(b) = c$, where both b and c are binary values. It's still possible that $f(1 - b) = c$, which implies that either $f(0) \oplus f(1) = 0$ or that $f(1 - b) = 1 - c$, which implies $f(0) \oplus f(1) = 1$. However, as we will explore in the next section, the quantum Deutsch algorithm can solve this problem with just one query to a black-box that implements a unitary operator closely related to the function f. This makes the Deutsch algorithm the first algorithm capable of accomplishing something impossible in the classical world. In doing so, it introduces crucial principles of quantum computing that are also leveraged in more advanced quantum algorithms.

5.1.2. The quantum version and its solution. To explain the quantum version of the Deutsch problem, we define the following unitary operator on $\mathbb{H}_2$:

$$(5.1.2) \qquad U_f : \mathbb{H}_2 \to \mathbb{H}_2, \quad |x\rangle |y\rangle \mapsto |x\rangle |f(x) \oplus y\rangle = |x\rangle X^{f(x)} |y\rangle.$$

Replacing f by U_f we obtain the quantum version of the Deutsch problem.

Problem 5.1.2 (Deutsch problem — quantum version).

Input: A black-box that implements the quantum operator U_f for a function $f : \{0, 1\} \to \{0, 1\}$.

Output: $f(0) \oplus f(1)$.

Exercise 5.1.3. Show that U_f is a unitary operator.

The quantum circuit that solves the Deutsch problem is shown in Figure 5.1.1. It uses important ingredients of quantum computing: *superposition, quantum parallelism, phase kickback,* and *quantum interference.* Now we describe the circuit step by step and explain these ingredients.

The input state is

$$(5.1.3) \qquad |\psi_0\rangle = |0\rangle |1\rangle.$$

In the first step, the circuit applies the Hadamard operator to the first and the second qubit which gives the state

$$(5.1.4) \qquad |\psi_1\rangle = |x_+\rangle |x_-\rangle = \frac{|0\rangle |x_-\rangle + |1\rangle |x_-\rangle}{\sqrt{2}}.$$

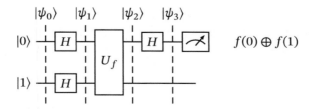

Figure 5.1.1. The quantum circuit that solves the Deutsch problem.

This is an equally weighted *superposition* of the states $|0\rangle|x_-\rangle$ and $|1\rangle|x_-\rangle$.

Next, the *phase kickback trick* is used. We note that

(5.1.5) $$U_f |0\rangle|x_-\rangle = |0\rangle X^{f(0)}|x_-\rangle = (-1)^{f(0)}|0\rangle|x_-\rangle$$

and

(5.1.6) $$U_f |1\rangle|x_-\rangle = |0\rangle X^{f(1)}|x_-\rangle = (-1)^{f(1)}|0\rangle|x_-\rangle.$$

These quantum states have the global phase factors $(-1)^{f(0}$ and $(-1)^{f(1)}$. But since global phase factors do not influence measurement outcomes, we cannot learn anything about $f(0)$ or $f(1)$ from measuring these states or evolutions of them. However, if we apply U_f to the superposition $|\psi_1\rangle$ we obtain

(5.1.7)
$$|\psi_2\rangle = U_f |x_+\rangle|x_-\rangle = U_f \frac{|0\rangle + |1\rangle}{\sqrt{2}}|x_-\rangle$$
$$= \frac{U_f |0\rangle|x_-\rangle + U_f |1\rangle|x_-\rangle}{\sqrt{2}} = \frac{(-1)^{f(0)}|0\rangle + (-1)^{f(1)}|1\rangle}{\sqrt{2}}|x_-\rangle.$$

In this operation the global phase factors $(-1)^{f(0)}$ and $(-1)^{f(1)}$ are *kicked back* to the amplitudes of $|0\rangle$ and $|1\rangle$ in the first qubit. As we will see, this opens up the possibility of gaining information about $f(0)$ and $f(1)$ through measurement of an evolution of $U_f |x_+\rangle|x_-\rangle$. Here, we also see *quantum parallelism in action*: one application of U_f changes both the amplitudes of $|0\rangle$ and $|1\rangle$.

Equation (5.1.7) implies

(5.1.8)
$$|\psi_2\rangle = (-1)^{f(0)}\frac{|0\rangle + (-1)^{f(0)\oplus f(1)}|1\rangle}{\sqrt{2}}|x_-\rangle$$
$$= \begin{cases} (-1)^{f(0)}|x_+\rangle|x_-\rangle & \text{if } f(0) \oplus f(1) = 0, \\ (-1)^{f(0)}|x_-\rangle|x_-\rangle & \text{if } f(0) \oplus f(1) = 1. \end{cases}$$

So up to the global phase factor $(-1)^{f(0)}$ the *quantum interference* of the two states $U_f |0\rangle|x_-\rangle$ and $U_f |1\rangle|x_-\rangle$ causes the amplitude of $|1\rangle$ in the first qubit to be $(-1)^{f(0)\oplus f(1)}$ while the amplitude of $|0\rangle$ in the first qubit is independent of this value. Measuring the first qubit in the basis $|x_+\rangle$ and $|x_-\rangle$ would give the desired result. Since measurement in the computational basis is used, the Hadamard operator is applied to the first qubit. This gives

(5.1.9) $$|\psi_3\rangle = (H \otimes I)|\psi_2\rangle = (-1)^{f(0)}|f(0) \oplus f(1)\rangle|x_-\rangle.$$

This state is separable with respect to the decomposition into the subsystems that contain the first and the second qubit, respectively. So it follows from Corollary 3.7.12 that measuring the first qubit of $|\psi_3\rangle$ in the computational basis gives $f(0) \oplus f(1)$ with probability 1.

We have thus proved the following theorem.

Theorem 5.1.4. *The quantum circuit in Figure 5.1.1 gives $f(0) \oplus f(1)$ with probability 1. It uses the black-box U_f once and, in addition, three Hadamard gates.*

So, while solving the classical Deutsch problem requires two applications of the function f, the Deutsch quantum circuit only needs one application of U_f. The next exercise generalizes the phase kickback trick.

Exercise 5.1.5. Consider a unitary single-qubit operator V and an eigenstate $|\psi\rangle$ of this operator.

(1) Show that applying the operator V to $|\psi\rangle$ means applying a global phase shift to this state.

(2) Show that applying the controlled-V operator $C(V)$ with the first qubit as a control to the state $|x_+\rangle|\psi\rangle$ kicks the global phase shift back to the amplitude of $|1\rangle$ in the first qubit.

(3) Find the spherical coordinates of the points on the Bloch sphere corresponding to the first qubit before and after the application of $C(V)$ to $|x_+\rangle|\psi\rangle$.

5.2. Oracle complexity

We note that specifying the Deutsch algorithm and analyzing its complexity necessitates a modification of the concepts of probabilistic algorithms and their complexity, as presented in Section 4.12. The only input required by the Deutsch algorithm is the oracle U_f. However, inputs of this kind are not accounted for in the quantum algorithm concept discussed thus far. As a result, the analysis of the algorithm's time complexity includes consideration of the number of calls to this oracle. We will adopt this approach for all other algorithms presented in this chapter.

5.3. The Deutsch-Jozsa algorithm

In 1992, David Deutsch and Richard Jozsa [**DJ92**] proposed a natural generalization of the Deutsch problem and a quantum algorithm to solve it. Here we present the improved version of the algorithm by Cleve et al. [**CEH$^+$98**] in 1998.

5.3.1. The classical version. Let

$$(5.3.1) \qquad\qquad f : \{0,1\}^n \to \{0,1\}$$

be a function: We call f *constant* if $f(\vec{x})$ is the same for all $\vec{x} \in \{0,1\}^n$ and we call f *balanced* if $f(\vec{x}) = 0$ for half of the arguments $\vec{x} \in \{0,1\}^n$ and $f(\vec{x}) = 1$ for the other half. The classical version of the Deutsch-Jozsa problem is the following.

Problem 5.3.1 (Deutsch-Jozsa problem — classical version).

Input: A black-box implementing a function $f : \{0,1\}^n \to \{0,1\}^n$ that is either constant or balanced.

Output: "constant" or "balanced", respectively, if f has this property.

Exercise 5.3.2. Show that every deterministic algorithm that solves the classical Deutsch-Jozsa problem requires $2^{n-1} + 1$ queries of the black-box implementing f.

As previously mentioned, the Deutsch-Jozsa problem represents a straightforward extension of the Deutsch problem. Its primary significance does not necessarily stem from direct practical applications but rather from the striking disparity in performance between the classical deterministic solution and the quantum solution, which is even more pronounced compared to the Deutsch problem. As demonstrated in Exercise 5.3.2, any deterministic classical algorithm necessitates a minimum of $2^{n-1}+1$ queries of f, whereas the quantum algorithm uses the operator U_f corresponding to the function f only once. However, it's worth noting that this is only part of the story, as the subsequent exercise reveals the existence of a significantly more efficient classical probabilistic algorithm for solving the Deutsch-Jozsa problem.

Exercise 5.3.3. Find a probabilistic algorithm which requires two evaluations of f and solves the Deutsch-Jozsa problem with a success probability of at least $\frac{1}{2}$.

5.3.2. The quantum version and its solution. The quantum version of the Deutsch-Jozsa problem uses the operator

$$(5.3.2) \qquad U_f : \mathbb{H}_n \otimes \mathbb{H}_1 \to \mathbb{H}_n \otimes \mathbb{H}_1, \quad |\vec{x}\rangle |y\rangle \mapsto |\vec{x}\rangle |f(\vec{x}) \oplus y\rangle = |\vec{x}\rangle X^{f(\vec{x})} |y\rangle$$

to find out whether f is constant or balanced.

Exercise 5.3.4. Prove that U_f defined in (5.3.2) is a unitary operator on $\mathbb{H}_n$.

The quantum version of the Deutsch-Jozsa problem is the following.

Problem 5.3.5 (Deutsch-Jozsa problem — quantum version).

Input: A positive integer n, a black-box implementing the operator U_f from (5.3.2) for a function $f : \{0,1\}^n \to \{0,1\}^n$ that is either constant or balanced.

Output: "constant" or "balanced", respectively, if f has this property.

The Deutsch-Josza problem is an example of a *promise problem* which in complexity theory refers to a decision problem that comes with a "promise" or guarantee about the possible answers it can have. Here, the two possible answers are "constant" or "balanced".

The quantum circuit that solves the Deutsch-Jozsa problem is shown in Figure 5.3.1. It also uses superposition, phase kickback, and interference and is similar to the quantum circuit in Figure 5.1.1 that solves the Deutsch problem.

In order to show that this circuit has the desired property, we use the following lemma.

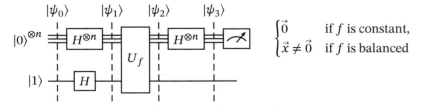

Figure 5.3.1. The quantum circuit $Q_{DJ}(n, U_f)$ that solves the Deutsch-Jozsa problem.

Lemma 5.3.6. *For all $\vec{x} \in \{0,1\}^n$ we have*

$$(5.3.3) \qquad H^{\otimes n} |\vec{x}\rangle = \frac{1}{\sqrt{2^n}} \sum_{\vec{z} \in \{0,1\}^n} (-1)^{\vec{x} \cdot \vec{z}} |\vec{z}\rangle$$

and

$$(5.3.4) \qquad |\vec{x}\rangle = \frac{1}{\sqrt{2^n}} \sum_{\vec{z} \in \{0,1\}^n} (-1)^{\vec{x} \cdot \vec{z}} H^{\otimes n} |\vec{z}\rangle.$$

Proof. We first note that for $x \in \{0,1\}$ we have

$$(5.3.5) \qquad H|x\rangle = \frac{1}{\sqrt{2}}(|0\rangle + (-1)^x |1\rangle) = \frac{1}{\sqrt{2}} \sum_{z \in \{0,1\}} (-1)^{xz} |z\rangle.$$

Hence, for $\vec{x} = (x_0, \ldots, x_{n-1}) \in \{0,1\}^n$ we have

$$
\begin{aligned}
H^{\otimes n} |\vec{x}\rangle &= \frac{1}{\sqrt{2^n}} \left(\sum_{z_0 \in \{0,1\}} (-1)^{x_0 z_0} |z\rangle \right) \otimes \cdots \otimes \left(\sum_{z_{n-1} \in \{0,1\}} (-1)^{x_{n-1} z_{n-1}} |z\rangle \right) \\
(5.3.6) \qquad &= \frac{1}{\sqrt{2^n}} \sum_{(z_0, \ldots, z_{n-1}) \in \{0,1\}^n} ((-1)^{x_0 z_0} |z_0\rangle) \otimes \cdots \otimes ((-1)^{x_{n-1} z_{n-1}} |z_{n-1}\rangle) \\
&= \frac{1}{\sqrt{2^n}} \sum_{\vec{z} \in \{0,1\}^n} (-1)^{\vec{x} \cdot \vec{z}} |\vec{z}\rangle.
\end{aligned}
$$

This proves (5.3.3). So $H^2 = I$ implies (5.3.4). $\qquad \square$

Now we determine the states $|\psi_i\rangle$, $0 \le i \le 3$, in the Deutsch-Jozsa circuit. The initial state is

$$(5.3.7) \qquad |\psi_0\rangle = |0\rangle^{\otimes n} |1\rangle.$$

The circuit applies $H^{\otimes(n+1)}$ to this state. It follows from (5.3.3) that this gives the state

$$(5.3.8) \qquad |\psi_1\rangle = H^{\otimes n} |0\rangle^{\otimes n} |x_-\rangle = \frac{1}{\sqrt{2^n}} \sum_{\vec{x} \in \{0,1\}^n} |\vec{x}\rangle |x_-\rangle.$$

This is the equally weighted *superposition* of the states $|\vec{x}\rangle |x_-\rangle$.

Next, the quantum circuit applies U_f to $|\psi_1\rangle$. We show that this is an application of the *phase kickback* trick. For all $\vec{x} \in \{0,1\}^n$ we have

$$(5.3.9) \qquad U_f |\vec{x}\rangle |x_-\rangle = (-1)^{f(\vec{x})} |\vec{x}\rangle |x_-\rangle.$$

Therefore, the application of U_f to $|\vec{x}\rangle |x_-\rangle$ modifies this state by the global phase factor $(-1)^{f(\vec{x})}$. It follows from (5.3.9) that

$$
\begin{aligned}
|\psi_2\rangle = U_f |\psi_1\rangle &= \frac{1}{\sqrt{2^n}} \sum_{\vec{x} \in \{0,1\}^n} U_f |\vec{x}\rangle |x_-\rangle \\
(5.3.10) \qquad &= \frac{1}{\sqrt{2^n}} \sum_{\vec{x} \in \{0,1\}^n} (-1)^{f(\vec{x})} |\vec{x}\rangle |x_-\rangle.
\end{aligned}
$$

Hence, applying U_f to the superposition $|\psi_1\rangle$ kicks the global phase shifts $(-1)^{\vec{x}}$ back to all the amplitudes of the states $|\vec{x}\rangle$ of the first n qubits. In order to extract information about the function f from this superposition, we note that by (5.3.4) we have

$$|\psi_2\rangle = \frac{(-1)^{f(\vec{0})}}{\sqrt{2^n}} \sum_{\vec{x}\in\{0,1\}^n} (-1)^{f(\vec{x})\oplus f(\vec{0})} |\vec{x}\rangle |x_-\rangle$$

(5.3.11)
$$= \frac{(-1)^{f(\vec{0})}}{2^n} \sum_{\vec{x}\in\{0,1\}^n} (-1)^{f(\vec{x})\oplus f(\vec{0})} \sum_{\vec{z}\in\{0,1\}^n} (-1)^{\vec{x}\cdot\vec{z}} H^{\otimes n} |\vec{z}\rangle |x_-\rangle$$

$$= \left(\frac{(-1)^{f(\vec{0})}}{2^n} \sum_{\vec{z}\in\{0,1\}^n} \left(\sum_{\vec{x}\in\{0,1\}^n} (-1)^{\vec{x}\cdot\vec{z}+f(\vec{x})\oplus f(\vec{0})} \right) H^{\otimes n} |\vec{z}\rangle \right) |x_-\rangle.$$

This is the tensor product of a quantum state in $\mathbb{H}_n$ with $|x_-\rangle$. The amplitude of the basis state $H^{\otimes n} |0\rangle_n$ in the state of the first n qubits is

(5.3.12)
$$\frac{(-1)^{f(\vec{0})}}{2^n} \sum_{x\in\{0,1\}^n} (-1)^{f(\vec{x})\oplus f(\vec{0})} = \begin{cases} (-1)^{f(\vec{0})} & \text{if } f \text{ is constant,} \\ 0 & \text{if } f \text{ is balanced.} \end{cases}$$

It follows from Corollary 3.7.12 that measuring the first n qubits in the basis $(H^{\otimes n} |\vec{z}\rangle)_{\vec{z}\in\{0,1\}^n}$ gives with probability 1 the information whether f is constant or balanced. To obtain this information by a measurement in the computational basis, the Deutsch-Jozsa circuit applies $H^{\otimes n}$ to the quantum state of the first n qubits of $|\psi_2\rangle$. By (5.3.11), this gives the final state

(5.3.13)
$$|\psi_3\rangle = \left(\frac{(-1)^{f(\vec{0})}}{2^n} \sum_{\vec{z}\in\{0,1\}^n} \left(\sum_{\vec{x}\in\{0,1\}^n} (-1)^{\vec{x}\cdot\vec{z}+f(\vec{x})\oplus f(\vec{0})} \right) |\vec{z}\rangle \right) |x_-\rangle.$$

Measuring the first n qubits of $|\psi_3\rangle$ in the computational basis gives $\vec{0}$ with probability 1 if f is constant and $\vec{z} \neq \vec{0}$ if f is balanced. As desired, this measurement distinguishes with probability 1 between constant and balanced functions f using $2n+1$ applications of the Hadamard operator H and one application of U_f. This can be considered as an exponential speedup of the best classical solution to the Deutsch problem.

Summarizing our discussion, we obtain the following theorem.

Theorem 5.3.7. *Let $n \in \mathbb{N}$, let $f : \{0,1\}^n \to \{0,1\}$ be a function that is constant or balanced, and let U_f be the unitary operator from (5.3.2). Then with probability 1 the quantum circuit Q_{DJ} returns $\vec{0}$ if f is constant and $\vec{x} \in \{0,1\}^n$, $\vec{x} \neq \vec{0}$, if f is balanced. It uses one U_f gate and $2n + 1$ Hadamard gates.*

So, compared to the best deterministic algorithm for solving the Deutsch-Jozsa problem, which by Exercise 5.3.2 requires $2^{n-1} + 1$ evaluations of the function f, the Deutsch-Jozsa algorithm represents a dramatic asymptotic speedup. However, compared to the probabilistic algorithm from Exercise 5.3.3, this advantage vanishes. This is different for Simon's algorithm, which is presented in the next section.

5.4. Simon's algorithm

This section focuses on *Simon's problem* and Simon's quantum algorithm for solving this problem. The algorithm was presented in 1994 by Daniel Simon [**Sim94, Sim97**] with the intention of showing quantum computing's supremacy over classical computing from the perspective of complexity theory. Simon's pioneering work demonstrated that the quantum variant of Simon's problem can be resolved exponentially faster compared to its classical counterpart. Notably, this marked the first instance of such an exponential acceleration, and it also laid the groundwork for inspiring the development of the Shor algorithm, a topic we will explore in Chapter 6.

5.4.1. The classical version.
Classically, Simon's problem is the following.

Problem 5.4.1 (Simon's problem — classical version).

Input: A black-box implementing a function $f : \{0,1\}^n \to \{0,1\}^n$ with the property that there is $\vec{s} \in \{0,1\}^n$, $\vec{s} \neq \vec{0}$, such that for all $\vec{x}, \vec{y} \in \{0,1\}^n$ we have $f(\vec{x}) = f(\vec{y})$ if and only if $\vec{x} = \vec{y}$ or $\vec{x} = \vec{y} \oplus \vec{s}$.

Output: The hidden string $\vec{s}$.

The next exercise gives a lower bound for solving Simon's problem using a classical deterministic algorithm.

Exercise 5.4.2. Let A be a classical deterministic algorithm that solves Simon's problem. Show that in the worst case, A must query the black-box implementing f at least $2^{n-1} + 1$ times.

We will see that the quantum algorithm for Simon's problem is much more efficient. But it is a probabilistic algorithm. Therefore, we must compare it with classical probabilistic algorithms. A lower bound for their performance is given in the next theorem, which was proved by Richard Cleve in [**Cle11**].

Theorem 5.4.3. *Any classical probabilistic algorithm that solves Simon's problem with probability at least 3/4 must make $\Omega(2^{n/2})$ queries to the black-box for f.*

5.4.2. The quantum version and its solution.
As in the quantum Deutsch problem, also in the quantum version of Simon's problem the function $f : \{0,1\}^n \to \{0,1\}^n$ is replaced by a unitary operator on $\mathbb{H}_n$. This operator is

$$(5.4.1) \qquad U_f : \mathbb{H}_n \otimes \mathbb{H}_n, \quad |\vec{x}\rangle |\vec{y}\rangle \mapsto |\vec{x}\rangle |f(\vec{x}) \oplus \vec{y}\rangle.$$

With this operator, the quantum version of Simon's problem can be stated as follows.

Problem 5.4.4 (Simon's problem — quantum version).

Input: A positive integer n, a black-box implementing the unitary operator U_f from (5.4.1) for a function $f : \{0,1\}^n \to \{0,1\}^n$ such that there is a vector $\vec{s} \in \{0,1\}^n$, $\vec{s} \neq \vec{0}$, with the property that for all $\vec{x}, \vec{y} \in \{0,1\}^n$ we have $f(\vec{x}) = f(\vec{y})$ if and only if $\vec{x} = \vec{y}$ or $\vec{x} = \vec{y} \oplus \vec{s}$.

Output: The hidden string $\vec{s}$.

Algorithm 5.4.5. Simon's algorithm

Input: A positive integer n and a black-box implementation of U_f as in (5.4.1) for a function $f : \{0,1\}^n \to \{0,1\}^n$ such that there is $\vec{s} \in \{0,1\}^n$, $\vec{s} \neq \vec{0}$, with the property that for all $\vec{x}, \vec{y} \in \{0,1\}^n$ we have $f(\vec{x}) = f(\vec{y})$ if and only if $\vec{x} = \vec{y}$ or $\vec{x} = \vec{y} \oplus \vec{s}$.

Output: The hidden string $\vec{s}$ from Simon's problem

1:   QSIMON(n, U_f)
2:      $W \leftarrow ()$
3:      **for** $j = 1$ to $n-1$ **do**
4:          $\vec{w}_j \leftarrow Q_{\text{Simon}}(n, U_f)$
5:          $W \leftarrow W \circ (\vec{w}_j)$
6:      **end for**
7:      $r \leftarrow$ rank W
8:      **if** $r = n-1$ **then**
9:          Find the unique nonzero solution $\vec{s}$ of the linear system $W^T\vec{x} = 0$
10:          **return** $\vec{s}$
11:      **else**
12:          **return** "Failure"
13:      **end if**
14: **end**

We explain the idea of Simon's Algorithm 5.4.5. The details and proofs are given below. Using the quantum circuit $Q_{\text{Simon}}(n, U_f)$ from Figure 5.4.1 the algorithm selects $n-1$ elements $\vec{w}_1, \ldots, \vec{w}_{n-1}$ in the orthogonal complement

$$(5.4.2) \qquad \vec{s}^{\perp} = \{\vec{w} \in \{0,1\}^n : \vec{w} \cdot \vec{s} = 0\}$$

of $\vec{s}$. If the matrix $W = (\vec{w}_1, \ldots, \vec{w}_{n-1})$ has rank $n-1$, then the algorithm computes the uniquely determined solution of the linear system $W^T\vec{x} = \vec{0}$ which happens to be the hidden string $\vec{s}$.

In the upcoming part of this section, we will adopt the notation from the quantum version of Simon's problem and prove the following theorem. In view of Theorem 5.4.3, it shows that Simon's algorithm offers an exponential speedup compared to any deterministic algorithm for Simon's problem.

Theorem 5.4.6. *Simon's Algorithm 5.4.5 returns the hidden string $\vec{s}$ from Simon's problem with probability at least 1/4. It requires $n-1$ applications of U_f and $O(n^3)$ other operations.*

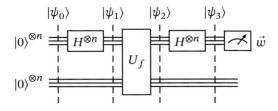

Figure 5.4.1. The quantum circuit $Q_{\text{Simon}}(n, U_f)$ used in Simon's algorithm.

We first prove that Simon's algorithm returns the correct result.

Proposition 5.4.7. *Let* $W = (\vec{w}_1, \dots, \vec{w}_{n-1})$ *be of dimension* $n-1$. *Then* $\vec{s}$ *is the uniquely determined solution of the linear system* $W^{\mathrm{T}}\vec{x} = \vec{0}$.

Proof. By Proposition 2.2.42, the orthogonal complement of $\vec{s}$ is a subspace of $\{0, 1\}^n$ of dimension $n-1$. By Proposition B.7.5, the dimension of the kernel of W^{T} is 1. Since $\vec{s}$ is in the kernel of W^{T} it follows that this kernel is $\{\vec{0}, \vec{s}\}$. □

It follows from Proposition 5.4.7 that Simon's algorithm returns the correct result if the quantum circuit $Q_{\mathrm{Simon}}(n, U_f)$ in Figure 5.4.1 returns elements of $\vec{s}^{\perp}$. This is what we will prove now. We need the following result.

Lemma 5.4.8. *Let* $\vec{s} \in \{0, 1\}^n$ *be nonzero. Then for all* $\vec{z} \in \{0, 1\}^n$ *we have*

$$(5.4.3) \qquad H^{\otimes n}\left(\frac{|\vec{z}\rangle + |\vec{z} \oplus \vec{s}\rangle}{\sqrt{2}}\right) = \frac{1}{\sqrt{2^{n-1}}} \sum_{\vec{w} \in \vec{s}^{\perp}} (-1)^{\vec{z}\cdot\vec{w}} |\vec{w}\rangle.$$

Proof. It follows from Lemma 5.3.6 that for all $\vec{z} \in \{0, 1\}^n$ we have

$$H^{\otimes n}\left(\frac{|\vec{z}\rangle + |\vec{z} \oplus \vec{s}\rangle}{\sqrt{2}}\right)$$

$$= \frac{1}{\sqrt{2^{n+1}}} \cdot \sum_{\vec{w} \in \{0,1\}^n} \left((-1)^{\vec{z}\cdot\vec{w}} + (-1)^{(\vec{z}\oplus\vec{s})\cdot\vec{w}}\right) |\vec{w}\rangle$$

$$(5.4.4) \qquad = \frac{1}{\sqrt{2^{n+1}}} \left(\sum_{\vec{w} \in \vec{s}^{\perp}} \left((-1)^{\vec{z}\cdot\vec{w}} + (-1)^{\vec{z}\cdot\vec{w}\oplus\vec{s}\cdot\vec{w}}\right) |\vec{w}\rangle \right.$$

$$\left. + \sum_{\vec{w} \in \{0,1\}^n \setminus \vec{s}^{\perp}} \left((-1)^{\vec{z}\cdot\vec{w}} + (-1)^{\vec{z}\cdot\vec{w}\oplus\vec{s}\cdot\vec{w}}\right) |\vec{w}\rangle \right).$$

If $\vec{w} \in \vec{s}^{\perp}$, then

$$(5.4.5) \qquad (-1)^{\vec{z}\cdot\vec{w}} + (-1)^{\vec{z}\cdot\vec{w}\oplus\vec{s}\cdot\vec{w}} = 2 \cdot (-1)^{\vec{z}\cdot\vec{w}}.$$

Also, if $\vec{w} \in \{0, 1\}^n \setminus \vec{s}^{\perp}$, then

$$(5.4.6) \qquad (-1)^{\vec{z}\cdot\vec{w}} + (-1)^{\vec{z}\cdot\vec{w}\oplus\vec{s}\cdot\vec{w}} = 0.$$

Hence, it follows from (5.4.4) that

$$(5.4.7) \quad H^{\otimes n}\left(\frac{|\vec{z}\rangle \oplus |\vec{z} + \vec{s}\rangle}{\sqrt{2}}\right) = \frac{2}{\sqrt{2^{n+1}}} \sum_{\vec{w} \in \vec{s}^{\perp}} (-1)^{\vec{z}\cdot\vec{w}} |\vec{z}\rangle = \frac{1}{\sqrt{2^{n-1}}} \sum_{\vec{w} \in \vec{s}^{\perp}} (-1)^{\vec{z}\cdot\vec{w}} |\vec{z}\rangle$$

as asserted. □

Now we can prove the following proposition.

Proposition 5.4.9. *The quantum circuit* $Q_{\mathrm{Simon}}(n, U_f)$ *from Figure 5.4.1 returns a uniformly distributed random element of* $\vec{s}^{\perp}$.

Proof. The quantum circuit $Q_{\text{Simon}}(n, U_f)$ operates on a quantum system that consists of two quantum registers of length n each of which is initialized to $|0\rangle^{\otimes n}$. So, we have

$$(5.4.8) \qquad |\psi_0\rangle = |0\rangle^{\otimes n} |0\rangle^{\otimes n}.$$

Then $H^{\otimes n}$ is applied to the first register. It follows from (5.3.3) that this gives the quantum state

$$(5.4.9) \qquad |\psi_1\rangle = \frac{1}{\sqrt{2^n}} \sum_{\vec{z} \in \{0,1\}^n} |\vec{z}\rangle |0\rangle^{\otimes n}.$$

It is an equally weighted superposition of the quantum states $|\vec{z}\rangle |0\rangle^{\otimes n}$. Next, the algorithm applies U_f to $|\psi_1\rangle$ and produces the state

$$(5.4.10) \qquad |\psi_2\rangle = U_f |\psi_1\rangle = \frac{1}{\sqrt{2^n}} \sum_{\vec{z} \in \{0,1\}^n} |\vec{z}\rangle |f(\vec{z})\rangle.$$

This is an instance of quantum parallelism: one application of the operator U_f gives a superposition of the states $|f(\vec{z})\rangle$, $\vec{z} \in \{0,1\}^n$. We will now show that this operation also leads to quantum interference which allows us to obtain $\vec{w} \in \vec{s}^\perp$. To see this, let I be a set of representatives of the elements of the quotient space $\{0,1\}^n / \{0, \vec{s}\}$. Then we have

$$(5.4.11) \qquad \{0,1\}^n = \bigsqcup_{\vec{z} \in I} \{\vec{z}, \vec{s} \oplus \vec{z}\}.$$

This implies that $|\psi_2\rangle$ can be rewritten as

$$(5.4.12) \qquad |\psi_2\rangle = \frac{1}{\sqrt{2^{n-1}}} \sum_{\vec{z} \in I} \frac{|\vec{z}\rangle + |\vec{z} \oplus \vec{s}\rangle}{\sqrt{2}} |f(\vec{z})\rangle.$$

From Lemma 5.4.8 we obtain

$$(5.4.13) \qquad |\psi_2\rangle = \frac{1}{2^{n-1}} \sum_{\vec{z} \in I} \sum_{\vec{w} \in \vec{s}^\perp} (-1)^{\vec{z} \cdot \vec{w}} H^{\otimes n} |\vec{w}\rangle |f(\vec{z})\rangle.$$

So quantum interference gives the equally weighted superposition of the quantum states $H^{\otimes n} |\vec{w}\rangle |f(\vec{z})\rangle$ with $\vec{w} \in \vec{s}^\perp$ and $\vec{z} \in I$. To allow extraction of some $\vec{w}$, the algorithm applies $H^{\otimes n}$ to the first quantum register. This gives the final state

$$(5.4.14) \qquad |\psi_3\rangle = \frac{1}{2^{n-1}} \sum_{\vec{z} \in I} \sum_{\vec{w} \in \vec{s}^\perp} (-1)^{\vec{z} \cdot \vec{w}} |\vec{w}\rangle |f(\vec{z})\rangle.$$

As shown in Exercise 5.4.10, measuring the first register of $|\psi_3\rangle$ in the computational basis of $\mathbb{H}_n$ gives every $\vec{w} \in \vec{s}^\perp$ with probability $1/2^{n-1}$. $\qquad\square$

Exercise 5.4.10. (1) Show that measuring the first register of $|\psi_3\rangle$ in (5.4.14) in the computational basis of $\mathbb{H}_n$ gives every $\vec{w} \in \vec{s}^\perp$ with probability $1/2^{n-1}$.

(2) Analyze the modification of Simon's algorithm where the second register is traced out before the measurement.

Next, we analyze Algorithm 5.4.5. The algorithm invokes the quantum circuit $Q_{Simon}(n, U_f)$ $n - 1$ times and constructs the matrix W. If the rank of the matrix W is $n - 1$, it follows from Proposition 5.4.7 that the hidden string $\vec{s}$ is the uniquely determined nonzero solution of the linear system $W^T\vec{x} = \vec{0}$. To estimate the success probability of the algorithm, we use the following lemma.

Lemma 5.4.11. *For all $n \in \mathbb{N}$ we have*

$$(5.4.15) \qquad \prod_{k=1}^{n-1}\left(1 - \frac{1}{2^k}\right) \geq \frac{1}{4}.$$

Proof. For any $x \in [0, 1/2]$ we have

$$(5.4.16) \qquad \log(1 - x) \geq -2x \log 2.$$

This is shown in Exercise 5.4.12 and implies

$$(5.4.17) \qquad \begin{aligned} \log\left(\prod_{k=1}^{n-1}\left(1 - \frac{1}{2^k}\right)\right) &= \sum_{k=1}^{n-1} \log\left(1 - \frac{1}{2^k}\right) \\ &\geq -2\log 2 \sum_{k=1}^{\infty} \frac{1}{2^k} \geq -2\log 2. \end{aligned}$$

This implies the assertion. $\square$

Exercise 5.4.12. Use elementary calculus to show that (5.4.16) holds.

Using Lemma 5.4.11 we now obtain the following estimate of the success probability of Simon's algorithm.

Proposition 5.4.13. *The success probability of Simon's algorithm is at least* $1/4$.

Proof. We show that for $1 \leq j \leq n - 1$ the matrix W_j computed by the algorithm in the jth iteration of the **for** loop has rank j with probability

$$(5.4.18) \qquad p_j = \prod_{k=n-j}^{n-1}\left(1 - \frac{1}{2^k}\right).$$

Using this equation for $j = n-1$ and Lemma 5.4.11 we see that the success probability of the Simon algorithm is at least $1/4$.

If $n = 1$, then (5.4.18) holds. So, let $n > 1$. We prove (5.4.18) be induction on j. In each iteration of the **for** loop, a random $\vec{w} \in \vec{s}^{\perp}$ is returned by $Q_{Simon}(n, U_j)$ according to the uniform distribution and is appended to the previous W. The matrix found in this way in the first iteration has rank 1 if $\vec{w}$ is nonzero. This happens with probability $(2^{n-1} - 1)/2^{n-1} = 1 - 1/2^{n-1} = p_1$. Now let $j \in N$, $1 < j \leq n - 1$, and assume that rank $W_{j-1} = j - 1$ and that this happens with probability p_{j-1}. We determine the probability that the vector $\vec{w}$ found in the jth iteration appends W_{j-1} to a matrix of rank j. Let $(\vec{b}_1, \ldots, \vec{b}_{n-1})$ be a basis of $\vec{s}^{\perp}$ such that $\vec{b}_1, \ldots, \vec{b}_{j-1}$ are the row vectors of W_{j-1}. Let $\vec{w} = \sum_{i=0}^{n-1} w_i \vec{b}_i$ be the vector returned in the jth iteration of the **for** loop by $Q_{Simon}(n, U_f)$. This vector appends W_{j-1} to a matrix of rank j if and only if at least one of the coefficients w_i of the basis elements $\vec{b}_i$ with $j \le i \le n-1$ is nonzero. This holds for $2^{n-1} - 2^{j-1}$ vectors in $\{0,1\}^n$. So, the probability of finding such a vector is

$$(5.4.19) \qquad p_{j-1}\left(1 - \frac{2^{j-1}}{2^{n-1}}\right) = p_{j-1}\left(1 - \frac{1}{2^{n-j}}\right) = p_j. \qquad \square$$

Finally, we analyze the complexity of the algorithm.

Proposition 5.4.14. *Simon's algorithm requires $n-1$ applications of U_f and $O(n^2)$ Hadamard gates and $O(n^3)$ additional operations.*

Proof. Clearly, the number of calls of $Q_{\text{Simon}}(n, U_f)$ is $n-1$. Since each application of $Q_{\text{Simon}}(n, U_n)$ uses $O(n)$ Hadamard gates, the total number of required Hadamard gates is $O(n^3)$. Also, by Proposition B.7.17 the linear system $W^T \vec{x} = \vec{0}$ can be solved using $O(n^2)$ operations. $\qquad \square$

Now Theorem 5.4.6 follows from Propositions 5.4.7, 5.4.9, 5.4.13, and 5.4.14.

5.5. Generalization of Simon's algorithm

We discuss a generalization of Simon's problem and a quantum algorithm for solving it. The idea is to replace the hidden string $\vec{s}$ by a hidden linear subspace S of the n-dimensional vector space $\{0,1\}^n$. The classical version of this generalization is the following.

Problem 5.5.1 (General Simon's problem — classical version).

Input: A black-box implementing a function $f : \{0,1\}^n \to \{0,1\}^n$ with the following property. There is a linear subspace S of $\{0,1\}^n$ such that for all $\vec{x}, \vec{y} \in \{0,1\}^n$ we have $f(\vec{x}) = f(\vec{y})$ if and only if $\vec{x} = \vec{y} \oplus \vec{s}$ for some $\vec{s} \in S$. The dimension m of S is also an input.

Output: A basis of S.

The subspace S in also referred to as a *hidden subgroup* of $\{0,1\}^n$. In Simon's original problem, we have $S = \{\vec{0}, s\}$. Finding a hidden subgroup is also the key step in Shor's factoring and discrete logarithm algorithms. They are discussed in Chapter 6.

The quantum version of the general Simon's problem is the following.

Problem 5.5.2 (General Simon's problem — quantum version).

Input: A black-box implementing the unitary operator U_f from (5.4.1) for a function $f : \{0,1\}^n \to \{0,1\}^n$ with the following property. There is a linear subspace S of $\{0,1\}^n$ such that for all $\vec{x}, \vec{y} \in \{0,1\}^n$ we have $f(\vec{x}) = f(\vec{y})$ if and only if $\vec{x} = \vec{y} \oplus \vec{s}$ for some $\vec{s} \in S$. The dimension m of S is also an input.

Output: A basis of S.

Algorithm 5.5.4, which is a small modification of Simon's Algorithm 5.4.5, solves the generalization of Simon's problem. The idea is the following. Using the quantum circuit $Q(U_f)$ from Figure 5.4.1 the algorithm selects m elements $\vec{w}_1, \ldots, \vec{w}_m$ in the orthogonal complement

(5.5.1) $$S^\perp = \{\vec{w} \in \{0,1\}^n : \vec{w} \cdot \vec{s} = 0 \text{ for all } \vec{s} \in S\}$$

of S randomly with the uniform distribution. This set is a linear subspace of $\{0,1\}^n$ of dimension $n - m$. If the matrix $W = (\vec{w}_1, \ldots, \vec{w}_m)$ has rank m, then the kernel of W is equal to the subspace S, and the algorithm returns a basis of this kernel.

In the remainder of this section, we will prove the following theorem.

Theorem 5.5.3. *Algorithm 5.5.4 returns a basis of the hidden subgroup S from the generalization of Simon's problem with probability at least $1/4$. It uses m applications of U_f and $O(n^3)$ other operations.*

Algorithm 5.5.4. General Simon's algorithm

Input: A black-box implementing U_f from (5.4.1) and the dimension m of the hidden subgroup S where f and S are as specified in the generalization of Simon's problem

Output: A basis of S

1:   GENERALSIMON(n, U_f, m)
2:       $W \leftarrow ()$
3:       **for** $j = 1$ to $n - m$ **do**
4:           $\vec{w}_j \leftarrow Q_{\text{Simon}}(n, U_f)$
5:           $W \leftarrow W \circ (\vec{w}_j)$
6:       **end for**
7:       $r \leftarrow \text{rank } W$
8:       **if** $r = m$ **then**
9:           Find a basis B of the kernel of W
10:          **return** B
11:      **else**
12:          **return** "Failure"
13:      **end if**
14: **end**

The proof of Theorem 5.5.3 is analogous to the proof of Theorem 5.4.6. Therefore, the proofs of the corresponding results are left to the reader as exercises. We start by determining the structure of $S^\perp$.

Proposition 5.5.5. *Let $W = (\vec{w}_1, \ldots, \vec{w}_{n-m})$ be of dimension $n - m$. Then S is the kernel of W^{T}.*

Exercise 5.5.6. Prove Proposition 5.5.5.

It follows from Proposition 5.5.5 that Algorithm 5.5.4 returns the correct result if the quantum circuit $Q_{\text{Simon}}(n, U_f)$ in Figure 5.4.1 returns elements of $S^\perp$ which we will prove now. For this, we need the following lemma.

Lemma 5.5.7. *Let $\vec{z} \in \{0,1\}^n$ and set*

(5.5.2) $$|\vec{z} \oplus S\rangle = \frac{1}{\sqrt{2^m}} \sum_{\vec{s} \in S} |\vec{z} \oplus \vec{s}\rangle.$$

Then we have

(5.5.3)
$$H^{\otimes n} |\vec{z} \oplus S\rangle = \frac{1}{\sqrt{2^{n-m}}} \sum_{\vec{w} \in S^{\perp}} (-1)^{\vec{z} \cdot \vec{w}} |\vec{z}\rangle .$$

Exercise 5.5.8. Prove Lemma 5.5.7.

Lemma 5.5.7 implies the following proposition.

Proposition 5.5.9. *The quantum circuit* $Q_{\text{Simon}}(n, U_f)$ *from Figure* 5.4.1 *returns a uniformly distributed random element of* $S^{\perp}$.

Exercise 5.5.10. Prove Proposition 5.5.9.

Finally, Theorem 5.5.3 is proved in the next exercise using Propositions 5.5.5 and 5.5.9.

Exercise 5.5.11. Prove Theorem 5.5.3.

Chapter 6

The Algorithms of Shor

In this chapter, we present the algorithms that Peter Shor first introduced in 1994 [**Sho94**], causing a significant stir in the cybersecurity domain. These are quantum algorithms that have the remarkable ability to compute integer factorizations and discrete logarithms in polynomial time. The intractability of these problems for large parameters forms the foundation of security in the most commonly used public-key cryptography, a pivotal pillar of overall cybersecurity, particularly internet security. Due to their profound significance in IT security, the Shor algorithms stand as the most renowned quantum algorithms. Their invention spurred the inception of the highly active research field known as Post-Quantum Cryptography.

The chapter starts by presenting the idea of the Shor factoring algorithm. Then, the most important tools used by Shors's algorithms, the Quantum Fourier Transform and quantum circuits for efficiently implementing it and its inverse, are explained. Subsequently, we show how the Quantum Fourier Transform is used to solve the quantum phase estimation problem which addresses the challenge of approximating the phase of the eigenvalue of a unitary operator when an associated eigenstate of this operator is known. Quantum phase estimation is then applied to finding the order of elements in the multiplicative group modulo a positive integer in polynomial time. For this, a quantum variant of the well-known fast exponentiation technique is essential. We then show how efficient order finding enables integer factorization in polynomial time. Also, we show how quantum phase estimation and quantum fast exponentiation lead to a polynomial time algorithm for discrete logarithms. We conclude the chapter by discussing the hidden subgroup problem and demonstrating that several computational problems in this and the previous section can be viewed as instances of this problem.

As usual, we identify linear operators on the state spaces $\mathbb{H}_n$, $n \in \mathbb{N}$, with their representation matrices with respect to the computational basis of $\mathbb{H}_n$. Furthermore, the complexity analyses of this chapter assume that all quantum circuits are constructed using the elementary quantum gates provided by the platform discussed in Section 4.12.2.

6.1. Idea of Shor's factoring algorithm

To enhance the clarity of Shor's factoring algorithm explanation, we provide a concise overview. Let us assume our goal is to discover a proper divisor of a composite number N. The algorithm's initial step involves selecting a random number a from $\mathbb{Z}_N$ with the uniform distribution. If $\gcd(a, N) > 1$, we have found a proper divisor of N, and our task is complete. Now, let us consider the scenario where $\gcd(N, a) = 1$. In this case, the algorithm proceeds to determine the order r of a modulo N. If this order happens to be even, then we can factorize $a^r - 1$ as $(a^{r/2} - 1)(a^{r/2} + 1)$, which is guaranteed to be divisible by N. If $a^{r/2} + 1$ is not divisible by N, then, as shown in Exercise 6.1.1, $\gcd(a^{r/2} - 1, N)$ is a proper divisor of N. The analysis of the algorithm will demonstrate that the probability of r being even and $a^{r/2} + 1$ not being divisible by N is sufficiently high.

Exercise 6.1.1. Let $N \in \mathbb{N}$ be a composite number and let $a \in \mathbb{Z}_N$ with $\gcd(a, N) = 1$. Assume that the order r of a modulo N is even and that $a^{r/2} + 1$ is not divisible by N. Show that $\gcd(a^{r/2} - 1, N)$ is a proper divisor of N.

So, our primary objective is now to determine the order of a modulo N. To achieve this, the Shor algorithm uses a precision parameter n and a unitary operator U_a on $\mathbb{H}_n$ with the following property. There is an orthonormal sequence $(|u_k\rangle)_{k \in \mathbb{Z}_r}$ of eigenstates with corresponding eigenvalue sequence $(e^{2\pi i \frac{k}{r}})_{k \in \mathbb{Z}_r}$. The Shor algorithm then finds an integer $x \in \mathbb{Z}_{2^n}$ such that $2\pi \frac{x}{2^n}$ is close to the phase $2\pi \frac{k}{r}$ of the eigenvalue associated to $|u_k\rangle$ for some $k \in \mathbb{Z}_r$. Then the continued fraction algorithm applied to $\frac{x}{2^n}$ is used to determine the denominator r which is the order of a modulo N.

To approximate one of the phases $2\pi \frac{k}{r}$, Shor's algorithm employs the quantum phase estimation algorithm. It can find an approximation to the phase of the eigenvalue of a given unitary operator U. However, it is important to note that this algorithm requires a corresponding eigenstate as an input. In general, preparing such an eigenstate efficiently is impossible. But we will demonstrate in Proposition 6.4.5 that $\sum_{k=0}^{r-1} |u_k\rangle = |1\rangle_n$ and, therefore, this superposition can be efficiently prepared. Consequently, the algorithm of Shor applies the quantum phase estimation algorithm to this superposition, obtaining an approximation for one of the phases $2\pi \frac{k}{r}$. Since the primary interest lies in determining the denominator r of these phases this is sufficient.

Before we can explain how quantum phase estimation operates, it is essential to introduce its primary component: the Quantum Fourier Transform.

6.2. The Quantum Fourier Transform

The most important tool used by Shor's algorithms is the Quantum Fourier Transform which we discuss in this section. As in Example 2.1.5, we use the bijection

$$(6.2.1) \qquad \text{stringToInt} : \{0, 1\}^n \to \mathbb{Z}_{2^n}, \quad \vec{x} = (x_1, \ldots, x_n) \mapsto \sum_{j=0}^{n-1} x_j 2^{n-j-1}$$

to identify the strings in $\{0, 1\}^n$ with the integers $x \in \mathbb{Z}_{2^n}$. Using this identification we write

(6.2.2)
$$|\vec{x}\rangle = |x\rangle_n$$

for the elements of the computational basis of $\mathbb{H}_n$.

To define the Quantum Fourier Transform, we use the following notation.

Definition 6.2.1. For any $\omega \in \mathbb{R}$ we set

(6.2.3)
$$|\psi_n(\omega)\rangle = \frac{1}{\sqrt{2^n}} \sum_{y=0}^{2^n-1} e^{2\pi i y \omega} |y\rangle_n .$$

Example 6.2.2. We have

$$\psi_2\left(\frac{1}{4}\right) = \frac{1}{4}\left(e^{2\pi i \cdot 0 \cdot \frac{1}{4}} |0\rangle_2 + e^{2\pi i \cdot 1 \cdot \frac{1}{4}} |1\rangle_2 + e^{2\pi i \cdot 2 \cdot \frac{1}{4}} |2\rangle_2 + e^{2\pi i \cdot 3 \cdot \frac{1}{4}} |3\rangle_2\right)$$
$$= \frac{1}{4}\left(|0\rangle_2 + i |1\rangle_2 - |2\rangle_2 - i |3\rangle_2\right).$$

Let $\omega \in \mathbb{R}$. Since

(6.2.4)
$$\frac{1}{2^n} \sum_{y=0}^{2^n-1} |e^{2\pi i \omega y}|^2 = 1$$

it follows that $|\psi_n(\omega)\rangle$ is a quantum state in $\mathbb{H}_n$. We now give another representation of this state.

Proposition 6.2.3. *Let $\omega \in \mathbb{R}$. Then we have*

(6.2.5)
$$|\psi_n(\omega)\rangle = \bigotimes_{j=0}^{n-1} \frac{|0\rangle + e^{2\pi i \cdot 2^{n-j-1}\omega} |1\rangle}{\sqrt{2}}.$$

Proof. We have

(6.2.6)
$$\bigotimes_{j=0}^{n-1} \frac{|0\rangle + e^{2\pi i \cdot 2^{n-j-1}\omega} |1\rangle}{\sqrt{2}}$$
$$= \frac{1}{\sqrt{2^n}} \sum_{\vec{y}=(y_0,\ldots,y_{n-1})\in\{0,1\}^n} \prod_{j=0}^{n-1} e^{2\pi i y_j 2^{n-j-1}\omega} |\vec{y}\rangle$$
$$= \frac{1}{\sqrt{2^n}} \sum_{\vec{y}=(y_0,\ldots,y_{n-1})\in\{0,1\}^n} e^{2\pi i (\sum_{j=0}^{n-1} y_j 2^{n-j-1})\omega} |\vec{y}\rangle$$
$$= \frac{1}{\sqrt{2^n}} \sum_{y=0}^{2^n-1} e^{2\pi i y \omega} |y\rangle_n$$
$$= |\psi_n(\omega)\rangle.$$

This proves the assertion. $\square$

Example 6.2.4. The alternative representation of the state $\psi_2\left(\frac{1}{4}\right)$, which was already considered in Example 6.2.2 ,is

$$\psi_2\left(\frac{1}{4}\right) = \frac{|0\rangle + e^{2\pi i \cdot 2^1 \cdot \frac{1}{4}}|1\rangle}{\sqrt{2}} \otimes \frac{|0\rangle + e^{2\pi i \cdot 2^0 \cdot \frac{1}{4}}|1\rangle}{\sqrt{2}}$$

$$= \frac{|0\rangle - |1\rangle}{\sqrt{2}} \otimes \frac{|0\rangle + i\,|1\rangle}{\sqrt{2}}.$$

Next, we define the Quantum Fourier Transform.

Definition 6.2.5. The *Quantum Fourier Transform* QFT_n is the linear operator on $\mathbb{H}_n$ that is defined in terms of the images of the computational basis states $|x\rangle_n$, $x \in \mathbb{Z}_{2^n}$, of $\mathbb{H}_n$ as follows:

$$(6.2.7) \qquad \mathrm{QFT}_n\,|x\rangle_n = \left|\psi_n\left(\frac{x}{2^n}\right)\right\rangle = \frac{1}{\sqrt{2^n}}\sum_{y=0}^{2^n-1} e^{2\pi i \frac{x}{2^n} y}\,|y\rangle_n.$$

Example 6.2.6. We have

$$\mathrm{QFT}_1 \frac{1}{\sqrt{2}}(|0\rangle + |1\rangle) = \frac{1}{\sqrt{2}}(\mathrm{QFT}_1\,|0\rangle + \mathrm{QFT}_1\,|1\rangle)$$

$$= \frac{1}{\sqrt{2}}\left(\frac{1}{\sqrt{2}}(|0\rangle + |1\rangle) + \frac{1}{\sqrt{2}}(|0\rangle - |1\rangle)\right) = |0\rangle.$$

Example 6.2.7. We have

$$(6.2.8) \qquad \mathrm{QFT}_n \left|\underbrace{00\cdots00}_{n}\right\rangle = \mathrm{QFT}\,|0\rangle_n = |\psi_n(0)\rangle = \frac{1}{\sqrt{2^n}}\sum_{y=0}^{2^n-1}|y\rangle_n.$$

So $\mathrm{QFT}_n\,|0\rangle_n$ is the equally weighted superposition of all computational basis states of $\mathbb{H}_n$.

Proposition 6.2.8. *The Quantum Fourier Transform* QFT_n *is a unitary operator on* $\mathbb{H}_n$*. Its inverse is the linear operator on* $\mathbb{H}_n$ *that is defined in terms of the images of the computational basis elements* $|x\rangle_n$*,* $x \in \mathbb{Z}_{2^n}$*, of* $\mathbb{H}_n$ *as follows:*

$$(6.2.9) \qquad \mathrm{QFT}_n^{-1}\,|x\rangle_n = \left|\psi_n\left(\frac{-x}{2^n}\right)\right\rangle = \frac{1}{\sqrt{2^n}}\sum_{y=0}^{2^n-1} e^{-2\pi i \frac{x}{2^n} y}\,|y\rangle_n.$$

Proof. The representation matrices of QFT_n and QFT_n^* with respect to the computational basis of $\mathbb{H}_n$ are

$$(6.2.10) \qquad\qquad \mathrm{QFT}_n = \left(\frac{e^{2\pi i \frac{x}{2^n} y}}{\sqrt{2^n}}\right)_{x,y\in\mathbb{Z}_{2^n}}$$

and

$$(6.2.11) \qquad\qquad \mathrm{QFT}_n^* = \left(\frac{e^{-2\pi i \frac{y}{2^n} z}}{\sqrt{2^n}}\right)_{y,z\in\mathbb{Z}_{2^n}}.$$

Let $x, z \in \mathbb{Z}_{2^n}$. Then the entry in row x and column z of the matrix products $\mathrm{QFT}_n \cdot \mathrm{QFT}_n^*$ and $\mathrm{QFT}_n^* \cdot \mathrm{QFT}_n$ is

$$(6.2.12) \qquad \frac{1}{2^n} \sum_{y=0}^{2^n-1} e^{2\pi i \frac{x}{2^n} y} e^{-2\pi i \frac{y}{2^n} z} = \begin{cases} 1 & \text{if } x = z, \\ \frac{1}{2^n} \frac{1 - e^{2\pi i (x-z)}}{1 - e^{2\pi i \frac{x-z}{2^n}}} = 0 & \text{if } x \neq z. \end{cases}$$

Hence, we have $\mathrm{QFT}_n \cdot \mathrm{QFT}_n^* = \mathrm{QFT}_n^* \cdot \mathrm{QFT}_n = I_n$. So QFT_n is unitary and (6.2.9) follows from (6.2.11). $\qquad \square$

We note that for all $x \in \mathbb{Z}_{2^n}$ we obtain from Proposition 6.2.3

$$(6.2.13) \qquad \mathrm{QFT}_n |x\rangle_n = \bigotimes_{j=0}^{n-1} \frac{|0\rangle + e^{2\pi i \cdot \frac{x}{2^{j+1}}} |1\rangle}{\sqrt{2}}.$$

In particular, we have

$$(6.2.14) \qquad \mathrm{QFT}_n |0\rangle_n = \bigotimes_{j=0}^{n-1} \frac{|0\rangle + |1\rangle}{\sqrt{2}} = H^{\otimes n} |0\rangle^{\otimes n}$$

which has already been shown in Example 6.2.7.

Exercise 6.2.9. Use equation (6.2.14) to show that

$$(6.2.15) \qquad H^{\otimes n} |0\rangle^{\otimes n} = \sum_{y=0}^{2^n-1} |y\rangle_n$$

and construct a quantum circuit that creates the equally weighted superposition of all computational basis states of $\mathbb{H}_n$.

We use Proposition 6.2.3 and (6.2.14) to give alternative formulas for QFT_n and QFT_n^{-1}. For $m \in \mathbb{N}$ and $(b_0, \ldots, b_{m-1}) \in \{0, 1\}^m$ we write

$$(6.2.16) \qquad 0.b_0 b_1 \cdots b_{m-1} = \sum_{i=0}^{m-1} b_i 2^{-i-1}.$$

Example 6.2.10. We have

$$0.1001 = 1 \cdot 2^{-1} + 0 \cdot 2^{-2} + 0 \cdot 2^{-3} + 1 \cdot 2^{-4} = 2^{-1} + 2^{-4}.$$

With this notation, we obtain the following result.

Proposition 6.2.11. Let $x \in \mathbb{Z}_{2^n}$ and let $\vec{x} = (x_0 x_1 x_2 \cdots x_{n-1}) \in \{0, 1\}^n$ such that $x = \sum_{i=0}^{n-1} x_i 2^{n-i-1}$. Then we have

$$(6.2.17) \qquad \mathrm{QFT}_n |x\rangle_n = \bigotimes_{j=0}^{n-1} \frac{|0\rangle + e^{2\pi i \cdot 0.x_{n-j-1} x_{n-j-2} \cdots x_{n-1}} |1\rangle}{\sqrt{2}}$$

and

$$(6.2.18) \qquad \mathrm{QFT}_n^{-1} |x\rangle_n = \bigotimes_{j=0}^{n-1} \frac{|0\rangle + e^{-2\pi i \cdot 0.x_{n-j-1} j x_{n-j-2} \cdots x_{n-1}} |1\rangle}{\sqrt{2}}.$$

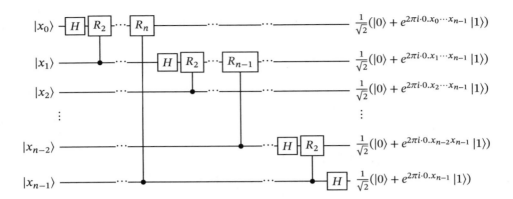

Figure 6.2.1. A quantum circuit that computes QFT$_n$ up to a permutation that reverses the order of the output qubits.

Proof. The assertion follows from equation (6.2.13).   □

Next, we present a quantum circuit that computes the Quantum Fourier Transform. Figure 6.2.1 shows this quantum circuit up to a permutation that reverses the order of the output qubits and Algorithm 6.2.12 is its complete specification.

Algorithm 6.2.12. Quantum Fourier Transform

Input: $|\psi\rangle \in \mathbb{H}_n$
Output: QFT$_n |\psi\rangle$
  1: Initialize the input register to $|\psi\rangle$
  2: **for** $j = 0$ to $n - 1$ **do**
  3:     Apply H to the jth qubit
  4:     **for** $k = 1$ to $n - j - 1$ **do**
  5:         Apply R_{k+1} to the jth qubit controlled by the $(j + k)$th qubit
  6:     **end for**
  7: **end for**
  8: Apply a permutation that reverses the order of the qubits in $|\psi\rangle$

The next theorem states the correctness of the QFT implementation.

Theorem 6.2.13. *The quantum circuit specified in Algorithm 6.2.12 computes* QFT$_n$ *and has size* $O(n^2)$.

Proof. Let $j \in \{0, \dots, n - 1\}$. We explain the evolution of the input qubit $|x_j\rangle$ in the quantum circuit shown in Figure 6.2.1. First, the quantum circuit applies the Hadamard operator to $|x_j\rangle$. This has the following effect on this qubit which is shown in Figure 6.2.2:

$$(6.2.19) \qquad\qquad H|x_j\rangle = \frac{1}{\sqrt{2}}\left(|0\rangle + e^{2\pi i \cdot 0.x_j}|1\rangle\right).$$

This operation does not change the other qubits.

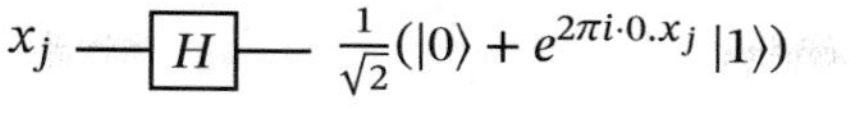

$$x_j \quad \boxed{H} \quad \frac{1}{\sqrt{2}}(|0\rangle + e^{2\pi i \cdot 0.x_j}|1\rangle)$$

Figure 6.2.2. Application of the Hadamard gate to $|x_j\rangle$.

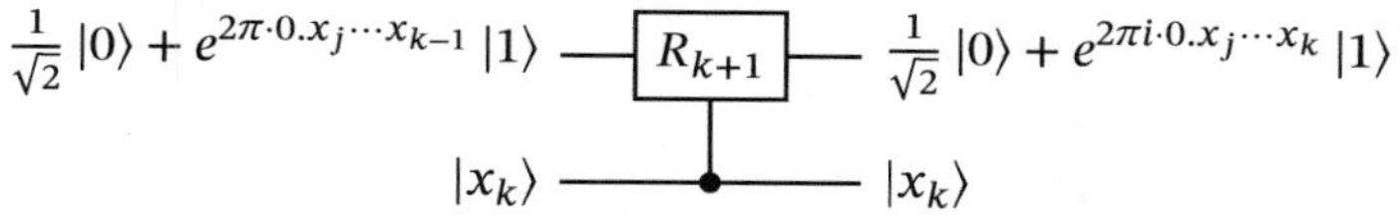

$$\frac{1}{\sqrt{2}}|0\rangle + e^{2\pi \cdot 0.x_j \cdots x_{k-1}}|1\rangle \quad \boxed{R_{k+1}} \quad \frac{1}{\sqrt{2}}|0\rangle + e^{2\pi i \cdot 0.x_j \cdots x_k}|1\rangle$$

$$|x_k\rangle \quad\longrightarrow\quad |x_k\rangle$$

Figure 6.2.3. The controlled-R_k operator kicks the phase $2\pi x_k/2^{k+1}$ back to the amplitude of $|1\rangle$.

Next, for $k = j+1, \ldots, n-1$ the quantum circuit applies the controlled-R_{k+1} operator to this qubit. As shown in Figure 6.2.3, this kicks the phase $2\pi \cdot 0.\underbrace{0 \cdots 0}_{k} x_k$ back to the amplitude of $|1\rangle$ in the jth qubit while the amplitude of $|0\rangle$ and the other qubits remain unchanged. So the final state of this qubit is

$$(6.2.20) \qquad \frac{1}{\sqrt{2}}\left(|0\rangle + e^{2\pi i \cdot 0.x_j \cdots x_{n-1}}|1\rangle\right).$$

We estimate the size of the quantum circuit. It can be seen in Figure 6.2.1 that $O(n^2)$ input qubits, elementary quantum gates, and controlled-R_k gates are used before the order of the output qubits is reversed. By Corollary 4.12.8, the implementation of the controlled-R_k gates requires $O(1)$ elementary gates. So, the size of this part of the quantum circuit is $O(n^2)$. By Proposition 4.5.2, the reversal of the output qubits requires another $O(n)$ elementary quantum gates. This shows that the total size of the circuit is $O(n^2)$. $\qquad\square$

Exercise 6.2.14. Find a quantum P-uniform circuit family (Q_n) such that for all $n \in \mathbb{N}$ and all $(b_0, \ldots, b_n) \in \{0,1\}^n$

$$(6.2.21) \qquad Q_n |b_0 \cdots b_{n-1}\rangle = |b_{n-1} \cdots b_0\rangle$$

and Q_n requires $O(n)$ CNOT gates.

Exercise 6.2.15. Verify that the quantum circuits in Figures 6.2.2 and 6.2.3 have the asserted outputs.

Theorem 6.2.13 implies the following corollary.

Corollary 6.2.16. *The quantum circuit in Figure 6.2.4 computes* QFT_n^{-1} *up to a permutation that reverses the order of the output qubits. It has size* $O(n^2)$.

Proof. The quantum circuit in Figure 6.2.4 is the inverse of the quantum circuit in Figure 6.2.1. $\qquad\square$

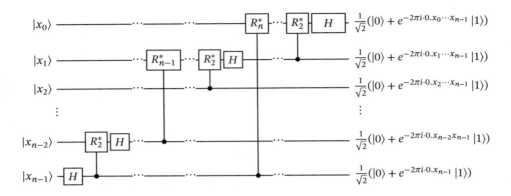

Figure 6.2.4. Quantum circuit that computes QFT_n^{-1} up to a permutation that reverses the order of the output bits.

6.3. Quantum phase estimation

In this section, we consider the following problem. Let $m \in \mathbb{N}$, let U be a unitary operator on $\mathbb{H}_m$, and let $|\psi\rangle$ be an eigenstate of U. Since U is unitary, its eigenvalues have absolute value 1 by Proposition 2.4.60. We may therefore write the eigenvalue associated to $|\psi\rangle$ as $e^{2\pi i \omega}$ with $\omega \in \mathbb{R}$. The phase $2\pi\omega$ of the eigenvalue is uniquely determined modulo 1. We will present a quantum algorithm that, given a precision parameter $n \in \mathbb{N}$, finds with high probability $x \in \mathbb{Z}_{2^n}$ such that $e^{2\pi i \frac{x}{n}}$ is close to the eigenvalue $e^{2\pi i \omega}$. The integer x will be obtained by measuring the final state of the quantum register produced by the quantum circuit in Figure 6.3.1.

6.3.1. The idea. We start by explaining the idea of the phase estimation algorithm which is shown in Figure 6.3.1. Using the Hadamard operator and the controlled-U operator, the algorithm constructs the state $|\psi_3\rangle = |\psi_n(\omega)\rangle$. If $\omega = \frac{x}{2^n}$ for some $x \in \mathbb{Z}_{2^n}$, then it follows from (6.2.7) that $|\psi_4\rangle = \mathrm{QFT}_n^{-1} |\psi_n(\omega)\rangle = |x\rangle$. So measuring this state gives x with probability 1. We will show that if ω is not of this form, then measuring $|\psi_4\rangle$ gives with high probability $x \in \mathbb{Z}_{2^n}$ such that $\frac{x}{2^n}$ is a good approximation to ω.

6.3.2. Approximating $|\psi_n(\omega)\rangle$. As discussed in Section 6.3.1, the phase estimation algorithm constructs and measures $\mathrm{QFT}_n^{-1} |\psi_n(\omega)\rangle$. The measurement outcome is $x \in \mathbb{Z}_{2^n}$ such that $\frac{x}{2^n}$ approximates the phase ω. It is important to note that $\frac{x}{2^n}$ falls within the interval $[0, 1[$, whereas ω may be any real number. However, adding integers to ω does not alter $|\psi_n(\omega)\rangle$. This justifies the subsequent definition, allowing us to quantify the accuracy of this approximation.

Definition 6.3.1. Let $\omega \in \mathbb{R}$, $n \in \mathbb{N}$, and $x \in \mathbb{Z}_{2^n}$. Then we set

$$(6.3.1) \qquad \Delta(\omega, n, x) = \omega - \frac{x}{2^n} - \left\lfloor \omega - \frac{x}{2^n} \right\rfloor.$$

The quantity $\Delta(\omega, n, x)$ from Definition 6.3.1 has the following properties.

Lemma 6.3.2. *Let* $\omega \in \mathbb{R}$, $n \in \mathbb{N}$, *and* $x \in \mathbb{Z}_{2^n}$. *Then the following hold.*

(1) $e^{2\pi i(\omega - \frac{x}{2^n})} = e^{2\pi i \Delta(\omega, n, x)}$.

(2) $-\frac{1}{2} < \Delta(\omega, n, x) \leq \frac{1}{2}$.

Proof. The first assertion follows from the periodicity of the function $f(y) = e^{2\pi i y}$ modulo 1. The second claim follows from the definition of $\Delta(\omega, n, x)$. $\qquad\square$

In several situations we are interested in estimation $\omega - \frac{x}{2^n}$ instead of $\Delta(\omega, n, x)$. The next lemma shows when the first expression can be replaced by the second.

Lemma 6.3.3. *Let* $\omega \in \mathbb{R}$, $n \in \mathbb{N}$, $x \in \mathbb{Z}_{2^n}$. *Assume that*

$$(6.3.2) \qquad |\Delta(\omega, n, x)| < \frac{1}{2^n} \quad \text{and} \quad 0 \leq \omega < 1 - \frac{1}{2^n}.$$

Then we have

$$(6.3.3) \qquad \Delta(\omega, n, x) = \omega - \frac{x}{2^n}.$$

Proof. Let

$$(6.3.4) \qquad z = \left\lfloor \omega - \frac{x}{2^n} \right\rfloor.$$

Then we have

$$(6.3.5) \qquad \Delta(\omega, n, x) = \omega - \frac{x}{2^n} - z$$

which implies

$$(6.3.6) \qquad z = \omega - \frac{x}{2^n} - \Delta(\omega, n, x).$$

So the first inequality in (6.3.2) implies

$$(6.3.7) \qquad \omega - \frac{x}{2^n} - \frac{1}{2^n} < z < \omega - \frac{x}{2^n} + \frac{1}{2^n}.$$

Using $x \in \mathbb{Z}_{2^n}$ and the inequalities for ω in (6.3.2) we obtain

$$(6.3.8) \qquad -1 \leq -\frac{1}{2^n} - \frac{x}{2^n} < z < 1 - \frac{x}{2^n}$$

which implies $z = 0$. $\qquad\square$

Finally, we prove the following statement which is crucial in the analysis of the phase estimation algorithm.

Proposition 6.3.4. *Let $n \in \mathbb{N}$ and $\omega \in \mathbb{R}$. For $x \in \mathbb{Z}_{2^n}$ denote by $p(x)$ the probability that x is the outcome of measuring $\mathrm{QFT}_n^{-1} |\psi_n(\omega)\rangle$ in the computational basis of $\mathbb{H}_n$. Then the following hold.*

(1) *If $2^n\omega \in \mathbb{Z}$, then for $x = 2^n\omega \bmod 2^n$ we have $p(x) = 1$ and $p(x) = 0$ for all other integers $x \in \mathbb{Z}_{2^n}$.*

(2) *If $2^n\omega \notin \mathbb{Z}$, then for all $x \in \mathbb{Z}_{2^n}$ we have*

$$(6.3.9) \qquad p(x) = \frac{1}{2^{2n}} \frac{\sin^2(2^n \pi \Delta(\omega, n, x))}{\sin^2(\pi \Delta(\omega, n, x))}.$$

Proof. Set $N = 2^n$ and $\Delta = \Delta(\omega, n, x)$. From Definition 6.2.1, Proposition 6.2.8, and Lemma 6.3.2 we obtain

$$\mathrm{QFT}_n^{-1} |\psi_n(\omega)\rangle = \frac{1}{\sqrt{N}} \sum_{y=0}^{N-1} e^{2\pi i \omega y} \mathrm{QFT}_n^{-1} |y\rangle_n$$

$$(6.3.10) \qquad = \frac{1}{\sqrt{N}} \sum_{y=0}^{N-1} e^{2\pi i \omega y} \frac{1}{\sqrt{N}} \sum_{x=0}^{N-1} e^{-2\pi i \frac{y}{N} x} |x\rangle_n$$

$$= \sum_{x=0}^{N-1} \frac{1}{N} \left(\sum_{y=0}^{N-1} e^{2\pi i \Delta y} \right) |x\rangle_n.$$

So for all $x \in \mathbb{Z}_N$ we have

$$(6.3.11) \qquad p(x) = \frac{1}{N^2} \left| \sum_{y=0}^{N-1} e^{2\pi i \Delta y} \right|^2.$$

Let $N\omega \in \mathbb{Z}$ and $x = N\omega \bmod N$. Then $\Delta = 0$ and $p(x) = 1$.

Assume that $N\omega \notin \mathbb{Z}$. Then evaluating the geometric series $\sum_{y=0}^{N-1} e^{2\pi i \Delta y}$ in (6.3.11) gives

$$(6.3.12) \qquad \mathrm{QFT}_n^{-1} |\psi_n(\omega)\rangle = \frac{1}{N} \sum_{x=0}^{N-1} \frac{1 - e^{2\pi i N \Delta}}{1 - e^{2\pi i \Delta}} |x\rangle_n.$$

Now for all $\theta \in \mathbb{R}$ we have

$$(6.3.13) \qquad |1 - e^{2\pi i \theta}| = |e^{-\pi i \theta} - e^{\pi i \theta}| = 2|\sin(\pi\theta)|.$$

It follows from equations (6.3.12) and (6.3.13) that

$$(6.3.14) \qquad p(x) = \left| \frac{1 - e^{2\pi i N \Delta}}{1 - e^{2\pi i \Delta}} \right|^2 = \frac{1}{N^2} \frac{\sin^2(N\pi\Delta)}{\sin^2 \pi\Delta}. \qquad \square$$

6.3.3. The problem and the algorithm. We now state the phase estimation problem which is also called the *eigenvalue estimation problem*.

Problem 6.3.5 (Phase estimation problem).

Input: Positive integers m and n, an implementation of the controlled-U operator for some unitary operator U on $\mathbb{H}_m$, and an eigenstate $|\psi\rangle$ of U.

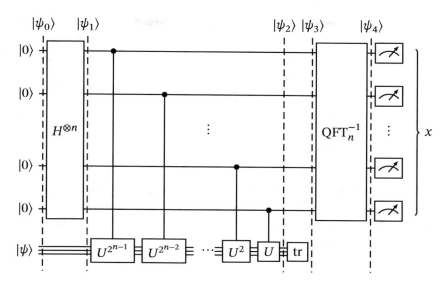

Figure 6.3.1. Quantum circuit for phase estimation.

Output: An integer $x \in \mathbb{Z}_{2^n}$ such that $|\Delta(\omega, n, x)| < 1/2^n$ where $\omega \in \mathbb{R}$ and $e^{2\pi i \omega}$ is the eigenvalue associated with $|\psi\rangle$.

A quantum circuit that solves this problem is shown in Figure 6.3.1 and specified in Algorithm 6.3.6.

Algorithm 6.3.6. Phase estimation algorithm

Input: Positive integers m, n, an implementation of the controlled-U operator for a unitary operator U on $\mathbb{H}_m$, $m \in \mathbb{N}$, an eigenstate $|\psi\rangle$ of U.

Output: $x \in \mathbb{Z}_{2^n}$

1:   PHASEESTIMATE$(m, n, U, |\psi\rangle)$
2:       Initialize the control register to $|0\rangle^{\otimes n}$
3:       Initialize the target register to $|\psi\rangle$
4:       Apply $H^{\otimes n}$ to the control register
5:       **for** $j = 1$ to $n - 1$ **do**
6:           Apply $U^{2^{n-j}}$ to the target register controlled by the jth qubit of the control register
7:       **end for**
8:       Trace out the target register
9:       Apply QFT_n^{-1} to the control register
10:      Measure the control register in the computational basis, the result being x
11:      **return** x
12: **end**

The next theorem states the correctness of the phase estimation algorithms and its success probability.

Theorem 6.3.7. *Let m, n be positive integers, and let U be a unitary operator on $\mathbb{H}_m$. Let $\omega \in \mathbb{R}$ be such that $e^{2\pi i \omega}$ is an eigenvalue of U and let $|\psi\rangle$ be a corresponding eigenstate. Also, let $x \in \mathbb{Z}_{2^n}$ be the return value of the phase estimation algorithm. Then the following hold.*

(1) *If $2^n \omega \in \mathbb{Z}$, then $x = 2^n \omega$ with probability 1.*

(2) *With probability at least $\frac{4}{\pi^2}$ we have $|\Delta(\omega, n, x)| \leq \frac{1}{2^{n+1}}$.*

(3) *With probability at least $\frac{8}{\pi^2}$ we have $|\Delta(\omega, n, x)| < \frac{1}{2^n}$.*

Proof. The quantum circuit operates on two quantum registers. The first is the control register. Its length is the precision parameter n and is initialized to $|0\rangle^{\otimes n}$. The second register is the target register. It is of length m and is initialized with the eigenvector $|\psi\rangle$ of U. So, the initial state of the algorithm is

$$(6.3.15) \qquad |\psi_0\rangle = |0\rangle^{\otimes n} |\psi\rangle.$$

The algorithm then applies $H^{\otimes n}$ to the control register. This gives the state

$$(6.3.16) \qquad |\psi_1\rangle = \left(\frac{|0\rangle + |1\rangle}{\sqrt{2}} \right)^{\otimes n} |\psi\rangle.$$

Next, we note that for $j = 0, \ldots, n - 1$ we have

$$(6.3.17) \qquad U^{2^{n-j-1}} |\psi\rangle = e^{2\pi i 2^{n-j-1} \omega} |\psi\rangle.$$

This is a global phase shift. These operators are applied to the target register controlled by the jth qubit of the control register which produces the new state

$$(6.3.18) \qquad |\psi_2\rangle = \bigotimes_{j=0}^{n-1} \frac{|0\rangle + e^{2\pi i 2^{n-j-1} j \omega} |1\rangle}{\sqrt{2}} |\psi\rangle = |\psi_n(\omega)\rangle |\psi\rangle.$$

This shows that the global phase shifts are kicked back to the amplitudes of $|1\rangle$ in the control qubits. Since the state $|\psi_2\rangle$ is separable with respect to the decomposition into the control and the target register, it follows from Corollary 3.7.12 that tracing out the target register yields the state

$$(6.3.19) \qquad |\psi_3\rangle = |\psi_n(\omega)\rangle.$$

Therefore, the final state is

$$(6.3.20) \qquad |\psi_4\rangle = \mathrm{QFT}_n^{-1} |\psi_n(\omega)\rangle.$$

This state is measured. So, the first assertion follows from Proposition 6.3.4.

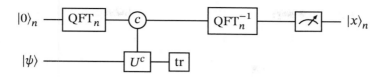

Figure 6.3.2. Simplified representation of the quantum circuit for eigenvalue estimation.

To prove the second assertion, we set

(6.3.21) $$N = 2^n, \quad x = \lfloor N\omega \rfloor \bmod N, \quad \theta = N\Delta(\omega, n, x).$$

Then we have

(6.3.22) $$|\theta| \le \frac{1}{2}.$$

From Proposition 6.3.4, Lemma A.5.3, Lemma A.5.5, and inequality (6.3.22) we obtain

(6.3.23) $$p(x) = \frac{1}{N^2} \frac{\sin^2(\pi\theta)}{\sin^2(\pi\theta/N)} \ge \frac{1}{N^2} \frac{(2\theta)^2}{(\pi\theta/N)^2} = \frac{4}{\pi^2}.$$

To prove the third assertion, we choose $x' \in \{x \pm 1\}$ such that

(6.3.24) $$N|\Delta(n, \omega, x')| = 1 - N|\Delta(n, \omega, x)| = 1 - \theta.$$

By Proposition 6.3.4 and Lemma A.5.5 the probability p for measuring x or $x+1$ satisfies

$$
\begin{aligned}
p &= \frac{1}{N^2} \frac{\sin^2(\pi\theta)}{\sin^2(\pi\theta/N)} + \frac{1}{N^2} \frac{\sin^2(\pi(1-\theta))}{\sin^2(\pi(1-\theta)/N)} \\
(6.3.25) \qquad &= \frac{\sin^2(\pi\theta)}{N^2} \left(\frac{1}{\sin^2(\pi\theta/N)} + \frac{1}{\sin^2(\pi(1-\theta)/N)} \right) \\
&\ge \frac{\sin^2(\pi\theta)}{\pi^2} \left(\frac{1}{\theta^2} + \frac{1}{(1-\theta)^2} \right).
\end{aligned}
$$

Since for $0 < \theta < 1$ the function

(6.3.26) $$f(x) = \sin^2(\pi\theta) \left(\frac{1}{\theta^2} + \frac{1}{(1-\theta)^2} \right)$$

attains its minimum 8 for $\theta = \frac{1}{2}$, the assertion follows. $\qquad\square$

Figure 6.3.2 is a simplified representation of the circuit in Figure 6.3.1. In this representation, we use a controlled-U^c operator which sends $|\psi\rangle$ to $U^c|\psi\rangle$ if the control register contains the state $|c\rangle_n$ for some $c \in \mathbb{Z}_{2^n}$.

6.4. Order finding

We will now present an important application of quantum phase estimation: a quantum algorithm that computes the order of an integer modulo another positive integer in polynomial time. This algorithm is the central building block in the integer factorization and discrete logarithm algorithms of Peter Shor, which we discuss in Sections 6.5 and 6.6.

6.4.1. The problem. The *order finding problem* is the following.

Problem 6.4.1 (Order finding problem).

Input: $N \in \mathbb{N}$ and $a \in \mathbb{Z}_N$ such that $\gcd(a, N) = 1$.

Output: The order r of a modulo N.

In the sequel, we explain a quantum polynomial time algorithm that solves the order problem. In the following, we let N, a, r be as in the order finding problem. We also let $n \in \mathbb{N}$.

6.4.2. The operator U_c. We introduce and discuss a unitary operator that is used in the order finding algorithm.

Definition 6.4.2. For any $c \in \mathbb{Z}$ with $\gcd(c, N) = 1$ we define the linear operator

$$(6.4.1) \qquad U_c : \mathbb{H}_n \to \mathbb{H}_n, \quad |x\rangle_n \mapsto \begin{cases} |cx \bmod N\rangle_n & \text{if } 0 \leq x < N, \\ |x\rangle_n & \text{if } N \leq x < 2^n. \end{cases}$$

Proposition 6.4.3. *For all $c \in \mathbb{Z}$ with $\gcd(c, N) = 1$, the operator U_c is unitary.*

Proof. By definition, the map U_c is linear. So it suffices to show that the map

$$(6.4.2) \qquad f_c : \mathbb{Z}_{2^n} \to \mathbb{Z}_{2^n}, \quad x \mapsto \begin{cases} cx \bmod N & \text{if } 0 \leq x < N, \\ c & \text{if } N \leq x < 2^n \end{cases}$$

is a bijection. For this, it suffices to show that f_c is surjective since the domain and the codomain of f_c are the same. Let $y \in \mathbb{Z}_{2^n}$. If $y \geq N$, then we have $f(y) = y$. If $y < N$, then we have $y = f(x)$ where $x \in \mathbb{Z}_N$ such that $cx \equiv y \bmod N$. This number x exists because $\gcd(c, N) = 1$. $\square$

Next, we determine the eigenstates of the operators U_{a^t} for all $t \in \mathbb{N}$. In the context of the order finding algorithm, only the case $t = 1$ is relevant. However, for the quantum discrete logarithm problem, we use $t > 1$.

Proposition 6.4.4. (1) *For every $k \in \mathbb{Z}$ and every $t \in \mathbb{N}$ the state*

$$(6.4.3) \qquad |u_k\rangle = \frac{1}{\sqrt{r}} \sum_{s=0}^{r-1} e^{-2\pi i \frac{k}{r} s} |a^s \bmod N\rangle_n$$

is an eigenstate of U_{a^t} with eigenvalue $e^{2\pi i \frac{tk}{r}}$.

(2) *The sequence $(|u_0\rangle, \ldots, |u_{r-1}\rangle)$ is an orthonormal basis of $\mathrm{Span}\{|a^s \bmod N\rangle_n : s \in \mathbb{Z}_r\}$.*

Proof. To prove the first assertion, let $k \in \mathbb{Z}$ and $t \in \mathbb{N}$. We note that the map

$$(6.4.4) \qquad \mathbb{Z}_r \to \mathbb{Z}_r, \quad s \mapsto (s + t) \bmod r$$

is a bijection. So we have

$$U_{a^t} |u_k\rangle = \frac{1}{\sqrt{r}} \sum_{s=0}^{r-1} e^{-2\pi i \frac{k}{r} s} U_{a^t} |a^s \bmod N\rangle_n$$

$$= \frac{1}{\sqrt{r}} \sum_{s=0}^{r-1} e^{-2\pi i \frac{k}{r} s} |a^{s+t} \bmod N\rangle_n$$

(6.4.5)
$$= e^{2\pi i \frac{kt}{r}} \frac{1}{\sqrt{r}} \sum_{s=0}^{r-1} e^{-2\pi i \frac{k}{r}(s+t) \bmod r} |a^{(s+t) \bmod r} \bmod N\rangle_n$$

$$= e^{2\pi \frac{kt}{r}} \frac{1}{\sqrt{r}} \sum_{s=0}^{r-1} e^{-2\pi i \frac{k}{r} s} |a^s \bmod N\rangle_n$$

$$= e^{2\pi \frac{kt}{r}} |u_k\rangle .$$

This concludes the proof of the first assertion.

Next, we turn to the second assertion. Since r is the order of a modulo N, the elements of the sequence $(|a^s \bmod N\rangle_n)_{0 \leq s < r}$ are pairwise different. Thus this sequence is a basis of $\mathrm{Span}\{|a^s \bmod N\rangle_n : 0 \leq s < r\}$. Also, for all $k, k' \in \mathbb{Z}_r$ we have

(6.4.6)
$$\langle u_k | u_{k'} \rangle = \frac{1}{r} \sum_{s=0}^{r-1} e^{2\pi i \frac{k-k'}{r} s}$$

$$= \begin{cases} 1 & \text{if } k = k', \\ \frac{1 - e^{2\pi i (k-k')s}}{1 - e^{2\pi i \frac{k-k'}{r} s}} = 0 & \text{if } k \neq k'. \end{cases}$$

Hence, the sequence $(|u_0\rangle, \ldots, |u_{r-1}\rangle)$ is an orthonormal basis of $\mathrm{Span}\{|a^s \bmod N\rangle_n : s \in \mathbb{Z}_r\}$, as claimed. $\qquad\square$

If we were to apply the phase estimation algorithm with the target register initialized to an eigenstate $|u_k\rangle$ of U_a, then we would obtain a rational approximation $\frac{x}{2^n}$ to $\frac{k}{r}$ and thus some information about the order $r \bmod N$. Unfortunately, this cannot be done because it is not known how to prepare the eigenstates of U_a. But the following proposition is of help.

Proposition 6.4.5. *We have*

(6.4.7)
$$\frac{1}{\sqrt{r}} \sum_{k=0}^{r-1} |u_k\rangle = |1\rangle_n .$$

Proof. Note that

(6.4.8)
$$\frac{1}{\sqrt{r}} \sum_{k=0}^{r-1} |u_k\rangle = \frac{1}{\sqrt{r}} \sum_{k=0}^{r-1} \frac{1}{\sqrt{r}} \sum_{s=0}^{r-1} e^{-2\pi i \frac{k}{r} s} |a^s \bmod N\rangle_n$$

$$= \frac{1}{r} \sum_{s=0}^{r-1} \left(\sum_{k=0}^{r-1} e^{-2\pi i \frac{k}{r} s} \right) |a^s \bmod N\rangle_n .$$

We determine the amplitude of $|1\rangle$ in the state in (6.4.8). Since $a^s \equiv 1 \bmod N$ if and only if $s \equiv 0 \bmod r$, it follows that this amplitude is

$$(6.4.9) \qquad \frac{1}{r} \sum_{k=0}^{r-1} e^{-2\pi i \frac{k}{r} \cdot 0} = \frac{r}{r} = 1.$$

Since $\frac{1}{\sqrt{r}} \sum_{k=0}^{r-1} |u_k\rangle$ is a quantum state, the amplitudes of the other basis states in this representation must be 0. This implies the assertion. $\qquad\square$

6.4.3. The algorithm. We now present and analyze a quantum order finding algorithm. The pseudocode is shown in Algorithm 6.4.6. It uses a variant of the phase estimation circuit, shown in Figure 6.4.1. This circuit differs from the quantum phase estimation circuit in that the target register is initialized to $|1\rangle_n$. This state can be prepared, and by Proposition 6.4.5 it is the equally weighted superposition of the eigenstates $|u_k\rangle$ of U_a for $0 \le k < r$. By Proposition 6.4.4, the corresponding eigenvalues are $e^{2\pi i \frac{k}{r}}$. Their phase contains information about the order r of a modulo N. As we will see in the proof of Theorem 6.4.8, the modified quantum phase estimation circuit can be used to determine r.

Algorithm 6.4.6. Quantum order finding algorithm

Input: $N \in \mathbb{N}$, $a \in \mathbb{Z}_N$ with $\gcd(a, N) = 1$, $n \in \mathbb{N}$ with $2r^2 \le 2^n \le 2N^2$ where r is the order a module N.

Output: r or "FAILURE"

1:    FINDORDER(N, a, n)
2:      **for** $j = 1, 2$ **do**
3:        Apply the quantum circuit Q_a from Figure 6.4.1 and obtain $x_j \in \mathbb{Z}_{2^n}$
4:        Apply the continued fraction algorithm to $\frac{x_j}{2^n}$
5:        **if** $|\frac{x_j}{2^n} - \frac{p}{q}| \le \frac{1}{2^n}$ for a convergent $\frac{p}{q}$ **then**
6:          Set $m_j \leftarrow p$ and $r_j \leftarrow q$
7:        **else**
8:          **return** "FAILURE"
9:        **end if**
10:      **end for**
11:      $r \leftarrow \text{lcm}(r_1, r_2)$
12:      **if** $a^r \equiv 1 \bmod N$ **then**
13:        **return** r
14:      **else**
15:        **return** "FAILURE"
16:      **end if**
17: **end**

In Algorithm 6.4.6 the precision parameter $n \in \mathbb{N}$ is chosen such that

$$(6.4.10) \qquad\qquad\qquad\qquad 2^n \ge 2r^2.$$

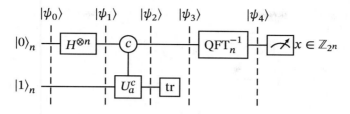

Figure 6.4.1. The modified phase estimation circuit Q_a used in the order finding algorithm.

For example, we may set

(6.4.11) $$n = \lceil 2\log_2 N \rceil + 1.$$

Then we have

(6.4.12) $$n \le 2\log_2 N + 2.$$

But if there is more information about N and r, we may be able to choose a smaller value for n. This is important because $2n$ is the number of qubits in the order finding algorithm. So, the smaller n is, the more efficient the order finding algorithm. For example, as shown in Exercise 6.4.7, for composite N we may choose $n = \lceil \log_2 N \rceil + 1$.

Exercise 6.4.7. (1) Show that (6.4.10) and (6.4.12) are satisfied for $n = \lceil 2\log_2 N \rceil + 1$.

(2) Let N be a composite number. Show that (6.4.10) and (6.4.12) are satisfied for $n = \lceil \log_2 N \rceil + 1$.

We will now prove the following theorem that states the correctness and the complexity of the algorithm.

Theorem 6.4.8. *On input of $N \in \mathbb{N}$ and $a \in \mathbb{Z}_N$ such that $\gcd(a, N) = 1$, Algorithm 6.4.6 computes the order r of a mod N with probability at least $\frac{3958924}{101761\pi^4} > 0.399$. The algorithm has running time $O((\log N)^3)$.*

For analyzing the success probability of the order finding algorithm, we need the following proposition in which we call a representation $r = \frac{p}{q}$ of a nonzero rational number r *reduced* if $q > 0$ and $\gcd(p, q) = 1$.

Proposition 6.4.9. *Denote by x the return value of the quantum circuit Q_a from Figure 6.4.1 and let $k \in \mathbb{Z}_r$. Then the following hold.*

(1) *With probability at least $\frac{8}{r\pi^2}$ we have*

(6.4.13) $$\left| \frac{x}{2^n} - \frac{k}{r} \right| < \frac{1}{2^n}.$$

(2) *If (6.4.13) holds, then $\frac{k}{r}$ is a convergent of the continued fraction expansion of $\frac{x}{2^n}$. It is the only convergent of this expansion whose reduced representation $\frac{p}{q}$ satisfies*

(6.4.14) $$\left| \frac{x}{2^n} - \frac{p}{q} \right| < \frac{1}{2^n} \quad and \quad q \le 2^{\frac{n-1}{2}}.$$

Proof. By Proposition 6.4.5, the initial state of Q_a is

$$(6.4.15) \qquad |\psi_0\rangle = |0\rangle_n |1\rangle_n = \frac{1}{\sqrt{r}} \sum_{k=0}^{r-1} |0\rangle_n |u_k\rangle.$$

Applying $H^{\otimes n}$ to the first register gives

$$(6.4.16) \qquad |\psi_1\rangle = \sum_{k=0}^{r-1} \left(\frac{|0\rangle + |1\rangle}{\sqrt{2}} \right)^{\otimes n} |u_k\rangle.$$

As seen in (6.3.18) we have

$$(6.4.17) \qquad |\psi_2\rangle = \frac{1}{\sqrt{r}} \sum_{k=0}^{r-1} \left| \psi_n \left(\frac{k}{r} \right) \right\rangle |u_k\rangle.$$

By Proposition 6.4.4, the sequence $(|u_k\rangle)$ is orthonormal. So it follows from Exercise 3.7.11 that tracing out the target register gives the mixed state

$$(6.4.18) \qquad |\psi_3\rangle = \left(\left(\frac{1}{r}, \left| \psi_n \left(\frac{0}{r} \right) \right\rangle \right), \ldots, \left(\frac{1}{r}, \left| \psi_n \left(\frac{r-1}{r} \right) \right\rangle \right) \right).$$

After applying QFT_n^{-1} the mixed state is

$$(6.4.19) \qquad |\psi_4\rangle = \left(\left(\frac{1}{r}, \mathrm{QFT}_n^{-1} \left| \psi_n \left(\frac{0}{r} \right) \right\rangle \right), \ldots, \left(\frac{1}{r}, \mathrm{QFT}_n^{-1} \left| \psi_n \left(\frac{r-1}{r} \right) \right\rangle \right) \right).$$

It follows from Theorem 6.3.7 that measuring this mixed state in the computational basis of $\mathbb{H}_n$ gives with probability at least $\frac{8}{r\pi^2}$ an integer $x \in \mathbb{Z}_{2^n}$ such that

$$(6.4.20) \qquad \left| \Delta \left(\frac{k}{r}, n, x \right) \right| < \frac{1}{2^n}.$$

In Exercise 6.4.10 it is shown that $\left| \Delta \left(\frac{k}{r}, n, x \right) \right| = \left| \frac{k}{r} - \frac{x}{2^n} \right|$. This concludes the proof of the first assertion.

Now assume that (6.4.13) holds. Since we have chosen n such that $2^n \geq 2r^2$ holds, we have

$$(6.4.21) \qquad r \leq 2^{\frac{n-1}{2}}$$

and

$$(6.4.22) \qquad \left| \frac{x}{2^n} - \frac{k}{r} \right| < \frac{1}{2r^2}.$$

So by Proposition A.3.35, the fraction $\frac{k}{r}$ is a convergent of the continued fraction expansion of $\frac{x}{2^n}$ and its reduced representation $\frac{p}{q}$ satisfies (6.4.14). To show the uniqueness of this convergent, let $\frac{p}{q}$ be another convergent of this continued fraction expansion that satisfies (6.4.14). Then we have

$$(6.4.23) \qquad |kq - pr| = rq \left| \frac{k}{r} - \frac{p}{q} \right| \leq 2^{n-1} \left(\left| \frac{x}{2^n} - \frac{k}{r} \right| + \left| \frac{x}{2^n} - \frac{p}{q} \right| \right) < \frac{2 \cdot 2^{n-1}}{2^n} = 1.$$

So we have $kq = pr$ which implies $\frac{k}{r} = \frac{p}{q}$. $\qquad\qquad\qquad\square$

Exercise 6.4.10. Use Lemma 6.3.3 to show that (6.4.20) implies $\left| \Delta \left(\frac{k}{r}, n, x \right) \right| = \left| \frac{k}{r} - \frac{x}{2^n} \right|$.

The next lemma provides a sufficient condition for Algorithm 6.4.6 to find the order r of a modulo N.

Lemma 6.4.11. *For $j = 1, 2$ let $k_j, m_j \in \mathbb{N}_0$ and $r_j \in \mathbb{N}$ with $\frac{k_j}{r} = \frac{m_j}{r_j}$, and assume that $\gcd(k_1, k_2, r) = \gcd(m_j, r_j) = 1$. Then $r = \operatorname{lcm}(r_1, r_2)$.*

Proof. Since $\frac{k_j}{r} = \frac{m_j}{r_j}$ and $\gcd(m_j, r_j) = 1$ for $j = 1, 2$, it follows that r_1 and r_2 are divisors of r. Therefore, $\operatorname{lcm}(r_1, r_2)$ is a divisor of r. This means that we can write

$$(6.4.24) \qquad r = u \cdot \operatorname{lcm}(r_1, r_2) = u u_1 r_1 = u u_2 r_2$$

with $u, u_1, u_2 \in \mathbb{N}$. So, we have

$$(6.4.25) \qquad \frac{m_j}{r_j} = \frac{k_j}{r} = \frac{k_j}{u u_j r_j}$$

for $j = 1, 2$. Hence u divides k_1, k_2, and r. Since $\gcd(k_1, k_2, r) = 1$, we have $u = 1$ and from (6.4.24) we obtain $r = \operatorname{lcm}(r_1, r_2)$. $\qquad \square$

The last statement in this section allows us to estimate the probability of the sufficient condition in Lemma 6.4.11 to occur.

Proposition 6.4.12. *Let Pr be a probability distribution on $\mathbb{Z}_r$ and let $c \in [0, 1]$ such that $\operatorname{Pr}(k) \geq \frac{c}{r}$ for all $k \in \mathbb{Z}_r$. Consider the experiment in which two integers k_1 and k_2 are independently chosen from $\mathbb{Z}_r$ according to the probability distribution Pr. Then the probability that the experiment gives a pair (k_1, k_2) such that $\gcd(k_1, k_2, r) = 1$ is at least $\frac{989731}{1628176} c^2 \geq 0.6 c^2$.*

Proof. First, note that the number of pairs (k_1, k_2) in $\mathbb{Z}_r^2$ with $\gcd(k_1, k_2, r) = 1$ is the same as the number of all such pairs in $\{1, \ldots, r\}^2$. This number is at least the number of coprime pairs in $\{1, \ldots, r\}^2$. But it is shown in [**Fon12**] that the latter number is at least $\frac{989731}{1628176} r^2 \geq 0.6 r^2$. Since, in the experiment, any such pair is chosen with probability at least $\frac{c^2}{r^2}$ it follows that the probability of choosing one such pair is at least $\frac{989731}{1628176} c^2$. $\qquad \square$

We can now prove the success probability stated in Theorem 6.4.8. Proposition 6.4.9 implies that for $j = 1, 2$ and each $k_j \in \mathbb{Z}_r$ the jth iteration of the **for** loop in Algorithm 6.4.6 finds with probability at least $\frac{8}{r\pi^2}$ integers $m_j, r_j \in \mathbb{Z}_{2^n}$ such that $\frac{m_j}{r_j} = \frac{k_j}{r}$. By Proposition 6.4.12, the probability that the two rounds of the **for** loop find such integers with $\gcd(k_1, k_2, r) = 1$ satisfies

$$(6.4.26) \qquad p \geq \frac{989731}{1628176} \frac{64}{\pi^4} = 0.399.$$

It remains to analyze the complexity of the order finding algorithm. Its bottleneck is the computation of the controlled-U_a^c operator which we discuss in the next section.

6.4.4. Modular exponentiation. By C-U_a we denote the controlled operator which for all $|c\rangle_n |t\rangle_n$ with $c, t \in \mathbb{Z}_{2^n}$ satisfies

(6.4.27)
$$\text{C-}U_a |c\rangle_n |t\rangle_n = \begin{cases} |c\rangle_n |a^c t \bmod N\rangle_n & \text{if } 0 \leq t < N, \\ |t\rangle_n & \text{if } N \leq t < 2^n. \end{cases}$$

We present a quantum circuit that implements C-U_a efficiently. For all $i \in \mathbb{Z}_n$ let

(6.4.28)
$$a_i = a^{2^i} \bmod N.$$

Then for $1 \leq i < n$ we have

(6.4.29)
$$a_i = a_{i-1}^2 \bmod N.$$

Let $c, t \in \mathbb{Z}_{2^n}$, $c = \sum_{i=0}^{n-1} c_i 2^{n-1-i}$ where $c_i \in \{0, 1\}$ for $0 \leq i < n$. For $0 \leq i \leq n$ set

(6.4.30)
$$t_i = \begin{cases} \prod_{l=0}^{i-1} a_{n-l-1}^{c_l} t \bmod N & \text{if } t < N, \\ t & \text{if } t \geq N. \end{cases}$$

Then we have

(6.4.31)
$$t_0 = t, \quad t_n = \begin{cases} a^c t \bmod N & \text{if } t < N, \\ t & \text{if } t \geq N. \end{cases}$$

Also, for $0 \leq i \leq n - 1$ we have

(6.4.32)
$$t_{i+1} = \begin{cases} a_{n-i}^{c_i} t_i \bmod n & \text{if } t < N, \\ t & \text{if } t \geq N. \end{cases}$$

Exercise 6.4.13. Verify (6.4.31) and (6.4.32).

The circuit implementing C-U_a uses the modular multiplication operator U_m that is defined by its effect on the computational basis states $|x\rangle |y\rangle$ of $\mathbb{H}_{2n}$, $x, y \in \mathbb{Z}_{2^n}$, as follows:

(6.4.33)
$$U_m |x\rangle |y\rangle = \begin{cases} |x\rangle |xy \bmod N\rangle & \text{if } (x, y) \in \mathbb{Z}_N^2 \wedge \gcd(y, N) = 1, \\ |x\rangle |y\rangle & \text{if } (x, y) \notin \mathbb{Z}_N^2 \vee \gcd(y, N) > 1. \end{cases}$$

This operator is unitary, because, as shown in Exercise 6.4.14, the map

(6.4.34)
$$(x, y) \mapsto \begin{cases} (x, xy \bmod N) & \text{if } (x, y) \in \mathbb{Z}_N^2 \wedge \gcd(y, N) = 1, \\ (x, y) & \text{if } (x, y) \notin \mathbb{Z}_N^2 \vee \gcd(y, N) > 1 \end{cases}$$

is a bijection of $\{0, 1\}^{2n}$.

Exercise 6.4.14. Show that the function in (6.4.34) is a bijection.

As shown in the overview of quantum circuits that implement modular multiplication [RC18], there exists a quantum circuit implementation of U_m with a size of $O(n^2)$.

The idea of a circuit implementation of C-U_a is to first compute $|a_i\rangle_n$ with a_i from (6.4.28) for $0 \leq i < n$ using (6.4.29). The circuit then computes $|t_i\rangle_n$ with t_i from (6.4.30) for $1 \leq i \leq n$ using (6.4.32). As seen in (6.4.31), the required result is $|t_n\rangle$.

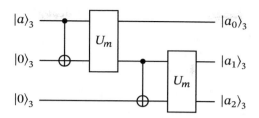

Figure 6.4.2. Circuit U_p for $n = 3$.

We now describe a quantum circuit U_p that constructs $|a_i\rangle$ for $0 \leq i < n$. It uses the bitwise CNOT operator which for $(x_0, \ldots, x_{n-1}), (y_0, \ldots, y_{n-1}) \in \{0,1\}^n$ gives

$$(6.4.35) \quad \begin{aligned} &\mathsf{CNOT}_n |x_0 \cdots x_{n-1}\rangle |y_0 \cdots y_{n-1}\rangle \\ &= |x_0 \cdots x_{n-1}\rangle X^{x_0} |y_0\rangle \cdots X^{x_{n-1}} |y_{n-1}\rangle. \end{aligned}$$

Note that we have

$$(6.4.36) \quad \mathsf{CNOT}_n |x_0 \cdots x_{n-1}\rangle |0 \cdots 0\rangle = |x_0 \cdots x_{n-1}\rangle |x_0 \cdots x_{n-1}\rangle.$$

Figure 6.4.2 shows the circuit U_p for $n = 3$. It works as follows. The input state is $|a\rangle_3 |0\rangle_3 |0\rangle_3$. In the first step, the circuit computes the new state

$$(6.4.37) \quad U_m \mathsf{CNOT}_n(|a\rangle_3 |0\rangle_3) |0\rangle_3 = U_m(|a_0\rangle_3 |a_0\rangle_3) |0\rangle_3 = |a_0\rangle_3 |a_1\rangle_3 |0\rangle_3.$$

In the second step, the circuit computes the new state

$$(6.4.38) \quad |a_0\rangle_3 U_m \mathsf{CNOT}_n(|a_1\rangle_3 |0\rangle_3) = |a_0\rangle_3 U_m(|a_1\rangle_3 |a_1\rangle_3) = |a_0\rangle_3 |a_1\rangle_3 |a_2\rangle_3.$$

The circuit specification for the general case is presented in Algorithm 6.4.15.

Algorithm 6.4.15. Circuit U_p for computing $|a_i\rangle_n = \left| a^{2^i} \bmod N \right\rangle_n$ for $0 \leq i < n$

Input: $n \in \mathbb{N}, N \in \mathbb{Z}_{2^n}, a \in \mathbb{Z}_N^*$
Output: $|a_0\rangle_n \cdots |a_{n-1}\rangle_n$ with a_i as in (6.4.28)
1:    $U_p(n, N, a)$
2:      /* The circuit operates on $|\psi\rangle = |\psi_0\rangle \cdots |\psi_{n-1}\rangle \in \mathbb{H}_n^{\otimes n}$ */
3:      $|\psi\rangle \leftarrow |a\rangle_n |0\rangle_n^{\otimes(n-1)}$
4:      **for** $i = 1, \ldots, n-1$ **do**
5:        $|\psi_{i-1}\rangle |\psi_i\rangle \leftarrow \mathsf{CNOT}_n |\psi_{i-1}\rangle |\psi_i\rangle$
6:        $|\psi_{i-1}\rangle |\psi_i\rangle \leftarrow U_m |\psi_{i-1}\rangle |\psi_i\rangle$
7:      **end for**
8: **end**

We now prove that the quantum circuit U_p has the required property and analyze its complexity.

Proposition 6.4.16. *We have*

$$(6.4.39) \quad U_p |a\rangle_n |0\rangle_n^{\otimes n-1} = |a_0\rangle_n \cdots |a_{n-1}\rangle_n.$$

Also, U_p has size $\mathrm{O}(n^3)$.

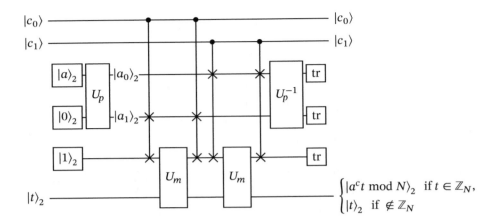

Figure 6.4.3. Quantum circuit that implements C-U_a for $n = 2$.

Proof. We use the notation of Algorithm 6.4.15 and show by induction on i that after the ith iteration of the **for** loop we have

$$(6.4.40) \qquad |\psi\rangle = |a_0\rangle_n \cdots |a_i\rangle_n |0\rangle_n \cdots |0\rangle_n .$$

The base case $i = 0$ follows from the choice of the input state. For the inductive step, assume that $0 \leq i < n - 1$ and that (6.4.40) holds. It follows from (6.4.29) that after the completion of the $(i + 1)$st iteration of the **for** loop we have

$$
\begin{aligned}
(6.4.41) \qquad |\psi\rangle &= |a_0\rangle_n \cdots |a_{i-1}\rangle\, U_m \mathsf{CNOT}_n(|a_i\rangle_n |0\rangle_n) |0\rangle_n \cdots |0\rangle_n \\
&= |a_0\rangle_n \cdots |a_{i-1}\rangle\, U_m(|a_i\rangle_n |a_i\rangle_n) |0\rangle_n \cdots |0\rangle_n \\
&= |a_0\rangle_n \cdots |a_{i-1}\rangle |a_i\rangle_n |a_i^2 \bmod N\rangle_n |0\rangle_n \cdots |0\rangle_n \\
&= |a_0\rangle_n \cdots |a_{i-1}\rangle |a_i\rangle_n |a_{i+1}\rangle_n |0\rangle_n \cdots |0\rangle_n .
\end{aligned}
$$

This proves that U_p gives $|a_0\rangle \cdots |a_{n-1}\rangle$. We estimate the size of the circuit. There are n iterations of the **for** loop. Each iteration executes CNOT_n and U_m. Both operations have complexity $O(n^2)$. So the total size is $O(n^3)$. $\qquad \square$

Next, we construct a quantum circuit that implements the unitary operator C-U_a from (6.4.32) which is shown in Figure 6.4.3 for $n = 2$. Its initial state is $|c\rangle_2 |t\rangle_2$ where $c, t \in \mathbb{Z}_4$. Then the ancilla qubits $|a\rangle_2$, $|0\rangle_2$, and $|1\rangle_2$ are inserted between $|c\rangle_2$ and $|t\rangle_2$. The operator U_p is applied to $|a\rangle_2 |0\rangle_2$. This gives the state $|c\rangle_2 |a_0\rangle_2 |a_1\rangle_2 |1\rangle_2 |t\rangle_2$. Next, $|a_1\rangle_2$ and $|1\rangle_2$ are swapped conditioned on $c_0 = 1$. Therefore, the input state to U_m is $|a_1^{c_0}\rangle_2 |t\rangle_2$. Applying U_m gives the state $|c\rangle_2 |a_0\rangle_2 |a_1^{1-c_0}\rangle_2 |a_1^{c_0}\rangle |t_1\rangle_2$. The next controlled swap changes the three ancillary registers back to $|a_0\rangle_2 |a_1\rangle_2 |1\rangle_2$. The circuit then applies another controlled swap, the operator U_m and the inverse swap. The result is the state $|c\rangle_2 |a_0\rangle_2 |a_1\rangle_2 |1\rangle_2 |t_2\rangle_2$. Tracing out the ancillary qubits, we obtain the required result.

Algorithm 6.4.17. Circuit that implements C-U_a

Input: $|c\rangle_n |t\rangle_n \in \mathbb{H}_{2n}$
Output: $|c\rangle_n |t_n\rangle_n$

1:    U_e
2:      */ The circuit operates on $|\psi\rangle = |\psi_0\rangle \cdots |\psi_{n+2}\rangle \in \mathbb{H}_n^{\otimes n(n+3)}*/$
3:      $|\psi\rangle \leftarrow |c\rangle_n |a\rangle_n |0\rangle_n^{\otimes(n-1)} |1\rangle_n |t\rangle_n$
4:      $|\psi_1\rangle \cdots |\psi_n\rangle \leftarrow U_p |\psi_1\rangle \cdots |\psi_n\rangle = |a_0\rangle_n \cdots |a_{n-1}\rangle_n$
5:      **for** $i = 1$ to n **do**
6:          $|\psi\rangle \leftarrow \mathsf{CSWAP}_{i-1} |\psi\rangle$
7:          $|\psi_{n+1}\rangle |\psi_{n+2}\rangle \leftarrow U_m |\psi_{n+1}\rangle |\psi_{n+2}\rangle$
8:          $|\psi\rangle \leftarrow \mathsf{CSWAP}_{i-1} |\psi\rangle$
9:      **end for**
10:      Trace out $|\psi_n\rangle \cdots |\psi_{2n}\rangle$
11: **end**

The general circuit that implements C-U_a is specified in Algorithm 6.4.17. It uses the controlled swap operator CSWAP_i for $i = 0, \ldots, n-1$. Its effect is seen in (6.4.42):

$$\mathsf{CSWAP}_i |c\rangle_n |\xi_0\rangle \cdots |\xi_{n-1}\rangle |\varphi_0\rangle |\varphi_1\rangle$$

$$(6.4.42) \qquad = \begin{cases} |c\rangle_n |\xi_0\rangle \cdots |\xi_{n-1}\rangle |\varphi_0\rangle |\varphi_1\rangle & \text{if } c_i = 0, \\ |c\rangle_n |\xi_0\rangle \cdots |\xi_{n-i-1}\rangle |\varphi_0\rangle |\xi_{n-i+1}\rangle \cdots |\xi_{n-i}\rangle |\varphi_1\rangle & \text{if } c_i = 1. \end{cases}$$

So CSWAP_i exchanges $|\xi_{n-i}\rangle$ and $|\varphi_0\rangle$ conditioned on c_i being 1. For example, the circuit in Figure 6.4.3 uses CSWAP_1. It swaps $|a_1\rangle_n$ and $|1\rangle_n$ conditioned on c_0 being 1.

We now prove the following result.

Proposition 6.4.18. *The circuit specified in Algorithm 6.4.17 implements the unitary operator* C-U_a *from (6.4.32) and has size* $O(n^3)$.

Proof. We prove by induction on i that after executing i iterations of the **for** loop we have

$$(6.4.43) \qquad |\psi\rangle = |c\rangle_n |a_0\rangle_n \cdots |a_{n-1}\rangle_n |1\rangle_n |t_i\rangle_n .$$

Applying this for $i = n$ it follows that the circuit implements C-U_a.

The base case follows by considering the instructions in lines 3 and 4 and Proposition 6.4.16.

For the inductive step, assume that $0 \leq i < n$ and that (6.4.43) holds. In the $(i+1)$st iteration of the **for** loop the instruction in line 6 swaps $|a_{n-i-1}\rangle_n$ and $|1\rangle_n$ conditioned on c_{i+1} being 1. This means that after this operation, we have

$$(6.4.44) \qquad |\psi_n\rangle |\psi_{n+1}\rangle = \left| a_{n-i-1}^{c_{i+1}} \right\rangle |t_i\rangle_n .$$

So the application of U_m to this quantum state gives

$$(6.4.45) \qquad \left| a_{n-i-1}^{c_{i+1}} \right\rangle |t_{i+1}\rangle_n .$$

Another application of CSWAP_i swaps $|a_{i+1}\rangle_n$ and $|1\rangle_n$ conditioned on c_{i+1} being 1. So, (6.4.43) holds.

Finally, we estimate the size of the circuit. The number of ancilla bits required by the circuit is $O(n^2)$. By Proposition 6.4.16, the circuit U_p has size $O(n^3)$. The number of iterations of the **for** loop is n. We analyze the complexity of the implementation of CSWAP_i. As seen in Figure 4.5.1, one quantum swap can be implemented using $O(1)$ elementary gates. So, swapping n qubits can be achieved using $O(n)$ elementary gates. Theorem 4.12.7 implies that CSWAP_i can also be implemented using $O(n)$ elementary quantum gates. Also, U_m requires $O(n^2)$ elementary quantum gates. So, the complexity of the **for** loop is $O(n^3)$ which concludes the proof. $\square$

Now we can prove the complexity statement of Theorem 6.4.8 as follows. The order finding algorithm uses the quantum circuit Q_a twice. By assumption, the precision parameter n used in this circuit satisfies $2^n \leq 2N^2$, which implies $n = O(\log N)$. So, by Proposition 6.4.18, the size of this circuit is $O((\log N)^3)$. By Proposition A.3.28, the application of the continued fraction algorithm requires time $O((\log N)^2)$. Furthermore, the calculation of the lcm, $a^r \bmod N$, and all other operations requires running time $O((\log N)^3)$. This implies the complexity statement of Theorem 6.4.8. If faster algorithms for integer multiplication, division with remainder, and lcm are used, the complexity of the order finding algorithm becomes $(\log N)^2(\log \log N)^{O(1)}$. Such algorithms are, for example, presented in [**AHU74**] and [**HvdH21**].

6.5. Integer factorization

In this section, we explain how to solve the following problem in quantum polynomial time using the order finding algorithm from Section 6.4.

Problem 6.5.1 (Integer factorization problem).

Input: A composite positive integer

Output: A proper divisor d of N

The best-known classical and fully analyzed Monte Carlo algorithm for this problem has subexponential complexity $e^{(1+o(1))(\log N \log \log N)^{1/2}}$ [**LP92**]. Furthermore, the best heuristic Monte Carlo algorithm for this problem has subexponential complexity $e^{(c+o(1))(\log N)^{1/3}(\log \log N)^{2/3}}$ where $c = \sqrt[3]{64/9}$ [**BLP93**]. The quantum factoring algorithm is Algorithm 6.5.2. Its idea has already been explained in Section 6.1. It selects $a \in \mathbb{Z}_N$ randomly with the uniform distribution and computes $d = \gcd(a, N)$. If $d > 1$, then d is a proper divisor of N; the algorithm returns this divisor and terminates. Otherwise, the algorithm calls $\mathsf{FindOrder}(N, a, n)$ with n from (6.4.11): By Theorem 6.4.8 it finds the order r of a modulo N with probability at least 0.399. If r is even, then

$$(6.5.1) \qquad\qquad (a^{r/2} - 1)(a^{r/2} + 1) \equiv 0 \bmod N.$$

Also, if N does not divide $a^{r/2} + 1$, then $\gcd(a^{r/2} - 1, N)$ is a proper divisor of N. This is what the algorithm tests.

Algorithm 6.5.2. Factoring using order finding

Input: $n \in \mathbb{N}$, an odd composite $N \in \mathbb{N}$
Output: A proper divisor d of N

```
 1:  FACTOR(N)
 2:      a ← randomInt(N)
 3:      d ← gcd(a, N)
 4:      if d > 0 then
 5:          return (d)
 6:      end if
 7:      n ← ⌈2 log₂ N⌉ + 1
 8:      r ← FindOrder(N, a, n)
 9:      if r ≠ "FAILURE" and r ≡ 0 mod 2 then
10:          b ← a^{r/2} mod N
11:          d ← gcd(b − 1, N)
12:          if d ≠ 1 then
13:              return d
14:          end if
15:      end if
16:      return "FAILURE"
17:  end
```

We present examples for a successful and an unsuccessful run of Algorithm 6.5.2.

Example 6.5.3. Let $N = 15$. Suppose that the factoring algorithm selects $a = 2$. Then $\gcd(a, N) = 1$. As can be easily verified, the order of 2 modulo 15 is 4. Suppose that the order finding algorithm returns $r = 4$. Then it computes $b = a^{r/2} \bmod N = 2^2 \bmod 15 = 4$ and $d = \gcd(b - 1, N) = \gcd(4 - 1, 15) = 3$ which is a proper divisor of 15.

Let $N = 15$. Now suppose that the factoring algorithm selects $a = 14$. Then $\gcd(a, N) = 1$. As can be easily verified, the order of 14 modulo 15 is 2. Suppose that the order finding algorithm returns $r = 2$. Then it computes $b = a^{r/2} \bmod N = 14$ and $d = \gcd(1 - 1, 15) = 15$, which is not a proper divisor of 15.

In the remainder of this section, we will prove the following theorem.

Theorem 6.5.4. *On input of an odd composite number $N \in \mathbb{N}$, Algorithm 6.5.2 returns a proper divisor of N with probability at least 0.199 and has running time $O((\log N)^3)$.*

We note that by employing the technique outlined in Section 1.3.4, it is possible to approach a success probability of 1 with increasing precision without altering the asymptotic complexity. This achievement can be accomplished through iterative application of the algorithm.

Now we analyze the success probability and the running time of the quantum factoring algorithm. It is successful if the order finding algorithm returns the order r of a, this order is even, and $a^{r/2} \not\equiv -1 \bmod N$. To obtain a lower bound for the probability of this happening, we need the following lemma.

Lemma 6.5.5. *Let p be an odd prime number and let $e \in \mathbb{N}$. Let $d \in \mathbb{N}$ be the exponent of 2 in the prime factor decomposition of $p-1$. Choose an integer $a \in \mathbb{Z}^*_{p^e}$ randomly with the uniform distribution. Then the probability for 2^d to divide the order of a modulo p^e is $1/2$.*

Proof. The order of $\mathbb{Z}^*_{p^e}$ is

$$(6.5.2) \qquad\qquad \varphi = \varphi(p^e) = (p-1)p^{e-1}.$$

Since p is odd, the exponent of 2 in the prime factorization of φ is d. Choose a primitive root $g \in \mathbb{Z}^*_{p^e}$ modulo p^e. Its order modulo p^e is φ and

$$(6.5.3) \qquad\qquad \mathbb{Z}_\varphi \to \mathbb{Z}^*_{p^e}, \quad k \mapsto g^k \bmod p^e$$

is a bijection. Select $k \in \mathbb{Z}_\varphi$ randomly with the uniform distribution. Due to the bijectivity of (6.5.3), the integer $a = g^k \bmod p^e$ is uniformly distributed in $\mathbb{Z}^*_{p^e}$. Also, the order of a is

$$(6.5.4) \qquad\qquad r = \frac{\varphi}{\gcd(k, \varphi)}.$$

Let d' be the exponent of 2 in the prime factor decomposition of r. Then (6.5.4) implies that $d' = d$ if k is odd and $d' < d$ if k is even. Since φ is even, half of the integers in $\mathbb{Z}_\varphi$ are even and half are odd. So, we have $d = d'$ with probability $1/2$. $\qquad\square$

The next proposition gives a lower bound for the conditional success probability of Algorithm 6.5.2 in the case where it selects an integer a that is coprime to N and the order finding algorithm returns the order of a modulo N.

Proposition 6.5.6. *Let N be an odd composite number with m different prime factors. Choose $a \in \mathbb{Z}^*_N$ randomly with the uniform distribution. Then the probability that the order r of a modulo N is even and that $a^{r/2} \not\equiv -1 \bmod N$ is at least $1 - 1/2^{m-1}$.*

Proof. We show that the probability that r is odd or that r is even and satisfies $a^{r/2} \equiv -1 \bmod N$ is at most $1/2^{m-1}$.

Let

$$(6.5.5) \qquad\qquad N = \prod_{i=1}^m p_i^{e_i}$$

be the prime factor decomposition of N. To choose $a \in \mathbb{Z}^*_N$ randomly with uniform distribution, select $a_i \in \mathbb{Z}^*_{p_i^{e_i}}$ randomly with the uniform distribution and apply Chinese remaindering (see [**Buc04**, Section 2.15]) to find $a \in \mathbb{Z}^*_N$ with $a \equiv a_i \bmod p_i^{e_i}$ for $1 \le i \le m$.

Let r be the order of a modulo N and for $1 \le i \le m$ let r_i be the order of a_i modulo $p_i^{e_i}$. Then by Exercise A.4.17, we have

$$(6.5.6) \qquad\qquad r = \mathrm{lcm}(r_1, \ldots, r_n).$$

Denote by f the exponent of 2 in the prime factor decomposition of r and denote by f_i the exponent of 2 in the prime factor decomposition of r_i modulo $p_i^{e_i}$. We show that if r is odd or r is even and satisfies $a^{r/2} \equiv -1 \bmod N$, then

$$(6.5.7) \qquad\qquad f = f_i \quad \text{for } 1 \le i \le m.$$

So let $i \in \{1, \ldots, m\}$. If r is odd, then $f = 0$ and (6.5.6) implies $f_i = 0$. Now assume that $f > 0$ and $a^{r/2} \equiv -1 \bmod N$. Since $a_i^r \equiv 1 \bmod p_i^{e_i}$, it follows that $r_i | r$ which implies $f_i \le f$. But since $a_i^{r/2} \equiv -1 \bmod p_i^{e_i}$, it follows that $f_i = f$.

From the above argument, it follows that the probability that r is odd or that r is even and satisfies $a^{r/2} \equiv -1 \bmod N$ is at most the probability that all f_i are equal. We show that this probability is at most $1/2^{m-1}$. We know that f_1 assumes some value with probability 1. It follows from Lemma 6.5.5 that for $2 \le i \le m$ we have $f_i = f_1$ with probability at most $1/2$. In fact, if f_1 is the exponent of 2 in the prime factorization of $p_i - 1$, then by Lemma 6.5.5 the probability that $f_i = f_1$ is $1/2$. And if f_1 is not this exponent, then by Lemma 6.5.5 the probability that $f_i = f_1$ is at most $1 - 1/2 = 1/2$. So the probability that all f_i are equal is at most $1/2^{m-1}$. $\qquad\square$

We can now prove Theorem 6.5.4.

The algorithm is at least in the following two cases successful. In the first case a is not coprime to N which occurs with probability $1 - 1/\varphi(N)$. In the second case (1) the algorithm finds the order r of a modulo N and (2) this order is even and $a^{r/2} \not\equiv -1 \bmod N$. By Theorem 6.4.8, (1) occurs with probability at least $\frac{3958924}{101761\pi^4}$. Furthermore, by Proposition 6.5.6, (2) happens with conditional probability at least $\frac{1}{2}$ since N has at least two prime factors. So the probability of (1) and (2) is at least

$$(6.5.8) \qquad\qquad \frac{3958924}{2 * 101761\pi^4} > 0.199.$$

Summarizing these results, the total success probability is at least

$$(6.5.9) \qquad\qquad \frac{N - \varphi(N) + 0.199\varphi(N)}{N} > 0.199.$$

By Theorem 6.4.8 and the choice of n in Algorithm 6.5.2, the running time of the algorithm is $O((\log N)^3)$.

Exercise 6.5.7. Find a polynomial time quantum algorithm that finds the prime factor decomposition of every positive integer.

Exercise 6.5.8. Let $N \in \mathbb{N}$, $a \in \mathbb{Z}_N^*$, and $r \in \mathbb{Z}_N$. Show how the quantum factoring algorithm can be used to check in polynomial time whether r is the order of a modulo N.

6.6. Discrete logarithms

In this section, we show how the order finding algorithm and the phase estimation algorithm can be used to solve the following problem in quantum polynomial time.

Problem 6.6.1 (Discrete Logarithm Problem).

Input: An odd integer N, $a, b \in \mathbb{Z}_N^*$ such that $b \equiv a^t \bmod N$ where $t \in \mathbb{Z}_r$ and r is the order of a modulo N

Output: The exponent t

The exponent t in the discrete logarithm problem is called the *discrete logarithm of b to base a modulo N*. The discrete logarithm problem is also referred to as the *DL problem*.

We will now present a quantum polynomial time DL algorithm. In the presentation, we will use the notation from the discrete logarithm problem.

We make a few preliminary remarks. Using the quantum factoring algorithm from the previous section, we can find the prime factorization of N in polynomial time, allowing us to compute the order $\varphi(N)$ of $\mathbb{Z}_N^*$ since

$$(6.6.1) \qquad\qquad \varphi(N) = N \prod_{p \mid N} \left(1 - \frac{1}{p} \right)$$

which is shown in [**Buc04**, Theorem 2.17.2]. By applying the quantum factoring algorithm, we can also determine the prime factorization of $\varphi(N)$. Subsequently, the *Pohlig-Hellman algorithm*, as described in Section 10.5 of [**Buc04**], provides a polynomial time reduction from the general DL problem to the problem of computing discrete logarithms of basis elements a whose order r is a known prime number. Therefore, to achieve a quantum polynomial time DL algorithm, we may assume that the order r of a modulo N is a prime number. Additionally, we assume that $t > 1$, as the cases $t = 0$ and $t = 1$ can be solved by inspection.

The idea of the quantum DL algorithm for this special case is the following. The algorithm selects an appropriate precision parameter $n \in \mathbb{N}$ and uses the unitary operators U_a and U_b that are specified in Definition 6.4.2. Since $b \equiv a^t \bmod N$, it follows from Proposition 6.4.4 that the eigenvalues of these operators are $e^{2\pi i \frac{k}{r}}$ and $e^{2\pi i \frac{tk}{r}}$ for $0 \le k < r$. Using quantum phase estimation, we can find $(x, y) \in \mathbb{Z}_{2^n}^2$ such that $\frac{x}{2^n}$ and $\frac{y}{2^n}$ are close enough to $\frac{k}{r}$ and $\frac{tk \bmod r}{r}$ for some $k \in \mathbb{Z}_r^*$ such that

$$(6.6.2) \qquad\qquad k = \left\lfloor \frac{rx}{2^n} \right\rceil \quad \text{and} \quad kt \bmod r = \left\lfloor \frac{ry}{2^n} \right\rceil.$$

Since r is a prime number, it follows that $\gcd(k, r) = 1$. So we can compute k' with $kk' \equiv 1 \bmod r$ and obtain

$$(6.6.3) \qquad\qquad t = k' \left\lfloor \frac{ry}{2^n} \right\rceil \bmod r.$$

The quantum discrete logarithm is Algorithm 6.6.2. In the remainder of this section, we will prove Theorem 6.6.3 which states that it is correct and runs in polynomial time.

Algorithm 6.6.2. Quantum discrete logarithm algorithm

Input: N, a, b, r such that r is the order of a modulo N, r is a prime number, and
$b \equiv a^t \bmod N$ for some $t \in \mathbb{Z}_r^*$.

Output: The discrete logarithm t of b to base a modulo N or "FAILURE"

1: $DL(N, a, b, r)$
2:      $n \leftarrow \lceil \log_2 r \rceil + 1$
3:      Apply the quantum circuit Q_{DL} from Figure 6.6.1 and obtain $(x, y) \in \mathbb{Z}_{2^n}^2$
4:      $k \leftarrow \lfloor xr/2^n \rceil \bmod r$
5:      **if** $k \neq 0$ **then**
6:          $l \leftarrow \lfloor yr/2^n \rceil \bmod r$
7:          $t \leftarrow lk^{-1} \bmod r$
8:          **return** t
9:      **end if**
10:     **return** "FAILURE"
11: **end**

Theorem 6.6.3. *On input of $N \in \mathbb{N}$, $a, b \in \mathbb{Z}_N^*$ such that the order of a modulo N is a prime number r and $b \equiv a^t \bmod N$ for some $t \in \mathbb{Z}_r^*$, Algorithm 6.6.2 returns t with probability at least $64(r-1)/r\pi^4 > 0.328$. Its running time is $O((\log N)^3)$.*

Algorithm 6.6.2 sets

$$(6.6.4) \qquad\qquad n = \lceil \log_2 \rceil + 1.$$

This implies

$$(6.6.5) \qquad\qquad 2^n > 2r.$$

Then it applies the quantum circuit from Figure 6.6.1. The next proposition describes the output of the circuit.

Proposition 6.6.4. *For all $k \in \mathbb{Z}_r^*$ the quantum circuit in Figure 6.6.1 gives with probability $64(r-1)/r\pi^4$ two integers $x, y \in \mathbb{Z}_{2^n}$ such that*

$$(6.6.6) \qquad\qquad k = \left\lfloor \frac{rx}{2^n} \right\rceil \quad and \quad kt \bmod r = \left\lfloor \frac{ry}{2^n} \right\rceil.$$

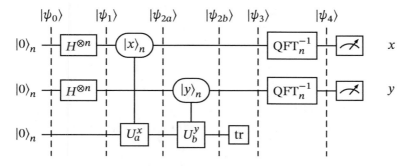

Figure 6.6.1. The quantum circuit Q_{DL} for discrete logarithm computation.

Proof. The circuit operates on the tensor product of three quantum registers of length n each of which is initialized to

(6.6.7)
$$|\psi_0\rangle = |0\rangle_n |0\rangle_n |1\rangle_n.$$

Then it applies $H^{\otimes n}$ to the first two quantum registers. This gives the state

(6.6.8)
$$\left(\frac{|0\rangle + |1\rangle}{\sqrt{2}} \right)^{\otimes n} \left(\frac{|0\rangle + |1\rangle}{\sqrt{2}} \right)^{\otimes n} |1\rangle_n.$$

Next, it applies the operator C-U_a from (6.4.27) to the first and the third quantum register. As in (6.4.17) this gives the state

(6.6.9)
$$|\psi_{2a}\rangle = \frac{1}{\sqrt{2^n r}} \sum_{k=0}^{r-1} \left(\left| \psi_n \left(\frac{k}{r} \right) \right\rangle \left(\sum_{y=0}^{2^n - 1} |y\rangle \right) |u_k\rangle \right).$$

Then the algorithm applies the operator C-U_b to the second and third quantum register. It follows from Proposition 6.4.4 that the $|u_k\rangle$ are eigenstates of U_b associated to the eigenvalues $e^{2\pi i \frac{kt}{r}}$. So, by the same argument, this gives the state

(6.6.10)
$$|\psi_{2b}\rangle = \frac{1}{r} \sum_{k=0}^{r-1} \left(\left| \psi_n \left(\frac{k}{r} \right) \right\rangle \left| \psi_n \left(\frac{kt}{r} \right) \right\rangle |u_k\rangle \right).$$

By Exercise 3.7.11, tracing out the third quantum register gives the mixed state

(6.6.11)
$$|\psi_3\rangle = \left(\left(\frac{1}{r}, \left| \psi_n \left(\frac{k}{r} \right) \right\rangle \left| \psi_n \left(\frac{kt}{r} \right) \right\rangle \right) \right)_{0 \le k < r}.$$

Now the algorithm applies QFT_n^{-1} to the first and second register. This gives the mixed state

(6.6.12)
$$|\psi_4\rangle = \left(\left(\frac{1}{r}, \mathrm{QFT}_n^{-1} \left| \psi_n \left(\frac{k}{r} \right) \right\rangle \mathrm{QFT}_n^{-1} \left| \psi_n \left(\frac{kt}{r} \right) \right\rangle \right) \right)_{0 \le k < r}.$$

Let $k \in \mathbb{Z}_r$. By Theorem 6.3.7 and (6.6.5) measuring these registers in the computational basis of $\mathbb{H}_n^{\otimes 2}$ gives with probability $64/\pi^4 r$ two integers $(x, y) \in \mathbb{Z}_{2^n}^2$ with

(6.6.13)
$$\left| \Delta \left(\frac{k}{r}, n, x \right) \right| < \frac{1}{2^n} < \frac{1}{2r}, \quad \left| \Delta \left(\frac{kt \bmod r}{r}, n, y \right) \right| < \frac{1}{2^n} < \frac{1}{2r}.$$

We note that

(6.6.14)
$$0 < \frac{k}{r} \le 1 - \frac{1}{r} < 1 - \frac{1}{2^n}, \quad \frac{kt \bmod r}{r} \le 1 - \frac{1}{r} < 1 - \frac{1}{2^n}.$$

So Lemma 6.3.3 and (6.6.13) imply

(6.6.15)
$$\left| \frac{k}{r} - \frac{x}{2^n} \right| < \frac{1}{2r}, \quad \left| \frac{kt \bmod r}{r} - \frac{y}{2^n} \right| < \frac{1}{2r}.$$

Consequently, we obtain

(6.6.16)
$$k = \left\lfloor \frac{rx}{2^n} \right\rceil \quad \text{and} \quad kt \bmod r = \left\lfloor \frac{ry}{2^n} \right\rceil$$

which concludes the proof. $\square$

It follows from Proposition 6.6.4 that with probability at least $64(r-1)/r\pi^4 \ge 32/\pi^4 > 0$ the quantum circuit in Figure 6.6.1 returns $x, y \in \mathbb{Z}_{2^n}$ that satisfy (6.6.6) for some $k \in \mathbb{Z}_r^*$. If this happens, then the algorithm computes $l = \lfloor yr/2^n \rceil$ which by Proposition 6.6.4 is $kt \bmod r$. So we have $t = lk^{-1} \bmod r$ which means that the algorithm produces the correct result. As in the analysis of the order finding algorithm, it can be seen that the complexity of the algorithm is $O((\log N)^3)$.

6.6.1. Discrete logarithms in other groups.
The presented quantum polynomial time algorithm effectively calculates the discrete logarithm in the multiplicative group modulo a positive integer. However, as it became apparent that this problem serves as the foundation for the security of public-key cryptography, a multitude of public-key cryptography algorithms emerged, which hinge upon the discrete logarithm problem in various groups. Notably, elliptic curve cryptography stands out from an application perspective. Its security relies on the intractability of computing discrete logarithms in the group of points of an elliptic curve over a finite field. Nevertheless, for all variants of the discrete logarithm problem, polynomial time quantum algorithms were uncovered. Consequently, none of the cryptographic algorithms based on discrete logarithms offer security against attacks from quantum computers. An exhaustive overview of these algorithms can be found in [**Jor**].

6.7. Relevance for cryptography

The discovery of Shor's algorithms, capable of factoring integers and computing discrete logarithms in polynomial time, has had a significant impact on cybersecurity. Traditional public-key cryptography, including the RSA public-key encryption and signature schemes and the digital signature algorithm, as described in [**Buc04**], becomes insecure in the face of these algorithms. These cryptographic systems are critical for cybersecurity, especially to secure the Internet.

Furthermore, variants of the Shor algorithm can also solve other discrete logarithm problems, such as those in elliptic curve groups over finite fields. As a result, all classical public-key cryptography becomes vulnerable when faced with sufficiently large quantum computers. In response, scientists have been actively working on developing post-quantum cryptography (see [**BLM17**], [**BLM18**]) that can resist quantum computer attacks.

6.8. The hidden subgroup problem

In this section, we discuss the hidden subgroup problem, which serves as a framework for addressing algorithmic problems in finite abelian groups. We will show that the algorithmic problems that we have studies so far can be viewed as hidden subgroup problems. However, there are several more instances of the hidden subgroup problem. For an overview see [**Wan10**].

6.8.1. The problem. To state the hidden subgroup problem, we need the following definition.

Definition 6.8.1. Let G be a group, let H be a subgroup of G, and let X be a set. We say that a function $f : G \to X$ hides the subgroup H if for all $g, g' \in G$ we have $f(g) = f(g')$ if and only if $gH = g'H$. In other words, the function f takes the same value for all elements of a coset of H, while it takes different values for elements of different cosets.

The hidden subgroup problem is the following.

Problem 6.8.2 (Hidden subgroup problem).

Input: A black-box that implements a function $f : G \to X$ where G is a group, X is a set, and f hides a finitely generated subgroup H of G.

Output: A finite generating system for H.

6.8.2. Hidden subgroup versions of the Deutsch and Simon problems. The Deutsch problem explained in Section 5.1 is to find out whether a function $f : \{0, 1\} \to \{0, 1\}$ is constant or balanced. So we can set $G = (\{0, 1\}, \oplus), X = \{0, 1\}$ and use the function f. If f is constant, then $f(x)$ is the same for all elements of G. Therefore, the function f hides $H = G$. Every generating system of the subgroup H contains the element 1. If f is balanced, then $f(0)$ and $f(1)$ are different. So f hides the subgroup $H = \{0\}$ with the cosets $0 \oplus H = \{0\}$ and $1 \oplus H = \{1\}$. In this case, the only generating system of H is (0). So we can tell whether f is constant or balanced by checking whether the generating system of H contains 1.

Exercise 6.8.3. Show that the Deutsch-Jozsa problem can be viewed as a hidden subgroup problem by finding G, X, f, and H as in Definition 6.8.1, and show that finding the hidden subgroup is equivalent to solving the Deutsch-Jozsa problem.

In Simon's problem from Section 5.4, a black-box implementation of a function $f : \{0, 1\}^n \to \{0, 1\}^n$ is given with the property that there is $\vec{s} \in \{0, 1\}^n, \vec{s} \neq 0$, such that for all $\vec{x}, \vec{y} \in \{0, 1\}^n$ we have $f(\vec{x}) = f(\vec{y})$ if and only if $\vec{y} = \vec{x} \oplus \vec{s}$. The task is to find $\vec{s}$. In the hidden subgroup version of Simon's problem, we can set $G = (\{0, 1\}^n, \oplus)$, $X = \{0, 1\}^n$ and use f from the original problem. The hidden subgroup is $H = \{\vec{0}, \vec{s}\}$. Its generating systems are $(\vec{s})$ and $(\vec{0}, \vec{s})$. Hence, if we find such a generating system, then we have solved Simon's problem. Conversely, the solution of Simon's problem gives a generating system of the hidden subgroup.

Exercise 6.8.4. Show that the generalization of Simon's problem can be viewed as a hidden subgroup problem by finding G, X, f, and H as in Definition 6.8.1, and show that finding the hidden subgroup is equivalent to solving the generalization of Simon's problem.

6.8.3. Hidden subgroup version of the order finding problem.

In the order finding problem from Section 6.4, an odd integer $N \in \mathbb{N}_{\geq 3}$ and $a \in \mathbb{Z}_N$ are given such that $\gcd(a, N) = 1$. The problem is to find the order r of a modulo N. To frame this problem as a hidden subgroup problem, we set $G = (\mathbb{Z}, +)$, $X = \mathbb{Z}_N$, $f : G \to X$, $j \mapsto a^j \bmod N$. The hidden subgroup is $H = r\mathbb{Z}$. It has the property that for all $j \in \mathbb{Z}$ we have $a^j \equiv 1 \bmod N$ if and only if $j \in H$.

We show that the problem of finding the order r of a modulo N is equivalent to finding a finite generating system of H. First, we note that if we find the order r of a modulo N, then we know the generating system (r) of H. To prove the converse, we need the following result.

Lemma 6.8.5. Let $r \in \mathbb{N}$ and let $m \in \mathbb{N}$ and $G = (r_0, \ldots, r_{m-1}) \in \mathbb{Z}^m$. Then G is a generating system of $r\mathbb{Z}$ if and only if $\gcd(r_0, \ldots, r_{m-1}) = r$.

Proof. The lemma follows from Theorem 1.7.5 in [**Buc04**]. $\square$

From Lemma 6.8.5 it follows that r can be determined as the gcd of every generating system of $r\mathbb{Z}$. Hence, finding a finite generating system of $H = r\mathbb{Z}$ allows one to find r.

6.8.4. Hidden subgroup version of the discrete logarithm problem.

Next, we show that the discrete logarithm problem can also be viewed as a hidden subgroup problem. In the discrete logarithm problem as we have phrased it above, we are given an odd integer N and $a, b \in \mathbb{Z}_N^*$ such that $b \equiv a^t$ for some $t \in \mathbb{Z}_r$ where r is the order of a modulo N which is also known. The task is to find the discrete logarithm t of b to the base a. We set $G = (\mathbb{Z}_r^2, +)$, $X = \mathbb{Z}_N$, and $f : G \to X$, $(x, y) \mapsto a^x b^y$. The hidden subgroup is $H = \mathbb{Z}_r(1, -t)$. We show that finding a finite generating system of H is equivalent to determining t. For this, we need the following lemma.

Lemma 6.8.6. Let $r, t, m \in \mathbb{N}$ and let $((x_0, y_0), \ldots, (x_{m-1}, y_{m-1}))$ be a generating system of $H = \mathbb{Z}_r(1, -t)$ where $(x_i, y_i) \in \mathbb{Z}_r^2$ for all $i \in \mathbb{Z}_m$. Then $d = \gcd(x_0, \ldots, x_{m-1})$ is coprime to r and $t = d' \gcd(y_0, \ldots, y_{m-1}) \bmod r$ where d' is the inverse of d modulo r.

Proof. Since $((x_0, y_0), \ldots, (x_{m-1}, y_{m-1}))$ is a generating system of $\mathbb{Z}_r(1, t)$, it follows that there are $u_0, \ldots, u_{m-1} \in \mathbb{Z}$ with

$$(6.8.1) \qquad \sum_{i=0}^{m-1} u_i(x_i, y_i) \equiv (1, -t) \bmod r.$$

Since $(x_i, y_i) \in \mathbb{Z}_r(1, -t)$ for all $i \in \mathbb{Z}_r$, (6.8.1) implies

$$(6.8.2) \qquad \sum_{i=0}^{m-1} u_i x_i(1, -t) \equiv (1, -t) \bmod r.$$

Let $d = \gcd(x_0, \ldots, x_{m-1})$ and let $x_i' = x_i/d$ for all $i \in \mathbb{Z}_r$. Then we obtain from (6.8.2)

$$(6.8.3) \qquad\qquad d \sum_{i=0}^{m-1} u_i x_i' \equiv 1 \quad \mod r.$$

So d is coprime to r. Since $y_i \equiv -tx_i \mod r$ for all $i \in \mathbb{Z}_m$, it follows that

$$(6.8.4) \qquad\qquad \gcd(y_0, \ldots, y_{m-1}) \equiv dt \mod r.$$

So if d' is the inverse of d modulo r, then we have

$$(6.8.5) \qquad\qquad d' \gcd(y_0, \ldots, y_{m-1}) \equiv t \mod r$$

which implies the assertion of the lemma. $\square$

Lemma 6.8.6 shows how to obtain the discrete logarithm t of b modulo N to base a from a generating system $((x_0, y_0), \ldots, (x_{m-1}, y_{m-1}))$ of $\mathbb{Z}_r(1, -t)$. We compute $d = \gcd(x_0, \ldots, x_{m-1})$, find the inverse d' of d modulo r, and determine $t = d' \gcd(y_0, \ldots, y_{m-1})$. Conversely, if we find the discrete logarithm t, then we know the generation system $(1, -t)$ of $\mathbb{Z}_r(1, -t)$.

Chapter 7

Quantum Search and Quantum Counting

This chapter explores the renowned search algorithm developed by Lov Grover [**Gro96**] and the related counting algorithms devised by Gilles Brassard, Peter Høyer, and Alain Tapp [**BHT98**]. These algorithms provide a quadratic acceleration of classical algorithms for unstructured search and counting problems. This refers to situations in which we are given black-box access to a function $f : \{0, 1\}^n \to \{0, 1\}$ and our objective is to discover an input $\vec{x} \in \{0, 1\}^n$ that satisfies $f(\vec{x}) = 1$ or to count the number of such inputs. Given the broad applicability of this problem, the algorithms elucidated in this section find utility across diverse fields, including cryptography and machine learning, and effectively amplify the efficiency of existing algorithms in these contexts.

The initial section of this chapter focuses on Grover's search algorithm. It shows that addressing the search problem necessitates only the measurement of a quantum state, which can be effectively prepared. However, the success probability of this approach proves to be inadequate. Here the crucial technique of amplitude amplification steps in. This technique serves to enhance the likelihood of obtaining a solution from a measurement, thus achieving quadratic acceleration. In the subsequent part of the chapter, the synergy between amplitude amplification and phase estimation, introduced in the preceding chapter, is explored. This synergy yields a quantum counting algorithm that, under certain conditions, also delivers a quadratic speedup.

In the complexity analyses of this chapter, we assume that all quantum circuits are constructed using the elementary quantum gates provided by the platform discussed in Section 4.12.2, along with implementations of operators U_f as specified in their respective contexts.

7.1. Quantum search

In this section, we present Grover's quantum search algorithm, featuring the fundamental technique of amplitude amplification, which is employed in numerous other quantum algorithms.

7.1.1. The classical search problem.

The algorithm of Grover solves the unstructured search problem. The classical version of this problem is as follows.

Problem 7.1.1 (Classical search problem).

Input: $n \in \mathbb{N}$ and a black-box that implements a function $f : \{0, 1\}^n \to \{0, 1\}$.

Output: A string $\vec{x} \in \{0, 1\}^n$ with $f(\vec{x}) = 1$.

This problem appears in innumerable applications. For example, we can think of $\{0, 1\}^n$ as the set of addresses of the entries in a database and we can think of the function f as marking the addresses x of the entries that meet a search criterion. This criterion is satisfied if $f(x) = 1$ and is not satisfied if $f(x) = 0$.

There are several variants of the search problem. For example, we may require that the search problem has a solution, "No Solution" may be a permitted output, and the number of solutions may be an input.

Requiring the domain of the function f to be $\{0, 1\}^n$ is not a restriction. To see this, identify $\{0, 1\}^n$ with $\mathbb{Z}_{2^n}$ as explained in Example 2.1.5. For any $N \in \mathbb{N}$ and any function $f : \mathbb{Z}_N \to \{0, 1\}$ the domain $\mathbb{Z}_N$ can be simply extended to the larger domain $\mathbb{Z}_{2^n}$ where $n = \lceil \log_2 N \rceil$ by assigning the value 0 to all inputs outside of $\mathbb{Z}_N$. Since $|\{0, 1\}^n| < 2N$, the extended search space is smaller than twice the size of the original search space.

On a classical computer, solving the search problem requires conducting 2^n evaluations of the function f in the worst case. However, as we will see in the sequel, Grover's algorithm is significantly faster, as it offers a quadratic speedup.

This has many applications. For example, Grover's algorithm can be used to search for a secret key in a symmetric encryption scheme, such as AES (e.g., see [**GLRS16**]). The quadratic speedup leads to the need to double the key length as protection against quantum computer attacks. Grover's algorithm can also be used to crack passwords stored in hashed form (e.g., see [**DGM+21**]). Another application of Grover's algorithm is quantum machine learning (e.g., see [**BWP+17**]).

7.1.2. The Grover search algorithm with a known number of solutions.

In this section, we present the first quantum search algorithm that requires the number of solutions of the search problem to be known. As in all quantum search problems presented in this chapter, the function f is replaced by the unitary operator

$$(7.1.1) \qquad\qquad U_f : \mathbb{H}_n \otimes \mathbb{H}_1 \to \mathbb{H}_n \otimes \mathbb{H}_1$$

that has been introduced in (5.3.2). For all $\vec{x} \in \{0, 1\}^n$ and $y \in \{0, 1\}$ it satisfies

$$(7.1.2) \qquad\qquad U_f |\vec{x}\rangle |y\rangle = |\vec{x}\rangle |f(\vec{x}) \oplus y\rangle = |\vec{x}\rangle X^{f(\vec{x})} |y\rangle .$$

The problem we aim to solve is the following.

Problem 7.1.2 (Quantum search problem with a known number of solutions).

Input: $n \in \mathbb{N}$, a black-box that implements U_f for a function $f : \{0,1\}^n \to \{0,1\}$, $M = |f^{-1}(1)|$. It is assumed that $M > 0$.

Output: A string $\vec{x} \in \{0,1\}^n$ with $f(\vec{x}) = 1$.

In the explanation of the Grover algorithm that solves Problem 7.1.2, we use the notation and assumptions of this problem and set $N = 2^n$.

First, we explain how the search problem can be solved by measuring the quantum state

$$(7.1.3) \qquad |s\rangle = \frac{1}{\sqrt{N}} \sum_{\vec{x} \in \{0,1\}^n} |\vec{x}\rangle.$$

Set

$$(7.1.4) \qquad |s_0\rangle = \frac{1}{\sqrt{N}} \sum_{\vec{x} \in \{0,1\}^n,\, f(\vec{x})=0} |\vec{x}\rangle, \quad |s_1\rangle = \frac{1}{\sqrt{N}} \sum_{\vec{x} \in \{0,1\}^n,\, f(\vec{x})=1} |\vec{x}\rangle,$$

and

$$(7.1.5) \qquad \theta = \arcsin \sqrt{\frac{M}{N}}.$$

Then the following proposition holds.

Proposition 7.1.3. *We have*

$$(7.1.6) \qquad |s\rangle = \sqrt{\frac{N-M}{N}} \, |s_0\rangle + \sqrt{\frac{M}{N}} \, |s_1\rangle = \cos\theta \, |s_0\rangle + \sin\theta \, |s_1\rangle.$$

Exercise 7.1.4. Prove Proposition 7.1.3.

If the quantum state $|s\rangle$ is measured in the computational basis of $\mathbb{H}_n$, then by Proposition 7.1.3, the probability of obtaining $\vec{x}$ with $f(\vec{x}) = 1$ is

$$(7.1.7) \qquad p = \sin^2 \theta = \frac{M}{N}.$$

This is exactly the probability of correctly guessing a solution to the search problem, and therefore this simple quantum strategy does not provide an advantage over classical solutions. To enhance the success probability of the quantum strategy, we employ a technique called *amplitude amplification* which significantly increases the amplitude of the state $|s_1\rangle$ and therefore the probability of finding a solution. Amplitude amplification uses a unitary operator G, called the *Grover iterator*. For every $\alpha \in \mathbb{R}$ it satisfies

$$(7.1.8) \qquad G(\cos\alpha \, |s_0\rangle + \sin\alpha \, |s_1\rangle) = \cos(\alpha + 2\theta) \, |s_0\rangle + \sin(\alpha + 2\theta) \, |s_1\rangle.$$

So, it follows from (7.1.6) that for every $k \in \mathbb{N}_0$ we have

$$(7.1.9) \qquad G^k \, |s\rangle = \cos(2k + 1)\theta \, |s_0\rangle + \sin(2k + 1)\theta \, |s_1\rangle.$$

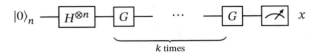

Figure 7.1.1. The quantum circuit for Grover's search algorithm.

The Grover quantum search algorithm is shown in Figure 7.1.1 and Algorithm 7.1.6. Its input is the state $|0\rangle_n$. Then the algorithm constructs

$$(7.1.10) \qquad\qquad |s\rangle = \mathbb{H}^{\otimes n} |0\rangle_n .$$

This equation holds by Lemma 5.3.6. Subsequently, the algorithm applies G^k to $|s\rangle$ and measures the resulting quantum state in the computational basis of $\mathbb{H}_n$. The number k is chosen so that $2(k + 1)\theta$ is as close as possible to $\frac{pi}{2}$. By (7.1.9), this maximizes the probability that the algorithm finds $\vec{x} \in \{0, 1\}^n$ with $f(\vec{x}) = 1$. In Theorem 7.1.21 we will estimate the number k of applications of G required in the search algorithm and the success probability of the algorithm. Before we state and prove this theorem, we construct the Grover iterator in the next section. Note that Algorithm 7.1.6 receives as input the black-box implementing U_f but applies the Grover iterator G. In Section 7.1.4, we will explain how G can be efficiently implemented using U_f.

7.1.3. The Grover iterator. We now explain the construction of the Grover iterator and prove its properties.

Definition 7.1.5. We define the following operators on $\mathbb{H}_n$:

$$(7.1.11) \qquad\qquad U_1 = I - 2 |s_1\rangle\langle s_1| , \quad U_s = 2 |s\rangle\langle s| - I$$

where I denotes the identity operator on $\mathbb{H}_n$. Then the *Grover iterator* is defined as the operator

$$(7.1.12) \qquad\qquad G = U_s U_1.$$

Algorithm 7.1.6. Grover algorithm for search problems with known number of solutions

---

Input: $n \in \mathbb{N}$, a black-box implementing U_f for some $f : \{0, 1\}^n \to \{0, 1\}$, and $M = |f^{-1}(1)|$. It is assumed that $M > 0$.
Output: $x \in \{0, 1\}^n$ such that $f(x) = 1$
   1:   QSEARCH(n, U_f, M)
   2:     $k \leftarrow \left\lfloor \frac{\pi}{4\theta} \right\rfloor$ where $\theta = \arcsin\left(\sqrt{\frac{M}{N}}\right)$
   3:     Apply the quantum circuit from Figure 7.1.1, the result being $\vec{x} \in \{0, 1\}^n$
   4: **end**

The operator U_s is sometimes also called the *Grover diffusion operator*. The next proposition states basic properties of the operators in Definition 7.1.5.

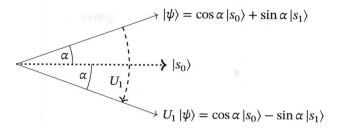

Figure 7.1.2. Applying U_1 to $|\psi\rangle = \cos\alpha\,|s_0\rangle + \sin\alpha\,|s_1\rangle$.

Proposition 7.1.7. (1) *The operators U_1 and U_s are unitary and Hermitian involutions.*

(2) *The Grover iterator G is unitary.*

Exercise 7.1.8. Prove Proposition 7.1.7.

Next, we present the geometric properties of U_1 and U_s. For this, we define the complex plane

$$(7.1.13) \qquad\qquad P = \mathbb{C}\,|s_0\rangle + \mathbb{C}\,|s_1\rangle.$$

We note that $(|s_0\rangle, |s_1\rangle)$ is an orthonormal basis of P. The next proposition states an important geometric property of U_1.

Proposition 7.1.9. *The operator U_1 acts as a reflection in the plane P across $|s_0\rangle$. In particular, for all $\alpha \in \mathbb{R}$ we have*

$$(7.1.14) \qquad U_1(\cos\alpha\,|s_0\rangle + \sin\alpha\,|s_1\rangle) = \cos\alpha\,|s_0\rangle - \sin\alpha\,|s_1\rangle.$$

Proposition 7.1.9 is illustrated in Figure 7.1.2 and proved in Exercise 7.1.10.

Exercise 7.1.10. Prove Proposition 7.1.9.

In order to describe the geometric meaning of U_s we define the quantum state

$$(7.1.15) \qquad\qquad |s^\perp\rangle = -\sin\theta\,|s_0\rangle + \cos\theta\,|s_1\rangle$$

which is in P and orthogonal to $|s\rangle$. We also define the matrix

$$(7.1.16) \qquad\qquad T = \begin{pmatrix} \cos\theta & -\sin\theta \\ \sin\theta & \cos\theta \end{pmatrix}.$$

As shown in Exercise 7.1.11, the matrix T is unitary, $(|s\rangle, |s^\perp\rangle)$ is another orthonormal basis of the plane P, and we have

$$(7.1.17) \qquad (|s\rangle, |s^\perp\rangle) = (|s_0\rangle, |s_1\rangle)\,T, \qquad (|s_0\rangle, |s_1\rangle) = (|s\rangle, |s^\perp\rangle)\,T^*.$$

Exercise 7.1.11. (1) Show that the matrix T is unitary.

(2) Prove that $(|s_1\rangle, |s_0\rangle)$ and $(|s\rangle, |s^\perp\rangle)$ are orthonormal bases of the plane P and verify (7.1.17).

The next proposition is the desired geometric interpretation of the operator U_s.

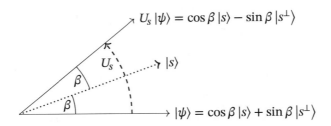

Figure 7.1.3. Applying U_s to $|\psi\rangle = \cos\beta\,|s\rangle + \sin\beta\,|s^\perp\rangle$.

Proposition 7.1.12. *The operator U_s acts as a reflection in the plane P across $|s\rangle$. In particular, for all $\alpha \in \mathbb{R}$ we have*

$$(7.1.18) \qquad U_s\left(\cos\alpha\,|s\rangle + \sin\alpha\,|s^\perp\rangle\right) = \cos\alpha\,|s\rangle - \sin\alpha\,|s^\perp\rangle.$$

The proposition is illustrated in Figure 7.1.3 and proved in Exercise 7.1.13.

Exercise 7.1.13. Prove Proposition 7.1.12.

Let $\alpha \in \mathbb{R}$. Using Propositions 7.1.9 and 7.1.12 we can describe the action of the Grover iterator on a quantum state $|\psi\rangle = \cos\alpha\,|s_0\rangle + \sin\alpha\,|s_1\rangle$ geometrically. This is illustrated in Figure 7.1.4. Since applying U_1 to $|\psi\rangle$ means reflecting $|\psi\rangle$ across $|s_0\rangle$, the angle between $|s_0\rangle$ and $U_1\,|\psi\rangle$ is α mod 2π. So, the angle between $|s\rangle$ and $U_1\,|\psi\rangle$ is $\alpha + \theta$ mod 2π. Next, applying U_s to $U_1\,|\psi\rangle$ means reflecting $U_1\,|\psi\rangle$ across $|s\rangle$. So the angle between $G\,|\psi\rangle = U_s U_1\,|\psi\rangle$ and $|s\rangle$ is $\alpha + \theta$ mod 2π and the angle between $|s_0\rangle$ and $G\,|\psi\rangle$ is $\alpha + 2\theta$ mod 2π. So we have

$$(7.1.19) \qquad G\,|\psi\rangle = \cos(\alpha + 2\theta)\,|s_0\rangle + \sin(\alpha + 2\theta)\,|s_1\rangle.$$

To verify (7.1.19) also algebraically, we need one more proposition.

Proposition 7.1.14. *Let $\alpha \in \mathbb{R}$, then we have*

$$(7.1.20) \qquad \cos\alpha\,|s\rangle + \sin\alpha\,|s^\perp\rangle = \cos(\alpha + \theta)\,|s_0\rangle + \sin(\alpha + \theta)\,|s_1\rangle$$

and

$$(7.1.21) \qquad \cos\alpha\,|s_0\rangle - \sin\alpha\,|s_1\rangle = \cos(\alpha + \theta)\,|s\rangle - \sin(\alpha + \theta)\,|s^\perp\rangle.$$

Exercise 7.1.15. Prove Proposition 7.1.14.

Here is the algebraic verification of (7.1.19).

Proposition 7.1.16. *Let $\alpha \in \mathbb{R}$. Then*

$$(7.1.22) \qquad G\left(\cos\alpha\,|s_0\rangle + \sin\alpha\,|s_1\rangle\right) = \cos(\alpha + 2\theta)\,|s_0\rangle + \sin(\alpha + 2\theta)\,|s_1\rangle.$$

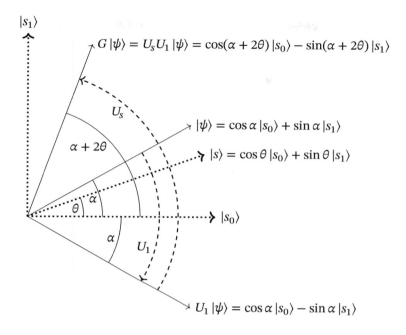

Figure 7.1.4. Applying the Grover iterator G to $|\psi\rangle = \cos\alpha\,|s_0\rangle + \sin\alpha\,|s_1\rangle$.

Proof. From Propositions 7.1.9, 7.1.12, and 7.1.14 we obtain

$$
\begin{aligned}
G\,(&\cos\alpha\,|s_0\rangle + \sin\alpha\,|s_1\rangle) \\
&= U_s U_1\,(\cos\alpha\,|s_0\rangle + \sin\alpha\,|s_1\rangle) \\
&= U_s(\cos\alpha\,|s_0\rangle - \sin\alpha\,|s_1\rangle) \\
&= U_s(\cos(\alpha+\theta)\,|s\rangle - \sin(\alpha+\theta)\,|s^\perp\rangle \\
&= \cos(\alpha+\theta)\,|s\rangle + \sin(\alpha+\theta)\,|s^\perp\rangle \\
&= \cos(\alpha+2\theta)\,|s_0\rangle + \sin(\alpha+2\theta)\,|s_1\rangle.
\end{aligned}
$$

(7.1.23)

$\square$

7.1.4. Implementation of the Grover iterator. The goal of this section is to show that the Grover G iterator can be efficiently implemented using the operator U_f. We use the quantum states $|s\rangle$, $|s^\perp\rangle$, $|s_0\rangle$, $|s_1\rangle$, the plane $P = \mathbb{C}\,|s\rangle + \mathbb{C}\,|s^\perp\rangle = \mathbb{C}\,|s_0\rangle + \mathbb{C}\,|s_1\rangle$, and the operators U_1 and U_s that were introduced in Section 7.1.3.

Figure 7.1.5 shows a quantum circuit that implements the operator U_1 on the plane P. Its correctness is stated in the next proposition.

Proposition 7.1.17. *The circuit in Figure 7.1.5 implements the operator U_1 in the plane P. It applies the black-box for U_f once and uses four additional elementary quantum gates.*

Proof. We will prove that the circuit computes $U_1\,|s_0\rangle$ and $U_1\,|s_1\rangle$ correctly. This suffices since U_1 is linear and $(|s_0\rangle, |s_1\rangle)$ is a basis of the plane P. Let $j \in \{0, 1\}$. First, we

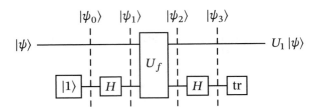

Figure 7.1.5. Implementation of U_1.

note that

$$U_1 |s_j\rangle = (I - 2 |s_1\rangle\langle s_1|) |s_j\rangle = (-1)^j |s_j\rangle,$$

$$U_f |s_j\rangle |0\rangle = \begin{cases} |s_j\rangle |0\rangle & \text{if } j = 0, \\ |s_j\rangle |1\rangle & \text{if } j = 1, \end{cases} \qquad U_f |s_j\rangle |1\rangle = \begin{cases} |s_j\rangle |1\rangle & \text{if } j = 0, \\ |s_j\rangle |0\rangle & \text{if } j = 1, \end{cases}$$

and therefore

$$(7.1.24) \quad U_f |s_j\rangle |x_-\rangle = \frac{U_f |s_j\rangle |0\rangle - U_f |s_j\rangle |1\rangle}{\sqrt{2}} = (-1)^j |s_j\rangle |x_-\rangle = (U_1 |s_j\rangle) |x_-\rangle.$$

This allows us to determine the intermediate states in the circuit. They are

$$|\psi_0\rangle = |b_j\rangle |1\rangle,$$
$$|\psi_1\rangle = |b_j\rangle |x_-\rangle,$$
$$|\psi_2\rangle = U_f |b_j\rangle |x_-\rangle = (U_1 |b_j\rangle) |x_-\rangle,$$
$$|\psi_3\rangle = (U_1 |b_j\rangle) |1\rangle.$$

This concludes the proof of the proposition.   □

Next, as Proposition 7.1.18 shows, Figure 7.1.6 shows an implementation of the operator $-U_s$. Since global phase factors do not change measurement outcomes, this is as good as implementing U_s.

Proposition 7.1.18. *The circuit in Figure 7.1.6 implements* $-U_s$ *on the plane P. It has size* $O(n)$.

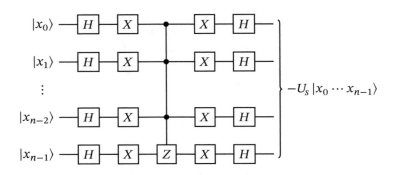

Figure 7.1.6. Implementation of $-U_s$.

Proof. By the definition of $|s\rangle$ and U_s we have

$$\text{(7.1.25)} \qquad U_s = H^{\otimes n}(2|0\rangle_n \langle 0|_n - I)H\otimes n.$$

Set

$$\text{(7.1.26)} \qquad V = 2|0\rangle_n \langle 0|_n - I.$$

To verify that the circuit implements $-U_s$, it suffices to show that

$$\text{(7.1.27)} \qquad V = -X^{\otimes n}C^{n-1}(Z)X^{\otimes n}.$$

In order to prove this equation, let $\vec{x} \in \{0,1\}^n$. Then

$$\text{(7.1.28)} \qquad V|\vec{x}\rangle = \begin{cases} -|\vec{x}\rangle & \text{if } \vec{x} \neq \vec{0}, \\ |\vec{x}\rangle & \text{if } \vec{x} = \vec{0} \end{cases}$$

We also have

$$\text{(7.1.29)} \qquad X^{\otimes n}|\vec{x}\rangle = |\neg\vec{x}\rangle$$

where $\neg\vec{x}$ denotes the string in $\{0,1\}^n$ that is obtained by negating all entries in $\vec{x}$. We now show that

$$\text{(7.1.30)} \qquad C^{n-1}(Z)X^{\otimes n}|\vec{x}\rangle = C^{n-1}(Z)|\neg\vec{x}\rangle = \begin{cases} -|\neg\vec{x}\rangle & \text{if } \vec{x} = \vec{0}, \\ |\neg\vec{x}\rangle & \text{if } \vec{x} \neq \vec{0}. \end{cases}$$

If $\vec{x} = \vec{0}$, then $\neg\vec{x} = (\underbrace{11\cdots 11}_{n})$. Hence $C^{n-1}(Z)|\neg\vec{x}\rangle$ applies the Pauli Z gate to the last qubit $|1\rangle$ which becomes $-|1\rangle$. This implies that the state $|\neg\vec{x}\rangle$ is changed to $-|\neg\vec{x}\rangle$. Assume that $\vec{x} \neq \vec{0}$. Then at least one of the entries of $\neg\vec{x}$ is 0. If one of the first $n-1$ entries is 0, then Z is not applied to the last qubit of $|\neg\vec{x}\rangle$, which means that $|\neg\vec{x}\rangle$ is not changed by applying $C^{n-1}(Z)$. But if the first $n-1$ entries of $\neg\vec{x}$ are 1, then its last entry is 0 and Z is applied to the last qubit $|0\rangle$ of $|\neg\vec{x}\rangle$ which does not change this qubit. So, again, $|\neg\vec{x}\rangle$ remains unchanged by applying $C^{n-1}(Z)$. In summary, (7.1.30) holds which together with (7.1.28) implies

$$\text{(7.1.31)} \qquad X^{\otimes n}C^{n-1}(Z)X^{\otimes n}|\vec{x}\rangle = \left.\begin{cases} -|\vec{x}\rangle & \text{if } \vec{x} = \vec{0}, \\ |\vec{x}\rangle & \text{if } \vec{x} \neq \vec{0} \end{cases}\right\} = V|\vec{x}\rangle.$$

We estimate the size of the circuit. It uses $O(n)$ Pauli X and Hadamard gates and one $C^{n-1}(Z)$ operator which by Proposition 4.9.11 and Corollary 4.12.8 can be implemented using the $O(n)$ elementary quantum gates. So in total, the circuit has size $O(n)$. $\qquad\square$

From Propositions 7.1.17 and 7.1.18 we obtain the following result.

Proposition 7.1.19. *The Grover iterator G can be implemented using one black-box for U_f and $O(n)$ additional elementary quantum gates.*

Exercise 7.1.20. Prove Proposition 7.1.19.

7.1.5. Analysis of the search algorithm with known number of solutions.
After the preparations of the previous sections, we can now estimate the success probability and the complexity of the Grover search algorithm with a known number of solutions.

Theorem 7.1.21. *Let $n \in \mathbb{N}$, $N = 2^n$, $f : \{0,1\}^n \to \{0,1\}$, $M = |f^{-1}(1)| > 0$. On input of n, a black-box that implements U_f and M, Algorithm 7.1.6 returns with probability at least $1 - \frac{M}{N}$ a string $\vec{x} \in \{0,1\}^n$ such that $f(\vec{x}) = 1$. The algorithm applies the black-box for U_f at most $\frac{\pi}{4}\sqrt{\frac{N}{M}}$ times and uses $O\left(\log N \sqrt{\frac{N}{M}}\right)$ additional elementary quantum gates.*

Proof. It follows from Proposition 7.1.16 that the final quantum state produced by the algorithm is

$$(7.1.32) \qquad |\psi\rangle = G^k |s\rangle = \cos(2k+1)\theta |s_0\rangle + \sin(2k+1)\theta |s_1\rangle$$

where

$$(7.1.33) \qquad \theta = \arcsin\sqrt{\frac{M}{N}} \quad \text{and} \quad k = \left\lfloor \frac{\pi}{4\theta} \right\rfloor.$$

Then the algorithm measures $|\psi\rangle$ in the computational basis of $\mathbb{H}_n$. It follows from the definition of $|s_0\rangle$ and $|s_1\rangle$ that this gives $\vec{x}$ such that $f(\vec{x}) = 1$ with probability

$$(7.1.34) \qquad p = \sin^2(2k+1)\theta.$$

To prove the theorem, we estimate k and p. It follows from Corollary A.5.6 and (7.1.33) that

$$(7.1.35) \qquad k \leq \frac{\pi}{4\theta} = \frac{\pi}{4\arcsin\sqrt{M/N}} \leq \frac{\pi}{4}\sqrt{\frac{N}{M}}.$$

To estimate p we observe that

$$(7.1.36) \qquad 0 < \theta \leq \frac{\pi}{2}.$$

We also set

$$(7.1.37) \qquad \tilde{k} = \frac{\pi}{4\theta} - \frac{1}{2}.$$

Then

$$(7.1.38) \qquad (2\tilde{k}+1)\theta = \left(\frac{\pi}{2} - \theta + \theta\right) = \frac{\pi}{2}.$$

Also, the choice of k in (7.1.33) implies

$$(7.1.39) \qquad 0 \leq \frac{\pi}{4\theta} - k < 1$$

and therefore

$$(7.1.40) \qquad -\frac{1}{2} \leq \frac{\pi}{4\theta} - \frac{1}{2} - k = \tilde{k} - k < \frac{1}{2}$$

which implies

$$(7.1.41) \qquad |k - \tilde{k}| \leq \frac{1}{2}.$$

It follows that

$$(7.1.42) \qquad |(2k+1)\theta - (2\tilde{k}+1)\theta| = |2(k-\tilde{k})\theta| \le \theta.$$

By (7.1.38) we have $\sin(2\tilde{k}+1)\theta = 1$ and $\cos(2\tilde{k}+1)\theta = 0$. These equations, the trigonometric identity (A.5.3), and equations (7.1.42) and (7.1.36) imply

$$(7.1.43) \qquad \begin{aligned} |\cos(2k+1)\theta| \\ = |\cos(2k+1)\theta \sin(2\tilde{k}+1)\theta - \cos(2\tilde{k}+1)\theta \sin(2k+1)\theta| \\ = |\sin\left((2k+1)\theta - 2(\tilde{k}+1)\theta\right)| \\ = \sin|(2k+1)\theta - 2(\tilde{k}+1)\theta| \le \sin\theta. \end{aligned}$$

Therefore, the failure probability of the Grover search algorithm after k iterations is

$$(7.1.44) \qquad \cos^2(2k+1)\theta \le \sin^2\theta = \frac{M}{N}$$

which implies the assertion about the success probability.

We estimate the complexity of the algorithm. It follows from (7.1.35) that the number of applications of the Grover iterator in the algorithm is bounded by $\frac{\pi}{4}\sqrt{\frac{N}{M}}$. So it follows from Proposition 7.1.19 that the algorithm invokes the black-box for U_f at most $\frac{\pi}{4}\sqrt{\frac{N}{M}}$ times and uses $O\left(\log N\sqrt{\frac{M}{N}}\right)$ additional elementary quantum gates. $\qquad \square$

7.1.6. A search algorithm with an unknown number of solutions. Again, let $f : \{0,1\}^n \to \{0,1\}$, $N = 2^n$, and assume that $M = |f^{-1}\{1\}| > 0$. We want to find $\vec{x} \in \{0,1\}^n$ with $f(\vec{x}) = 1$. In this section, we present and analyze Algorithm 7.1.22 that solves this search problem in the case where the number of solutions M is not known. This algorithm is a Las Vegas algorithm. It repeatedly computes and measures $G^k |s\rangle$ for a randomly chosen k from $\mathbb{Z}_m$ where m increases exponentially until a solution of the search problem is found.

Algorithm 7.1.22. Quantum search when the number of solutions is unknown

---

Input: $n \in \mathbb{N}$, a black-box implementing U_f for some function $f : \{0,1\}^n \to \{0,1\}$
Output: $\vec{x} \in \{0,1\}^n$ such that $f(\vec{x}) = 1$
  1: QSEARCH(n, U_f)
  2:     $m \leftarrow 1$
  3:     $\lambda \leftarrow 6/5$
  4:     **repeat**
  5:         $k \leftarrow \text{randomInt}(m)$
  6:         Apply the quantum circuit from Figure 7.1.1, the result being $\vec{x} \in \{0,1\}^n$
  7:         $m \leftarrow \left\lceil \min\{\lambda m, \sqrt{N}\} \right\rceil$
  8:     **until** $f(\vec{x}) = 1$
  9:     **return** $\vec{x}$
10: **end**

Our goal is to prove the following theorem.

Theorem 7.1.23. *Assume that $M \leq \frac{3N}{4}$. Then the expected number of applications of the Grover iterator and thus of the operator U_f required by Algorithm 7.1.22 to find a solution of the search problem is at most $9\sqrt{\frac{N}{M}}$. The expected running time of the algorithm is $\left(\sqrt{\frac{N}{M}}\right)^{1+o(1)}$.*

We note that the condition $M \leq \frac{3N}{4}$ is not a restriction, since for $M > \frac{3N}{4}$ guessing a solution of the search problem has success probability at least $\frac{3}{4}$.

In the proof of Theorem 7.1.23, we again use the angle

$$(7.1.45) \qquad\qquad \theta = \arcsin\sqrt{\frac{M}{N}}.$$

Our proof requires the following two auxiliary results.

Lemma 7.1.24. *For any $\alpha \in \mathbb{R}$ and $m \in \mathbb{N}$ we have*

$$(7.1.46) \qquad\qquad 2\sin\alpha \sum_{k=0}^{m-1} \cos(2k+1)\alpha = \sin 2m\alpha.$$

Proof. We prove the assertion by induction on m. For $m = 1$, the trigonometric identity (A.5.4) gives

$$(7.1.47) \qquad\qquad 2\sin\alpha \sum_{k=0}^{m-1} \cos((2k+1)\alpha) = 2\sin\alpha\cos\alpha = \sin 2\alpha.$$

Now let $m \geq 1$ and assume that (7.1.46) holds. Then this equation and the trigonometric identities (A.5.2) and (A.5.3) imply

$$
\begin{aligned}
2\sin\alpha \sum_{k=0}^{m} \cos(2k+1)\alpha \\
= 2\sin\alpha\left(\sum_{k=0}^{m-1} \cos(2k+1)\alpha + \cos(2m+1)\alpha\right) \\
= \sin 2m\alpha + 2\sin\alpha\cos(2m+1)\alpha \\
= \sin 2m\alpha + \sin\alpha\cos(2m+1)\alpha - \cos\alpha\sin(2m+1)\alpha \\
+ \sin\alpha\cos(2m+1)\alpha + \cos\alpha\sin(2m+1)\alpha \\
= \sin 2m\alpha - \sin 2m\alpha + \sin 2(m+1)\alpha \\
= \sin 2(m+1)\alpha.
\end{aligned}
$$

(7.1.48)

$\square$

Lemma 7.1.25. *Let $m \in \mathbb{N}$ and assume that k is chosen randomly with the uniform distribution from $\mathbb{Z}_m$. Then measuring $G^k |s\rangle$ gives a solution of the search problem with probability*

$$p_m = \frac{1}{2} - \frac{\sin 4m\theta}{4m \sin 2\theta}. \tag{7.1.49}$$

In particular, we have $p_m \geq \frac{1}{4}$ when $m \geq \frac{1}{\sin 2\theta}$.

Proof. By (7.1.9) the probability of obtaining a solution of the search problem when measuring $G^k |s\rangle$ for some $k \in \mathbb{N}_0$ is $\sin^2(2k+1)\theta$. So if k is chosen randomly from $\mathbb{Z}_m$ for some $m \in \mathbb{N}$, then equation (7.1.32), the trigonometric identity (A.5.7), and Lemma 7.1.24 imply that this probability is

$$\begin{aligned}
p_m &= \frac{1}{m} \sum_{k=0}^{m-1} \sin^2(2k+1)\theta \\
&= \frac{1}{2m} \sum_{k=0}^{m-1} (1 - \cos(2k+1)2\theta) \\
&= \frac{1}{2} - \frac{\sin 4m\theta}{4m \sin 2\theta}.
\end{aligned} \tag{7.1.50}$$

If $m \geq \frac{1}{\sin 2\theta}$, then

$$\frac{\sin 4m\theta}{4m \sin 2\theta} \leq \frac{\sin 4m\theta}{4} \leq \frac{1}{4}. \tag{7.1.51} \qquad \square$$

We now prove Theorem 7.1.23. Set

$$m_0 = \frac{1}{\sin 2\theta}. \tag{7.1.52}$$

Since $\sin \theta = \sqrt{\frac{M}{N}}$ and $\cos \theta = \sqrt{\frac{N-M}{N}}$, it follows from (A.5.4) and $M \leq \frac{3N}{4}$ that

$$m_0 = \frac{1}{2 \sin \theta \cos \theta} = \frac{N}{2\sqrt{(N-M)M}} \leq \sqrt{\frac{N}{M}}. \tag{7.1.53}$$

In the jth iteration of the loop in Algorithm 7.1.22 we have

$$m = \left\lceil \min\{\lambda^{j-1}, \sqrt{N}\} \right\rceil \tag{7.1.54}$$

with $\lambda = \frac{6}{5}$. Also, the expected number of applications of the Grover iterator in this loop is bounded as follows:

$$E_j = \frac{m}{2} \leq \min \frac{1}{2}\{\lambda^{j-1}, \sqrt{N}\}. \tag{7.1.55}$$

We say that the algorithm reaches the critical stage if for the first time $m \geq m_0$. This happens when in line 7 of the algorithm we have $j = \lceil \log_\lambda m_0 \rceil$. From (7.1.55) and $\lambda = \frac{6}{5}$ it follows that the expected number of applications of the Grover iterator before the algorithm finds a solution or reaches the critical stage is at most

$$\frac{1}{2} \sum_{j=1}^{\lceil \log_\lambda m_0 \rceil} \lambda^{j-1} = \frac{\lambda^{\lceil \log_\lambda m_0 \rceil} - 1}{2(\lambda - 1)} < \frac{\lambda}{2(\lambda - 1)} m_0 = 3m_0. \tag{7.1.56}$$

If the critical stage is reached, then in every iteration of the **repeat** loop in the algorithm from this point on, we have $m \geq m_0 = 1/\sin 2\theta$. By Lemma 7.1.24, the success probability in each of these iterations is at least $\frac{1}{4}$. So for all $u \geq 1$ the probability that the algorithm is successful in the $(\lceil \log_\lambda m_0 \rceil + u)$th iteration of the loop is at most $\frac{3}{4}^{u-1}$. Therefore, the expected number of applications of the Grover iterator needed to succeed in the critical stage is at most

$$(7.1.57) \qquad \frac{\lambda^{\lceil \log_\lambda m_0 \rceil}}{2} \sum_{u=0}^{\infty} \left(\frac{3\lambda}{4} \right)^u < \frac{3m_0}{5} \sum_{u=0}^{\infty} \left(\frac{9}{10} \right)^u = 6m_0.$$

Therefore, the total expected number of applications of the Grover iterator in the algorithm is bounded by $9m_0$ which by (7.1.52) is bounded by $9\sqrt{\frac{N}{M}}$. The estimate of the expected running time of the algorithm is derived from Propositions 7.1.17 and 7.1.18.

7.2. Quantum counting

A problem closely related to the search problem is the following. For $n \in \mathbb{N}$, $f : \{0, 1\}^n \rightarrow \{0, 1\}$ a function, determine the number $M = |f^{-1}(1)|$ of solutions of the search problem. In this section, we describe quantum algorithms to find approximations of M or even to determine M exactly.

These algorithms utilize the quantum states $|s\rangle$, $|s^\perp\rangle$, $|s_0\rangle$, and $|s_1\rangle$, the plane $P = \mathbb{C}|s\rangle + \mathbb{C}|s^\perp\rangle = \mathbb{C}|s_0\rangle + \mathbb{C}|s_1\rangle$, the operators U_f, U_1, and U_s, $N = 2^n$, and the angle $\theta = \arcsin \left(\sqrt{\frac{M}{N}} \right)$ which were previously discussed in this chapter.

7.2.1. Implementing the controlled Grover iterator. When evaluating the time complexity of the quantum counting algorithms, we consider the frequency of using the U_f operator. This enables us to draw a comparison with classical counting algorithms, where the crucial information lies in the number of evaluations of the function f required for the counting process.

To determine the frequency of using U_f, we need to take into account that the algorithms discussed in this section utilize quantum phase estimation to approximate the eigenvalues of the Grover iterator G. This technique is explained in detail in Section 6.3, and its implementation involves the quantum circuit shown in Figure 6.3.1. This implementation utilizes controlled-G^{2^i} operators for $0 \leq i < l$, where l represents the precision parameter.

In Section 6.4.4, we demonstrate an efficient approach to implement the controlled-U_a^c operators. However, this simplification is not applicable in our current situation, since G is the Grover operator for an arbitrary function f and does not possess special properties as U_a. As a result, we need an alternative method to implement the controlled-G^{2^i} operators, which can be achieved as follows.

Recall from equation (7.1.12) that the Grover iterator is expressed as $G = U_s U_1$. Figure 7.1.6 presents an implementation of U_s, which, as per Proposition 7.1.18, requires $\mathrm{O}(n)$ elementary quantum gates. It follows from Theorem 4.12.7 that the controlled-U_s operator can also be implemented by a quantum circuit using $O(n)$ elementary quantum gates. Next, in Figure 7.1.5, a quantum circuit implementation of U_1 is presented, utilizing the U_f operator and $O(1)$ elementary quantum gates. To transform it into an implementation of the controlled-U_1 operator, we require the controlled-U_f operator. If an implementation of U_f is available that exclusively uses elementary quantum gates, then, by following the method described in the proof of Theorem 4.12.7, a quantum circuit for the controlled-U_f operator can be constructed using only elementary quantum gates. We assume that the controlled-U_f operator is provided in this manner or through some other means. Then, by employing the method from the proof of Theorem 4.12.7, a quantum circuit implementing the controlled-U_1 operator can be constructed that uses $O(1)$ elementary quantum gates and one controlled-U_f gate. Combining these results, we obtain the following result.

Proposition 7.2.1. *There is an implementation of the controlled Grover iterator that requires one controlled-U_f gate and $O(n)$ additional elementary quantum gates.*

In the complexity analyses presented in the following sections, we consider using one controlled-U_f gate as equivalent to using one U_f gate. This is because if the circuits implementing these gates are constructed from elementary gates, their sizes are proportional.

7.2.2. An approximate quantum counting algorithm. We begin by presenting an approximate quantum counting algorithm that also serves to demonstrate the principles employed in the other counting algorithms covered in this chapter.

The following proposition reveals the eigenvalues of the restricted Grover iterator $G|_P$ within the plane P from (7.1.13). It demonstrates that the counting problem can be effectively addressed through quantum phase estimation. To achieve this, we introduce the quantum states:

$$(7.2.1) \qquad |s_+\rangle = \frac{|s_1\rangle + i\,|s_0\rangle}{\sqrt{2}}, \quad |s_-\rangle = \frac{|s_1\rangle - i\,|s_0\rangle}{\sqrt{2}}.$$

Proposition 7.2.2. *The pair $(|s_+\rangle, |s_-\rangle)$ is an orthonormal basis of eigenstates of the restriction $G|_P$ of the Grover iterator to the plain P. The corresponding eigenvalues are $e^{2i\theta}$ and $e^{-2i\theta i}$ with θ from (7.1.45), and we have*

$$(7.2.2) \qquad |s\rangle = \frac{-i}{\sqrt{2}}\left(e^{i\theta}\,|s_+\rangle - e^{-i\theta}\,|s_-\rangle\right).$$

Proof. By Exercise 7.2.3, the pair $(|s_+\rangle, |s_-\rangle)$ is an orthonormal basis of P. Also, we know from Proposition 7.1.16 that for all $\alpha \in \mathbb{R}$ we have

$$(7.2.3) \qquad G(\cos\alpha\,|s_0\rangle + \sin\alpha\,|s_1\rangle) = \cos(\alpha + 2\theta)\,|s_0\rangle + \sin(\alpha + 2\theta)\,|s_1\rangle.$$

As shown in Exercise 7.2.3, this implies

$$(7.2.4) \qquad G\,|s_0\rangle = \cos 2\theta\,|s_0\rangle + \sin 2\theta\,|s_1\rangle, \quad G\,|s_1\rangle = -\sin 2\theta\,|s_0\rangle + \cos 2\theta\,|s_1\rangle$$

and therefore

$$(7.2.5) \qquad G\,|s_+\rangle = e^{2i\theta}\,|s_+\rangle, \quad G\,|s_-\rangle = e^{-2i\theta}\,|s_-\rangle.$$

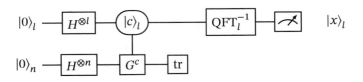

Figure 7.2.1. The approximate quantum counting algorithm.

So $|s_+\rangle$ and $|s_-\rangle$ are eigenstates of $G|_P$ associated with the eigenvalues $e^{2i\theta}$ and $e^{-2i\theta}$, respectively. Equation (7.2.2) is also proved in Exercise 7.2.3.   □

Exercise 7.2.3.   (1) Show that $(|s_+\rangle, |s_-\rangle)$ is an orthonormal basis of P.

(2) Verify equations (7.2.4), (7.2.5), and (7.2.2).

Proposition 7.2.2 demonstrates the feasibility of estimating the value of $M = N \sin^2 \theta$ by approximating one of the phases $\pm 2\theta$ associated with the eigenvalues of the Grover iterator G. To achieve this approximation, we employ Algorithm 7.2.4, which utilizes the quantum circuit depicted in Figure 7.2.1 as its main component.

The idea of this algorithm is the following. By Exercise 6.2.9 and (7.2.2) we have

$$(7.2.6) \qquad H^{\otimes n} |0\rangle^{\otimes n} = |s\rangle = \frac{-i}{\sqrt{2}} \left(e^{i\theta} |s_+\rangle - e^{-i\theta} |s_-\rangle \right).$$

This indicates that we can efficiently create an equally weighted superposition of the two eigenstates of $G|_P$. This is done in the second register of the quantum circuit depicted in Figure 7.2.1. The circuit proceeds to perform quantum phase estimation with a precision parameter $l \in \mathbb{N}$ on this superposition to identify a value $x \in \mathbb{Z}_L$, where $L = 2^l$, such that $\frac{x}{L}$ is an approximation for one of the real numbers $\frac{\pm\theta}{\pi}$. As a result, the computed value $\tilde{M} = N \sin^2 \frac{\pi x}{2^l}$ provides an approximation of $M = \sin^2 \theta$. Algorithm 7.2.4 performs these calculations and returns the obtained approximation. It is essential to note that the returned value $\tilde{M}$ is a real number and the algorithm can only provide a rational approximation to this number. Therefore, implementations of the algorithm must ensure that the precision of the approximation is sufficient for the specific application's requirements.

Algorithm 7.2.4. Approximate quantum counting algorithm

Input: $n \in \mathbb{N}$, U_f for some $f : \{0,1\}^n \to \{0,1\}$, and a precision parameter $l \in \mathbb{N}$
Output: An approximation $\tilde{M}$ to $M = |f^{-1}(1)|$

 1: QCOUNT(n, U_f, l)
 2:       Apply the quantum circuit from Figure 7.2.1, the result being $x \in \mathbb{Z}_L$ where $L = 2^l$
 3:       **return** $\tilde{M} \leftarrow N \sin^2 \frac{\pi x}{L}$
 4: **end**

The following theorem establishes the correctness of Algorithm 7.2.4 and provides insight into its computational complexity.

Theorem 7.2.5. *Assume that the input of Algorithm 7.2.4 is n, U_f, l as specified in the algorithm and let $L = 2^l$. Denote by $\tilde{M}$ the output of the algorithm. Then the following are true.*

(1) *With probability at least $\frac{8}{\pi^2}$ we have*

$$(7.2.7) \qquad |\tilde{M} - M| \leq 2\pi \frac{\sqrt{M(N-M)}}{L} + \frac{\pi^2 N}{L^2}.$$

(2) *The algorithm requires $O(L)$ applications of U_f and $O(Ln^2)$ additional elementary operations.*

Proof. From (7.2.2) and (6.3.18) it follows that before tracing the target register, the state of the quantum circuit is

$$(7.2.8) \qquad |\varphi\rangle = -\frac{-i}{\sqrt{2}} \left(e^{i\theta} \psi_l \left(\frac{\theta}{\pi}\right) |s_+\rangle - e^{-i\theta} \psi_l \left(\frac{-\theta}{\pi}\right) |s_-\rangle \right).$$

Since $(|s_+\rangle, |s_-\rangle)$ is an orthonormal basis of the plane P, Corollary 3.7.12 implies that after tracing out the target register, the control register is in the mixed state

$$(7.2.9) \qquad \left(\left(\frac{1}{2}, \psi_l\left(\frac{\theta}{\pi}\right)\right), \left(\frac{1}{2}, \psi_l\left(-\frac{\theta}{\pi}\right)\right) \right).$$

So it follows from Theorem 6.3.7 and Lemma 6.3.3 that with probability at least $\frac{8}{\pi^2}$ the measurement result x in Algorithm 7.2.4 satisfies

$$(7.2.10) \qquad \left| \frac{x}{L} - \frac{\theta}{\pi} \right| < \frac{1}{L}.$$

Set

$$(7.2.11) \qquad p = \sin^2 \theta = \frac{M}{N}.$$

Then we have

$$(7.2.12) \qquad \sin \theta = \sqrt{p}, \quad \cos \theta = \sqrt{1-p},$$

and

$$(7.2.13) \qquad M = Np.$$

So if we set

$$(7.2.14) \qquad \tilde{\theta} = \frac{\pi x}{L}, \quad \tilde{p} = \sin^2 \tilde{\theta},$$

then the return value of Algorithm 7.2.4 is

$$(7.2.15) \qquad \tilde{M} = N\tilde{p}.$$

We will now prove that

$$(7.2.16) \qquad |\tilde{p} - p| < 2\pi \frac{\sqrt{p(1-p)}}{L} + \frac{\pi^2}{L^2}.$$

Multiplying this equation by N, we obtain the assertion of the theorem. We set

$$(7.2.17) \qquad \varepsilon = \tilde{\theta} - \theta.$$

Then (7.2.10) implies

(7.2.18)
$$|\varepsilon| < \frac{\pi}{L}.$$

First, assume that $\tilde{p} \geq p$ and note that by the trigonometric identites (A.5.2) and (A.5.7) we have

(7.2.19)
$$\sin\theta\cos\theta\sin 2\varepsilon + (1 - 2\sin^2\theta)\sin^2\varepsilon$$

$$= 2\sin\theta\cos\theta\sin\varepsilon\cos\varepsilon + (\cos^2\theta - \sin^2\theta)\sin^2\varepsilon$$

$$= \sin^2\theta\cos^2\varepsilon + 2\sin\theta\cos\theta\sin\varepsilon\cos\varepsilon + \cos^2\theta\sin^2\varepsilon - \sin^2\theta(\sin^2\varepsilon + \cos^2\varepsilon)$$

$$= \sin^2(\theta + \varepsilon) - \sin^2\theta.$$

Hence, using Lemma A.5.3 and equations (7.2.12), and (7.2.18) we obtain

$$|\tilde{p} - p| = \tilde{p} - p = \sin^2(\theta + \varepsilon) - \sin^2\theta$$

$$= \sin\theta\cos\theta\sin 2\varepsilon + (1 - 2\sin^2\theta)\sin^2\varepsilon$$

(7.2.20)
$$= \sqrt{p(1-p)}\sin 2\varepsilon + (1-2p)\sin^2\varepsilon$$

$$\leq 2\varepsilon\sqrt{p(1-p)} + \varepsilon^2 < 2\pi\frac{\sqrt{p(1-p)}}{L} + \frac{\pi^2}{L}.$$

Next, let $\tilde{p} \leq p$. Then using analogous arguments we obtain

$$|\tilde{p} - p| = p - \tilde{p} = \sin^2\theta - \sin^2(\theta - \varepsilon)$$

$$= \sin\theta\cos\theta\sin 2\varepsilon - (1 - 2\sin^2\theta)\sin^2\varepsilon$$

(7.2.21)
$$= \sqrt{p(1-p)}\sin 2\varepsilon + (2p-1)\sin^2\varepsilon$$

$$\leq 2\varepsilon\sqrt{p(1-p)} + \varepsilon^2 \leq 2\pi\frac{\sqrt{p(1-p)}}{L} + \frac{\pi^2}{L^2}.$$

This concludes the proof of (7.2.16) which implies (7.2.7).

We will now establish the complexity statement. The implementation of the quantum phase estimation circuit necessitates the utilization of controlled-G^{2^i} operators for $0 \leq i < l$. It follows from Proposition 7.2.1 that these, in turn, require a total of $\sum_{i=0}^{l-1} 2^l = 2^l = L - 1$ applications of the U_f operator and an additional $O(Ln^2)$ elementary quantum gates. $\qquad\square$

7.2.3. A quantum counting algorithm with pre-selected error.
Algorithm 7.2.6 is a modification of the approximate quantum counting algorithm from the previous section. Given an error parameter $\varepsilon \in \mathbb{R}$, $0 < \varepsilon < 1$, it computes an approximation $\tilde{M}$ of $M = f^{-1}(1)$ such that

(7.2.22)
$$|M - \tilde{M}| \leq \varepsilon M.$$

We explain the idea of the algorithm. For increasing values of $l = 0, 1, 2, \ldots$, it calls QCount(n, U_f, l) until the return value is different from 0 for the first time or $2^l \geq 2\sqrt{N}$. Denote the l-value of the first occurrence of a nonzero return value by $l_{\max}$.

In the analysis of the algorithm, it is demonstrated that with probability at least $\cos^2 \frac{2}{5}$, we have

(7.2.23)
$$2^{l_{max}} \geq \frac{2}{5\pi}\sqrt{\frac{N}{M}}.$$

Then the number l_{max} of QCount calls that return 0 provides crucial information about the magnitude of the solution count M. A larger value of l_{max} implies a smaller value of M. Furthermore, it is proven that for $l = l_{max} + \left\lceil \log_2 \frac{20\pi^2}{\varepsilon} \right\rceil$, where ε denotes the chosen precision, the call $QCount(n, U_f, l)$ provides the desired approximation $\tilde{M}$ with a probability of at least $\frac{8}{\pi^2}$. Therefore, the total success probability of the algorithm is at least $\frac{8}{\pi^2} \cos^2 \frac{2}{5} > \frac{2}{3}$.

Algorithm 7.2.6. Approximate counting with pre-selected approximation precision

Input: $n \in \mathbb{N}$, U_f with $f : \{0,1\}^n \to \{0,1\}$, a parameter $\varepsilon \in \mathbb{R}$ with $0 < \varepsilon < 1$
Output: An approximation $\hat{M} \in \mathbb{N}_0$ to M such that $|M - \hat{M}| \leq \varepsilon M$

1:   ApproxQCount(n, U_f, ε)
2:      $l \leftarrow 0$
3:      **repeat**
4:         $l \leftarrow l + 1$
5:         $\tilde{M} \leftarrow QCount(n, U_f, l)$
6:      **until** $\tilde{M} \neq 0$ or $2^l \geq 2\sqrt{N}$
7:      $l \leftarrow l + \left\lceil \log_2 \frac{20\pi^2}{\varepsilon} \right\rceil$
8:      $\tilde{M} \leftarrow QCount(n, U_f, l)$
9:      **return** $\hat{M} \leftarrow \lfloor \tilde{M} \rceil$
10: **end**

The following theorem establishes both the correctness and the computational complexity of Algorithm 7.2.6.

Theorem 7.2.7. *Let n, U_f, ε be the input of Algorithm 7.2.6. Denote by $\hat{M}$ the return value of the algorithm. Then the following are true.*

(1) *With probability at least $\frac{2}{3}$ we have*

(7.2.24)
$$|\hat{M} - M| < \varepsilon M.$$

(2) *The algorithm requires $O\left(\frac{\sqrt{N}}{\varepsilon}\right)$ applications of U_f and $O\left(\frac{n^2\sqrt{N}}{\varepsilon}\right)$ additional elementary operations.*

Proof. Let

(7.2.25)
$$\theta = \arcsin\sqrt{\frac{M}{N}}, \quad k = \left\lceil \log_2 \frac{1}{5\theta} \right\rceil.$$

Then we have

(7.2.26)
$$2^k \geq 2^{\log_2 \frac{1}{5\theta}} = \frac{1}{5\theta}.$$

Also, Corollary A.5.6 implies

$$(7.2.27) \qquad 2^k \le 2^{\log_2 \frac{1}{5\theta}+1} = \frac{2}{5\theta} = \frac{2}{5\arcsin\sqrt{\frac{M}{N}}} \le \frac{2}{5}\sqrt{\frac{N}{M}}.$$

In line 6 of the algorithm we obtain $l = k$ if the call QCount(n, U_f, l) in line 5 of the algorithm returns 0 on k occasions. This happens with probability

$$(7.2.28)$$
$$p \underset{(1)}{=} \prod_{l=1}^{k} \frac{1}{2^{2l}} \frac{\sin^2(2^l\theta)}{\sin^2\theta} \underset{(2)}{\ge} \prod_{l=1}^{k} \cos^2(2^l\theta)$$
$$\underset{(3)}{=} \frac{\sin^2(2^{k+1}\theta)}{2^{2k}\sin^2(2\theta)} \underset{(4)}{=} \cos^2(2^{k+1}\theta) \underset{(5)}{\ge} \cos^2\frac{2}{5}.$$

Here, equation (1) follows from Proposition 6.3.4, inequality (2) is a consequence of Lemma A.5.10 with $x = 2^l$ and $\alpha = \theta$ which is applicable because of (7.2.26), equation (3) is obtained from Lemma A.5.11, for equation (4) we use Lemma A.5.10 again with $x = 2^k$ and $\alpha = 2\theta$, and inequality (5) uses Lemma A.5.7 $\cos x \ge x$, and (7.2.26).

Assume that the maximum value l_{max} for l assumed in the **repeat** loop is at least k. As in line 7 of the algorithm set

$$(7.2.29) \qquad l = l_{max} + \left\lceil \log_2 \frac{20\pi^2}{\varepsilon} \right\rceil.$$

Corollary A.5.4 implies

$$(7.2.30) \qquad \theta = \arcsin\sqrt{\frac{M}{N}} \le \frac{\pi}{2}\sqrt{\frac{M}{N}}.$$

So with $L = 2^l$ we obtain from (7.2.29), (7.2.26), and (7.2.30)

$$(7.2.31) \qquad \frac{1}{L} \le 5\theta \frac{\varepsilon}{20\pi^2} \le \frac{\varepsilon}{8\pi}\sqrt{\frac{M}{N}}.$$

This implies that QCount(n, U_f, l) returns with probability at least $\frac{8}{\pi^2}$ a real number $\tilde{M}$ with

$$(7.2.32)$$
$$|M - \tilde{M}| \le 2\pi \frac{\sqrt{M(N-M)}}{L} + \pi^2 \frac{N}{L^2}$$
$$\le \frac{\varepsilon}{4}M\sqrt{\frac{N-M}{N}} + \frac{\varepsilon^2}{64}$$
$$\le \varepsilon M\left(\frac{1}{4} + \frac{1}{64}\right)M < \frac{\varepsilon M}{2}.$$

Set $\hat{M} = \lfloor \tilde{M} \rceil$. If $\varepsilon M < 1$, then (7.2.32) implies $|\hat{M} - M| < \frac{1}{2}$. So we have $\hat{M} = M$ and $|\hat{M} - M| = 0$. If $\varepsilon M \ge 1$, then we have $|\hat{M} - \tilde{M}| \le \frac{1}{2} \le \varepsilon \frac{M}{2}$. Together with (7.2.32) this implies $|\hat{M} - M| \le |\hat{M} - \tilde{M}| + |\tilde{M} - M| < \varepsilon M$. The total success probability is

$$(7.2.33) \qquad \frac{8}{\pi^2}\cos^2\frac{2}{5} \ge \frac{2}{3}.$$

The complexity statement can be seen as follows. According to Theorem 7.2.5, the call to QCount(n, U_f, l) in line 5 necessitates $O(2^l)$ applications of U_f and $O(n^2 2^l)$

additional elementary operations. Furthermore, due to the condition in line 6, the maximum value of 2^l in the **repeat** loop is bounded by $O(\sqrt{N})$. As a result, this **repeat** loop requires $O(\sqrt{N})$ applications of U_f and $O(n^2\sqrt{N})$ additional operations. After the assignment in line 7, the value of l becomes $O\left(\frac{\sqrt{N}}{\varepsilon}\right)$. Another application of Theorem 7.2.5 completes the proof. $\qquad\qquad\qquad\square$

7.2.4. Exact counting. The two algorithms, QCount and ApproxQCount, can be effectively utilized to count the number of solutions $M = f^{-1}(1)$ exactly. The approach involves employing ApproxQCount$(n, U_f, \frac{1}{2})$ to obtain a reliable approximation $\tilde{M}_1$ of M. Then, using this approximation, we find an appropriate value of $l \in \mathbb{N}$ such that QCount(n, U_f, l) provides $\tilde{M}_2$ that satisfies $|M - \tilde{M}_2| < \frac{1}{2}$. Consequently, M is the nearest integer value of $\tilde{M}_2$. This entire process is implemented in Algorithm 7.2.8.

Algorithm 7.2.8. Exact counting

Input: $n \in \mathbb{N}$, U_f with $f : \{0,1\}^n \to \{0,1\}$
Output: $M = |f^{-1}(1)|$

  1: ExactQCount(n, U_f)

  2:     $\tilde{M}_1 \leftarrow$ ApproxQCount$\left(n, U_f, \frac{1}{2}\right)$

  3:     $l \leftarrow \left\lceil \log_2 26\sqrt{\tilde{M}_1 N} \right\rceil$

  4:     $\tilde{M}_2 \leftarrow$ QCount(n, U_f, l)

  5:     **return** $M \leftarrow \lfloor \tilde{M}_2 \rceil$

  6: **end**

Theorem 7.2.9. *On input of $n \in \mathbb{N}$ and U_f for some function $f : \{0,1\}^n \to \{0,1\}$, Algorithm 7.2.8 returns $M = |f^{-1}(1)|$ with probability at least $\frac{1}{2}$. The algorithm requires $O(\sqrt{MN})$ applications of U_f and $O(n^2\sqrt{MN})$ additional elementary operations.*

Proof. We have $|\tilde{M}_1 - M| \leq \frac{M}{2}$ and, therefore, $\tilde{M}_1 \geq \frac{M}{2}$. Choose $l = \left\lceil \log_2 26\sqrt{\tilde{M}_1 N} \right\rceil$ as in line 3 and $L = 2^l$. Then we have

(7.2.34)
$$\frac{1}{L} \leq \frac{1}{26\sqrt{\tilde{M}_1 N}}.$$

This implies

(7.2.35)
$$|M - \tilde{M}_2| \leq \frac{2\pi}{26}\sqrt{\frac{MN}{\tilde{M}_1 N}} + \frac{\pi^2 N}{26^2 \tilde{M}_1 N}$$
$$\leq \frac{4\pi}{26} + \frac{\pi^2}{26^2} < \frac{1}{2}.$$

So it follows that the algorithm returns the correct M.

The complexity statement follows from Theorems 7.2.5 and 7.2.7. $\qquad\qquad\square$

Chapter 8

The HHL Algorithm

In the previous chapters, we explored early quantum algorithms and fundamental techniques, such as phase estimation and amplitude amplification. Shor's algorithms for integer factorization and computing discrete logarithms stand out for their significant speedups compared to classical algorithms. They solve these problems in polynomial time, which is classically only possible in subexponential time. Additionally, the Grover search algorithm offers a quadratic speedup, which is also very impressive. These advancements have inspired researchers to seek quantum algorithms with similar advantages across various computing domains. Many other quantum algorithms have been discovered, building upon the techniques we have presented. For a comprehensive overview of the presently known quantum algorithms, refer to [**Jor**].

In this chapter, we focus on a fascinating and more recent quantum algorithm, the HHL algorithm, proposed by Aram W. Harrow, Avinatan Hassidim, and Seth Lloyd in 2008 [**HHL08**]. The HHL algorithm addresses a crucial problem encountered in countless applications in science and engineering: solving linear systems over $\mathbb{C}$. When certain conditions are met and the problem is appropriately formulated, the HHL algorithm achieves exponential speedup compared to the best classical algorithms. Its theoretical significance and practical applications make it an intriguing subject of study.

Our presentation is based on the description of the algorithm in [**DHM**$^+$**18**] and aims to provide an impression of the algorithm and its analysis. However, a detailed explanation goes beyond the scope of this book.

8.1. The problem

One of the most significant challenges in algorithmic linear algebra is known as the *Linear Systems Problem* (*LSP*), which we also discuss in Section B.7.5. Here, we focus on a specific case of LSP, involving the parameters $M \in \mathbb{N}$, $A \in \mathsf{GL}(M, \mathbb{C})$, and $\vec{b} \in \mathbb{C}^M$. The objective is to compute $\vec{x} = A^{-1}\vec{b}$. In Section B.7.5, we demonstrate that Gaussian elimination can find $\vec{x}$ using operations $\mathrm{O}(M^3)$ in $\mathbb{C}$. However, since computers can only handle rational approximations to complex numbers, several algorithms have been developed to efficiently find good approximations to $\vec{x}$ using polynomial time in M.

The HHL algorithm addresses the Quantum Linear System Problem (QLSP). As in LSP, it uses $M \in \mathbb{N}$, $A \in GL(M, \mathbb{C})$, $\vec{b} \in \mathbb{C}^M$, and $\vec{x} = A^{-1}\vec{b}$. To simplify the description of the HHL algorithm, the following assumptions are made.

(1) $M = 2^m$ with $m \in \mathbb{N}$.

(2) A is Hermitian; hence, $A \in GL(M, \mathbb{C})$ and Proposition 2.4.60 imply that the eigenvalues of A are nonzero real numbers.

(3) The eigenvalues of A are in $[0, 2\pi[$.

(4) $\left\| \vec{b} \right\| = 1$.

If these assumptions are not satisfied, it can be ensured by appropriately modifying the HHL parameters M, A, and $\vec{b}$ that they are satisfied, and therefore the HHL algorithm still works. This modification is presented in the next exercise.

Exercise 8.1.1. Let $A \in GL(M, \mathbb{C})$, $\vec{b}, \vec{x} \in \mathbb{C}^M$ with $A\vec{b} = \vec{x}$. Show that $A' = \begin{pmatrix} 0 & A \\ A^* & 0 \end{pmatrix}$ is Hermitian and for $\vec{b}' = (\vec{b}, \vec{0})$ and $\vec{x}' = (\vec{0}, \vec{x})$ we have $A'\vec{b}' = \vec{x}'$.

Let $\vec{b} = (b_0, \ldots, b_{M-1})$. Since $\left\| \vec{b} \right\| = 1$,

$$(8.1.1) \qquad\qquad |b\rangle = \sum_{i \in \mathbb{Z}_M} b_i |i\rangle_m$$

is a quantum state in $\mathbb{H}_m$. So the number $m = \log_2 M$ of qubits required to represent $\vec{b}$ is logarithmic in the dimension M of the linear system to be solved. This opens the possibility for the HHL algorithm to find

$$(8.1.2) \qquad\qquad |x\rangle = \sum_{i \in \mathbb{Z}_M} x_i |i\rangle_m$$

where $\vec{x} = (x_0, \ldots, x_{M-1})$ in time polylogarithmic in M which would be an exponential advantage over all classical LSP algorithms. However, note that $|x\rangle$ may not be a quantum state, since the Euclidean length of $\vec{x}$ may not be 1.

In the upcoming section, we provide an overview of the HHL algorithm, and in Section 8.3, we delve into the conditions under which the algorithm achieves its exponential speedup. Here, we highlight some essential considerations. To achieve exponential speedup, the algorithm cannot directly read all components of $\vec{x}$ since this vector has length M. Instead, it can extract certain properties of $\vec{x}$ by measuring $|x\rangle$ with respect to an observable of $\mathbb{H}_m$. In several application contexts, this is sufficient. Additionally, the input of the algorithm cannot simply be given as A and $\vec{b}$ because representing these objects in a standard form requires a time complexity of $\Omega(M^2)$. Therefore, both A and $\vec{b}$ must be very sparse, and there must exist an efficient method to access the entries of A and $\vec{b}$.

8.2. Overview

Let $m, M, A, \vec{b}, \vec{x}, |b\rangle$, and $|x\rangle$ be as specified in the previous section for the HHL problem. In the following, we give an overview of the HHL algorithm.

Since A is Hermitian, it follows from Theorem 2.4.53 that we can choose an orthonormal basis $(|u_0\rangle, \dots, |u_{M-1}\rangle)$ of eigenstates of A. Denote by $\lambda_0, \dots, \lambda_M$ the corresponding eigenvalues. They are nonzero real numbers in $[0, 2\pi[$ by assumption. It follows from the Spectral Theorem 2.4.56 that

$$(8.2.1) \qquad A = \sum_{j=0}^{M-1} \lambda_j |u_j\rangle\langle u_j|.$$

So, the inverse of A is

$$(8.2.2) \qquad A^{-1} = \sum_{j=0}^{M-1} \frac{1}{\lambda_j} |u_j\rangle\langle u_j|.$$

The quantum state $|b\rangle$ can be written as

$$(8.2.3) \qquad |b\rangle = \sum_{j=0}^{M-1} \beta_j |u_j\rangle$$

with $\beta_j \in \mathbb{C}$ for $j \in \mathbb{Z}_M$. With this notation, we have

$$(8.2.4) \qquad |x\rangle = A^{-1} |b\rangle = \sum_{j=0}^{M-1} \frac{\beta_j}{\lambda_j} |u_j\rangle.$$

The HHL circuit shown in Figure 8.2.1 uses this identity to approximate $|x\rangle$. We explain how this works by determining the intermediate states $|\psi_0\rangle, \dots, |\psi_4\rangle$.

The HHL circuit operates on a quantum register which is composed of three smaller quantum registers. The first is the *ancilla register*. It contains one ancillary qubit. The second is the *clock register*. It is of length $n \in \mathbb{N}$ which is a precision constant. The third register is the *b-register*. It is of length m. To simplify the explanation of the algorithm,

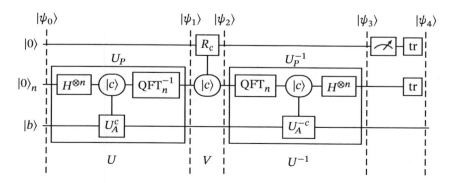

Figure 8.2.1. The HHL circuit.

we assume that the eigenvalues of A can be written as

(8.2.5) $$\lambda_j = \frac{2\pi c_j}{2^n} \quad \text{with } c_j \in \mathbb{Z}_{2^n} \text{ for } 0 \le j < M.$$

This means that all λ_j have a finite binary expansion of length at most n and can therefore be precisely determined by quantum phase estimation. We will show that the final state $|\psi_4\rangle$ is proportional to $|x\rangle$ if the measurement of the first qubit gives 1. If (8.2.5) is not true, the algorithm finds an approximation of a quantum state proportional to $|x\rangle$.

The initial state of the HHL algorithm is

(8.2.6) $$|\psi_0\rangle = |0\rangle\,|0\rangle_n\,|b\rangle .$$

Next, we note that

(8.2.7) $$|\psi_1\rangle = U\,|\psi_0\rangle = |0\rangle\,U_P\,|0\rangle_n\,|b\rangle$$

with U and U_P from Figure 8.2.1. Here, U_P is the phase estimation circuit introduced in Section 6.3. It is used to estimate the eigenvalues of

(8.2.8) $$U_A = e^{iA} = \sum_{j \in \mathbb{Z}_M} e^{i\lambda_j}\,|u_j\rangle\langle u_j|$$

introduced in Definition 2.4.69. By Theorem 2.4.72, this operator is unitary since A is Hermitian. In addition, (8.2.8) shows that $(|u_0\rangle, \dots, |u_{M-1}\rangle)$ is an orthonormal basis of the eigenstates of U_A and

(8.2.9) $$e^{i\lambda_0} = e^{2\pi i \frac{c_0}{2^n}}, \dots, e^{i\lambda_{M-1}} = e^{2\pi i \frac{c_{M-1}}{2^n}}$$

are the eigenvalues corresponding to the basis elements. It follows from equation (6.3.20) in the analysis of the phase estimation algorithm, Definition 6.2.5, and the invertibility of QFT_n shown in Proposition 6.2.8 that

$$|\psi_1\rangle = |0\rangle \sum_{j=0}^{M-1} \beta_j U_P\,|0\rangle_n\,|u_j\rangle$$

(8.2.10) $$= |0\rangle \sum_{j=0}^{M-1} \beta_j \mathrm{QFT}_n^{-1}\left|\psi_n\left(\frac{c_j}{2^n}\right)\right\rangle |u_j\rangle$$

$$= |0\rangle \sum_{j=0}^{M-1} \beta_j\,|c_j\rangle_n\,|u_j\rangle .$$

In order to obtain $|\psi_2\rangle$, the HHL circuit applies the operator V to $|\psi_1\rangle$ which is also shown in Figure 8.2.1. This operator acts as the rotation

(8.2.11) $$R_c = R_{\hat{y}}(-2\theta(c))$$

on the ancilla register controlled by the clock register $|c\rangle_n$, $c \in \mathbb{Z}_{2^n}$, and does not change the clock and the b register. Here, $R_{\hat{y}}$ is from Definition 4.3.7,

(8.2.12) $$\theta(c) = \arcsin\frac{C}{\lambda(c)} \quad \text{with } \lambda(c) = \frac{C}{2^n},$$

and the constant $C \in \mathbb{R}$ is chosen such that $\theta(c)$ in (8.2.12) is defined, in the interval $[0, \frac{\pi}{2}]$, and the success probability of the algorithm is maximized. From (4.3.9), we obtain the following:

$$(8.2.13) \qquad R_c \, |0\rangle = \cos \theta(c) \, |0\rangle + \sin \theta(c) \, |0\rangle = \sqrt{1 - \frac{C^2}{\lambda(c)^2}} \, |0\rangle + \frac{C}{\lambda(c)} \, |1\rangle.$$

This implies

$$|\psi_2\rangle = V \, |\psi_1\rangle = \sum_{j=0}^{M-1} \beta_j R_c \, |0\rangle \, |c_j\rangle_n \, |u_j\rangle$$

$$(8.2.14)$$

$$= |0\rangle \sum_{j=0}^{M-1} \beta_j \sqrt{1 - \frac{C^2}{\lambda_j}} \, |c_j\rangle_n \, |u_j\rangle + |1\rangle \sum_{j=0}^{M-1} \beta_j \frac{C}{\lambda_j} \, |c_j\rangle_n \, |u_j\rangle.$$

Exercise 8.2.1. Show that the operator V is unitary.

As in (8.2.10), we see that

$$|\psi_3\rangle = |0\rangle \sum_{j=0}^{M-1} \beta_j \sqrt{1 - \frac{C^2}{\lambda_j}} \, |0\rangle_n \, |u_j\rangle + |1\rangle \sum_{j=0}^{M-1} \beta_j \frac{C}{\lambda_j} \, |0\rangle_n \, |u_j\rangle$$

$$(8.2.15)$$

$$= |0\rangle \, |0\rangle_n \sum_{j=0}^{M-1} \beta_j \sqrt{1 - \frac{C^2}{\lambda_j}} \, |u_j\rangle + |1\rangle \, |0\rangle_n \, C \, |x\rangle.$$

Exercise 8.2.2. Verify (8.2.15)

So we obtain the following result.

Theorem 8.2.3. *Measuring the first qubit of $|\psi_3\rangle$ gives $|1\rangle$ with probability $C \left\| \vec{x} \right\|$. If $|1\rangle$ is measured, then the final state in the HHL circuit is*

$$(8.2.16) \qquad\qquad |\psi_4\rangle = \frac{C}{\left\| \vec{x} \right\|} \, |x\rangle.$$

Proof. Measuring the first qubit of $|\psi_3\rangle$ means measuring the observable $O = (|0\rangle \langle 0| + |1\rangle \langle 1|) I_B$ where B is the quantum system comprising the second and third quantum registers. Therefore, the probability of measuring $|1\rangle$ is $C \left\| \vec{x} \right\|$ and if $|1\rangle$ is measured, then (8.2.16) holds. $\qquad\square$

We note that the proportionality factor $\frac{C}{\left\| \vec{x} \right\|}$ can be obtained from C and the probability of measuring $|1\rangle$. Also, if the measurement of the first qubit gives $|1\rangle$ but (8.2.5) does not hold, which in general is the case, then the final state is

$$(8.2.17) \qquad\qquad |\psi_4\rangle = \frac{C}{\left\| \vec{x}' \right\|} \, |x'\rangle$$

where $\vec{x}' = (x_0', \ldots, x_{M-1}')$ is an approximation of $\vec{x}$ and $|x'\rangle = \sum_{j=0}^{M-1} x_j' \, |u_j\rangle$.

8.3. Analysis and applications

This section presents the results of the complexity analysis of the HHL algorithm. As in the two previous chapters, we assume in the following that all quantum circuits are constructed using the elementary quantum gates from the platform specified in Section 4.12.2.

To state the complexity result, we need some further notations and assumptions. We use $M, A, \lambda_j, j \in \mathbb{Z}_M, \vec{b}, \vec{x}, \vec{x}', |b\rangle, |x\rangle$, and $|x'\rangle$ as introduced in the previous sections.

The *condition number* of A is

$$(8.3.1) \qquad \kappa = \frac{\max\{|\lambda_i| : i \in \mathbb{Z}_M\}}{\min\{|\lambda_i| : i \in \mathbb{Z}_M\}}.$$

It is assumed that the matrix A is s-sparse and efficiently row computable, which means that there are at most s nonzero entries per row, and, given a row index, the entries of the corresponding row can be computed in time $O(s)$.

The parameters C and n are chosen in such a way that $\left\| \vec{x} - \vec{x}' \right\| < \varepsilon$ for some error parameter $\varepsilon \in \mathbb{R}_{>0}$.

With the given notations and assumptions, the time complexity of the HHL algorithm can be expressed as

$$(8.3.2) \qquad \mathcal{O}\left(\frac{\log M \cdot s^2 \cdot \kappa^2}{\varepsilon} \right).$$

So, the complexity is only polylogarithmic in M as long as $s\kappa$ has this property. This means that if A is sparse and the condition number is small, the algorithm provides an exponential speedup compared to the best-known classical algorithms. However, the end result is a quantum state representing $\vec{x}'$, and as previously mentioned in Section 8.1, obtaining all M components of $\vec{x}'$ would take at least time M and thus undermine exponential speedup. But within the time complexity (8.3.2), it is possible to measure the final state of the HHL algorithm and obtain information about $\vec{x}$. This has interesting applications in various domains such as machine learning, data analysis, and optimization.

Foundations

To understand quantum algorithms and their underlying principles, a solid mathematical foundation is indispensable. To ensure the comprehensiveness of this book, this appendix provides important mathematical concepts and results used throughout.

The initial part of this appendix comprises fundamental notions such as numbers, relations, functions, and operations. Subsequently, we explain directed graphs, used to model Boolean circuits, and the asymptotic notation, which is an indispensable tool for algorithm analysis.

To comprehend the algorithms devised by Peter Shor, as discussed in Chapter 6, some number theory becomes essential. This is addressed in the next section and includes the exploration of the utility of continued fractions in determining good rational approximations, which is crucial in the order finding algorithm.

We also present fundamental concepts from algebra, encompassing groups, rings, and fields. In addition, many algorithms require familiarity with trigonometric identities and inequalities. These are presented in the concluding section.

A.1. Basics

A.1.1. Numbers.
We denote the usual sets of numbers as follows.

- $\mathbb{N}$ is the set of *natural numbers*; i.e., $\mathbb{N} = \{1, 2, \ldots\}$.
- $\mathbb{N}_0$ is the set of *natural numbers* including 0; i.e., $\mathbb{N}_0 = \{0, 1, 2, \ldots\}$.
- $\mathbb{Z}$ is the set of *integers*; i.e., $\mathbb{Z} = \{0, \pm 1, \pm 2, \ldots\}$.
- $\mathbb{Q}$ is the set of *rational numbers*; i.e., $\mathbb{Q} = \{\frac{p}{q} : p \in \mathbb{Z}, q \in \mathbb{N}\}$.
- $\mathbb{R}$ is the set of *real numbers*, i.e., the set of all numbers that can be represented by infinite decimals, like $\sqrt{2} = 1.414\ldots$ or $\pi = 3.14159\ldots$.
- $\mathbb{C}$ is the set of *complex numbers*, i.e., the set of all numbers $\gamma = \alpha + i\beta$ where α, β are real numbers and i is a square root of -1; i.e., $i^2 = -1$. In this representation,

α is called the *real part* of γ and is denoted by $\Re\gamma$ and β is called the imaginary part of γ and is denoted by $\Im\gamma$.

We note that

(A.1.1) $$\mathbb{N} \subset \mathbb{N}_0 \subset \mathbb{Z} \subset \mathbb{Q} \subset \mathbb{R} \subset \mathbb{C}.$$

For every $k \in \mathbb{N}$ we write

(A.1.2) $$\mathbb{Z}_k = \{0, 1, \ldots, k-1\}.$$

Furthermore, if $l \in \mathbb{Z}$, we denote by $l \bmod k$ the remainder of the division of l by k.

Example A.1.1. We have $\mathbb{Z}_5 = \{0, 1, 2, 3, 4\}$ and $123 \bmod 5 = 3$.

Next, we use the following notation.

Definition A.1.2. Let r be a real number. Then we set

(1) $\lfloor r \rfloor = \max\{z \in \mathbb{Z} : z \le r\}$,

(2) $\lceil r \rceil = \min\{z \in \mathbb{Z} : z > r\}$, and

(3) $\lfloor r \rceil$ to the uniquely determined integer z with $-\frac{1}{2} \le r - z < \frac{1}{2}$.

Example A.1.3. We have $\lfloor 1.3 \rfloor = 1$, $\lceil 1.3 \rceil = 2$, $\lfloor 1.3 \rceil = 1$, $\lfloor -1.3 \rfloor = -2$, $\lceil -1.3 \rceil = -1$, $\lfloor -1.3 \rceil = -1$.

A.1.2. Relations.

Definition A.1.4. (1) Let S and T be sets. The *Cartesian product* $S \times T$ of S and T is the set of all pairs (s, t) with $s \in S$ and $t \in T$; that is,

(A.1.3) $$S \times T = \{(s, t) : s \in S, t \in T\}.$$

(2) More generally, if $k \in \mathbb{N}$ and $S_1, \ldots, S_k$ are sets, then the *Cartesian product* $S_0 \times \cdots \times S_{k-1}$ of these sets is the set of all tuples $(s_1, \ldots, s_k)$ where $s_i \in S_i$, $i \in \mathbb{Z}_k$; i.e.,

(A.1.4) $$S_0 \times \cdots \times S_{k-1} = \{(s_0, \ldots, s_{k-1}) : s_i \in S_i, i \in \mathbb{Z}_k\}.$$

We also write $\prod_{i=0}^{k-1}$ for this Cartesian product.

Definition A.1.5. Let S and T be sets. A *relation* between S and T is a subset R of the Cartesian product $S \times T$. If $S = T$, then R is called a relation on S.

Example A.1.6. Consider the two sets $S = \{\text{"odd", "even"}\}$, $T = \mathbb{Z}$. Then "is the parity of" is a relation between S and T. Denote it by R. A pair (s, t) is in R if and only if s is the parity of t. For example, ("even", 2) is in R. Also, ("odd", -3) is in R. However, ("odd", 0) is not in R.

We introduce a few important notions regarding relations on a single set.

Definition A.1.7. Let S be a set and let $R \subset S \times S$ be a relation on S.

(1) The relation R is called *reflexive* if $(s, s) \in R$ for all $s \in S$.

(2) The relation R is called *symmetric* if for any pair (s, t) in S, the pair (t, s) is also in R.

(3) The relation R is called *antisymmetric* if $(s,t) \in R$ and $(t,s) \in R$ implies $s = t$ for all $s, t \in S$.

(4) The relation R is called *transitive* if for all $s, t, u \in S$ such that both (s,t) and (t,u) are in R, the pair (s,u) is also in R.

(5) The relation R is called an *equivalence relation* if it is reflexive, symmetric, and transitive.

Example A.1.8. Consider the relation $\leq$ on $\mathbb{Z}$. To be more explicit, this relation is defined as

(A.1.5) $$R = \{(s,t) : s, t \in \mathbb{Z}, s \leq t\}.$$

This relation is reflexive since $s \leq s$ for all $s \in \mathbb{Z}$. It is antisymmetric since $s \leq t$ and $t \leq s$ implies $s = t$ for all $s, t \in \mathbb{Z}$. The relation is also transitive, since $s \leq t$ and $t \leq u$ implies $s \leq u$ for all $s, t, u \in \mathbb{Z}$.

Definition A.1.9. Let S be a set, and let $R \subset S \times S$ be an equivalence relation on S.

(1) The *equivalence class* of an element $s \in S$ with respect to the relation R is the set $[s]_R = \{t \in S : (s,t) \in R\}$.

(2) The set of all equivalence classes of S with respect to R is written as S/R. An element of an equivalence class is called a *representative* of this equivalence class.

Theorem A.1.10. *If S is a set and R is an equivalence relation on S, then the equivalence classes of two elements S are either equal or disjoint. In other words, S is the disjoint union of the equivalence classes in S/R.*

Exercise A.1.11. Prove Theorem A.1.10.

Example A.3.3 shows an equivalence relation.

A.1.3. Functions. In this section, we introduce and discuss functions.

Definition A.1.12. A function is a triplet $f = (S, T, R)$ where S and T are sets and R is a relation between S and T that associates every element of S with exactly one element of T. This means that for every $s \in S$ there is exactly one $t \in T$ such that $(s,t) \in R$. This element t is denoted by $f(s)$. We will write the function as

(A.1.6) $$f : S \to T$$

or, more explicitly, as

(A.1.7) $$f : S \to T, \quad s \mapsto f(s).$$

Such a function is also called a *map* or *mapping* from S to T. The set of all functions (S, T, f) is denoted by S^T.

We introduce more terminology regarding functions.

Definition A.1.13. Let

(A.1.8) $$f : S \to T$$

be a function.

(1) The set S is called the *domain* and T is called the *codomain* of f.

(2) Any $s \in S$ is called an *argument* or *input* of f and $f(s)$ is the *value* or *image* of f on input x. We also call $f(s)$ the *image* of s under f and say that f *maps* s *to* $f(s)$.

(3) If $S' \subset S$, then we denote by $f(S')$ the set of the images of all arguments in the subset S'; i.e.,

(A.1.9) $$f(S') = \{f(s) : s \in S'\}.$$

We call $f(S')$ the *image* of S' under f.

(4) If $s \in S$ and $t \in T$ with $f(s) = t$, then we call s an *inverse image* of t under f.

(5) Let $T' \subset T$. The set of all inverse images of the elements of T' is denoted by $f^{-1}(T')$. Furthermore, the inverse image $f^{-1}(\{t\})$ of a single element $t \in T$ is denoted by $f^{-1}(t)$.

Definition A.1.14. (1) The *identity function* on a set S is the function $I_S : S \to S$, $s \mapsto s$. This function is also called the *identity map*, *identity mapping*, or *identity relation*.

(2) For sets A, B, C with $B \subset A$ and a map $f : A \to C$ we denote by $f|_B$ the *restriction* of f to B, i.e., the map $f|_B : B \to C, b \mapsto f(b)$.

Definition A.1.15. Let $f : S \to T$ be a function.

(1) The function f is called *injective, one-to-one*, or an *injection* if for all $s \in S$ there is exactly one inverse image of $f(s)$ under f, namely s.

(2) The function f is called *surjective, onto*, or a *surjection* if $f(S) = T$; i.e., for all $t \in T$ there is an argument $s \in S$ such that $f(s) = t$.

(3) The function f is called *bijective* or a *bijection* if it is injective and surjective.

(4) If f is a bijection and $S = T$, then f is called a *permutation* of S.

(5) If f is a bijection, then we denote by f^{-1} the function that maps $t \in T$ to its uniquely determined inverse image $s \in S$ under f. This function is called the *inverse* of f.

Example A.1.16. (1) Consider the function

(A.1.10) $$f : \mathbb{Z} \to \{\text{"even", "odd"}\}, \quad s \mapsto \text{parity of } s.$$

For example, we have $f(2) = \text{"even"}$ and $f(-3) = \text{"odd"}$. This function is not injective because there are many even and odd integers. However, the function is surjective because there exist even and odd integers.

(2) Next, consider the function

(A.1.11) $$f : \mathbb{Z} \to \mathbb{Z}, \quad s \mapsto s \bmod 11.$$

This function is neither injective nor surjective. For example, $f(0) = f(11) = 0$. Therefore, f is not injective. In addition, f is not surjective since $f(\mathbb{Z}) = \mathbb{Z}_{11}$. However, if we restrict the domain of f to $\mathbb{Z}_{11}$, then f becomes injective. Also, if we restrict the codomain of f to $\mathbb{Z}_{11}$, then f becomes surjective. In fact, the function $f : \mathbb{Z}_{11} \to \mathbb{Z}_{11}$, $s \mapsto s \bmod 11$ is the identity map on $\mathbb{Z}_{11}$. Another bijection is

$$(A.1.12) \qquad f : \mathbb{Z}_{11} \to \mathbb{Z}_{11}, \quad s \mapsto (s+1) \bmod 11.$$

This function is not the identity map.

Next, we introduce the composition of functions.

Definition A.1.17. Let S, T, U be sets and let

$$(A.1.13) \qquad f : T \to U, \quad g : S \to T$$

be functions. Then the *composition* of f and g is the map

$$(A.1.14) \qquad f \circ g : S \to U, \quad s \mapsto f(g(x)).$$

Example A.1.18. Consider the functions

$$(A.1.15) \qquad f : \mathbb{Z}_6 \to \mathbb{Z}_5, \quad x \mapsto x \bmod 5$$

and

$$(A.1.16) \qquad g : \mathbb{Z} \to \mathbb{Z}_6, \quad x \mapsto x \bmod 6.$$

Then

$$(A.1.17) \qquad f \circ g : \mathbb{Z} \to \mathbb{Z}_5, \quad x \mapsto f(g(x)) = (x \bmod 6) \bmod 5.$$

For instance, we have

$$(A.1.18) \qquad (f \circ g)(11) = f(g(11)) = (11 \bmod 6) \bmod 5 = 5 \bmod 5 = 0.$$

A.1.4. Operations. In order to be able to define algebraic structures, we introduce operations on a nonempty set S.

Definition A.1.19. A *binary operation* on S is a map

$$(A.1.19) \qquad \circ : S \times S \to S.$$

We write the image $\circ(s, s')$ of (s, s') under this map as $s \circ s'$.

Definition A.1.20. Let $\circ$ be an operation on S.

(1) The operation $\circ$ is called *associative* if $(a \circ b) \circ c = a \circ (b \circ c)$ for all $a, b, c \in S$.

(2) The operation $\circ$ is called *commutative* if $a \circ b = b \circ a$ for all $a, b \in S$.

(3) An element i of S is called the *identity element* with respect to $\circ$ if $a \circ i = i \circ a = a$ for all $a \in S$.

Example A.1.21. Addition and multiplication are binary operations on the sets of natural numbers $\mathbb{N}$ and on the set of integers $\mathbb{Z}$.

Let m be a positive integer. We define the binary operations addition and multiplication on $\mathbb{Z}_k$ as follows:

$$(A.1.20) \qquad \begin{aligned} +_m &: \mathbb{Z}_m \times \mathbb{Z}_m, \quad (a,b) \mapsto a +_m b = (a+b) \bmod m, \\ \cdot_m &: \mathbb{Z}_m \times \mathbb{Z}_m, \quad (a,b) \mapsto a \cdot_m b = (a \cdot b) \bmod m. \end{aligned}$$

A.1.5. Directed graphs. To define Boolean circuits in Chapter 1, directed graphs are required, which we introduce now.

Definition A.1.22. A directed graph is a pair $G = (V, E)$ where V is a nonempty set and E is a subset of V^2. An element of V is called a *vertex* of G and an element E is called an edge of G.

Figure A.1.1 gives an example of a directed graph.

Definition A.1.23. Let $G = (V, E)$ be a directed graph.

(1) An edge (u, v) of G is called an *outgoing edge* of u and an *incoming edge* of v.

(2) Let $k \in \mathbb{N}$. A sequence $(v_0, \ldots, v_k)$ is called a *path* in G if $(v_i, v_{i+1}) \in E$ for $0 \leq i < k$. The *length* of such a path is k and is called a *cycle* if $k > 0$ and $v_0 = v_k$.

(3) The graph G is called *acyclic* if it has no cycles.

Exercise A.1.24. (1) Find the incoming and outgoing edges of all vertices of the graph G in Figure A.1.1.

(2) Remove a minimum number of edges from G such that the graph becomes acyclic.

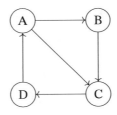

Figure A.1.1. Example of a directed graph.

A.2. The asymptotic notation

In order to compare the asymptotic behavior of functions, the following notation is used. It is especially useful in the complexity analysis of algorithms.

Definition A.2.1. Let $X \subset \mathbb{R}_{\geq 0}$ and let $f, g : X \to \mathbb{R}_{\geq 0}$ be functions. Then we write the following.

(1) $f = o(g)$ if for all $\varepsilon > 0$ there is $x_0 > 0$ such that for all $x > x_0$ we have $f(x) \leq \varepsilon g(x)$.

(2) $f = \omega(g)$ if $g = o(f)$.

(3) $f = O(g)$ if there are $C > 0$ and $x_0 > 0$ such that for all $x > x_0$ we have $f(x) \le Cg(x)$.

(4) $f = \Omega(g)$ if $g = O(f)$.

(5) $f = \Theta(g)$ if $f = O(g)$ and $g = O(f)$.

This terminology can be asymptotically interpreted as follows: If $f = O(g)$, then f grows not faster than g. If $f = o(g)$, then f grows much slower than g. If $f = \Theta(g)$, then these functions grow equally fast.

Example A.2.2. We have $n^2 = o(n^3 + 1)$, $n^2 = O(2n^2 + n + 1)$, $n^2 = \Theta(2n^2 + n + 1)$.

Exercise A.2.3. Prove the statements in Example A.2.2.

A.3. Number theory

We present the concepts and results from number theory that are used in this book.

A.3.1. Divisibility. We begin by discussing divisibility in $\mathbb{Z}$.

Definition A.3.1. We say that an integer m *divides* an integer a if there is an integer n with $a = nm$. If m divides a, then m is called a *divisor* of a, a is called a *multiple* of m, and we write $m \mid a$. We also say that a is *divisible* by m. If m is not a divisor of a, then we write $m \nmid a$.

Example A.3.2. We have $13 \mid 182$ because $182 = 14 * 13$. Likewise, we have $-5 \mid 30$ because $30 = (-6) * (-5)$. The divisors of 30 are $\pm 1, \pm 2, \pm 3, \pm 5, \pm 6, \pm 10, \pm 15, \pm 30$.

We note that any integer m divides 0 because $0 = m * 0$. The only integer that is divisible by 0 is 0 because $a = 0 * m$ implies $a = 0$.

We also use divisibility in $\mathbb{Z}$ in the following example of an equivalence relation.

Example A.3.3.

Let s and t be integers, and let m be a positive integer. We write

(A.3.1) $$s \equiv t \bmod m$$

if m divides $t - s$. Consider the relation

(A.3.2) $$R = \{(s, t) : s, t \in \mathbb{Z}, s \equiv t \bmod m\}.$$

It is called a *congruence relation*. It is reflexive because m divides $s - s = 0$ for all $s \in \mathbb{Z}$. This relation is symmetric since m is a divisor of $t - s$ if and only if m is a divisor of $s - t$. Finally, the relation is transitive. To see this, we let s, t, u be integers. Suppose that

(A.3.3) $$s \equiv t \bmod m, \quad t \equiv u \bmod m.$$

Then m divides $t - s$ and $u - t$. So we can write

(A.3.4) $$t - s = xm, \quad u - t = ym$$

with two integers x, y. Therefore, we have the following:

(A.3.5) $$u - s = (u - t) + (t - s) = xm + ym = (x + y)m.$$

This implies that m divides $u - s$ which means that

(A.3.6) $$s \equiv u \bmod m.$$

Since the congruence relation is reflexive, symmetric, and transitive, it is an equivalence relation and we have

(A.3.7) $$\mathbb{Z}/R = \{\{i + m\mathbb{Z}\} : 0 \le i < m\}.$$

For instance, if $m = 3$, then there are the three equivalence classes $0 + 3\mathbb{Z} = \{0, \pm 3, \pm 6, \dots\}, 1 + 3\mathbb{Z} = \{\dots, -5, -2, 1, 4, 7, \dots\}$, and $2 + 3\mathbb{Z} = \{\dots, -4, -1, 2, 5, 8, \dots\}$. Typically, $\mathbb{Z}/R$ is written as $\mathbb{Z}/m\mathbb{Z}$.

A.3.2. Greatest common divisor. Our next topic is the greatest common divisor of two integers and we will explain that the Euclidean algorithm computes it efficiently. The proofs of all the results presented in this section can be found in [**Buc04**, Section 1.10].

Definition A.3.4. A *common divisor* of a and b is an integer that divides both a and b.

Proposition A.3.5. *Among all common divisors of two integers a and b, which are not both zero, there is exactly one greatest divisor (with respect to $\le$). It is called the* greatest common divisor (gcd) *of a and b.*

For completeness, we set the greatest common divisor of 0 and 0 to 0 (that is, $\gcd(0, 0) = 0$). Therefore, the greatest common divisor of two numbers is never negative.

We present another useful characterization of the greatest common divisor.

Proposition A.3.6. *There is exactly one nonnegative common divisor of a and b, which is divisible by all other common divisors of a and b, namely the greatest common divisor of a and b.*

Example A.3.7. The greatest common divisor of 18 and 30 is 6. The greatest common divisor of -10 and 20 is 10. The greatest common divisor of -20 and -14 is 2. The highest common divisor of 12 and 0 is 12.

An important property of the greatest common divisor is that it can be computed very efficiently by the Euclidean Algorithm 1.1.16. The next proposition states its complexity.

Proposition A.3.8. *The Euclidean algorithm uses time $O((\text{bitLength } a)(\text{bitLength } b))$ and space $O(\text{bitLength } a + \text{bitLength } b)$ to compute $\gcd(a, b)$.*

In several contexts, we use Euler's totient function which is defined next.

Definition A.3.9. Let $N \in \mathbb{N}$.

(1) By $\mathbb{Z}_m^*$ we denote the set of all $a \in \mathbb{Z}_m$ with $\gcd(a, m) = 1$.

(2) We write $\varphi(m) = |\mathbb{Z}_m^*|$.

(3) The function that sends $m \in \mathbb{N}$ to $\varphi(m)$ is referred to as *Euler's totient function*.

Example A.3.10. We have $\mathbb{Z}_1^* = \emptyset$, $\varphi(1) = 0$, $\mathbb{Z}_2^* = \{1\}$, $\varphi(2) = 1$, $\mathbb{Z}_{15}^* = \{1, 2, 4, 7, 8, 11, 13, 14\}$, and $\varphi(15) = 8$.

Exercise A.3.11. Let $m \in \mathbb{N}$, $m > 1$. Prove that $\mathbb{Z}_m^*$ is a group with respect to multiplication modulo m.

A.3.3. Least common multiple. We also require the least common multiple of two integers.

Definition A.3.12. Let $n \in \mathbb{N}$ and let $a_0, \ldots, a_{n-1}$ be nonzero integers. Then the *least common multiple* of these integers is the smallest positive integer that is a multiple of all a_i. It is denoted by $\mathrm{lcm}(a_0, \ldots, a_{n-1})$.

Example A.3.13. The least common multiple of $2, 3, 4$ is $\mathrm{lcm}(2, 3, 4) = 12$.

The next exercise justifies the definition of the least common multiple and finds an algorithm for computing it.

Exercise A.3.14. (1) Prove the existence and uniqueness of the least common multiple of finitely many nonzero integers. Why do these numbers have to be nonzero?

(2) Utilize the Euclidean algorithm to devise an algorithm for computing the least common multiple in quadratic running time.

A.3.4. Prime factor decomposition. A famous quantum algorithm by Peter Shor, which we discuss in Chapter 6, can compute the prime factor decomposition of a positive integer in polynomial time. We now explain its mathematical foundation.

Definition A.3.15. An integer $p > 1$ is called a *prime number* if it has exactly two positive divisors, namely 1 and p. Instead of "prime number" we also simply say "prime". An integer $a > 1$ that is not a prime is called *composite* . If the prime p divides the integer a, then p is called the *prime divisor* of a.

Example A.3.16. The first eight prime numbers are $2, 3, 5, 7, 11, 13, 17, 19$. As of 2023, the largest known prime number is $2^{82,589,933} - 1$.

We state the fundamental theorem of arithmetic. It is also called the unique factorization theorem and goes back to Euclid.

Theorem A.3.17. *Every integer $a > 1$ can be written as the product of prime numbers. Up to permutation, the factors in this product are uniquely determined.*

Example A.3.18. The French mathematician Pierre de Fermat (1601–1665) thought that all of the so-called *Fermat numbers*

$$F_i = 2^{2^i} + 1$$

are primes. In fact, $F_0 = 3$, $F_1 = 5$, $F_2 = 17$, $F_3 = 257$, and $F_4 = 65537$ are prime numbers. However, in 1732 Euler discovered that $F_5 = 641 * 6700417$ is composite. Both factors in this decomposition are primes. F_6, F_7, F_8, and F_9 are also composite. The factorization of F_6 was found in 1880 by Landry and Le Lasseur. The factorization of F_7 was found in 1970 by Brillhart and Morrison. The factorization of F_8 was computed in 1980 by Brent and Pollard, and F_9 was factored in 1990 by Lenstra, Lenstra, Manasse, and Pollard (see [**LLMP93**] where also references for the other results mentioned in this example can be found). This shows the difficulty of the factoring problem. But on the other hand, we also see that there is considerable progress. It took until 1970 to factor the 39-digit number F_7, but only 20 years later the 155-digit number F_9 was factored.

A.3.5. The continued fraction algorithm. In this section, we present the continued fraction algorithm (CFA) and its properties. It is used in Shor's order finding algorithm to compute good rational approximations.

We start with an example.

Example A.3.19. The continued fraction $[a_0, a_1, a_2] = [1, 2, 3]$ represents the rational number

$$(A.3.8) \qquad [1, 2, 3] = 1 + \cfrac{1}{2 + \cfrac{1}{3}}.$$

This rational number is

$$(A.3.9) \qquad [1, 2, 3] = 1 + \cfrac{1}{2 + \cfrac{1}{3}} = 1 + \cfrac{1}{\frac{7}{3}} = 1 + \frac{3}{7} = \frac{10}{7}.$$

Definition A.3.20. Let $n \in \mathbb{N}_0$ and let $(a_0, a_1, \dots, a_n) \in \mathbb{Q}_{\geq 0} \times \mathbb{Q}_{>0}^n$. Then we set

$$(A.3.10) \qquad [a_0, a_1, \dots, a_n] = a_0 + \cfrac{1}{a_1 + \cfrac{1}{a_2 + \cfrac{}{\ddots + \cfrac{1}{a_{n-1} + \cfrac{1}{a_n}}}}}.$$

This defines the map

$$(A.3.11) \qquad \mathbb{Q}_{\geq 0} \times \mathbb{Q}_{>0}^n \to \mathbb{Q} : (a_0, a_1, \dots, a_n) \mapsto [a_0, \dots, a_n].$$

We make some remarks concerning this definition. Let $n \in \mathbb{N}_0$ and $(a_0, \dots, a_n) \in \mathbb{Q}_{\geq 0} \times \mathbb{Q}_{>0}^n$. Then we will denote by $[a_0, \dots, a_n]$ the rational number from Definition A.3.20 but also the sequence $(a_0, \dots, a_n)$. This will make our conversation easier. However, for each rational number, there are multiple sequences that can be used to express it in this manner. For example, for $n > 0$ we have

$$(A.3.12) \qquad [\underbrace{a_0, \dots, a_n}_{n}] = [\underbrace{a_0, \dots, a_{n-1} + \frac{1}{a_n}}_{n-1}].$$

Note that the sequence on the left side of (A.3.12) has length n, while the sequence on the right side has length $n - 1$. The following lemma generalizes this observation.

Lemma A.3.21. *Let $n, k \in \mathbb{N}_0$, $(a_0, \dots, a_n) \in \mathbb{Q}_{\geq 0} \times \mathbb{Q}_{>0}^n$, and $(b_0, \dots, b_k) \in \mathbb{Q}_{>0}^{k+1}$. Then we have*

$$(A.3.13) \qquad [a_0, \dots, a_n, [b_0, \dots, b_k]] = [a_0, \dots, a_n, b_0, \dots, b_k].$$

Proof. We prove the assertion by induction on k. For $k = 0$, it follows from the fact that $[b_0] = b_0$. Now, assume that the assertion holds for $k - 1$. The induction hypothesis and (A.3.12) imply

$$[a_0, \ldots, a_n, [b_0, \ldots, b_k]]$$

$$= \left[a_0, \ldots, a_n, \left[b_0, \ldots, b_{k-1} + \frac{1}{b_k}\right]\right]$$

(A.3.14)
$$= \left[a_0, \ldots, a_n, b_0, \ldots, b_{k-1} + \frac{1}{b_k}\right]$$

$$= [a_0, \ldots, a_n, b_0, \ldots, b_k]. \qquad \square$$

Now we define finite simple continued fractions.

Definition A.3.22. A *finite simple continued fraction* is a finite sequence $(a_0, a_1, \ldots, a_n)$ $\in \mathbb{N}_0 \times \mathbb{N}^n$. It has *length n* and *represents the rational number* $[a_0, \ldots, a_n]$.

As in the general case, we write $[a_0, \ldots, a_n]$ for a continued fraction $(a_0, \ldots a_n)$; i.e., we write it in the same way as the rational number represented by it. Example A.3.19 shows a finite simple continued fraction.

Next, we present an algorithm that, given a nonnegative rational number, finds a continued fraction that represents it.

Algorithm A.3.23. Continued fraction algorithm

Input: $p \in \mathbb{N}_0, q \in \mathbb{N}$
Output: A continued fraction $[a_0, a_1, \ldots, a_n] \in \mathbb{N}^{n+1}$ that represents $\frac{p}{q}$

  1:   $\text{CFA}(p, q)$
  2:     $r_{-1} \leftarrow p$
  3:     $r_0 \leftarrow q$
  4:     $i \leftarrow -1$
  5:     **repeat**
  6:       $i \leftarrow i + 1$
  7:       $a_i \leftarrow \lfloor r_{i-1}/r_i \rfloor$
  8:       $r_{i+1} \leftarrow r_{i-1} - a_i r_i$
  9:     **until** $r_{i+1} = 0$
10:     $n \leftarrow i$
11:     **return** $([a_0, \ldots, a_n])$
12: **end**

Example A.3.24. Let $p = 15$ and $q = 13$. The sequence of values r_i and a_i from Algorithm A.3.23 is shown in Table A.3.1. We verify that $[1, 6, 2] = \frac{15}{13}$. We have

(A.3.15)
$$[1, 6, 2] = 1 + \cfrac{1}{6 + \cfrac{1}{2}} = 1 + \cfrac{1}{\cfrac{13}{2}} = 1 + \frac{2}{13} = \frac{15}{13}.$$

Table A.3.1. Run of the continued fraction algorithm with input $p = 15, q = 13$.

i	-1	0	1	2
r_i	15	13	2	1
a_i		1	6	2

As shown in Exercise A.3.25, the continued fraction algorithm is a variant of the Euclidean algorithm that outputs the sequence $(a_0, \ldots, a_n)$ of quotients that are computed in this algorithm.

Exercise A.3.25. Use the notation from Algorithm A.3.23 and show that $r_n = \gcd(p, q)$.

The next proposition shows that the continued fraction algorithm yields the correct result.

Proposition A.3.26. *On input of $p, q \in \mathbb{N}$, Algorithm A.3.23 computes a continued fraction $[a_0, a_1, \ldots, a_n]$ for some $n \in \mathbb{N}_0$ that represents $\frac{p}{q}$ and satisfies $a_n > 1$ if $n > 0$. It is the only continued fraction with these properties.*

Proof. We use the notation of Algorithm A.3.23. After each iteration of the **repeat** loop, we have

$$(A.3.16) \qquad r_{i-1} = a_i r_i + r_{i+1} \quad \text{and} \quad 0 \leq r_{i+1} < r_i \quad \text{for } 0 \leq i \leq n.$$

This shows that the sequence $r_0, r_1, \ldots$ is strictly decreasing. Therefore, the algorithm terminates.

For $0 \leq i \leq n$ let $\alpha_i = [a_i, \ldots, a_n]$. Then we have

$$(A.3.17) \qquad \alpha_n = a_n \quad \text{and} \quad \alpha_i = a_i + \frac{1}{\alpha_{i+1}} \quad \text{for } 0 \leq i < n.$$

We show by induction on $i = n, n - 1, \ldots, 0$ that

$$(A.3.18) \qquad \alpha_i = \frac{r_{i-1}}{r_i} \quad \text{and} \quad a_i = \lfloor \alpha_i \rfloor.$$

For $i = 0$, this shows that $[a_0, \ldots, a_n] = \frac{p}{q}$.

The base case where $i = n$ is obtained from $r_{n-1} = a_n r_n$, which follows from (A.3.16) and $r_{n+1} = 0$. For the inductive step, let $i \in \mathbb{N}, n \geq i > 0$ and assume that (A.3.18) holds. Then (A.3.16) implies

$$(A.3.19) \qquad \frac{r_{i-2}}{r_{i-1}} = a_{i-1} + \frac{1}{\frac{r_{i-1}}{r_i}} = a_{i-1} + \frac{1}{\alpha_i} = \alpha_{i-1}.$$

Since $\alpha_i = \frac{r_{i-1}}{r_i} > 1$ by (A.3.16), this implies $a_{i-1} = \lfloor \alpha_i \rfloor$. This completes the induction proof.

Next, we assume that $n > 0$ and show that $a_n > 1$. Since $r_{n+1} = 0$, we have $r_{n-1} = a_n r_n$. So $a_n = 1$ would imply $r_{n-1} = r_n$ which contradicts (A.3.16).

Finally, we prove the uniqueness of $[a_0, \ldots, a_n]$. Let $k \in \mathbb{N}, k \geq n$, and let $[b_0, \ldots, b_k]$ be a continuous fraction that represents $\frac{p}{q}$ with $b_k > 1$ for $k > 0$. If $k = 0$, then we have $n = 0$ and $\frac{p}{q} = a_0 = b_0$. Assume that $k > 0$. For $0 \leq i \leq k$, let $\beta_i = [b_i, \ldots, b_k]$. Then we have

(A.3.20) $$\beta_k = b_k \quad \text{and} \quad \beta_i = b_i + \frac{1}{\beta_{i+1}} \quad \text{for } 0 \leq i < k.$$

Since $b_k > 1$, it follows from (A.3.20) that

(A.3.21) $$\beta_i > 1 \quad \text{for } 0 \leq i \leq k.$$

From (A.3.20) and (A.3.21) we obtain

(A.3.22) $$b_i = \lfloor \beta_i \rfloor \quad \text{for } 1 \leq i \leq k.$$

We show by induction on $i = 0, \ldots, n$ that

(A.3.23) $$\alpha_i = \beta_i \quad \text{and} \quad a_i = b_i.$$

By assumption, we have $\alpha_0 = [a_0, a_1, \ldots, a_n] = \frac{r_0}{r_{i-1}} = [b_0, \ldots, b_k] = \beta_0$ and by (A.3.18) and (A.3.22) $b_0 = a_0$. Now assume that $0 \leq i < n$ and that (A.3.23) holds. Then it follows from (A.3.17) and (A.3.20) that

(A.3.24) $$\beta_{i+1} = \frac{1}{b_i - \beta_i} = \frac{1}{a_i - \alpha_i} = \alpha_{i+1}.$$

So (A.3.18) and (A.3.22) imply

(A.3.25) $$b_{i+1} = \lfloor \beta_{i+1} \rfloor = \lfloor \alpha_{i+1} \rfloor = a_{i+1}.$$

Finally, since by (A.3.20) we have $\beta_i > b_i$ for $1 \leq i < k$, it follows from $\beta_n = \alpha_n = a_n = b_n$ that $n = k$. $\square$

Definition A.3.27. Let $\alpha \in \mathbb{Q}_{\geq 0}$. The uniquely determined continued fraction from Proposition A.3.26 that represents α is called the *continued fraction expansion* of α.

We estimate the running time of the continued fraction algorithm.

Proposition A.3.28. *Let $p \in \mathbb{N}_0$, $q \in \mathbb{N}$ and let $l = \max\{\text{bitLength}(p), \text{bitLength}(q)\}$. Then the number of iterations in the continued fraction algorithm is $O(l)$ and its running time and space requirement is $O(l^2)$.*

Proof. The statement can be proved using the techniques from Section 1.6.4 in **[Buc04]**. Note that the space is required to represent continued fractions. $\square$

We now introduce convergents of continued fractions.

Definition A.3.29. Let $[a_0, \ldots, a_n]$ be a continued fraction. Then for $0 \leq i \leq n$ the rational numbers $[a_0, \ldots, a_i]$ are called its *convergents*.

Example A.3.30. The convergents of the continued fraction expansion $[1, 6, 2]$ of $\frac{15}{13}$ are $[1] = 1$, $[1, 6] = 1 + \frac{1}{6} = \frac{7}{6}$, and $[1, 6, 2] = \frac{15}{13}$.

The following proposition provides a method for computing convergents.

Proposition A.3.31. *Let $n \in \mathbb{N}_0$ and $(a_0, \dots, a_n) \in \mathbb{Q}_{\geq 0} \times Q_{>0}^n$. Set*

$$\text{(A.3.26)} \qquad p_{-2} = 0, \quad p_{-1} = 1, \quad q_{-2} = 1, \quad q_{-1} = 0$$

and for $0 \leq i \leq n$ let

$$\text{(A.3.27)} \qquad p_i = a_i p_{i-1} + p_{i-2}, \quad q_i = a_i q_{i-1} + q_{i-1}.$$

Then

$$\text{(A.3.28)} \qquad [a_0, \dots, a_i] = \frac{p_i}{q_i} \quad \text{for } 0 \leq i \leq n.$$

Proof. We prove the assertion by induction on n. For the base case, we note that

$$\text{(A.3.29)} \qquad \frac{p_0}{q_0} = \frac{a_0}{1} = [a_0]$$

and

$$\text{(A.3.30)} \qquad \frac{p_1}{q_1} = \frac{a_1 a_0 + 1}{a_1} = a_0 + \frac{1}{a_1} = [a_0, a_1].$$

For the inductive step, let $i \in \mathbb{N}_0$, $0 \leq i < n$ and assume that the assertion of the proposition holds for i. The induction hypothesis gives

$$
\begin{aligned}
[a_0, \dots, a_{i+1}] &= \left[a_0, \dots, a_i + \frac{1}{a_{i+1}} \right] \\
&= \frac{\left(a_i + \frac{1}{a_{i+1}} \right) p_{i-1} + p_{i-2}}{\left(a_i + \frac{1}{a_{i+1}} \right) q_{i-1} + q_{i-2}} \\
\text{(A.3.31)} \qquad &= \frac{a_i p_{i-1} + p_{i-2} + \frac{p_{i-1}}{a_{i+1}}}{a_i q_{i-1} + q_{i-2} + \frac{q_{i-1}}{a_{i+1}}} = \frac{p_i + \frac{p_{i-1}}{a_{i+1}}}{q_i + \frac{q_{i-1}}{a_{i+1}}} \\
&= \frac{a_{i+1} p_i + p_{i-1}}{a_{i+1} q_i + q_{i-1}} = \frac{p_{i+1}}{q_{i+1}}. \qquad \square
\end{aligned}
$$

Exercise A.3.32. Let $[a_0, \dots, a_n]$ be a finite simple continued fraction. Use the notation from Proposition A.3.31 and show that $p_{i-1} q_i + p_i q_{i-1} = (-1)^i$ for $-1 \leq i \leq n$ and $\gcd(p_i, q_i) = 1$ for $-2 \leq i \leq n$.

Now we show that there are exactly two finite simple continued fractions that represent a given positive rational number.

Proposition A.3.33. *Let $\alpha \in \mathbb{Q}_{>0}$ and let $[a_0, \dots, a_n]$ be the continued fraction expansion of α. Then $[a_0, \dots, a_{n-1}, a_n - 1, 1]$ is the only other finite simple continued fraction that represents α.*

Proof. For $-2 \leq i \leq n$ denote by p_i, q_i the integers from Proposition A.3.31 for the continued fraction $[a_0, \dots, a_n]$, and for $-2 \leq i \leq n+1$ denote by p_i', q_i' the corresponding integers for the continued fraction $[a_0, \dots, a_n - 1, 1]$. Then we have

$$\text{(A.3.32)} \quad p_n' = (a_n - 1) p_{n-1} + p_{n-2}, \quad p_{n+1}' = p_n' + p_{n-1} = a_n p_{n-1} + p_{n-2} = p_n.$$

In the same way, it can be verified that $q'_{n+1} = q_n$. This shows that $[a_0, \ldots, a_n - 1, 1] = \alpha$.

The uniqueness is proved in Exercise A.3.34. $\qquad\square$

Exercise A.3.34. Verify the uniqueness claim in Proposition A.3.33.

Finally, we prove that sufficiently good approximations to a positive rational number are convergents of its continued fraction expansion.

Proposition A.3.35.

$$\left| \frac{p}{q} - \alpha \right| \le \frac{1}{2q^2}. \tag{A.3.33}$$

Then $\frac{p}{q}$ is a convergent of the continued expansion of the fraction of α.

Proof. If $\alpha = 0$, then $p = 0$ and $\frac{p}{q}$ is the only convergent of α. Let $\alpha \ne 0$. Let δ be the rational number with

$$\alpha = \frac{p}{q} + \frac{\delta}{2q^2}. \tag{A.3.34}$$

Then the assumption of the proposition implies $|\delta| \le 1$. By Proposition A.3.33, we can choose a simple continued fraction $[a_0, \ldots, a_n]$ that represents $\frac{p}{q}$ such that

$$\operatorname{sign} \delta = (-1)^n. \tag{A.3.35}$$

For $-2 \le i \le n$, define p_i, q_i as in Proposition A.3.31. Then we have $\frac{p}{q} = \frac{p_n}{q_n}$. Set

$$\lambda = \frac{2}{|\delta|} - \frac{q_{n-1}}{q_n}. \tag{A.3.36}$$

Then $q_{n-1} > q_n$ and $|\delta| \le 1$ imply

$$\lambda > 2 - 1 = 1. \tag{A.3.37}$$

Also, we have

$$\lambda p_n + p_{n-1} \frac{2p_n}{|\delta|} - \frac{p_n q_{n-1}}{q_n} + p_{n-1} \tag{A.3.38}$$

and

$$\lambda q_n + q_{n-1} = \frac{2q_n}{|\delta|}. \tag{A.3.39}$$

It follows from Exercise A.3.32 and (A.3.35) that

$$\frac{\lambda p_n + p_{n-1}}{\lambda q_n + q_{n-1}} = \frac{p_n}{q_n} - \frac{(p_n q_{n-1} - p_{n-1} q_n)|\delta|}{2q_n^2} = \frac{p_n}{q_n} - \frac{\delta}{2q_n^2} = \alpha. \tag{A.3.40}$$

So it follows from Proposition A.3.31 that

$$\alpha = [a_0, \ldots, a_{n-1}, \lambda]. \tag{A.3.41}$$

If $[b_0, \ldots, b_k]$ is the continued fraction expansion of λ, then (A.3.37) implies $b_0 > 0$. Therefore, it follows from Lemma A.3.21 that $\alpha = [a_0, \ldots, a_n, b_0, \ldots, b_k]$. Since $b_k > 1$, this is the continued fraction expansion of α. So $\frac{p_n}{q_n} = [a_0, \ldots, a_n]$ is a convergent of α. $\qquad\square$

Example A.3.36. We note that

$$(A.3.42) \qquad \left| \frac{15}{13} - \frac{7}{6} \right| = \frac{1}{78} \leq \frac{1}{2*6^2} = \frac{1}{72}.$$

So Proposition A.3.35 predicts that $\frac{7}{6}$ is a convergent of the continued fraction expansion of $\frac{15}{13}$ which we have shown in Example A.3.30.

A.4. Algebra

We introduce a few basic concepts of algebra that are required in this book.

A.4.1. Semigroups.

Definition A.4.1. A *semigroup* is a pair $(S, \circ)$ where S is a nonempty set and $\circ$ is an associative binary operation on S. If it is clear from the context which operation $\circ$ we refer to, then we also write S instead of $(S, \circ)$.

Next, we introduce some notions concerning semigroups.

Definition A.4.2. Let $(S, \circ)$ be a semigroup.

(1) The semigroup is called *abelian* or *commutative* if the operation $\circ$ is commutative.

(2) The semigroup is called a *monoid* if S contains an identity element with respect to $\circ$.

Exercise A.4.3. Prove that $\mathbb{N}$ is not a monoid with respect to addition.

We show that the identity element and the inverses in monoids are uniquely determined.

Proposition A.4.4. *A monoid has exactly one identity element.*

Proof. Let i and i' be identity elements in S. Then $i = i \circ i' = i'$. $\qquad\square$

Definition A.4.5. An element u of a monoid $(S.\circ)$ with unit element 1 is called *invertible* of S if there is $u' \in S$ with $u \circ u' = u' \circ u = 1$.

Proposition A.4.6. *Every invertible element of a monoid has a uniquely determined inverse.*

Proof. Suppose that u is a unit of a monoid $(S, \circ)$ with the unit element 1 and that u' and u'' are inverses of u. Then $u' = u \circ 1 = u' \circ (u \circ u'') = (u' \circ u) \circ u'' = 1 \circ u'' = u''$. $\qquad\square$

A.4.2. Groups. We now define groups.

Definition A.4.7. (1) A *group* is a monoid in which every element is invertible.

(2) A *commutative* or *abelian* group is a group in which the group operation is commutative.

Proposition A.4.8. *If $(S, \circ)$ is a monoid and U is the set of all invertible elements in S, then $(U, \circ)$ is a group, called the* unit group *of $(S, \circ)$.*

Exercise A.4.9. Prove Proposition A.4.8.

We give a few concrete examples of semigroups, monoids, and groups.

Example A.4.10. The claims in this example are shown in Exercise A.4.11.

(1) $(\mathbb{N}, +)$ is an abelian semigroup. But $(\mathbb{N}, +)$ is not a monoid since there is no identity element in this semigroup.

(2) $(\mathbb{Z}, +), (\mathbb{Q}, +), (\mathbb{R}, +)$, and $(\mathbb{C}, +)$ are abelian groups with identity element 0.

(3) $(\mathbb{N}, \cdot)$ and $(\mathbb{Z}, \cdot)$ are abelian monoids with identity element 1. But they are not groups. In fact, the only unit in $(\mathbb{N}, \cdot)$ is 1 and the only units in $(\mathbb{Z}, \cdot)$ are ± 1.

(4) $(\mathbb{Q} \setminus \{0\}, \cdot), (\mathbb{R} \setminus \{0\}, \cdot)$, and $(\mathbb{C} \setminus \{0\}, \cdot)$ are abelian groups with identity element 1.

(5) If $m \in \mathbb{N}$, then $(\mathbb{Z}_m, +_m)$ is a finite abelian group.

(6) If $m \in \mathbb{N}$, then $(\mathbb{Z}_m^*, \cdot_m)$ is a finite abelian group.

Exercise A.4.11. Show that the claims in Example A.4.10 are correct.

Exercise A.4.12. Let m be a positive integer. Determine if each of $(\mathbb{Z}_m, +_m)$ and $(\mathbb{Z}_m, \cdot_m)$ is a semigroup (abelian?), a monoid, or a group. If applicable, determine the respective identity elements and unit groups.

We also define the order of elements of finite groups. This notion is motivated by the following observation.

Proposition A.4.13. *Let G be a finite group and let $g \in G$. Then there is $n \in \mathbb{N}$ such that $g^n = 1$.*

Exercise A.4.14. Prove Proposition A.4.13.

Here is the definition of element orders and the order of an integer modulo another integer.

Definition A.4.15. (1) Let G be a finite group and let $g \in G$. Then the smallest positive integer n with $g^n = 1$ is called the *order* of g in G.

(2) Let $N \in N$ and let $a \in \mathbb{Z}_N^*$. Then the order of a in the multiplicative group $\mathbb{Z}_N^*$ is called the *order of a modulo N*.

Example A.4.16. The order of 3 modulo 4 is 3 because $2^2 \equiv 4 \bmod 7$ and $2^3 \equiv 8 \equiv 1 \bmod 7$.

For a further discussion of the order of elements of finite abelian groups we refer to [**Buc04**, Sections 2.9 and 2.14].

Exercise A.4.17. Let $m \in \mathbb{N}$, let $p_1, \ldots, p_m$ be prime numbers, and let $N = \prod_{i=1}^m p_i^{e_i}$ where $e_1, \ldots, e_m$ are positive integers. Also, let $a \in \mathbb{Z}$ such that $\gcd(a, N) = 1$. Show that the order of a modulo N is the least common multiple of the orders of a modulo $p_i^{e_i}$ for $1 \le i \le n$.

A.4.3. The symmetric group. In many contexts, for example, in Section 4.5, symmetric groups play an important role. We discuss them in this section.

Proposition A.4.18. *Let S be a nonempty set. Then the following hold.*

(1) *The set of all permutations of S is a group with respect to composition. It is called the* symmetric group on S.

(2) *If S is finite, then the symmetric group on S has order* $|S|!$.

Exercise A.4.19. Prove Proposition A.4.18.

In the next definition, a special symmetric group is introduced.

Definition A.4.20. Let $n \in \mathbb{N}$. The *symmetric group of degree n* is the group of permutations of $\mathbb{Z}_n$. It is denoted by S_n. Also, if $\pi \in S_n$, then we write π as

$$(A.4.1) \qquad \begin{pmatrix} 0 & 1 & 2 & \dots & n-1 \\ \pi(0) & \pi(1) & \pi(2) & \dots & \pi(n-1) \end{pmatrix}$$

or simply as the sequence $(\pi(0), \pi(1), \dots, \pi(n-1))$.

Exercise A.4.21. Let $n \in \mathbb{N}$. Show that $S_n = n! = 2 \cdot 3 \cdot 4 \cdots n$.

Example A.4.22. There are $3! = 6$ permutations of degree 3. They are

$$(A.4.2) \qquad \begin{pmatrix} 0 & 1 & 2 \\ 0 & 1 & 2 \end{pmatrix}, \quad \begin{pmatrix} 0 & 1 & 2 \\ 0 & 2 & 1 \end{pmatrix}, \quad \begin{pmatrix} 0 & 1 & 2 \\ 1 & 0 & 2 \end{pmatrix},$$
$$\begin{pmatrix} 0 & 1 & 2 \\ 1 & 2 & 0 \end{pmatrix}, \quad \begin{pmatrix} 0 & 1 & 2 \\ 2 & 0 & 1 \end{pmatrix}, \quad \begin{pmatrix} 0 & 1 & 2 \\ 2 & 2 & 0 \end{pmatrix}.$$

We introduce transpositions.

Definition A.4.23. A *transposition* is a permutation in a symmetric group over some set S that exchanges two elements of S and does not change the other elements. If a and b are the exchanged elements, then we write the transposition as (a, b).

Example A.4.24. The transpositions in S_3 are the permutations

$$(A.4.3) \qquad \begin{pmatrix} 0 & 1 & 2 \\ 0 & 2 & 1 \end{pmatrix}, \quad \begin{pmatrix} 0 & 1 & 2 \\ 1 & 0 & 2 \end{pmatrix}, \quad \begin{pmatrix} 0 & 1 & 2 \\ 2 & 1 & 0 \end{pmatrix}.$$

They can also be written as $(1, 2)$, $(0, 1)$, and $(0, 2)$.

We prove an important representation theorem for the symmetric group of degree n.

Theorem A.4.25. *Let $n \in \mathbb{N}$. Then every element of the symmetric group of degree n can be written as a composition of at most $n - 1$ transpositions.*

Proof. We prove the theorem by induction on n. For $n = 1$, the assertion is valid because S_1 does not contain transpositions. Assume that the assertion holds for n and let $\pi \in S_{n+1}$.

First, assume that $\pi(n) = n$. Then the map $\pi' = \pi|_{\mathbb{Z}_n}$ is in S_n. By the induction hypothesis, π' can be written as the composition of at most $n - 1$ transpositions in S_n that are also transpositions in S_{n+1}. Also, since $\pi(n) = n$, the permutation π is the composition of the same transpositions.

Now assume $\pi(n) = j$ with $j < n$. Then the map $\pi' = \pi \circ (j, n)|_{\mathbb{Z}_n}$ is in S_n. By the induction hypothesis, π' can be written as the composition of at most $n - 1$ transpositions in S_n that are also transpositions in S_{n+1}. Therefore, the permutation π is the composition of the same transpositions and (j, n). These are at most n transpositions. □

Example A.4.26. Consider the permutation $\pi = (3, 2, 1, 0)$ in S_4. We can write

$$(A.4.4) \qquad \pi = \begin{pmatrix} 0 & 1 & 2 & 3 \\ 3 & 2 & 1 & 0 \end{pmatrix} = \begin{pmatrix} 0 & 1 & 2 & 3 \\ 0 & 2 & 1 & 3 \end{pmatrix} \circ (0, 3) = (1, 2) \circ (0, 3).$$

Exercise A.4.27. Find a polynomial time algorithm that computes the representation of a permutation in S_n as a product of at most $n - 1$ transpositions in S_n.

We define the sign of the elements of the symmetric group.

Definition A.4.28. Let $k \in \mathbb{N}$ and let $\sigma \in S_k$. Then the *sign* of σ is defined as

$$(A.4.5) \qquad \text{sign}(\sigma) = |\{(u, v) \in \mathbb{Z}_k^2 : u < v \text{ and } \sigma(u) > \sigma(v)\}|.$$

Example A.4.29. Consider the permutation $(0, 2, 1) \in S_3$. Then we have $\sigma(0) = 0$, $\sigma(1) = 2$, and $\sigma(2) = 1$. So $(u, v) = (1, 2)$ is the only pair in $\mathbb{Z}_3^2$ such that $u < v$ and $\sigma(u) > \sigma(v)$ and, therefore, $\text{sign}(\sigma) = 1$.

We show how to determine the sign of a permutation.

Proposition A.4.30. *Let $k, m \in \mathbb{N}$ and suppose that $\sigma \in S_k$ can be written as a composition of m transpositions. Then $\text{sign}\, \sigma = (-1)^m$.*

Proof. We prove the assertion by induction on m. If $m = 1$, then σ is equal to a transposition (u, v) with $u, v \in \mathbb{Z}_k$. This implies $\text{sign}\, \sigma = -1$. Now assume that $m \geq 1$ and that the assertion holds for m. Also, suppose that $\sigma \in S_k$ can be represented as a composition of $m + 1$ transpositions. Then we can write $\sigma = \tau \circ (u, v)$ where $\tau \in S_k$ is the composition of m transpositions and (u, v) is another transposition with $u, v \in \mathbb{Z}_k$. This representation and the induction hypothesis imply $\text{sign}\, \sigma = -\text{sign}\, \tau = (-1)^{m+1}$. □

Corollary A.4.31. *Let $k \in \mathbb{N}$. Then the map $S_k \to \{\pm 1\}$, $\sigma \mapsto \text{sign}\, \sigma$ is a surjective homomorphism.*

Exercise A.4.32. Prove Corollary A.4.31.

A.4.4. Subgroups. We also introduce the notion of a subgroup.

Definition A.4.33. Let $(G, \circ)$ be a group, and let H be a subset of G. Then H is called a *subgroup* of G if $(H, \circ)$ is a group.

Example A.4.34. $(2\mathbb{Z}, +)$ is a subgroup of $(\mathbb{Z}, +)$ where $2\mathbb{Z} = \{2z : z \in \mathbb{Z}\}$ is the set of all even integers.

Next, we discuss quotient groups. We will confine our discussion to commutative groups, as this is the only aspect required for the context of this book.

Proposition A.4.35. *Let $(G, \circ)$ be a commutative group with identity element e and let H be a subgroup of G. Let $G/H = \{gH : g \in G\}$. Then*

$$(A.4.6) \qquad \circ : G/H \times G/H \to G/H, \quad (g_0 H, g_1 H) \to (g_0 \circ g_1)H$$

is a well-defined operation on G/H and $(G/, \circ)$ is a commutative group with identity element eH. It is called the quotient *of G and H.*

Exercise A.4.36. Prove Proposition A.4.35.

Example A.4.37. The set $5\mathbb{Z}$ of all integer multiples of 5 is a subgroup of the commutative group $\mathbb{Z}$ with respect to addition. The corresponding quotient group is $\mathbb{Z}/5\mathbb{Z} = \{a + 5\mathbb{Z}, 0 \le a < 5\}$. Its identity element is $0 + 5\mathbb{Z} = 5\mathbb{Z}$.

A.4.5. Rings and fields. Another basic notion in algebra is that of a ring, which we define now.

Definition A.4.38. A ring is a triple $(R, +, \cdot)$ where R is a nonempty set and $+$ and $\cdot$ are binary operations on R called addition and multiplication. They satisfy the following conditions.

(1) $(R, +)$ is an abelian group.
(2) $(R, \cdot)$ is a monoid.
(3) Multiplication is *distributive* with respect to addition, meaning that
- $a \cdot (b + c) = a \cdot b + a \cdot c$ for all $a, b, c \in R$ (*left distributivity*),
- $(b + c) \cdot a = b \cdot a + c \cdot a$ for all $a, b, c \in R$ (*right distributivity*).

Definition A.4.39. Let $(R, +, \cdot)$ be a ring.

(1) The ring R is called *commutative* if the semigroup $(R, \cdot)$ is commutative.
(2) The *unit group* of the ring R is the unit group of the monoid $(R, \cdot)$.
(3) A *zero divisor* in R is an element $a \in R$ such that there are $x, y \in R$ with $xa = ay = 0$.
(4) The ring R is called a *field* if $(R \setminus \{0\}, \cdot)$ is an abelian group.

Theorem A.4.40. *Let $(R, +, \cdot)$ be a ring. Then the set of units and the set of zero divisors in this ring are disjoint.*

Exercise A.4.41. Prove Theorem A.4.40.

We give a few examples of rings, their unit groups, and zero divisors.

Example A.4.42. The claims of this example are verified in Exercise A.4.43.

(1) The integers, equipped with the usual addition and multiplication, are a commutative ring without zero divisors.

(2) If k is a positive integer, then $(\mathbb{Z}_k, +_k, \cdot_k)$ is a commutative ring where $+_k$ and $\cdot_k$ are defined as explained in Example A.1.21. The units in this ring are all integers $a \in \mathbb{Z}_k$ such that $\gcd(a, k) = 1$. The zero divisors in this ring are all integers $a \in \mathbb{Z}_k$ such that $\gcd(a, k) > 1$.

(3) The rational numbers, real numbers, and complex numbers equipped with the usual addition and multiplication are fields.

Exercise A.4.43. Verify the claims in Example A.4.42.

As explained in [**Buc04**, Section 2.20], for all prime numbers p and all positive integers e, there is a finite field with $q = p^e$ elements. It is uniquely determined up to isomorphism and is denoted by $\mathbb{F}_q$. These are all the finite fields that exist.

A.4.6. Polynomial rings. In this section, we assume that R is a commutative ring with unit element 1 and we discuss polynomials over R.

Definition A.4.44. A *polynomial* in one *variable* x over R is an expression

(A.4.7) $$f(x) = a_n x^n + a_{n-1} x^{n-1} + \cdots + a_1 x + a_0$$

where $a_i \in R$ for $0 \leq i \leq n$ and x is not an element of R. We also use the following notation.

(1) The ring elements a_i are called the *coefficients* of f.

(2) The set of all polynomials over R in the variable x is denoted by $R[x]$.

(3) If in (A.4.7) the coefficient a_n is not zero, then n is called the *degree* of the polynomial f, we write $n = \deg f$, and a_n is called the *leading coefficient* of f.

(4) If all the coefficients of f except the leading one are zero, then f is called a *monomial*.

(5) If $f = a_0$, then f is called a *constant polynomial* or simply a *constant*.

(6) If $r \in R$, then we write $f(r) = a_n r^n + \cdots + a_0$.

(7) If $r \in R$ with $f(r) = 0$, then r is called the *zero* or the *root* of f.

Example A.4.45. The polynomials $2x^3 + x + 1$, x, 1 are elements of $\mathbb{Z}[x]$. The first polynomial has degree 3, the second has degree 1, and the third has degree 0. Also, the second and third polynomials are monomials. The first polynomial is not a monomial.

Example A.4.46. Consider the polynomial $x^2 + 1$ in $\mathbb{Z}_2[x]$. This polynomial has the only zero 1.

We define sums and products of polynomials over R. Let

$$g(x) = b_m x^m + \cdots + b_0$$

be another polynomial over R and let $n \geq m$. The *sum* of the polynomials f and g is the polynomial

$$(f + g)(x) = (a_n + b_n)x^n + \cdots + (a_0 + b_0).$$

Here, the undefined coefficients are set to zero.

Example A.4.47. Let $f(x) = x^3 + 2x^2 + x + 2$ and $g(x) = x^2 + x + 1$. Then $(f+g)(x) = x^3 + 3x^2 + 2x + 3$.

The *product* of the polynomials f and g is

$$(fg)(x) = c_{n+m}x^{n+m} + \cdots + c_0,$$

where

$$c_k = \sum_{i=0}^{k} a_i b_{k-i}, \quad 0 \leq k \leq n + m.$$

In this formula, the undefined coefficients a_i and b_i are set to 0.

Example A.4.48. Let $f(x) = x^3 + 2x^2 + x + 2$ and $g(x) = x^2 + x + 1 \in \mathbb{Z}[x]$. Then

$$\begin{aligned}
(fg)(x) &= (x^3 + 2x^2 + x + 2)(x^2 + x + 1) \\
&= x^5 + (2+1)x^4 + (1+2+1)x^3 + (2+1+2)x^2 + (2+1)x + 2 \\
&= x^5 + 3x^4 + 4x^3 + 5x^2 + 3x + 2.
\end{aligned}$$

In the proof of Proposition 2.2.25 we use the discriminant of quadratic polynomials which are introduced in the next exercise.

Exercise A.4.49. Let $f(x) = ax^2 + bx + c \in R[x]$. Then the discriminant of f is $\Delta(f) = b^2 - 4ac$. Prove the following. Let $R = \mathbb{R}$. If $\delta(f) > 0$, then f has two distinct real zeros. If $\Delta(f) = 0$, then f has two identical real zeros. If $\Delta(f) < 0$, the f has two complex nonreal zeros which are the complex conjugates of each other.

Proposition A.4.50. *Let F be a field and let $f \in F[x]$ be a polynomial of degree n. If $\alpha \in F$ is a zero of f, then $n \geq 1$ and we can write $f = (x-\alpha)g$ with a polynomial $g \in F[x]$ of degree $n - 1$.*

Exercise A.4.51. Prove Proposition A.4.50.

Example A.4.52. The polynomial $f(x) = x^2 - 1 \in \mathbb{Q}[x]$ has the zero 1 and can be written as $f(x) = (x - 1)(x + 1)$.

Finally we state the fundamental theorem of algebra. It can be found in [**FK03**, 31.18 Theorem].

Theorem A.4.53. *If $f \in \mathbb{C}[x]$ is a polynomial of degree n. Then f can be written as $f = \prod_{i=0}^{n-1}(x - \alpha_i)$ with complex zeros α_i for $0 \leq i < n$.*

Exercise A.4.54. Let $f \in \mathbb{R}[x]$ be a polynomial of degree $n \in \mathbb{N}$. Show that there are nonnegative integers s and t such that $n = s + 2t$ and f has s real zeros and $2t$ pairs of complex conjugate zeros.

A.5. Trigonometric identities and inequalities

We present trigonometric identities and inequalities that are necessary for several proofs. We start with a few identities that can be found in [**Abr72**], and the equation numbers from [**Abr72**] are provided after each identity:

(A.5.1) $\sin^2 x + \cos^2 x = 1$,   4.3.10,

(A.5.2) $\sin(x + y) = \sin x \cos y + \cos x \sin y$,   4.3.16,

(A.5.3) $\sin(x - y) = \sin x \cos y - \cos x \sin y$,   4.3.13, 4.3.14, 4.3.16,

(A.5.4) $\sin 2x = 2 \sin x \cos x$,   4.3.24,

(A.5.5) $\cos(x + y) = \cos x \cos y - \sin x \sin y$,   4.3.17,

(A.5.6) $\cos(x - y) = \cos x \cos y + \sin x \sin y$,   4.3.13, 4.3.14, 4.3.17,

(A.5.7) $\cos 2x = \cos^2 x - \sin^2 x = 2 \cos^2 x - 1 = 1 - 2 \sin^2 x$,   4.3.25

(A.5.8) $\cos((n + 1)y) + \cos((n - 1)y) = 2 \cos(y) \cos(ny)$ for all $n \in \mathbb{N}_0$

Exercise A.5.1. Prove (A.5.8).

We now demonstrate some necessary trigonometric inequalities.

Lemma A.5.2. *For all* $x, y \in \mathbb{R}$ *we have*

(A.5.9) $|\sin x - \sin y| \leq |x - y|$, $|\cos x - \cos y| \leq |x - y|$.

Proof. Let $x, y \in \mathbb{R}$. If $x = y$, then the statement is valid. So, let $x \neq y$. By the mean value theorem, there is $z \in \mathbb{R}$ with

(A.5.10) $\sin(x) - \sin(y) = \cos(z)(x - y)$.

This implies

(A.5.11) $|\sin(x) - \sin(y)| = |\cos(z)||x - y| \leq |x - y|$.

Likewise, the mean value theorem implies

(A.5.12) $\cos(x) - \cos(y) = -\sin(z)(x - y)$.

This implies

(A.5.13) $|\cos(x) - \cos(y)| = |\sin(z)||x - y| \leq |x - y|$.   □

Lemma A.5.3. *For all* $x \in [0, \frac{1}{2}]$ *we have* $\sin \pi x \geq 2x$.

Proof. [**Abr72**, 4.3.79].   □

Corollary A.5.4. *For all* $y \in [0, 1]$ *we have* $\arcsin y \leq \frac{\pi y}{2}$.

Lemma A.5.5. *For all* $x \in \mathbb{R}$ *we have* $|\sin x| \leq |x|$.

Proof. [**Abr72**, 4.3.80].   □

Corollary A.5.6. *For all* $y \in [-1, 1]$ *we have* $|\arcsin y| \geq |y|$.

Lemma A.5.7. *For all* $x \in [0, \pi/2]$ *we have* $\cos x \geq x$.

Proof. Consider the function $f(x) = \cos x - x$. We have $f(0) = 1$, $f(\pi/2) = 0$, and $f'(x) = -\sin x - 1 < 0$ for all $x \in [0, \pi/2]$.   □

Lemma A.5.8. *For $x, y \in \mathbb{R}$ we have*

(A.5.14) $$\sin^2(x + y) - \sin^2 x = \sin x \cos x \sin(2y) + (1 - 2\sin^2 x)\sin^2 y,$$

(A.5.15) $$\sin^2 x - \sin^2(x - y) = \sin x \cos x \sin(2y) - (1 - 2\sin^2 x)\sin^2 y.$$

Exercise A.5.9. Prove Lemma A.5.8 using the appropriate trigonometric identities.

Lemma A.5.10. *Let $x, \alpha \in \mathbb{R}_{>0}$ with $0 \le x\alpha < \frac{\pi}{2}$. Then we have*

(A.5.16) $$\frac{\sin x\alpha}{x \sin \alpha} \ge \cos x\alpha.$$

Proof. Since $0 \le x\alpha < \pi/2$ we have

(A.5.17) $$\frac{\sin x\alpha}{\cos x\alpha} = \tan x\alpha \ge x\alpha \ge \alpha \sin \alpha. \qquad \square$$

Lemma A.5.11. *For all $\alpha \in \left[0, \frac{\pi}{2}\right[$ and $k \in \mathbb{N}$ we have*

(A.5.18) $$\prod_{l=1}^{k} \cos(2^l \alpha) = \frac{\sin(2^{k+1}\alpha)}{2^k \sin 2\alpha}.$$

Proof. We prove the assertion by induction on k. For $k = 0$ the assertion holds since both sides of (A.5.18) are equal to 1. So assume that $k \ge 0$ and that the assertion is true for k. Then we have

(A.5.19)
$$\prod_{l=1}^{k+1} \cos(2^l \alpha) = \cos(2^{k+1}\alpha) \prod_{l=0}^{k} \cos(2^l \alpha)$$
$$= \cos(2^{k+1}\alpha) \frac{\sin(2^{k+1}\alpha)}{2^k \sin(2\alpha)} = \frac{\sin(2^{k+2}\alpha)}{2^{k+1} \sin(2\alpha)}$$

where the last equality follows from the trigonometric identity (A.5.4). $\qquad \square$

Linear Algebra

Linear algebra plays an essential role in modeling phenomena in diverse scientific disciplines. Its efficiency in algorithmic solutions empowers the resolution of computational challenges and the formulation of concrete predictions in various scientific domains.

In the context of this book, linear algebra assumes particular importance, since it includes the theory of Hilbert spaces, which serves as a framework for modeling quantum mechanics. To comprehend this theory fully, it becomes necessary to establish a foundation in linear algebra, which we provide in this appendix.

The appendix is divided into two parts. In the initial part, which includes Sections B.1 to B.7, we provide a brief overview of fundamental concepts commonly found in introductory linear algebra courses. We assume the reader's familiarity with these concepts and present them as reference points, omitting proofs, examples, and exercises. The topics encompass vectors, matrices, modules over rings, vector spaces, linear maps, characteristic polynomials, eigenvalues, and eigenspaces. This section also covers the Gaussian elimination algorithm, its complexity, and its applications in determining bases for linear map images and kernels, as well as solving linear systems.

The subsequent part of this appendix, Section B.8, presents tensor products of modules and vector spaces, as well as the concept of the partial trace. This area of study typically falls outside the scope of introductory linear algebra courses but is crucial for modeling quantum mechanics mathematically. As this topic may be unfamiliar or entirely new to readers, we include comprehensive explanations, proofs, examples, and exercises to facilitate understanding.

Throughout this chapter, we use the following notation. By k, l, m, and n we denote positive integers, $(R, +, \cdot)$ is a commutative ring, and $(F, +, \cdot)$ is a field. We write 0 for the identity elements with respect to addition in R and F and we write 1 for the identity elements with respect to multiplication in R and F. If $r \in R$ or $r \in F$ is invertible with respect to multiplication, we write r^{-1} for its multiplicative inverse in R or F, respectively.

B.1. Vectors

Definition B.1.1. (1) A *vector* over a nonempty set S is a sequence $\vec{v} = (v_0, \dots, v_{k-1})$ of elements in S. The positive integer k is called the *length* of $\vec{v}$. The elements v_i are called the *entries* or *components* of $\vec{v}$.

(2) The set of all vectors of length k over S is denoted by S^k.

(3) For $\vec{v} = (v_0, \dots, v_{k-1}) \in S^k$ we also write $\vec{v} = (v_i)_{0 \le i < k}$ or $\vec{v} = (v_i)_{i \in \mathbb{Z}_k}$.

(4) If $\vec{v} \in S^l$ and $\vec{w} \in S^k$, then $\vec{v} \| \vec{w}$ denotes the concatenation of $\vec{v}$ and $\vec{w}$ which is an element of S^{k+l}.

Usually, we will start by numbering the entries of a vector by 0, but we may also number the entries differently. We do not distinguish between row vectors and column vectors. However, an analogous distinction is introduced in Section B.4.1.

Proposition B.1.2. *Let S be finite and let $k \in \mathbb{N}$. Then $|S^k| = |S|^k$.*

B.1.1. Vector operations. For vectors over the ring R, we define the following operations.

Definition B.1.3. Let $r \in R$ and $\vec{v} = (v_0, \dots, v_{k-1}), \vec{w} = (w_0, \dots, w_{k-1}) \in R^k$.

(1) The *scalar product* of r with $\vec{v}$ is defined as

$$(\text{B.1.1}) \qquad r\vec{v} = r \cdot \vec{v} = (rv_0, rv_1, \dots, rv_{k-1}).$$

(2) The *sum* of $\vec{v}$ and $\vec{w}$ is defined as

$$(\text{B.1.2}) \qquad \vec{v} + \vec{w} = (v_0 + w_0, v_1 + w_1, \dots, v_{k-1} + w_{k-1}).$$

(3) We write $-\vec{w} = (-w_0, \dots, -w_{k-1})$ and $\vec{v} - \vec{w}$ for $\vec{v} + (-\vec{w})$.

(4) The *dot product* of $\vec{v}$ and $\vec{w}$ is an element of R which is defined as

$$(\text{B.1.3}) \qquad \vec{v} \cdot \vec{w} = \sum_{i=0}^{k-1} v_i w_i.$$

B.2. Modules and vector spaces

Definition B.2.1. An *R-module* or *module over R* is a triplet $(M, +, \cdot)$ where M is a nonempty set and

$$(\text{B.2.1}) \qquad \begin{aligned} + &: M \times M \to M, (\vec{v}, \vec{w}) \mapsto \vec{v} + \vec{w}, \\ \cdot &: R \times M \to M, (r, \vec{w}) \mapsto r\vec{w} = r \cdot \vec{w} \end{aligned}$$

are maps, called *addition* and *scalar multiplication*, which satisfy the following conditions.

(1) $(M, +)$ is an abelian group.

(2) Scalar multiplication is *associative*; that is, $(r \cdot s) \cdot \vec{v} = r \cdot (s \cdot \vec{v})$ for all $r, s \in R$ and all $\vec{v} \in M$.

(3) The identity element 1 of R is also an *identity element* with respect to scalar multiplication; that is, $1 \cdot \vec{v} = \vec{v}$ for all $\vec{v} \in M$.

(4) Scalar multiplication is *distributive* with respect to addition in M; that is, $r \cdot (\vec{v} + \vec{w}) = r \cdot \vec{v} + r \cdot \vec{w}$ for all $r \in R$ and all $\vec{v}, \vec{w} \in M$.

(5) Scalar multiplication is *distributive* with respect to addition in R; that is, $(r+s) \cdot \vec{v} = r \cdot \vec{v} + s \cdot \vec{v}$ for all $r, s \in R$ and all $\vec{v} \in M$.

Definition B.2.2. Any module over the field F is called a *vector space over F* or an *F-vector space*.

Proposition B.2.3. (1) $(R^k, +, \cdot)$ *is an R-module, where "+" and "$\cdot$" denote addition and scalar multiplication in $\mathbb{R}^k$, respectively.*

(2) $(F^k, +, \cdot)$ *is an F-vector space, where "+" and "$\cdot$" denote addition and scalar multiplication in $\mathbb{F}^k$, respectively.*

(3) *For all $\vec{v} = (v_0, \ldots, v_{k-1}) \in R^k$, the element $-\vec{v} = (-v_0, \ldots, -v_{k-1})$ is the additive inverse of $\vec{v}$.*

(4) *The* zero vector $\vec{0} = (0, \ldots, 0)$ *is the neutral element in $\mathbb{R}^k$ with respect to addition.*

To simplify our notation, we denote an R-module $(M, +, \cdot)$ or an F-vector space $(V, +, \cdot)$ also by M and V, respectively, if it is clear what is meant by addition and scalar multiplication. Unless otherwise specified, module addition is always denoted by $+$ and scalar multiplication by $\cdot$.

B.2.1. Submodules.

Definition B.2.4. (1) An *R-submodule* of M is a nonempty subset N of M such that N is a subgroup of M with respect to addition and N is *closed under scalar multiplication*; that is, $r\vec{v} \in N$ for all $\vec{v} \in N$ and all $r \in R$. If it is clear from the context what is meant by the ring R, we call N a *submodule* of M.

(2) An *F-subspace* W of V is an *F-submodule* W of V. If it is clear from the context what is meant by the field F, we call W a *subspace* of M.

Proposition B.2.5. (1) *Every R-submodule of M is an R-module with the same addition and scalar multiplication as in M.*

(2) *Every F-subspace of V is an F-vector space with the same addition and scalar multiplication as in V.*

Definition B.2.6. (1) Let $\vec{v}_0, \ldots, \vec{v}_{l-1}, \vec{v}$ be vectors in M. We say that $\vec{v}$ is a *linear combination* of the vectors $\vec{v}_0, \ldots, \vec{v}_{l-1}$ if $\vec{v}$ can be written as

$$(B.2.2) \qquad \vec{v} = r_0 \vec{v}_0 + \cdots + r_{l-1} \vec{v}_{l-1} = \sum_{j=0}^{l-1} r_j \vec{v}_j$$

with $r_j \in R$ for $0 \le j < l$. The ring elements r_j are called the *coefficients* of the linear combination (B.2.2).

(2) The *linear combination of the empty sequence* in M is defined to be $\vec{0}$.

(3) For any subset S of M we define the *span* of S as the set of all linear combinations of finitely many elements of S including the empty set. We write it as $\mathrm{Span}(S)$. So, we have

(B.2.3)
$$\mathrm{Span}(S) = \left\{ \sum_{j=0}^{l-1} r_j \vec{v}_j \; : \; l \in \mathbb{N}_0, r_j \in R, \vec{v}_j \in S, \text{ for all } j \in \mathbb{Z}_l \right\}.$$

In particular, the span of the empty set is $\{\vec{0}\}$.

(4) We say that M is *finitely generated* if $M = \mathrm{Span}(S)$ for a finite subset S of M.

Proposition B.2.7. *Let S be a subset of M. Then the span of S is an R-module, and it is the (with respect to inclusion) smallest submodule of M that contains S. It is called the submodule generated by S.*

Proposition B.2.8. *Let X be a set of submodules of M. Then the set*

(B.2.4)
$$\sum_{N \in X} N = \left\{ \sum_{N \in X} \vec{v}_N \; : \; \vec{v}_N \in N, \text{ finitely many } \vec{v}_N \text{ are nonzero} \right\}$$

is a submodule of M. It is called the sum *of the submodules in X.*

Definition B.2.9. Let X be a set of submodules of M.

(1) The sum $\sum_{N \in X} N$ of the submodules in X is called *direct* if all of its nonzero elements $\vec{v}$ have a uniquely determined representation

(B.2.5)
$$\vec{v} = \sum_{N \in X} \vec{v}_N$$

where $\vec{v}_N \in N$ and only finitely many of these elements are nonzero.

(2) If $\sum_{N \in X} N$ is direct, then the module $P = \sum_{N \in X} N$ is called the *direct sum* of the submodules in X.

Definition B.2.10. A submodule W of an F-vector space V is called an *F-linear subspace* of V or simply a *subspace* of V. All notions concerning modules transfer analogously to subspaces.

B.2.2. Direct product of modules.

Proposition B.2.11. *Let $M_0, \ldots, M_{l-1}$ be R-modules. On the Cartesian product $\prod_{i=0}^{l-1} M_i$ we define componentwise addition and scalar multiplication as follows. If $(\vec{v}_0, \ldots, \vec{v}_{l-1})$, $(\vec{w}_0, \ldots, \vec{w}_{l-1}) \in \prod_{i=0}^{l-1} M_i$ and $r \in R$, then we set*

(B.2.6)
$$(\vec{v}_0, \ldots, \vec{v}_{l-1}) + (\vec{w}_0, \ldots, \vec{w}_{l-1}) = (\vec{v}_0 + \vec{w}_0, \ldots, \vec{v}_{l-1} + \vec{w}_{l-1}),$$
$$r \cdot (\vec{v}_0, \ldots, \vec{v}_{l-1}) = (r \cdot \vec{v}_0, \ldots, r \cdot \vec{v}_{l-1}).$$

Then $\left(\prod_{i=0}^{l-1} M_i, +, \cdot \right)$ is an R-module. It is called the direct product *of the R-modules $M_0, \ldots, M_{l-1}$ and is also written as $\prod_{i=0}^{l-1} M_i$.*

B.2.3. Quotient modules.

Proposition B.2.12. *Let N be a submodule of an R-module M. On the quotient group M/N we define the scalar product*

(B.2.7) $\qquad \cdot : R \times M/N \to M/N, \quad (r, \vec{v} + N) \mapsto r \cdot (\vec{v} + N) = r\vec{v} + N.$

Then $(M/N, +, \cdot)$ is an R-module. It is called a quotient module *or, more precisely, the* quotient *of M and N.*

B.2.4. Free modules.

Definition B.2.13. Let M be an R-module and let I be a nonempty set. Let $B = (\vec{b}_i)_{i \in I}$ be a family of elements $\vec{b}_i \in M$.

(1) B is called a *generating system* of M if $M = \sum_{i \in I} R\vec{b}_i$.

(2) B is called *linearly independent* if B is the direct sum of the submodules $R\vec{b}_i$, $i \in I$.

(3) B is called *linearly dependent* if B is not linearly independent.

(4) B is called a *basis* of M if B is a linearly independent generating system of M.

(5) If M has a basis, then M is called a *free module*.

Theorem B.2.14. *Every vector space V over a field F is free.*

B.3. Linear maps between modules

Definition B.3.1. (1) A map

(B.3.1) $\qquad\qquad\qquad\qquad f : M \to N$

is called *R-linear* or an *R-module homomorphism* if it preserves the operations of addition and scalar multiplication; that is, we have
   (a) $f(\vec{v} + \vec{w}) = f(\vec{v}) + f(\vec{w})$ for all $\vec{v}, \vec{w} \in M$,
   (b) $f(r\vec{v}) = rf(\vec{v})$ for all $r \in R$ and all $\vec{v} \in V$.

(2) The set of all R-module homomorphisms $f : M \to N$ is denoted by $\text{Hom}_R(M, N)$. If it is clear from the context which ring R we refer to, then we simply write $\text{Hom}(M, N)$.

(3) A bijective R-module homomorphism $f : M \to N$ is called an *R-module isomorphism* or an *isomorphism* between M and N.

(4) The R-modules M and N are called *isomorphic* if there is an isomorphism between M and N.

The notions from Definition B.3.1 transfer analogously to vector spaces.

Definition B.3.2. Let

(B.3.2) $\qquad\qquad\qquad\qquad f, g : M \to N$

be functions and let $r \in R$.

(1) The *sum* of f and g is the function

(B.3.3) $\qquad\qquad f + g : M \to N, \quad \vec{v} \mapsto f(\vec{v}) + g(\vec{v}).$

(2) The *scalar product* of r with f is the function

(B.3.4) $$rf = r \cdot f : M \to N, \quad \vec{v} \mapsto rf(\vec{v}).$$

Proposition B.3.3. $(\mathrm{Hom}(M, N), +, \cdot)$ *is an R-module.*

Proposition B.3.4. *For every* $\vec{v} \in M$ *define the map*

(B.3.5) $$f_{\vec{v}} : R \to M, \quad r \mapsto r\vec{v}.$$

Then $f_{\vec{v}} \in \mathrm{Hom}(R, M)$ *and the map*

(B.3.6) $$M \to \mathrm{Hom}(R, M), \quad \vec{v} \mapsto f_{\vec{v}}$$

is an isomorphism of R-modules.

B.3.1. Kernel and image.

Definition B.3.5. Let $f \in \mathrm{Hom}_R(M, N)$.

(1) The image of f is defined as

(B.3.7) $$\mathrm{im}(f) = \{f(\vec{u}) : \vec{u} \in M\}.$$

(2) The *kernel* of f is defined as

(B.3.8) $$\ker(f) = \{\vec{u} \in M : f(\vec{u}) = \vec{0}\}.$$

The following theorem is called the *fundamental homomorphism theorem* for R-modules.

Theorem B.3.6. *Let* $f \in \mathrm{Hom}_R(M, N)$. *Then the kernel of* f *is an R-submodule of* M, *the image of* f *is an R-submodule of* N, *and the map*

(B.3.9) $$g : M/\ker(f) \to \mathrm{im}(f), \quad \vec{u} + \ker(f) \mapsto f(\vec{u})$$

is an isomorphism of R-modules.

B.3.2. The dual module.

Definition B.3.7. (1) The set $\mathrm{Hom}_R(M, R)$ of homomorphisms between M and the ring R is called the *dual module* of M. It is denoted by M^*.

(2) If V is an F-vector space, then V^* is called the *dual vector space* of M.

B.3.3. R-algebras.

Definition B.3.8. An *R-algebra* is a tuple $(M, +, \cdot, \circ)$ which has the following properties.

(1) $(M, +, \cdot)$ is an R-module.

(2) $(M, +, \circ)$ is a ring.

(3) The scalar multiplication of the R-module M is associative with respect to the $\circ$- operation; that is,

(B.3.10) $$r \cdot (A \circ B) = (r \cdot A) \circ B = A \circ (r \cdot B)$$

for all $A, B \in M$ and $r \in R$.

Definition B.3.9. Let $(M, +, \cdot, \circ)$ and $(N, +, \cdot, \circ)$ be R-algebras. A function

(B.3.11) $$f : M \to N$$

is called *linear* if it is an R-module homomorphism and also preserves the operation $\circ$; that is, we have

(B.3.12) $$f(A \circ B) = f(A) \circ f(B)$$

for every $A, B \in M$ and

(B.3.13) $$f(I_M) = I_N$$

where I_M and I_N are the identity elements of M and N, respectively. Such a linear map is also called an *R-algebra homomorphism*. The set of all such homomorphisms is denoted by $\operatorname{Hom}_R(M, N)$. The homomorphism f is called an *R-algebra isomorphism* if it is bijective.

B.3.4. Endomorphisms.

Definition B.3.10. (1) An R-module homomorphism that maps M to itself is called an *R-module endomorphism* or simply an *endomorphism of M*.

(2) The set of all endomorphisms of M is denoted by $\operatorname{End}_R(M)$ or by $\operatorname{End}(M)$ if it is clear to which ring R we refer.

(3) A bijective endomorphism of M is called an *automorphism of M*.

(4) The set of all automorphisms of M is denoted by $\operatorname{Aut}_R(M)$ or simply by $\operatorname{Aut}(M)$ if it is clear which ring R we refer to.

Proposition B.3.11. *The quadruple* $(\operatorname{End}(M), +, \cdot, \circ)$ *is an R-algebra where $+$ and $\cdot$ denote addition and scalar product as specified in Definition* B.3.2 *and $\circ$ means composition of functions. It is called the* endomorphism algebra of M. *Also,* $(M, +, \circ)$ *is called the* endomorphism ring of M.

If M is an R-algebra, then the notions of an R-algebra endomorphism and R-algebra automorphism are defined analogously to the corresponding notions for R-modules. Furthermore, the sets of all such endomorphisms and automorphisms are denoted by $\operatorname{End}_R(M)$ and $\operatorname{Aut}_R(M)$, respectively.

B.4. Matrices

In this section, we introduce matrices which play a very important role in linear algebra as representations of module homomorphisms. We let S be a nonempty set.

Definition B.4.1. (1) A $k \times l$ *matrix* over S is a two-dimensional schema

(B.4.1) $$A = \begin{pmatrix} a_{0,0} & a_{0,1} & \cdots & a_{0,l-1} \\ a_{1,0} & a_{1,1} & \cdots & a_{1,l-1} \\ \vdots & \vdots & & \vdots \\ a_{k,0} & a_{k,1} & \cdots & a_{k,l-1} \end{pmatrix}$$

where $a_{i,j} \in S$ for $0 \leq i < k$, $0 \leq j < l$. The elements $a_{i,j}$ are called the *entries* of the matrix A. This matrix can also be written as $A = (a_{i,j})_{0 \leq i < k, 0 \leq j < l}$ or $A = (a_{i,j})_{i \in \mathbb{Z}_k, j \in \mathbb{Z}_l}$. If the ranges of k and l are clear from the context, we also write $A = (a_{i,j})$.

(2) The set of all $k \times l$ matrices over S is denoted by $S^{(k,l)}$.

Before we give examples of matrices, we need the following definition.

Definition B.4.2. Let

$$(B.4.2) \qquad A = \begin{pmatrix} a_{0,0} & a_{0,1} & \cdots & a_{0,l-1} \\ a_{1,0} & a_{1,1} & \cdots & a_{1,l-1} \\ \vdots & \vdots & & \vdots \\ a_{k,0} & a_{k,1} & \cdots & a_{k,l-1} \end{pmatrix} \in S^{(k,l)}.$$

(1) We call $(a_{i,0}, a_{i,1}, \ldots, a_{i,l-1})$, $0 \leq i < k$, the *row vectors* or *rows* of A and $(a_{0,j}, a_{1,j}, \ldots, a_{k-1,j})$, $0 \leq j < l$, the *column vectors* or *columns* of A.

(2) The *transpose* of A is the matrix

$$(B.4.3) \qquad A^{\mathrm{T}} = (a_{j,i})_{j \in \mathbb{Z}_l, i \in \mathbb{Z}_k} = \begin{pmatrix} a_{0,0} & a_{1,0} & \cdots & a_{k-1,0} \\ a_{0,1} & a_{1,1} & \cdots & a_{k-1,1} \\ \vdots & \vdots & & \vdots \\ a_{0,l-1} & a_{1,l-1} & \cdots & a_{k-1,l-1} \end{pmatrix} \in S^{(l,k)}.$$

So, the rows of A^{T} are the columns of A and vice versa.

Definition B.4.3. Let $\vec{a}_0, \ldots, \vec{a}_{l-1} \in S^k$. Then we write

$$(B.4.4) \qquad A = (\vec{a}_0, \ldots, \vec{a}_{l-1})$$

for the matrix in $S^{(k,l)}$ with column vectors $\vec{a}_0, \ldots, \vec{a}_{l-1}$ and

$$(B.4.5) \qquad A = \begin{pmatrix} \vec{a}_0 \\ \vdots \\ \vec{a}_{l-1} \end{pmatrix}$$

for the matrix in $S^{(l,k)}$ with row vectors $\vec{a}_0, \ldots, \vec{a}_{l-1}$.

B.4.1. Vectors as matrices. In some contexts, it is useful to identify vectors with matrices. We do this in the following way. Let S be a nonempty set, let $k \in \mathbb{N}$, and let $\vec{a} = (a_0, \ldots, a_{k-1}) \in S^k$. Then we identify $\vec{a}$ with the matrix

$$(B.4.6) \qquad \vec{a} = \begin{pmatrix} a_0 \\ a_1 \\ \vdots \\ a_{k-1} \end{pmatrix} \in S^{(k,1)}.$$

This matrix has $\vec{a}$ as its only column vector.

Using this identification, we define the transpose of $\vec{a}$ to be the matrix with $\vec{a}$ as its only row vector; that is,

$$(B.4.7) \qquad \vec{a}^{\mathrm{T}} = \begin{pmatrix} a_0 & a_1 & \cdots & a_{k-1} \end{pmatrix} \in S^{(1,k)}.$$

B.4.2. Matrix operations.

Definition B.4.4. Let $r \in R$, $A = (a_{i,j})$, $B = (b_{i,j}) \in R^{(l,k)}$.

(1) The *scalar product* of r with A is the matrix $r \cdot A = rA = (ra_{i,j}) \in R^{(k,l)}$. This operation is called *scalar multiplication*.

(2) The *sum* of A and B is

(B.4.8) $$A + B = (a_{i,j} + b_{i,j}) \in R^{(k,l)}.$$

This operation is called the *(componentwise) addition* of matrices.

(3) We write $-B = (-b_{i,j})$ and $A - B$ for $A + (-B)$.

Proposition B.4.5. (1) *$(R^{(k,l)}, +, \cdot)$ is an R-module where "+" denotes matrix addition and "$\cdot$" stands for scalar multiplication on $R^{(k,l)}$.*

(2) *For all $A \in R^{(k,l)}$ the matrix $-A$ is the additive inverse of A.*

(3) *The neutral element of the group $(R^{(k,l)}, +)$ is the zero matrix in $R^{(k,l)}$ all of whose entries are zero. We denote it by $0_{k,l}$ or as 0 if it is clear from the context what is meant by k, l.*

Definition B.4.6. Let $A = (a_{i,j}) \in R^{(k,l)}$ and $B = (b_{i,j}) \in R^{(l,m)}$. Then we define the product of A and B as

(B.4.9) $$A \cdot B = \left(\sum_{u=0}^{l-1} a_{i,u} b_{u,j} \right)_{i \in \mathbb{Z}_k, j \in \mathbb{Z}_m}.$$

Instead of $A \cdot B$ we also write AB.

Proposition B.4.7. *Matrix multiplication is associative in the following sense. If $A \in R^{(k,l)}$, $B \in R^{(l,m)}$, and $C \in R^{(m,n)}$, then we have*

(B.4.10) $$(A \cdot B) \cdot C = A \cdot (B \cdot C).$$

Definition B.4.8. Let $A \in R^{(k,l)}$ with column vectors $\vec{a}_0, \ldots, \vec{a}_{l-1}$ and let $\vec{v} = (v_0, \ldots, v_{l-1}) \in R^l$. Then we define the product of A with $\vec{v}$ as

(B.4.11) $$A \cdot \vec{v} = A\vec{v} = \sum_{j=0}^{l-1} v_i \vec{a}_i.$$

Note that the product of a matrix A with a vector $\vec{v}$ is the same as the product of A with the matrix corresponding to $\vec{v}$.

B.5. Square matrices

Square matrices, that is, matrices with the same number of rows and columns, are of special interest in linear algebra. In this section, we discuss the structure and properties of the set $R^{(k,k)}$ of all $k \times k$ square matrices over R.

Definition B.5.1. (1) Let $A \in S^{(k,k)}$,

$$(B.5.1) \qquad A = \begin{pmatrix} \boldsymbol{a_{0,0}} & a_{0,1} & a_{0,2} & \cdots & a_{0,k-1} \\ a_{1,0} & \boldsymbol{a_{1,1}} & a_{1,2} & \cdots & a_{1,k-1} \\ a_{2,0} & a_{2,1} & \boldsymbol{a_{2,2}} & \cdots & a_{2,k-1} \\ \vdots & \vdots & \vdots & \ddots & \vdots \\ a_{k-1,0} & a_{k-1,1} & a_{k-1,2} & \cdots & \boldsymbol{a_{k-1,k-1}} \end{pmatrix}.$$

Then the entries $a_{i,i}$, $0 \leq i < k$, are called the *diagonal elements* of A (highlighted in (B.5.1)). The other entries are called the *off-diagonal elements* of A.

(2) The *zero matrix of order k* over R is the matrix in $R^{(k,k)}$ all of whose entries are 0. We denote it by 0_k or simply by 0 if it is clear from the context what is meant by k.

(3) The *identity matrix of order k* over R is the following square matrix in $R^{(k,k)}$:

$$(B.5.2) \qquad I_k = \begin{pmatrix} 1 & 0 & 0 & \cdots & 0 \\ 0 & 1 & 0 & \cdots & 0 \\ 0 & 0 & 1 & \cdots & 0 \\ \vdots & \vdots & \vdots & \ddots & \vdots \\ 0 & 0 & 0 & \cdots & 1 \end{pmatrix}.$$

All of its diagonal elements are 1 and the off-diagonal elements are 0. The matrix I_k is also called the *unit matrix* of order and is denoted by I if k is clear from the context.

Definition B.5.2. Let $A = (a_{i,j}) \in S^{(k,k)}$.

(1) We say that A is an *upper triangular matrix* or in *upper triangular form* if A is of the form

$$(B.5.3) \qquad A = \begin{pmatrix} a_{0,0} & a_{0,1} & a_{0,2} & \cdots & a_{0,k-1} \\ 0 & a_{1,1} & a_{1,2} & \cdots & a_{1,k-1} \\ 0 & 0 & a_{2,2} & \cdots & a_{2,k-1} \\ \vdots & \vdots & \vdots & \ddots & \vdots \\ 0 & 0 & 0 & \cdots & a_{k-1,k-1} \end{pmatrix};$$

that is, $a_{i,j} = 0$ for $0 \leq j < i < k$.

(2) We say that A is a *lower triangular matrix* or in *lower triangular form* if A is of the form

$$(B.5.4) \qquad A = \begin{pmatrix} a_{0,0} & 0 & 0 & \cdots & 0 \\ a_{1,0} & a_{1,1} & 0 & \cdots & 0 \\ a_{2,0} & a_{2,1} & a_{2,2} & \cdots & 0 \\ \vdots & \vdots & \vdots & \ddots & \vdots \\ a_{k-1,0} & a_{k-1,1} & a_{k-1,2} & \cdots & a_{k-1,k-1} \end{pmatrix};$$

that is, $a_{i,j} = 0$ for $0 \leq i < j < k$.

(3) We say that A is a *diagonal matrix* or in *diagonal form* if A is of the form

(B.5.5)
$$A = \begin{pmatrix} a_{0,0} & 0 & 0 & \cdots & 0 \\ 0 & a_{1,1} & 0 & \cdots & 0 \\ 0 & 0 & a_{2,2} & \cdots & 0 \\ \vdots & \vdots & \vdots & \ddots & \vdots \\ 0 & 0 & 0 & \cdots & a_{k-1,k-1} \end{pmatrix};$$

that is, $a_{i,j} = 0$ for $0 \le i, j < k$, $i \ne j$. For such a matrix, we also write

(B.5.6)
$$A = \operatorname{diag}(a_{0,0}, \dots, a_{k-1,k-1}).$$

B.5.1. Algebraic structure of $R^{(k,k)}$.

Proposition B.5.3. *The set* $(R^{(k,k)}, +, \cdot, \cdot)$ *is an R-algebra where "+" is matrix addition, the first "·" means scalar multiplication, and the second "·" stands for matrix multiplication. The neutral element with respect to addition is the zero matrix* 0_k *of order k. The identity element with respect to multiplication is the identity matrix* I_k *of order k.*

Definition B.5.4. (1) A matrix $A \in R^{(k,k)}$ is called invertible if it has an inverse in the multiplicative semigroup $R^{(k,k)}$; that is, there is a matrix $B \in R^{(k,k)}$ such hat $AB = BA = I_k$.

(2) If $A \in R^{(k,k)}$ is invertible, then we denote by A^{-1} the multiplicative inverse of A. This matrix is also called the *inverse of A*.

(3) The unit group of $R^{(k,k)}$, that is, the set of all invertible $k \times k$ matrices with entries from R, is called the *general linear group of degree k over R* and is denoted by $GL(k, R)$.

Definition B.5.4 uses the fact that the inverse of an element of a semigroup is uniquely determined. We will show in Corollary B.5.21 that for a matrix $A \in R^{(k,k)}$ to be invertible, it suffices that A has a left or right inverse, respectively; that is, there is a matrix $B \in R^{(k,k)}$ such that $BA = I_k$ or $AB = I_k$, respectively. This right or left inverse of A is its inverse.

B.5.2. Permutation matrices.
Permutation matrices are obtained by permuting the row vectors of the identity matrix.

Definition B.5.5. Let $\sigma \in S_k$. Then the *permutation matrix* P_σ associated with σ is the matrix in $R^{(k,k)}$ with row vectors $\vec{e}_{\sigma(0)}, \dots, \vec{e}_{\sigma(k-1)}$, in this order.

We provide two other representations of permutation matrices. For this, we recall the Kronecker delta which is the following function:

(B.5.7)
$$\mathbb{N}_0^2 \to \{0, 1\}, \quad (i, j) \mapsto \delta_{i,j} = \begin{cases} 1 & \text{if } i = j, \\ 0 & \text{if } i \ne j. \end{cases}$$

Proposition B.5.6. *Let* $\sigma \in S_k$. *Then the following hold.*

(1) $P_\sigma = (\delta_{\sigma(i),j})_{i,j \in \mathbb{Z}_k} = (\delta_{i,\sigma^{-1}(j)})_{i,j \in \mathbb{Z}_k}$.

(2) P_σ *is the matrix with column vectors* $\vec{e}_{\sigma^{-1}(0)}, \dots, \vec{e}_{\sigma^{-1}(k-1)}$, *in this order.*

We prove two important properties of permutation matrices.

Proposition B.5.7. (1) *For all* $\sigma, \tau \in S_k$ *we have* $P_{\sigma \circ \tau} = P_\tau P_\sigma$.

(2) *For all* $\sigma \in S_k$ *the matrix* P_σ *is invertible, and we have* $P_{\sigma^{-1}} = P_\sigma^{-1} = P_\sigma^T$.

Proof. Let $\sigma, \tau \in S_k$. Then Proposition B.5.6 implies

$$
P_\tau P_\sigma = \left(\sum_{u=0}^{k-1} \delta_{\tau(i),u} \delta_{\sigma(u),j} \right)_{i,j \in \mathbb{Z}_k}
$$
$$
= (\delta_{\sigma(\tau(i)),j})_{i,j \in \mathbb{Z}_k}
$$
$$
= P_{\sigma \circ \tau}.
$$

This proves the first assertion and also implies

(B.5.8) $I_k = P_{\sigma^{-1}} P_\sigma = P_\sigma P_{\sigma^{-1}}.$

This shows that P_σ is invertible and $P_{\sigma^{-1}} = (P_\sigma)^{-1}$. Also, by Proposition B.5.6 we have $P_\sigma^T = (\delta_{i,\sigma(j)})$ and $P_{\sigma^{-1}} = (\delta_{i,\sigma(j)})$ which proves that these two matrices are the same. □

From Proposition B.5.6, we obtain the following corollary.

Corollary B.5.8. *The set of permutation matrices in $R^{(k,k)}$ is a subgroup of* $\mathrm{GL}(k, R)$.

We also determine the effect of multiplying matrices by permutation matrices from the left and right.

Proposition B.5.9. (1) *For all $A \in R^{(k,l)}$ with row vectors $\vec{a}_0, \ldots, \vec{a}_{k-1}$ and all $\sigma \in S_k$ the product $P_\sigma A$ is the matrix in $R^{(k,l)}$ with row vectors $\vec{a}_{\sigma(0)}, \ldots, \vec{a}_{\sigma(k-1)}$, in this order.*

(2) *For all $A \in R^{(l,k)}$ with column vectors $\vec{a}_0, \ldots, \vec{a}_{k-1}$ and all $\sigma \in S_l$ the product $A P_\sigma$ is the matrix in $R^{(l,k)}$ with column vectors $\vec{a}_{\sigma^{-1}(0)}, \ldots, \vec{a}_{\sigma^{-1}(l-1)}$, in this order.*

Proof. Let $A \in R^{(k,l)}$ and $\sigma \in S_k$. Then, for all $i \in \mathbb{Z}_k$, the product $\vec{e}_{\sigma(i)} A$ is the $\sigma(i)$th row vector of A. Together with the definition of P_σ this implies the first assertion. Now, let $\sigma \in S_l$. Let $j \in \mathbb{Z}_l$. By Proposition B.5.6, the jth column vector of P_σ is $\vec{e}_{\sigma^{-1}(j)}$ and the product $A \vec{e}_{\sigma(j)}$ is the $\sigma^{-1}(j)$th column vector of A. This implies the second assertion. □

B.5.3. Determinants.

Definition B.5.10. Consider a map

(B.5.9) $\det : R^{(k,k)} \to R.$

(1) The map f is called *multilinear* if it has the following two properties. For all $A \in R^{(k,k)}$ with column vectors $\vec{a}_0, \ldots, \vec{a}_{k-1}$, all $\vec{b} \in R^k$, all $j \in \mathbb{Z}_k$, and all $r \in R$ we

have

$$
\det(\vec{a}_0, \ldots, \vec{a}_{j-1}, \vec{a}_j + \vec{b}, \vec{a}_{j+1}, \ldots, \vec{a}_{k-1})
$$

(B.5.10)
$$
= \det(\vec{a}_0, \ldots, \vec{a}_{j-1}, \vec{a}_j, \vec{a}_{j+1}, \ldots, \vec{a}_{k-1})
$$
$$
+ \det(\vec{a}_0, \ldots, \vec{a}_{j-1}, \vec{b}, \vec{a}_{j+1}, \ldots, \vec{a}_{k-1})
$$

and

(B.5.11)
$$
\det(\vec{a}_0, \ldots, \vec{a}_{j-1}, r\vec{a}_j, \vec{a}_{j+1}, \ldots, \vec{a}_{k-1})
$$
$$
= r \cdot \det(\vec{a}_0, \ldots, \vec{a}_{j-1}, \vec{a}_j, \vec{a}_{j+1}, \ldots, \vec{a}_{k-1}).
$$

(2) The map det is called *alternating* if for every matrix $A \in R^{(k,k)}$ which has two equal columns we have $\det(A) = 0$.

(3) The map det is called *normalized* if $\det(I_k) = 1$.

(4) The map det is called a *determinant function* if it is multilinear, alternating, and normalized.

Proposition B.5.11. *Let* det $: R^{(k,k)} \to R$ *be multilinear and alternating. Then, for all* $A \in R^{(k,k)}$ *with column vectors* $\vec{a}_0, \ldots, \vec{a}_{k-1}$ *the following hold.*

(1) *Adding a multiple of one column to another column of A does not change* $\det(A)$; *that is, for all all* $r \in R$ *and all* $i, j \in \mathbb{Z}_k$ *with* $i \neq j$ *we have*

(B.5.12)
$$
\det(\vec{a}_0, \ldots, \vec{a}_{j-1}, \vec{a}_j + r\vec{a}_i, \vec{a}_{j+1} \ldots, \vec{a}_{k-1}) = \det A.
$$

(2) *Swapping two columns of A changes the sign of* $\det(A)$; *that is, for all* $i, j \in \mathbb{Z}_k$ *with* $i \neq j$ *we have*

(B.5.13)
$$
\det(\vec{a}_0, \ldots, \vec{a}_j, \ldots, \vec{a}_i \ldots, \vec{a}_{k-1}) = -\det A.
$$

(3) *If one column of A is zero, then* $\det A = 0$.

Theorem B.5.12. *The map*

(B.5.14)
$$
\det : R^{(k,k)} \to R, \quad A \mapsto \det(A) = \sum_{\sigma \in S_k} \text{sign}(\sigma) \prod_{j=0}^{k-1} a_{\sigma(j),j}
$$

is a determinant function and it is the only determinant function that maps $R^{(k,k)}$ *to* R.

Definition B.5.13. *For* $A \in R^{(k,k)}$ *the value* $\det(A)$ *is called the* determinant *of A.*

The formula (B.5.14) is called the *Leibniz formula* for evaluating determinants.

Proposition B.5.14. (1) *The determinant of a square matrix over R and its transpose are the same; that is, for all* $A \in R^{(k,k)}$ *we have*

(B.5.15)
$$
\det(A) = \det(A^{\mathsf{T}}).
$$

(2) *The determinant is linear with respect to matrix multiplication; that is, for all* $A, B \in R^{(k,k)}$ *we have*

(B.5.16)
$$
\det(AB) = \det(A) \det(B).
$$

(3) *The determinant of a triangular matrix (upper or lower) is the product of its diagonal entries.*

Definition B.5.15. Let $A \in R^{(k,k)}$ and assume that $k > 1$. Also, let $i, j \in \mathbb{Z}_k$. Then the minor $A_{i,j}$ is the matrix in $R^{(k-1,k-1)}$ that is obtained by deleting the ith row and jth column in A.

Here is the *Laplace expansion formula* for determinants.

Theorem B.5.16. *Let $k > 1$ and let $A = (a_{i,j}) \in R^{(k,k)}$. Then for every $i \in \mathbb{Z}_k$ we have*

$$(B.5.17) \qquad \det A = \sum_{j=0}^{k-1} (-1)^{i+j} a_{i,j} \det A_{i,j}$$

and for every $j \in \mathbb{Z}_k$ we have

$$(B.5.18) \qquad \det A = \sum_{i=0}^{k-1} (-1)^{i+j} a_{i,j} \det A_{i,j}.$$

Proposition B.5.17. *Let $A \in R^{(k,k)}$ in upper or lower triangular form. Then the determinant is the product of the diagonal elements of A.*

B.5.4. The unit group of $R^{(k,k)}$.

Definition B.5.18. Let $A = (a_{i,j}) \in R^{(k,k)}$. Then the *adjugate* of A is defined as the matrix

$$(B.5.19) \qquad \mathrm{adj}(A) = \left((-1)^{i+j} \det A_{j,i}\right)_{i,j \in \mathbb{Z}_k} \in R^{(k,k)}$$

where $A_{i,j}$ are the minors of A. We also write $\mathrm{adj}\, A$ instead of $\mathrm{adj}(A)$.

The adjugate of a square matrix has the following property that allows us to compute inverses of square matrices.

Proposition B.5.19. *Let $A \in R^{(k,k)}$. Then we have*

$$(B.5.20) \qquad (\mathrm{adj}\, A)A = A(\mathrm{adj}\, A) = \det A \cdot I_k.$$

Now we can characterize the elements of $\mathrm{GL}(k, R)$ and show how to compute the inverses of square matrices.

Theorem B.5.20.   (1) *A matrix $A \in R^{(k,k)}$ is invertible if and only if $\det(A)$ is a unit in R; that is,*

$$(B.5.21) \qquad \mathrm{GL}(k, R) = \{A \in R^{(k,k)} : \det A \in U(R)\}.$$

   (2) *Let $A \in \mathrm{GL}(k, R)$. Then we have*

$$(B.5.22) \qquad \det(A^{-1}) = (\det A)^{-1}$$

   and the inverse of A is

$$(B.5.23) \qquad A^{-1} = \det(A)^{-1} \mathrm{adj}(A).$$

Corollary B.5.21. *Let $A, B \in R^{(k,k)}$ with $AB = I_k$. Then $A, B \in \mathrm{GL}(k, R)$, $B = A^{-1}$, and $A = B^{-1}$.*

Corollary B.5.22. *If F is a field, then we have*

$$(B.5.24) \qquad \mathrm{GL}(k, F) = \{A \in F^{(k,k)} : \det A \neq 0\}.$$

Lemma B.5.23. *Let $A, B \in \mathrm{GL}(k, R)$. Then we have $(AB)^{-1} = B^{-1}A^{-1}$.*

B.5.5. Trace.

Definition B.5.24. The *trace* of a square matrix A over R is the sum of its diagonal elements. It is denoted by $\mathrm{tr}(A)$ or $\mathrm{tr}\,A$.

Proposition B.5.25. (1) *The trace map* $\mathrm{tr}\ :\ R^{(k,k)}\ \to\ R$ *is R-linear; that is,* $\mathrm{tr}(aA + bB) = a\,\mathrm{tr}(A) + b\,\mathrm{tr}(B)$ *for all $a, b \in R$ and $A, B \in R^{(k,k)}$.*

(2) $\mathrm{tr}(A^{\mathrm{T}}) = \mathrm{tr}(A)$ *for all $A \in R^{(k,k)}$.*

(3) $\mathrm{tr}(AB) = \mathrm{tr}(BA)$ *for all $A, B \in R^{(k,k)}$.*

B.5.6. Characteristic polynomials.

Definition B.5.26. The *characteristic polynomial* of a matrix $A \in F^{(k,k)}$ is the polynomial $p_A(x) = \det(xI_k - A) \in R[x]$.

Proposition B.5.27. *Let $A \in R^{(k,k)}$ and let*

$$(\mathrm{B.5.25}) \qquad p_A(x) = x^k + \sum_{i=0}^{k-1} r_i x^i$$

with $r_i \in R$ for all $i \in \mathbb{Z}_k$. Then we have

$$(\mathrm{B.5.26}) \qquad r_{k-1} = -\,\mathrm{tr}(A)$$

and

$$(\mathrm{B.5.27}) \qquad r_0 = (-1)^k \det(A).$$

Corollary B.5.28. *Let $A \in R^{(k,k)}$ and assume that the characteristic polynomial of A can be written as*

$$(\mathrm{B.5.28}) \qquad p_A(x) = \prod_{i=0}^{k-1} (x - \lambda_i)$$

with $\lambda_i \in R$ for $0 \le i < k$. Then we have

$$(\mathrm{B.5.29}) \qquad \mathrm{tr}(A) = \sum_{i=0}^{k-1} \lambda_i$$

and

$$(\mathrm{B.5.30}) \qquad \det(A) = \prod_{i=0}^{k-1} \lambda_i.$$

B.5.7. Similar matrices.

Definition B.5.29. Two matrices $A, B \in R^{(k,k)}$ are called *similar* if there is a matrix $U \in \mathrm{GL}(k, R)$ such that $B = U^{-1}AU$.

Proposition B.5.30. *Similar matrices in $R^{(k,k)}$ have the same characteristic polynomial, trace, and determinant.*

B.6. Free modules of finite dimension

In this section, we discuss free R-modules with finite bases. Let M be an R-module.

B.6.1. Operations on M^k.

Definition B.6.1. Let $B = (\vec{b}_0, \ldots, \vec{b}_{k-1}) \in M^k$.

(1) We define the product of B with a vector $\vec{x} = (x_0, \ldots, x_{k-1}) \in R^k$ as

$$\text{(B.6.1)} \qquad B\vec{x} = B \cdot \vec{x} = \sum_{i=0}^{k-1} x_i \vec{b}_i.$$

(2) Let $T \in R^{(k,l)}$ with column vectors $\vec{t}_0, \ldots, \vec{t}_{l-1}$. Then we define the product of B with T as

$$\text{(B.6.2)} \qquad BT = B \cdot T = (B\vec{t}_0, \ldots, B\vec{t}_{l-1}).$$

Proposition B.6.2. *Let* $B = (\vec{b}_0, \ldots, \vec{b}_{k-1}) \in M^k$.

(1) *For all* $r \in R$ *and all* $\vec{x}, \vec{y} \in R^k$ *we have* $(rB)\vec{x} = B(r\vec{x})$ *and* $B(\vec{x} + \vec{y}) = B\vec{x} + B\vec{y}$.

(2) *For all* $r \in R$ *and* $X \in R^{(k,l)}$ *and* $Y \in R^{(k,l)}$ *we have* $(rB)X = B(rX)$ *and* $B(X+Y) = BX + BY$.

(3) *For all* $X \in R^{(k,l)}$ *and* $Y \in R^{(l,m)}$ *we have* $(BX)Y = B(XY)$.

B.6.2. Bases and dimension.

Proposition B.6.3. *Let* $B = (\vec{b}_0, \ldots, \vec{b}_{m-1})$ *be a sequence of elements in* M. *Then* B *is linearly independent if and only if it follows from* $\sum_{j=0} r_j \vec{b}_j = 0$ *with* $r_j \in R$ *for* $0 \leq j < m$ *that* $r_0, \ldots, r_{m-1} = 0$.

Theorem B.6.4. *Let* M *be finitely generated. Then the following hold.*

(1) *If* M *is free, then all bases of* M *are finite and have the same length which is called the* dimension *of* M.

(2) *If* M *has a finite basis* B *of length* k, *then every basis of* M *can be obtained as* BT *with* $T \in \mathsf{GL}(k, R)$; *that is, the set of all bases of* M *is* $B\mathsf{GL}(k, R)$.

Corollary B.6.5. *Let* F *be a field, and let* V *be a finitely generated* F-*vector space. Then* V *has a finite basis* B, *all bases of* V *have the same length, which is called the* dimension *of* V, *and for any basis* B *of* V, *the set of all bases of* V *is* $B\mathsf{GL}(k, F)$.

Definition B.6.6. Let $B = (\vec{b}_0, \ldots, \vec{b}_{k-1})$ be a basis of M and let $\vec{v} \in M$. Then the uniquely determined vector $\vec{x} \in R^k$ such that $\vec{v} = B\vec{x}$ is called the *coefficient vector* of $\vec{v}$ with respect to the basis B. We denote it by $\vec{v}_B$.

Theorem B.6.7. *Let* $B = (\vec{b}_0, \ldots, \vec{b}_{k-1})$ *be a basis of* M. *Then the map*

$$\text{(B.6.3)} \qquad M \to R^k, \quad \vec{v} \mapsto \vec{v}_B,$$

that sends an element $\vec{v} \in M$ *to its coefficient vector with respect to the basis* B, *is an isomorphism of* R-*modules.*

Proposition B.6.8. *Let B be a finite basis of M of length k and let $T \in \mathrm{GL}(k, R)$. Then for all $\vec{v} \in M$ we have*

(B.6.4) $$\vec{v}_B = T\vec{v}_{BT}.$$

B.6.3. Linear maps. Let M, N be free R-modules of dimensions k and l, respectively. In this section, we construct isomorphisms between the R-modules $\mathrm{Hom}_R(M, N)$ and $R^{(k,l)}$ and between the R-algebras $\mathrm{End}_R(M)$ and $R^{(k,k)}$. This shows that we can identify homomorphisms between finite-dimensional R-modules with matrices over R. However, we will see that these identifications depend on the choice of bases of M and N.

Let $B = (\vec{b}_0, \dots, \vec{b}_{k-1})$ and $C = (\vec{c}_0, \dots, \vec{c}_{l-1})$ be R-bases of M and N, respectively. Recall that for any $\vec{v} \in M$ we denote by $\vec{v}_B$ the coefficient vector of $\vec{v}$ with respect to the basis B. Also, for any $\vec{w} \in N$ we denote by $\vec{w}_C$ the coefficient vector of $\vec{w}$ with respect to the basis C. By Theorem B.6.7 the maps $M \to R^l, \vec{v} \mapsto \vec{v}_B$ and $N \to R^k, \vec{w} \mapsto \vec{w}_C$ are R-module isomorphisms. We now define the representation matrices of linear maps from M to N.

Definition B.6.9. (1) For $f \in \mathrm{Hom}_R(M, N)$, we define $\mathrm{Mat}_{B,C}(f)$ as the matrix in $R^{(l,k)}$ with column vectors $f(\vec{b}_0)_C, \dots, f(\vec{b}_{k-1})_C$. This matrix is called the *representation matrix* of f with respect to the bases B and C.

(2) For $f \in \mathrm{End}(M)$ we write $\mathrm{Mat}_B(f)$ for $\mathrm{Mat}_{B,B}(f)$ and call this matrix the *representation matrix* of f with respect the basis B.

(3) Let T be in $R^{(l,k)}$. Then we define the map

(B.6.5) $$f_{T,B.C} : M \to N, \quad \vec{v} \mapsto CT\vec{v}_B.$$

This map is in $\mathrm{Hom}_R(M, N)$.

(4) Let $M = N$ and let T be in $R^{(k,k)}$. Then we write $f_{T,B}$ for $f_{T,B,B}$. This map is in $\mathrm{End}(M)$.

Theorem B.6.10. (1) *The map*

(B.6.6) $$\mathrm{Hom}_R(M, N) \to R^{(l,k)}, \quad f \mapsto \mathrm{Mat}_{B,C}(f)$$

is an R-module isomorphism. The inverse of this isomorphism sends $T \in R^{(l,k)}$ to $f_{T,B,C}$.

(2) *The map*

(B.6.7) $$\mathrm{End}_R(M) \to R^{(k,k)}, \quad f \mapsto \mathrm{Mat}_B(f)$$

is an R-algebra isomorphism.

Proposition B.6.11. *Let $S \in \mathrm{GL}(k, R)$ and $T \in \mathrm{GL}(l, R)$. Then for all $f \in \mathrm{Hom}_R(M, N)$ we have*

(B.6.8) $$\mathrm{Mat}(f)_{BS,CT} = T^{-1}\mathrm{Mat}(f)_{B,C}S.$$

Proof. We apply Proposition B.6.8 and obtain

(B.6.9) $$f(\vec{v}) = C\mathrm{Mat}_{B,C}(f)\vec{v}_B = CTT^{-1}\mathrm{Mat}_{B,C}(f)SS^{-1}\vec{v}_B.$$

This implies the assertion. $\qquad\square$

B.6.4. Endomorphisms. Let M be a free R-module of finite dimension k and let $B = (\vec{b}_0, \ldots, \vec{b}_{k-1})$ be a basis of M. The next theorem follows immediately from Theorem B.6.10.

Theorem B.6.12. *The map*

(B.6.10) $$\mathrm{End}_R(M) \to R^{(k,k)}, \quad f \mapsto \mathrm{Mat}_B(f)$$

is an isomorphism of R-algebras. The inverse of this isomorphism is

(B.6.11) $$R^{(k,k)} \to \mathrm{End}_R(M), \quad T \mapsto f_{T,B}.$$

Theorem B.6.13. (1) *An endomorphism f of M is an automorphism if and only if* $\mathrm{Mat}_B(f) \in \mathrm{GL}(R, k)$.

(2) *For all $f \in \mathrm{Aut}_R(M)$ we have*

(B.6.12) $$\mathrm{Mat}_B(f^{-1}) = \mathrm{Mat}_B(f)^{-1}.$$

Proposition B.6.14. *For all $U \in \mathrm{GL}(k, R)$ we have $M_{BU}(f) = U^{-1} M_B(f) U$.*

Corollary B.6.15. *The set of representation matrices of an endomorphism f of M is the equivalence class of all matrices in $R^{(k,k)}$ that are similar to $\mathrm{Mat}_B(f)$.*

Corollary B.6.16. *The characteristic polynomials, traces, and determinants of all matrix representations of an endomorphism of a k-dimensional R-module M are the same.*

This result justifies the following definition.

Definition B.6.17. The characteristic polynomial, determinant, and trace of an endomorphism of a finitely generated free R-module is the characteristic polynomial, trace, and determinant of any of its representation matrices in $R^{(k,k)}$, respectively.

B.6.5. Dual modules. Let M be a free R-module of finite dimension k and let $B = (\vec{b}_0, \ldots, \vec{b}_{k-1})$ be a basis of M.

Theorem B.6.18. *The dual module M^* is isomorphic to M as an R-module. In particular, M^* is a finitely generated free module and its dimension is the dimension of M.*

B.7. Finite-dimensional vector spaces

Most of the linear algebra used to model quantum algorithms refers to vector spaces of finite dimensions. This section discusses this topic. We let V, W be F-vector spaces of dimensions k and l, respectively.

B.7.1. Bases and generating systems. We know from Theorem B.6.4 that all bases of V have the same length k and that for every basis B of V, the set of all bases of V is the coset $\mathrm{GL}(k, F)B$. We now state the *Steinitz Exchange Lemma* which allows us to obtain more results for bases and generating systems of V.

Lemma B.7.1. *Let $m, n \in \mathbb{N}$, let $U = (\vec{u}_0, \ldots, \vec{u}_{m-1}) \in V^m$ be linearly independent, and let $G \in V^n$ be a generating system of V. Then $m \le n$ and there are elements $\vec{u}_m, \ldots, \vec{u}_{n-1}$ in G such that $(\vec{u}_0, \ldots, \vec{u}_{n-1})$ is a generating system of V.*

Theorem B.7.2. (1) *Linearly independent sequences in V have length $\leq k$.*

(2) *A linearly independent sequence in V is a basis of V if and only if its length is k.*

(3) *Every linearly independent system can be extended to a basis of V.*

(4) *Generating systems of V have length $\geq k$.*

(5) *A generating system of V is a basis of V if and only if its length is k.*

B.7.2. The rank of a matrix.

Proposition B.7.3. *The vector spaces generated by the rows and columns of a matrix A over F, respectively, have the same dimension. This dimension is called the* rank *of A. It is denoted by* rank(A) *or* rank A.

Example B.7.4. Consider the matrix

$$(B.7.1) \qquad A = \begin{pmatrix} 1 & 0 \\ 0 & 1 \\ 1 & 1 \end{pmatrix}$$

over $\mathbb{F}_2$. The rank of A is 2. Indeed, the column rank of A is 2 because the two column vectors of A are linearly independent. Also, the row rank of this matrix is 2 because the first two row vectors of A are linearly independent and the third row vector is a linear combination of the first two.

The next proposition establishes a connection between the kernel and the image of a linear map from V to W and the rank of a representation matrix of this map.

Proposition B.7.5. *Let $f \in \text{Hom}_F(V, W)$. Then the rank r of all representation matrices of f is the same and the following hold.*

(1) *The dimension of the image of f is r.*

(2) *The dimension of the kernel of A is $k - r$.*

Definition B.7.6. Let $A \in F^{(k,l)}$.

(1) The *kernel* of A is defined as the kernel of the map $R^l \to R^k, \vec{x} \mapsto A\vec{x}$.

(2) The *image* of A is defined as the image of the map $R^l \to R^k, \vec{x} \mapsto A\vec{x}$.

Definition B.7.7. We call a matrix $A \in F^{(k,k)}$ *singular* if $\det A = 0$ and we call it *nonsingular* otherwise.

Proposition B.7.8. *Let $A \in F^{(k,k)}$. Then the following statements are equivalent.*

(1) *A is nonsingular.*

(2) *The rank of A is k.*

(3) *The columns of A form a basis of F^k.*

(4) *The rows of A form a basis of F^k.*

B.7.3. Row and column echelon form.

Definition B.7.9. Let $A = (a_{i,j}) \in R^{(l,k)}$ with row vectors $\vec{a}_0, \ldots, \vec{a}_{l-1}$.

(1) We say that A is in *row echelon form* if the following conditions are satisfied.
   (a) All rows of A that have only zero entries are at the bottom of A; that is, if $u, v \in \mathbb{Z}_k$ such that $\vec{a}_u \neq \vec{0}$ and $\vec{a}_v = \vec{0}$, then $u < v$.
   (b) If $u > 0$ and $\vec{a}_u$ is nonzero, then the first nonzero entry in $\vec{a}_u$ is strictly to the right of the first nonzero entry in $\vec{a}_{u-1}$; that is,

(B.7.2) $$\min\{j \in \mathbb{Z}_l : a_{u,j} \neq 0\} > \min\{j \in \mathbb{Z}_l : a_{u-1,j} \neq 0\}.$$

(2) We say that A is in *reduced row echelon form* if A is in row echelon form and the first nonzero element in each nonzero row is 1.

Definition B.7.10. We say that a matrix $A \in R^{(k,l)}$ is in *column echelon form* if A^T is in row echelon form. Also, we say that A is in *reduced column echelon form* if A^T is in reduced row echelon form.

We note that a square matrix in row echelon form is an upper triangular matrix. Also, a square matrix in column echelon form is a lower triangular matrix.

B.7.4. The Gauss elimination algorithm.

In this section, we explain the Gauss Elimination Algorithm B.7.13 that transforms a matrix $A \in F^{(l,k)}$ into column echelon form. Despite its name, this algorithm was already known in China in the second century. Since the algorithm uses division by nonzero elements, the algorithm is only guaranteed to work over fields but, in general, not over rings.

The correctness of the algorithm is stated in the next theorem.

Theorem B.7.11. *On input of $k, l \in \mathbb{N}$ and $A \in F^{(l,k)}$, Algorithm B.7.13 returns $A' F^{(l,k)}$, $S \in \mathsf{GL}(k, F)$, $v \in \mathbb{N}$, and $w \in \mathbb{N}_0$ such that A' is in column echolon form, $A' = AS$, v is the number of nonzero columns in A', and $\det S = (-1)^w$.*

The name "Gaussian elimination algorithm" derives from the fact that in the **for** loop starting at line 15, the entries $a_{u,j}$ are "eliminated" for $v < j < k$.

Algorithm B.7.13 can also be used to transform $A \in F^{(k,l)}$ into row echelon form as follows. We apply Algorithm B.7.13 to the transpose of A. The algorithm returns A', S, v, w. We replace A', S by their transposes. Then A' is in row echelon form and we have $A' = SA$, v is the number of nonzero rows of A', and $\det S = (-1)^w$.

Theorem B.7.12. *Let $k, l \in \mathbb{N}$ and $A \in F^{(l,k)}$ be the input of Algorithm B.7.13 and let $n = \max\{k, l\}$. The algorithm then uses $O(n^3)$ operations in F and space for $O(n^2)$ elements of F.*

The Gauss elimination algorithm also requires time and space to initialize and increment the loop variables u, v, and w. However, these time and space requirements are dominated by the complexity of the operations in the field F. Therefore, we do not mention them explicitly.

Algorithm B.7.13. Gaussian elimination

Input: $k, l \in \mathbb{N}, A \in F^{(l,k)}$

Output: $A' \in F^{(l,k)}, S \in \mathsf{GL}(k,F), v, w \in \mathbb{N}_0$, such that A' is in column echolon form, $A' = AS$, v is the number of nonzero columns in A', and $\det S = (-1)^w$.

1: COLUMNECHOLON(k, l, A)
2:     $u, v, w \leftarrow 0$
3:     $A' \leftarrow A$
4:     /* The entries of A' are $a'_{i,j}$ */
5:     $S \leftarrow I_k$
6:     /*The columns of A', S are $\vec{a}'_j$ and $\vec{s}_j$, respectively.
7:     **while** $u < l$ and $v < k$ **do**
8:         **if** one of $a'_{u,v}, a'_{u,v+1}, \ldots, a'_{u,k-1}$ is nonzero **then**
9:             Select $v_{\text{pivot}} \in \{v, \ldots, k-1\}$ such that $a'_{u,v_{\text{pivot}}} \neq 0$
10:            $\mathbf{if}$ $v \neq v_{\text{pivot}}$ **then**
11:                Swap $\vec{a}'_v$ and $\vec{a}'_{v_{\text{pivot}}}$
12:                Swap $\vec{s}_v$ and $\vec{s}_{v_{\text{pivot}}}$
13:                $w \leftarrow w + 1$
14:            **end if**
15:            **for** $j = v+1, \ldots, k-1$ **do**
16:                $\alpha \leftarrow a_{u,j}/a_{u,v}$
17:                $\vec{a}_j \leftarrow \vec{a}_j - \alpha \vec{a}_v$
18:                $\vec{s}_j \leftarrow \vec{s}_j - \alpha \vec{s}_v$
19:            **end for**
20:            $v \leftarrow v + 1$
21:         **end if**
22:         $u \leftarrow u + 1$
23:     **end while**
24:     **return** A', S, v, w
25: **end**

Several modifications of Algorithm B.7.13 are possible. Depending on the desired output, we can omit the computation of S or w which simplifies the algorithm and improves its performance. When F is the field of real or complex numbers, the algorithm can only use approximations of the entries of the matrix A. Then, a good selection of the pivot element is crucial for keeping error propagation under control.

B.7.5. Applications of the Gauss elimination algorithm.

Theorem B.7.14. *Let $l, k \in \mathbb{N}$ and $A \in F^{(l,k)}$ be the input of the Gauss elimination algorithm and let $A' \in F^{(l,k)}, S \in \mathsf{GL}(k,F), v, w \in \mathbb{N}$ be its output. Then the following hold.*

(1) *The rank of A is v.*

(2) *The sequence consisting of the first v column vectors of A' is a basis of the image of A.*

(3) *The sequence consisting of the last $v - k$ columns of S is a basis of the kernel of A.*

(4) *If $k = l$, then $(-1)^w \det A$ is the product of the diagonal elements of A'.*

Also, with $n = \max\{k, l\}$ the computation of these objects requires $\mathrm{O}(k^3)$ operations in F and space for $\mathrm{O}(k^2)$ elements of F.

Next, we discuss the problem of solving *linear systems of equations*. By this we mean the following. Let $A \in F^{(l,k)}$ and $\vec{b} \in F^l$. The goal is to find all $\vec{x} \in F^k$ such that

(B.7.3) $A\vec{x} = \vec{b}.$

If $\vec{x} \in F^l$ satisfies (B.7.3), then $\vec{x}$ is called a *solution of the linear system* (B.7.3). We first characterize the solutions of linear systems.

Proposition B.7.15. *Let $A \in F^{(l,k)}$ and let $b \in F^l$. Then the set of all the solutions of the linear system $A\vec{x} = \vec{b}$ is empty or a coset of the kernel of A, i.e., of the form $\vec{x} + \ker(A)$ where $\vec{x}$ is any of the solutions of the linear system.*

Proposition B.7.15 shows how to find the set of all solutions of the linear system (B.7.3). First, decide whether the linear system has a solution and if this is the case, find one. Second, determine the basis of the kernel of A. We have already explained how the second task can be achieved using the Gauss algorithm. So, it remains to solve the first task. This is done in Algorithm B.7.16.

Algorithm B.7.16. Solving a linear system

Input: $k, l \in \mathbb{N}, A \in F^{(l,k)}, \vec{b} \in F^l$
Output: $\vec{x} \in F^k$ with $A\vec{x} = \vec{b}$ or "No solution"

1: $\mathrm{SOLVE}(k, l, A, \vec{b})$
2:      $(A', S, v, w) \leftarrow \mathrm{ColumnEcholon}(k, l, A)$
3:      /* The entries of A' are $a'_{i.j}$. The entries of $\vec{b}$ are b_i /*
4:      **for** $j = 0, \ldots, v - 1$ **do**
5:          $u = \min\{i \in \mathbb{Z}_k : a'_{i,j} \neq 0\}$
6:          $y_j = \left(b_u - \sum_{i=0}^{j-1} x_i a'_{u,i}\right) / a'_{u,j}$
7:      **end for**
8:      $\vec{y} \leftarrow (y_0, \ldots, y_{v-1}, \underbrace{0, \ldots, 0}_{k-v})$
9:      **if** $A'\vec{y} = b$ **then**
10:          $\vec{x} \leftarrow S\vec{y}$
11:          **return** $\vec{x}$
12:     **else**
13:          **return** "No solution"
14:     **end if**
15: **end**

Proposition B.7.17. *Let $k, l \in \mathbb{N}$, $A \in F^{(l,k)}$, and $\vec{b} \in F^l$ be the input of Algorithm B.7.16. If the algorithm returns $\vec{x} \in F^k$, then this vector satisfies $A\vec{x} = \vec{b}$. If the algorithm returns "No solution", then the linear system $A\vec{x} = \vec{b}$ has no solution. The algorithm requires $\mathrm{O}(n^3)$ operations in F and space for $\mathrm{O}(n^2)$ elements of F.*

B.7.6. Eigenvalues, eigenvectors, and eigenspaces.

Definition B.7.18. (1) Let $A \in \mathrm{End}(V)$. An *eigenvalue* of A is a field element $\lambda \in F$ such that $A\vec{v} = \lambda\vec{v}$ for some nonzero vector $\vec{v} \in V$. Such a vector $\vec{v}$ is called an *eigenvector of A corresponding to the eigenvalue λ* or simply an *eigenvector* of A.

(2) Let $A \in F^{(k,k)}$. An eigenvalue or eigenvector of A is defined as an eigenvalue or eigenvector of the endomorphism $F^k \to F^k, \vec{v} \mapsto A\vec{v}$, respectively.

Proposition B.7.19. *Let $f \in \mathrm{End}(V)$ and let $\vec{v} \in V$ be an eigenvector of f. Then there is exactly one eigenvalue $\lambda \in F$ of f such that $f(\vec{v}) = \lambda\vec{v}$. It is called the eigenvalue associated with the eigenvector $\vec{v}$.*

Proposition B.7.20. *If λ is an eigenvalue of an endomorphism $f \in \mathrm{End}(V)$, then the set of all eigenvectors corresponding to this eigenvalue is a subspace of V. It is called the eigenspace of f associated with the eigenvalue λ.*

Proposition B.7.21. *A field element $\lambda \in F$ is an eigenvalue of an endomorphism $f \in \mathrm{End}(V)$ if and only if $p_f(\lambda) = 0$.*

Corollary B.7.22. *If $A \in F^{(k,k)}$ is in upper or lower triangular form, then the eigenvalues of A are its diagonal elements.*

Definition B.7.23. If λ is an eigenvalue of $f \in \mathrm{End}(V)$, then the dimension of the eigenspace associated with λ is called the *geometric multiplicity* of λ. Furthermore, the *algebraic multiplicity* of λ is the power to which $(x - \lambda)$ divides $p_A(x)$.

Proposition B.7.24. *Let $f \in \mathrm{End}(V)$, let $\lambda_0, \dots, \lambda_{l-1}$ be the eigenvalues of f, let $m_0, \dots, m_{l-1}$ be their geometric multiplicities, and let $E_0, \dots, E_{l-1}$ be the corresponding eigenspaces. Then the following hold.*

(1) *The sum of the eigenspaces $E_0, \dots, E_{l-1}$ is direct.*

(2) *There is a basis $(\vec{b}_0, \dots, \vec{b}_{k-1})$ of V such that for all $j \in \mathbb{Z}_l$ the sequence $(\vec{b}_{M_j}, \dots, \vec{b}_{M_j + m_j - 1})$ is a basis of E_j where $M_j = \sum_{i=0}^{j-1} m_i$.*

Corollary B.7.25. *Let $A \in F^{(k,k)}$, let $\lambda_0, \dots, \lambda_{l-1}$ be the eigenvalues of A, and let $m_0, \dots, m_{l-1}$ be their geometric multiplicities. Then A is similar to a matrix of the form*

(B.7.4)
$$A = \begin{pmatrix} A_1 & A_2 \\ \mathbf{0} & A_3 \end{pmatrix}$$

where

(B.7.5)
$$A_1 = \mathrm{diag}(\underbrace{\lambda_0, \dots, \lambda_0}_{m_0}, \dots, \underbrace{\lambda_{l-1}, \dots, \lambda_{l-1}}_{m_{l-1}}),$$

$\mathbf{0}$ *stands for the matrix in $F^{(k-l,k)}$ with only zero entries, $A_2 \in F^{(l,k-l)}$, and $A_3 \in F^{(l,k-l)}$.*

Corollary B.7.26. *Let $f \in \mathrm{End}(V)$. Then the geometric multiplicities of the eigenvalues of f are less than or equal to the corresponding algebraic multiplicities.*

B.7.7. Diagonizable matrices.

Definition B.7.27. A matrix $A \in R^{(k,k)}$ is called *diagonizable* if A is similar to a diagonal matrix.

Theorem B.7.28. *Let $A \in \mathbb{F}^{(k,k)}$. Then the following statements are equivalent.*

(1) *A is diagonizable.*

(2) *The characteristic polynomial $p_A(x)$ of A is a product of linear factors, and the geometric multiplicity of each eigenvalue is equal to its algebraic multiplicity.*

(3) *$\mathbb{F}^k$ is the direct sum of the eigenspaces of A.*

Corollary B.7.29. *Let $A \in \mathbb{F}^{(k,k)}$ be diagonizable, let $\lambda_0, \ldots, \lambda_{l-1}$ be the distinct eigenvalues of A, and let $m_0, \ldots, m_{l-1}$ be their algebraic multiplicities. Then there is a basis $(\vec{b}_0, \ldots, \vec{b}_{k-1})$ of eigenvectors of A such that the first m_0 of them are eigenvectors for the eigenvalue λ_0, the next m_1 of them are eigenvectors for the eigenvalue λ_1, etc. Also, we have*

$$(B.7.6) \qquad B^{-1}AB = \mathrm{diag}(\underbrace{\lambda_0, \ldots, \lambda_0}_{m_0}, \underbrace{\lambda_1, \ldots, \lambda_1}_{m_1}, \ldots, \underbrace{\lambda_{l-1}, \ldots, \lambda_{l-1}}_{m_{l-1}})$$

where B is the matrix with column vectors $\vec{b}_0, \ldots, \vec{b}_{k-1}$.

B.8. Tensor products

Let $M_0, \ldots, M_{m-1}, P$ be modules over a ring R where $m \in \mathbb{N}$. We discuss the tensor product of the modules $M_0, \ldots, M_{m-1}$. We let M be the direct product of the modules $M_0, \ldots, M_{m-1}$.

B.8.1. Idea and characterization.
A tensor product of the modules $M_0, \ldots, M_{m-1}$ combines these modules to a larger module T that respects the original module structures and their respective operations. Such a construction is useful to model combinations of systems that are individually modeled as R-modules, for example the combination of state spaces in quantum mechanics. For the definition of tensor products, multilinear maps are required.

Definition B.8.1. A function

$$(B.8.1) \qquad\qquad\qquad f : M \to P$$

is called *multilinear* if for all $(\vec{v}_0, \ldots, \vec{v}_{m-1}) \in M$ and all $j \in \mathbb{Z}_m$ the functions

$$(B.8.2) \qquad M_j \to P, \quad \vec{v} \mapsto f(\vec{v}_0, \ldots, \vec{v}_{j-1}, \vec{v}, \vec{v}_{j+1}, \ldots, \vec{v}_{m-1})$$

are R-module homomorphisms. If $m = 2$, then f is called *bilinear*.

Example B.8.2. Let $m = 3$, $R, M_0, M_1, M_2, P = \mathbb{Z}$. The map

$$(B.8.3) \qquad\qquad f : \mathbb{Z}^3 \to \mathbb{Z}, \quad (x_0, x_1, x_2) \mapsto x_0 x_1 x_2$$

is multilinear. To see this, we note that for $(x_0, x_1, x_2) \in \mathbb{Z}^3$ and $r, x \in \mathbb{Z}$ we have $f(x_0 + x, x_1, x_2) = (x_0 + x)x_1, x_2 = x_0 x_1 x_2 + x x_1 x_2 = f(x_0, x_1, x_2) + f(x, x_1, x_2)$, and $f(rx_0, x_1, x_2) = rx_0 x_1 x_2 = rf(x_0, x_1 x_2)$. Therefore, f is linear in its first argument. In the same way, it can be shown that f is linear in the other two arguments.

We present a condition for the value of a multilinear function to be 0.

Lemma B.8.3. *Let $f : M \to P$ be multilinear. Then for all $(\vec{v}_0, \ldots, \vec{v}_{m-1}) \in M$ and all $j \in \mathbb{Z}_m$ we have*

$$(B.8.4) \qquad f(\vec{v}_0, \ldots, \vec{v}_{j-1}, \vec{0}, \vec{v}_{j+1}, \ldots, \vec{v}_{m-1}) = 0.$$

Proof. Let $(\vec{v}_0, \ldots, \vec{v}_{m-1}) \in M$ and let $j \in \mathbb{Z}_m$. Since f is multilinear, we have

$$
\begin{aligned}
(B.8.5) \qquad & f(\vec{v}_0, \ldots, \vec{v}_{j-1}, \vec{0}, \vec{v}_{j+1}, \ldots, \vec{v}_{m-1}) \\
& = f(\vec{v}_0, \ldots, \vec{v}_{j-1}, 0 \cdot \vec{0}, \vec{v}_{j+1}, \ldots, \vec{v}_{m-1}) \\
& = 0 \cdot f(\vec{v}_0, \ldots, \vec{v}_{j-1}, \vec{0}, \vec{v}_{j+1}, \ldots, \vec{v}_{m-1}) = 0.
\end{aligned}
$$

This concludes the proof. $\qquad\qquad\square$

Now we define tensor products.

Definition B.8.4. A *tensor product* of $M_0, \ldots, M_{m-1}$ over R is a pair (T, θ) where T is an R-module and

$$(B.8.6) \qquad \theta : M \to T$$

is a multilinear map that has the following properties.

(1) The image $\theta(M)$ of M under θ spans T.

(2) Universal property: For every R-module P and every multilinear map

$$(B.8.7) \qquad \phi : M \to P$$

there is $\Phi \in \mathrm{Hom}_R(T, P)$ such that

$$(B.8.8) \qquad \phi = \Phi \circ \theta.$$

Example B.8.5. Let $m = 3$, $R, M_0, M_1, M_2, T = \mathbb{Z}$. Define the map $\theta : \mathbb{Z}^3 \to \mathbb{Z}$, $(x_0, x_1, x_2) \mapsto x_0 x_1 x_2$. We claim that $(\mathbb{Z}, \theta)$ is a tensor product of M_0, M_1, and M_2. We have seen in Example B.8.2 that θ is multilinear. Also, since for every $t \in \mathbb{Z}$ we have $t = \theta(t, 1, 1)$, it follows that $\theta(\mathbb{Z}^3) = \mathbb{Z}$.

To prove the universal property, let $\phi : \mathbb{Z}^3 \to \mathbb{Z}$ be a linear map. Define

$$(B.8.9) \qquad \Phi : \mathbb{Z} \to \mathbb{Z}, \quad t \mapsto \phi(t, 1, 1).$$

This map is well-defined. Also, if $z, t, t' \in Z$, then the multilinearity of ϕ implies

$$(B.8.10) \qquad \Phi(t + t') = \phi(t + t', 1, 1) = \phi(t, 1, 1) + \phi(t', 1, 1) = \Phi(t) + \Phi(t')$$

and

$$(B.8.11) \qquad \Phi(zt) = \phi(zt, 1, 1) = z\phi(t, 1, 1) = t\Phi(t).$$

So Φ is linear, and we have $\phi = \Phi \circ \theta$. We also show that Φ is the only linear map between T and P such that $\Phi \circ \theta = \phi$. So let $\Phi' : \mathbb{Z} \to \mathbb{Z}$ be another linear map such that $\phi = \Phi' \circ \theta$. Then for all $t \in \mathbb{Z}$ we have

$$(B.8.12) \qquad \Phi(t) = \phi(t, 1, 1) = \Phi'(\theta(t, 1, 1)) = \Phi'(t).$$

Exercise B.8.6. Let $R = \mathbb{F}_2$, $M_0 = M_1 = \mathbb{F}_2^2$. Find a tensor product of M_0 and M_1.

We will now generalize Example B.8.5 and show how to construct the map Φ from Definition B.8.4 and prove that it is uniquely determined by ϕ in this definition.

Lemma B.8.7. *Let T be an R-module and let $\theta : M \to T$ be a multilinear map with* Span $\theta(M) = T$. *Then the following statements are equivalent.*

(1) *The pair (T, θ) has the universal property.*

(2) *For every R-module P, every multilinear map $\phi : M \to P$, every $n \in \mathbb{N}$, all $(\vec{v}_0, \ldots, \vec{v}_{n-1}) \in M$, and all $r_0, \ldots, r_{n-1} \in R$ with $\sum_{j=0}^{n-1} r_j \theta(\vec{v}_j) = 0$ we have $\sum_{j=0}^{n-1} r_j \phi(\vec{v}_j) = 0$.*

Proof. Let P be an R-module and let $\phi : M \to P$ be multilinear. Assume that (θ, T) has the universal property. Then there is a linear map $\Phi : T \to P$ with $\phi = \Phi \circ \theta$. Let $n \in \mathbb{N}$, $\vec{v}_0, \ldots, \vec{v}_{n-1} \in M$, and $r_0, \ldots, r_{n-1} \in R$ with $\sum_{j=0}^{n-1} r_j \theta(\vec{v}_j) = 0$. Then the linearity of Φ implies

$$(B.8.13) \qquad \sum_{j=0}^{n-1} r_j \phi(\vec{v}_j) = \sum_{j=0}^{n-1} r_j \Phi \circ \theta(\vec{v}_j) = \Phi\left(\sum_{j=0}^{n-1} r_j \theta(\vec{v}_j)\right) = \Phi(0) = 0.$$

Conversely, assume that the second condition holds. Consider the map

$$(B.8.14) \qquad \Phi : T \to P, \quad \sum_{j=0}^{n-1} r_j \theta(\vec{v}_j) \mapsto \sum_{j=0}^{n-1} r_j \phi(\vec{v}_j)$$

for all $n \in \mathbb{N}$, $\vec{v}_j \in M$ for $0 \le j < n$. We show that Φ is well-defined. Since Span $\theta(M) = T$, every $t \in T$ can be written as

$$(B.8.15) \qquad t = \sum_{j=0}^{m-1} r_j \theta(\vec{v}_j)$$

where $n \in \mathbb{N}$, $\vec{v}_j \in M$, and $r_j \in R$ for $0 \le j < n$. So Φ is defined for all $t \in T$ and we must show that the image of t under Φ is independent of the representation of t. Let

$$(B.8.16) \qquad t = \sum_{i=0}^{n'-1} r_i' \theta(\vec{v}_i')$$

be another representation where $n' \in \mathbb{N}$, $\vec{v}_i' \in M$, and $r_i' \in R$ for $0 \le i < n'$. By inserting summands with coefficients 0 in the sums on the right sides of (B.8.15) and (B.8.16) and changing the order of the terms in these sums, we achieve $n = n'$ and $\vec{v}_j = \vec{v}_j'$ for all $j \in \mathbb{Z}_n$. So we have

$$(B.8.17) \qquad 0 = \sum_{j=0}^{n-1} (r_j - r_j') \theta(\vec{v}_j).$$

Therefore, the second condition implies

$$(B.8.18) \qquad \sum_{j=0}^{n-1} r_i \phi(\vec{v}_j) - \sum_{j=0}^{n-1} r_j' \phi(\vec{v}_j) = \sum_{j=0}^{n-1} (r_j - r_i') \phi(\vec{v}_j) = 0.$$

This shows that Φ is, in fact, well-defined. The proof of the linearity is left to the reader as Exercise B.8.8. $\qquad\square$

Exercise B.8.8. Prove the linearity of the map defined in (B.8.14).

From the proof of Lemma B.8.7 we obtain the following result.

Proposition B.8.9. *Let (T, θ) be a tensor product of $M_0, \ldots, M_{m-1}$, let P be an R-module, and let $\phi : T \to P$ be a linear map. Then the map*

$$(B.8.19) \qquad \Phi : T \to P, \quad \sum_{i=0}^{n-1} r_i \theta(\vec{v}_i) \mapsto \sum_{i=0}^{n-1} r_i \phi(\vec{v}_i)$$

is a well-defined homomorphism that satisfies $\phi = \Phi \circ \theta$ and it is the only linear map with this property.

Proof. We have shown in the proof of Lemma B.8.7 that Φ is a well-defined homomorphism with $\phi = \Phi \circ \theta$. To prove the uniqueness, let $\Phi' : T \to P$ be another linear map with these properties. Also, let $t \in T$ with a representation as in (B.8.15). Then we have $\Phi'(t) = \Phi'(\sum_{i=0}^{n-1} r_i \theta(\vec{v}_i)) = \sum_{i=0}^{n-1} r_i \Phi' \circ \theta(\vec{v}_i) = \sum_{i=0}^{n-1} r_i \phi(\vec{v}_i) = \Phi(t)$. $\qquad\square$

Example B.8.10. We use the tensor product $(\mathbb{Z}, \theta)$ from Example B.8.5 and consider the map $\phi : \mathbb{Z}^3 \to \mathbb{Z}, (x_0, x_1, x_2) \mapsto 2x_1$. It is multilinear. The uniquely determined linear map Φ from Proposition B.8.9 satisfies

$$(B.8.20) \qquad \Phi(t) = \phi(1, t, 1) = 2t$$

and this equation completely determines Φ.

B.8.2. Uniqueness. We show that a tensor product of $M_0, \ldots, M_{m-1}$ is uniquely determined up to tensor product isomorphism.

Definition B.8.11. Let (T, θ) and (T', θ') be tensor products of $M_0, \ldots, M_{m-1}$ over R. A *tensor product isomorphism* between (T, θ) and (T', θ') is an R-module isomorphism $\Theta : T' \to T$ that satisfies $\theta = \Theta \circ \theta'$.

Theorem B.8.12. *Tensor products of $M_0, \ldots, M_{m-1}$ over R are uniquely determined up to tensor product isomorphism. Furthermore, for two tensor products of $M_0, \ldots, M_{m-1}$ over R, the isomorphism between them is uniquely determined.*

Proof. Let (T, θ) and (T', θ') be tensor products of $M_0, \ldots, M_{m-1}$. Then it follows from the universal property of tensor products that there are linear maps $\Theta : T' \to T$ and $\Theta' : T \to T'$ such that

$$(B.8.21) \qquad \theta' = \Theta' \circ \theta \quad \text{and} \quad \theta = \Theta \circ \theta'.$$

This implies

$$(B.8.22) \qquad \theta' = (\Theta' \circ \Theta) \circ \theta' \quad \text{and} \quad \theta = (\Theta \circ \Theta') \circ \theta.$$

Equation (B.8.22) implies

$$(B.8.23) \qquad \Theta' \circ \Theta|_{\theta'(M)} = I_{\theta'(M)} \quad \text{and} \quad \Theta \circ \Theta'|_{\theta(M)} = I_{\theta(M)}.$$

Since Θ and Θ' are linear transformations and since $\operatorname{Span}(\theta(M)) = T$ and $\operatorname{Span}(\theta'(M)) = T'$ we obtain from (B.8.23)

(B.8.24) $$\Theta' \circ \Theta = \Theta' \circ \Theta|_{\operatorname{Span}(\theta'(M))} = I_{\operatorname{Span}(\theta'(M))} = I_{T'}$$

and

(B.8.25) $$\Theta \circ \Theta' = \Theta \circ \Theta'|_{\operatorname{Span}(\theta(M))} = I_{\operatorname{Span}(\theta(M))} = I_T.$$

So Θ is an isomorphism between T' and T. The uniqueness of Θ follows from Proposition B.8.9 $\qquad\Box$

B.8.3. Construction. We construct a tensor product of $M_0, \dots, M_{m-1}$ over R. For this, let L be the set of all formal linear combinations

(B.8.26) $$\sum_{i=0}^{k-1} r_i(\vec{v}_{i,0}, \dots, \vec{v}_{i,m-1})$$

where $k \in \mathbb{N}_0$, $r_i \in R$, $(\vec{v}_{i,0}, \dots, \vec{v}_{i,m-1}) \in \prod_{j=0}^{m-1} M_j$ for $0 \le i < k$ such that the tuples $(\vec{v}_{i,0}, \dots, \vec{v}_{i,m-1})$ are pairwise different and also nonzero. For $k = 0$, the sum in (B.8.26) is the empty linear combination which we denote by $\vec{0}$.

On L we define addition in the obvious way. For all $k \in \mathbb{N}$, $\vec{v}_0, \dots, \vec{v}_{k-1} \in M$, and $r, r_0, \dots, r_{k-1}, s_0, \dots, s_{k-1} \in R$ we set

(B.8.27) $$\sum_{i=0}^{k-1} r_i \vec{v}_i + \sum_{i=0}^{k-1} s_i \vec{v}_i = \sum_{i=0}^{k-1} (r_i + s_i) \vec{v}_i$$

and

(B.8.28) $$r \sum_{i=0}^{k-1} r_i \vec{v}_i = \sum_{i=0}^{k-1} (r r_i) \vec{v}_i.$$

From these rules, we also obtain formulas for adding two linear combinations in L. For this, we write both as linear combinations of the same elements of M by inserting summands with coefficients zero and changing the order of the terms in the sum if necessary. As shown in Exercise B.8.13, L is an R-module.

Exercise B.8.13. Verify that L is an R-module.

Example B.8.14. Let $M_0 = M_1 = \mathbb{Z}$. Then the module L consists of all formal sums $\sum_{j=0}^{k-1} r_i(v_i, w_i)$ where $k \in \mathbb{N}_0$ and $r_i, v_i, w_i \in \mathbb{Z}$ for $0 \le i < k$ such that the tuples (v_i, w_i) are pairwise different and different from $(0, 0)$. For example $(3, 2) - 2 \cdot (1, 2)$ and $(1, 2)$ are two different elements of L.

We note that a sequence of nonzero and pairwise different elements of M is by definition linearly independent in L.

Let S be the submodule of L which is generated by all elements of L of the form

(B.8.29)
$$\begin{aligned}
&(\vec{v}_0, \dots, \vec{v}_{i-1}, \vec{v}, \vec{v}_{i+1}, \dots, \vec{v}_{m-1}) \\
&+ (\vec{v}_0, \dots, \vec{v}_{i-1}, \vec{w}, \vec{v}_{i+1}, \dots, \vec{v}_{m-1}) \\
&- (\vec{v}_0, \dots, \vec{v}_{i-1}, \vec{v} + \vec{w}, \vec{v}_{i+1}, \dots, \vec{v}_{m-1})
\end{aligned}$$

and

$$(B.8.30) \qquad \begin{aligned} & r(\vec{v}_0, \ldots, \vec{v}_{i-1}, \vec{v}, \vec{v}_{i+1}, \ldots, \vec{v}_{m-1}) \\ & - (\vec{v}_0, \ldots, \vec{v}_{i-1}, r\vec{v}, \vec{v}_{i+1}, \ldots, \vec{v}_{m-1}) \end{aligned}$$

where $r \in R$, $(\vec{v}_0, \ldots, \vec{v}_{m-1}) \in M$, $i \in \mathbb{Z}_m$, $\vec{v}, \vec{w} \in M_i$. For $\vec{v} = (\vec{v}_0, \ldots, \vec{v}_{m-1}) \in M$ we denote the residue class of $\vec{v}$ modulo S by

$$(B.8.31) \qquad \vec{v}_0 \otimes_R \cdots \otimes_R \vec{v}_{m-1}.$$

If the ring R is understood, then we also write this residue class as

$$(B.8.32) \qquad \vec{v}_0 \otimes \cdots \otimes \vec{v}_{m-1} = \bigotimes_{j=0}^{m-1} \vec{v}_j.$$

Also, we write the quotient module L/S as

$$(B.8.33) \qquad M_0 \otimes_R \cdots \otimes_R M_{m-1}.$$

If the ring R is understood, then we also write it as

$$(B.8.34) \qquad M_0 \otimes \cdots \otimes M_{m-1} = \bigotimes_{j=0}^{m-1} M_i.$$

So we have defined the map

$$(B.8.35) \qquad \bigotimes : \prod_{j=0}^{m-1} M_j \to \bigotimes_{j=0}^{m-1} M_j, \quad (\vec{v}_0, \ldots, \vec{v}_{m-1}) \mapsto \bigotimes_{i=0}^{m-1} \vec{v}_i.$$

It follows from the definition of S that the following relations hold for all $r \in R$, $(\vec{v}_0, \ldots, \vec{v}_{m-1}) \in M$, $i \in \mathbb{Z}_m$, and $\vec{v}, \vec{w} \in M_i$:

$$(B.8.36) \qquad \begin{aligned} & \vec{v}_0 \otimes \cdots \otimes \vec{v}_{i-1} \otimes \vec{v} \otimes \vec{v}_{i+1} \otimes \cdots \otimes \vec{v}_{m-1} \\ & + \vec{v}_0 \otimes \cdots \otimes \vec{v}_{i-1} \otimes \vec{w} \otimes \vec{v}_{i+1} \otimes \cdots \otimes \vec{v}_{m-1} \\ & = \vec{v}_0 \otimes \cdots \otimes \vec{v}_{i-1} \otimes \vec{v} + \vec{w} \otimes \vec{v}_{i+1} \otimes \cdots \otimes \vec{v}_{m-1}, \end{aligned}$$

and

$$(B.8.37) \qquad \begin{aligned} & r\vec{v}_0 \otimes \cdots \otimes \vec{v}_i \otimes \cdots \otimes \vec{v}_{m-1} \\ & = \vec{v}_0 \otimes \cdots \otimes r\vec{v}_i \otimes \cdots \otimes \vec{v}_{m-1}. \end{aligned}$$

Example B.8.15. As in Example B.8.14, consider the $\mathbb{Z}$-modules $M_0 = M_1 = \mathbb{Z}$. In this example, we have presented the two different elements $(3, 2) - 2 \cdot (1, 2)$ and $(1, 2)$ of L. But applying (B.8.36) and (B.8.37) and using the multilinearity of $\otimes$ we find that $3 \otimes_{\mathbb{Z}} 2 - 2 \cdot (1 \otimes_{\mathbb{Z}} 2) = 3 \otimes_{\mathbb{Z}} 2 - 2 \otimes_{\mathbb{Z}} 2 = 1 \otimes_{\mathbb{Z}} 2$. Therefore, the corresponding elements in $M_0 \otimes_{\mathbb{Z}} M_1$ are the same.

Theorem B.8.16. *The pair* $(\bigotimes_{j=0}^{m-1} M_j, \otimes)$ *is a tensor product of* $M_0, \ldots, M_{m-1}$ *over R.*

Proof. By definition, the map

(B.8.38)
$$\otimes : \prod_{j=0}^{m-1} M_j \to \bigotimes_{j=0}^{m-1} M_j, \quad (\vec{v}_0, \ldots, \vec{v}_{m-1}) \mapsto \bigotimes_{j=0}^{m-1} \vec{v}_j$$

is bilinear and its image spans $\bigotimes_{j=0}^{m-1} M_j$.

We prove the universal property by verifying the second condition in Lemma B.8.7. By the definition of $\otimes$, an element in $\bigotimes_{j=0}^{m-1} M_j$ is zero if and only if it is a linear combination of elements $\theta(\vec{v})$ with $\vec{v} \in S$. Hence, it suffices to show that the second condition of Lemma B.8.7 holds for the generators of S shown in (B.8.29) and (B.8.30). But this follows from the multilinearity of ϕ. $\qquad\square$

The uniqueness of the tensor product shown in Theorem B.8.12 justifies the following definition.

Definition B.8.17. The pair $(\bigotimes_{j=0}^{m-1} M_j, \otimes)$ is called *the tensor product of* $M_0, \ldots, M_{m-1}$ *over R*. We simply write it as $M_0 \otimes_R \cdots \otimes_R M_{m-1}$ or as $M_0 \otimes \cdots \otimes M_{m-1} = \bigotimes_{i=0}^{m-1} M_i$ if R is understood.

We make the following remark. Let $n \in \mathbb{N}$ and let $N_0, \ldots, N_{n-1}$ be R-modules. Then the map

(B.8.39)
$$\left(\bigotimes_{i=0}^{m-1} M_i \right) \otimes \left(\bigotimes_{i=0}^{n-1} N_i \right) \to M_0 \otimes \cdots \otimes M_{m-1} \otimes N_0 \otimes \cdots \otimes N_{n-1},$$
$$\left(\bigotimes_{i=0}^{m-1} \vec{v}_i \right) \otimes \left(\bigotimes_{i=0}^{n-1} \vec{w}_i \right) \mapsto \vec{v}_0 \otimes \cdots \otimes \vec{v}_{m-1} \otimes \vec{w}_0 \otimes \cdots \otimes \vec{w}_{n-1}$$

induces an isomorphism of tensor products. Using this isomorphism, we identify the domain and image of this map. For an R-module M and $k \in \mathbb{N}$, we also write

(B.8.40)
$$M^{\otimes k} = \bigotimes_{i=0}^{k-1} M$$

and for $\vec{v} \in M$

(B.8.41)
$$\vec{v}^{\otimes k} = \bigotimes_{i=0}^{k-1} \vec{v}.$$

Example B.8.18. We construct the tensor product $\mathbb{Z}^{\otimes 3}$. Its elements are the linear combinations of $x_0 \otimes x_1 \otimes x_2$ with integer coefficients where $x_i \in \mathbb{Z}$. We claim that

(B.8.42)
$$\mathbb{Z}^{\otimes 3} = \mathbb{Z} \cdot 1^{\otimes 3}.$$

To verify (B.8.42) we first note that $\mathbb{Z} \cdot 1^{\otimes 3} \subset \mathbb{Z}^{\otimes 3}$. To show the reverse inclusion, let $x_0 \otimes x_1 \otimes x_2 \in \mathbb{Z}^{\otimes 3}$ with $x_0, x_1, x_2 \in \mathbb{Z}$. Due to the multilinearity of the tensor product, we have $x_0 \otimes x_1 \otimes x_2 = x \cdot 1^{\otimes 3}$ where $x = x_0 x_1 x_2$. So $x_0 \otimes x_1 \otimes x_2 \in \mathbb{Z} \cdot 1^{\otimes 3}$.

B.8.4. Homomorphisms. We show that homomorphisms of R-modules can be combined in the obvious way to tensor products of homomorphisms. For this, let $N_0, \ldots, N_{m-1}$ be further R-modules and set

$$(B.8.43) \qquad M = \bigotimes_{j=0}^{m-1} M_j \quad \text{and} \quad N = \bigotimes_{j=0}^{m-1} N_i.$$

The next proposition shows that with each element of the tensor product $\bigotimes_{j=0}^{m-1} \operatorname{Hom}(M_j, N_j)$ we can associate a homomorphisms in $\operatorname{Hom}(M, N)$.

Proposition B.8.19. *For $0 \le j < m$ let $f_j \in \operatorname{Hom}(M_j, N_j)$. Then the map*

$$(B.8.44) \qquad M \to N, \quad \bigotimes_{j=0}^{m-1} \vec{v}_j \mapsto \bigotimes_{j=0}^{m-1} f_j(\vec{v}_j)$$

defines a map in $\operatorname{Hom}(M, N)$. We refer to this homomorphism as $\bigotimes_{j=0}^{m-1} f_j$.

Proof. The map

$$(B.8.45) \qquad M \to N, \quad \bigotimes_{j=0}^{m-1} \vec{v}_j \mapsto \bigotimes_{j=0}^{m-1} f_j(\vec{v}_j)$$

is multilinear because the maps f_j are linear. Since $(\bigotimes_{j=0}^{m-1} M_j, \otimes)$ is a tensor product of $M_0, \ldots, M_{m-1}$, the assertion follows from Proposition B.8.9. $\qquad \square$

We note that, in general, the map that sends the tensor product of elements in $\operatorname{Hom}(M_j, N_j)$ to the corresponding homomorphism in $\operatorname{Hom}(M, N)$ is not injective. Therefore, several such tensor products may be associated with the same homomorphism in $\operatorname{Hom}(M, N)$. But as we will see in Section B.9.3, the map is an R-module isomorphism if the modules M_j and N_j are finite-dimensional vector spaces.

Example B.8.20. Let $R, M_0, M_1, N_0, N_1 = \mathbb{Z}_4$. Let $f : \mathbb{Z}_4 \to \mathbb{Z}_4$, $v \mapsto 2v$ mod 4. We determine the homomorphism in $\operatorname{End}(\mathbb{Z}_4^{\otimes 2})$ associated with $f^{\otimes 2}$. It sends $x \otimes y \in \mathbb{Z}_4^{\otimes 2}$ to $(2x \bmod 4) \otimes (2y \bmod 4) = 4(x \otimes y) = 0 \otimes y\mathbb{Z}_4 = \vec{0}$. So, it is the zero map that can also be represented as the tensor product of the zero map in $\operatorname{End}(\mathbb{Z}_4)$ with itself.

B.9. Tensor products of finite-dimensional vector spaces

Let $m, k_0, \ldots, k_{m-1} \in \mathbb{N}$. For $0 \le j < m$ let V_j be an F-vector space of dimension k_j and let $B_j = (\vec{b}_{0,j}, \ldots, \vec{b}_{k_j-1,j})$ be an F-bases of V_j.

We discuss the properties of the tensor product $V_0 \otimes_F \cdots \otimes_F V_{m-1}$. To simplify our notation, we write $\otimes$ for $\otimes_F$.

B.9.1. Representation. We explain how to represent the tensor product

$$(B.9.1) \qquad\qquad V = \bigotimes_{j=0} V_i$$

as an F-vector space of multi-dimensional matrices over F. For this, we set

$$(B.9.2) \qquad \vec{k} = (k_0, \ldots, k_{m-1}), \quad \mathbb{Z}_{\vec{k}} = \prod_{j=0}^{m-1} \mathbb{Z}_{k_j}, \quad k = \prod_{j=0}^{m-1} k_j.$$

Definition B.9.1. (1) By $F^{\vec{k}}$ we mean the set of all *m-dimensional matrices* $(\alpha_{\vec{i}})_{\vec{i} \in \mathbb{Z}_{\vec{k}}}$ with entries $\alpha_{\vec{i}} \in F$.

(2) Let $\vec{i} \in \mathbb{Z}_{\vec{k}}$. The *standard unit matrices* in $F^{\vec{k}}$ are the matrices $E_{\vec{i}}$ in $F^{\vec{k}}$ such that the entry with index $\vec{i}$ is 1 and it is the only nonzero entry of $E_{\vec{i}}$.

Proposition B.9.2. *The set $F^{\vec{k}}$ equipped with componentwise addition and scalar multiplication is a k-dimensional F-vector space and $(E_{\vec{i}})_{\vec{i} \in \mathbb{Z}}$ is a basis of $F^{\vec{k}}$.*

Exercise B.9.3. Prove Proposition B.9.2.

Example B.9.4. Let $F = \mathbb{F}_2$, $m = 3$, $k_0 = k_1 = k_2 = 2$. Then we have $\vec{k} = (2, 2, 2)$, $\mathbb{Z}_{\vec{k}} = \mathbb{Z}_2^3$, $k = 8$. The set $F^{\vec{k}} = \mathbb{F}_2^{(2,2,2)}$ contains 2^8 three-dimensional matrices with 8 entries each, which can be 0 or 1. For $\vec{i} = (i_0, i_1, i_2) \in \mathbb{Z}_{(2,2,2)} = \mathbb{Z}_2^3$ the standard unit matrix $E_{\vec{i}}$ is the matrix in $\mathbb{F}_2^{(2,2,2)}$ such that the entry with index $\vec{i}$ is 1 and all other entries are 0. These matrices form a basis of the eight-dimensional $\mathbb{F}_2$-vector space $\mathbb{F}_2^{(2,2,2)}$.

Let $(\vec{v}_0, \ldots, \vec{v}_{m-1}) \in \prod_{j=0}^{m-1} V_j$. For all $j \in \mathbb{Z}_m$, write the coefficient vector of $\vec{v}_j$ with respect to the basis B_j of V_j as

$$(B.9.3) \qquad\qquad (\vec{v}_j)_{B_j} = (v_{0,j}, \ldots, v_{k_j, j}).$$

Also, define the m-dimensional matrix

$$(B.9.4) \qquad \mathrm{Mat}(\vec{v}_0, \ldots, \vec{v}_{m-1}) = \left(\prod_{j=0}^{m-1} v_{i_j, j} \right)_{(i_0, \ldots, i_{m-1}) \in \mathbb{Z}_{\vec{k}}}.$$

Note that this matrix depends on the choice of the bases of the vector spaces V_j. If we want to make this dependence explicit, we write $\mathrm{Mat}_{B_0, \ldots, B_{m-1}}(\vec{v}_0, \ldots, \vec{v}_{m-1})$.

Example B.9.5. Let $F = \mathbb{Q}$, $m = 3$, $k_0 = k_1 = k_2 = 2$. The three-dimensional matrix $M((1,1), (1,-1), (1,0))$ is presented in Table B.9.1.

So we have defined a map

$$(B.9.5) \qquad \mathrm{Mat} : \prod_{j=0}^{m-1} V_j \to F^{\vec{k}}, \quad (\vec{v}_0, \ldots, \vec{v}_{m-1}) \mapsto \mathrm{Mat}(\vec{v}_0, \ldots, \vec{v}_{m-1}).$$

Proposition B.9.6. *The pair $(F^{\vec{k}}, \mathrm{Mat})$ is a tensor product of $V_0, \ldots, V_{m-1}$.*

Table B.9.1. Mat$((1,1),(1,-1),(1,0))$.

index (i_0, i_1, i_2)	entry $\prod_{j=0}^{m-1} v_{i_j,j}$
$(0,0,0)$	$v_{0,0}v_{0,1}v_{0,2} = 1$
$(0,0,1)$	$v_{0,0}v_{0,1}v_{1,2} = 0$
$(0,1,0)$	$v_{0,0}v_{1,1}v_{0,2} = -1$
$(0,1,1)$	$v_{0,0}v_{1,1}v_{1,2} = 0$
$(1,0,0)$	$v_{1,0}v_{0,1}v_{0,2} = 1$
$(1,0,1)$	$v_{1,0}v_{0,1}v_{1,2} = 0$
$(1,1,0)$	$v_{1,0}v_{1,1}v_{0,2} = -1$
$(1,1,1)$	$v_{1,0}v_{1,1}v_{1,2} = 0$

Proof. The map Mat is well-defined since the B_i are bases of the V_i. Also, it is easy to verify that Mat is multilinear. Next, we note that for all $\vec{i} = (i_0, \dots, i_{m-1}) \in \mathbb{Z}_{\vec{k}}$ we have

$$(B.9.6) \qquad \mathrm{Mat}(\vec{b}_{i_0}, \dots, \vec{b}_{i_{m-1}}) = E_{\vec{i}}.$$

Hence, Proposition B.9.2 implies that $\mathrm{Span}(\mathrm{Mat}(\prod_{i=0}^{m-1} V_i)) = F^{\vec{k}}$.

To prove the universal property, let P be an F-vector space and let $\phi : \prod_{i=0}^{m-1} V_i \to P$ be multilinear. Define the map

$$(B.9.7) \qquad \Phi : F^{\vec{k}} \to P, \quad E_{(i_0, \dots, i_{m-1})} \mapsto \phi(\vec{b}_{i_0,0}, \dots, \vec{b}_{i_{m-1}, m-1}).$$

This map is well-defined since $(E_{\vec{i}})_{\vec{i} \in \mathbb{Z}_{\vec{k}}}$ is a basis of $F^{\vec{k}}$. It is linear by definition, and as shown in Exercise B.9.7 it follows from the multilinearity of ϕ that $\phi = \Phi \circ \mathrm{Mat}$. $\square$

Exercise B.9.7. Show that in the proof of Proposition B.9.6 we have $\phi = \Phi \circ \mathrm{Mat}$.

From Proposition B.9.6 and Theorem B.8.12 we obtain the following corollary.

Corollary B.9.8. *The map*

$$(B.9.8) \qquad \bigotimes_{i=0}^{m-1} V_i \to F^{\vec{k}}, \quad \bigotimes_{i=0}^{m-1} \vec{v}_i \mapsto \mathrm{Mat}(\vec{v}_0, \dots, \vec{v}_{m-1})$$

is the uniquely determined isomorphism between the tensor products $(\bigotimes_{i=0}^{m-1} V_i, \bigotimes)$ and $(F^{\vec{k}}, \mathrm{Mat})$.

Corollary B.9.8 justifies the following definition.

Definition B.9.9. We use the map

$$(B.9.9) \qquad \bigotimes_{i=0}^{m-1} F^{k_i} \to F^{\vec{k}}, \quad \vec{v}_0 \otimes \cdots \otimes \vec{v}_{k-1} \mapsto \mathrm{Mat}(\vec{v}_0, \dots, \vec{v}_k)$$

to identify the tensor product $\bigotimes_{j=0}^{m-1} F^{k_j}$ with $\mathbb{F}^{\vec{k}}$.

Next, we show that the tensor product of the B_j is a basis of V. For this, we need the following definition and result.

Definition B.9.10. For $0 \leq j < m$ let $d_j \in \mathbb{N}$ and let $D_j = (\vec{v}_{0,j}, \ldots, \vec{v}_{d_j-1,j})$ be finite sequences in V_j. Define

$$(B.9.10) \qquad D_0 \otimes \cdots \otimes D_{m-1} = \left(\bigotimes_{j=0}^{m-1} \vec{v}_{i_j,j} \right)_{i_j \in \mathbb{Z}_{d_j}, j \in \mathbb{Z}_m} .$$

Proposition B.9.11. *In the situation of Definition B.9.10 we have*

(1) $\bigotimes_{j=0}^{m-1} \mathrm{Span}(D_j) = \mathrm{Span}\left(\bigotimes_{j=0}^{m-1} D_j \right)$.

(2) $D_0, \ldots, D_{m-1}$ *are linearly independent if and only if* $\bigotimes_{j=0}^{m-1} D_i$ *is linearly independent.*

Proof. The first assertion follows from the fact that $\bigotimes_{j=0}^{m-1} \mathrm{Span}(D_j)$ is the set of linear combinations of the elements of $\bigotimes_{j=0}^{m-1} D_j$ with coefficients in F.

We prove the second assertion. Let $D_0, \ldots, D_{m-1}$ be linearly independent. It follows from Corollary B.9.8 that $\bigotimes_{j=0}^{m-1} \mathrm{Span}(D_j)$ is an F-vector space of dimension $d = \prod_{j=0}^{m-1} d_j$. By the first assertion, $\bigotimes_{j=0}^{m-1} D_j$ is a generating system of this tensor product with d elements. So $\bigotimes_{j=0}^{m-1} D_j$ must be a basis of this tensor product. The converse is left to the reader as Exercise B.9.12. $\qquad\square$

Exercise B.9.12. In the situation of Proposition B.9.11 show that the linear independence of $\bigotimes_{j=0}^{m-1} D_j$ implies the linear independence of D_j for $0 \leq j < m$.

B.9.2. Tensor product of matrices. Let $m, n, u, v \in \mathbb{N}$. We define a map

$$(B.9.11) \qquad \theta : F^{(m,n)} \times F^{(u,v)} \to F^{(mu,nv)}$$

as follows. Let $A \in F^{(m,n)}$ and $B \in F^{(u,v)}$ with $A = (a_{i,j})$ and $B = (b_{k,l})$. Then we set

$$(B.9.12) \qquad \theta(A,B) = \begin{pmatrix} a_{0,0}B & a_{0,1}B & \cdots & a_{0,n-1}B \\ a_{1,0}B & a_{1,1}B & \cdots & a_{1,n-1}B \\ \vdots & & & \vdots \\ a_{m-1,0}B & a_{m-1,1}B & \cdots & a_{m-1,n-1}B \end{pmatrix} \in F^{(mu,nv)}.$$

If we write

$$(B.9.13) \qquad \theta(A,B) = (c_{i,j})_{i \in \mathbb{Z}_{mu}, j \in \mathbb{Z}_{nv}},$$

then we have

$$(B.9.14) \qquad c_{pu+q,rv+s} = a_{p,q} b_{r,s}$$

for all $p \in \mathbb{Z}_m, q \in \mathbb{Z}_u, r \in \mathbb{Z}_n, s \in \mathbb{Z}_v$.

Exercise B.9.13. Verify (B.9.14).

We explain the meaning of $\theta(A,B)$. For this, we assume that we have modified the representation in Definition B.9.9 such that the matrices in $F^{(m,n)}$ become vectors in F^{mn} and matrices in $F^{(u,v)}$ become vectors in F^{uv}. The details are worked out in Exercise B.9.15.

Proposition B.9.14. *Let f_A, f_B be the linear maps in $\mathrm{Hom}(F^n, F^m)$ and $\mathrm{Hom}(F^v, F^u)$, respectively, that have the representation matrices A and B with respect to the standard bases of F^m, F^n, F^u, and F^v. Then $\theta(A, B)$ is the representation matrix of $f_A \otimes f_B$ with respect to the standard bases of F^{mu} and F^{nv}.*

Exercise B.9.15. Prove Proposition B.9.14.

Example B.9.16. Let

$$(B.9.15) \qquad A = \begin{pmatrix} 0 & 1 \\ 0 & 0 \end{pmatrix} \quad \text{and} \quad B = \begin{pmatrix} 1 & 0 \\ 0 & 0 \end{pmatrix}.$$

Then

$$(B.9.16) \qquad A \otimes B = \begin{pmatrix} 0 \cdot B & 1 \cdot B \\ 0 \cdot B & 0 \cdot B \end{pmatrix} = \begin{pmatrix} 0 & 0 & 1 & 0 \\ 0 & 0 & 0 & 0 \\ 0 & 0 & 0 & 0 \\ 0 & 0 & 0 & 0 \end{pmatrix}.$$

The next proposition allows us to identify $A \otimes B$ with the matrix $\theta(A, B)$.

Proposition B.9.17. (1) *The pair $(F^{(mu,nv)}, \theta)$ is a tensor product of $F^{(m,n)}$ and $F^{(u,v)}$.*

(2) *The uniquely determined isomorphism between the tensor products $F^{(m,n)} \otimes F^{(u,v)}$ and $F^{(mu,nv)}$ of $F^{(m,n)}$ and $F^{(u,v)}$ is*

$$(B.9.17) \qquad F^{(m,n)} \otimes F^{(u,v)} \to F^{(mu,nv)}, \quad A \otimes B \mapsto \theta(A, B).$$

Proof. The map θ is multilinear by definition. Next, we show that

$$(B.9.18) \qquad \mathrm{Span}\, \theta(F^{(m,n)} \times F^{(u,v)}) = F^{(mu,nv)}.$$

For $p \in \mathbb{Z}_m, q \in \mathbb{Z}_n$ let $A_{p,q} = (a_{i,j}) \in F^{(m,n)}$ with

$$(B.9.19) \qquad a_{i,j} = \begin{cases} 1 & \text{if } i = p, j = q, \\ 0 & \text{otherwise.} \end{cases}$$

Also denote by $B_{r,s}$ the analogous matrix in $F^{(u,v)}$. Then it follows from (B.9.14) that $\theta(A_{p,q}, B_{r,s})$ is the matrix $(c_{i,j}) \in F^{(uv,mn)}$ with

$$(B.9.20) \qquad c_{i,j} = \begin{cases} 1 & \text{if } i = pu + q, j = rv + s, \\ 0 & \text{otherwise.} \end{cases}$$

So $(\theta(A_{p,q}, B_{r,s}))$ is a basis of $F^{(mu,nv)}$ which is in $\theta(F^{(m,n)} \times F^{(u,v)})$. This implies (B.9.18). To show the universal property, let P be an F-vector space and let $\phi : F^{(m,n)} \times F^{(u,v)} \to P$ be a multilinear map. Since $(A_{p,q})$ is a basis of $F^{(m,n)}$ and $(B_{r,s})$ is a basis of $F^{(u,v)}$, it follows from the multilinearity of ϕ that

$$(B.9.21) \qquad \Phi : F^{(mu,nv)} \to P, \quad \theta(M_{p,q}, N_{r,s}) \mapsto \phi(M_{p,q}, N_{r,s})$$

is a well-defined homomorphism with the property that $\phi = \Phi \circ \theta$. This proves the first assertion. The second assertion follows from Theorem B.8.12. $\qquad\square$

Proposition B.9.17 justifies the following definition.

Definition B.9.18. Let $A \in F^{(m,n)}$ and $B \in F^{(u,v)}$. Then we identify $A \otimes B$ with the matrix $\theta(A, B)$ from (B.9.12) and we call this matrix the *tensor product of A and B*.

B.9.3. Tensor product of homomorphisms. Let $W_0, \ldots, W_{m-1}$ be further F-vector spaces of finite dimensions $l_0, \ldots, l_{m-1} \in \mathbb{N}$. Set $l = \prod_{j=0}^{m-1} l_j$,

$$(B.9.22) \qquad V = \bigotimes_{i=0}^{m-1} V_i, \quad \text{and} \quad W = \bigotimes_{i=0}^{m-1} W_i.$$

In this situation, we can strengthen the assertion in Proposition B.8.19 as follows.

Proposition B.9.19. *The map*

$$(B.9.23) \qquad \bigotimes_{j=0}^{m-1} \mathrm{Hom}(V_j, W_j) \to \mathrm{Hom}(V, W), \quad \bigotimes_{j=0}^{m-1} f_j \mapsto \bigotimes_{j=0}^{m-1} f_j$$

defines an isomorphism of F-vector spaces. In this definition, $\bigotimes_{j=0}^{m-1} f_j$ means two different things: a map in $\bigotimes_{j=0}^{m-1} \mathrm{Hom}(V_j, W_j)$ and the corresponding map in $\mathrm{Hom}(V, W)$ defined in Proposition B.8.19.

Proof. We prove the assertion by induction on m. For $m = 1$, the assertion clearly holds. Assume that $m > 1$ and that the assertion holds for $m - 1$. We set

$$(B.9.24) \qquad P = \bigotimes_{j=0}^{m-2} V_j, \quad Q = \bigotimes_{j=0}^{m-2} W_j, \quad R = V_{m-1}, \quad S = W_{m-1}.$$

It follows from the induction hypothesis that the map

$$(B.9.25) \qquad \bigotimes_{j=0}^{m-2} \mathrm{Hom}(V_j, W_j) \to \mathrm{Hom}(P, R), \quad \bigotimes_{j=0}^{m-2} f_j \mapsto \bigotimes_{j=0}^{m-2} f_j$$

is an isomorphism. Denote the dimensions of P, Q, R, S by n, m, v, u, respectively. Then we can identify $\mathrm{Hom}(P, R)$ with $F^{(m,n)}$, $\mathrm{Hom}(Q, S)$ with $F^{(u,v)}$, and $\mathrm{Hom}(P \otimes Q, Q \otimes R)$ with $F^{(mu,nv)}$. It follows from Proposition B.9.14 and Proposition B.9.17 that the map

$$(B.9.26) \qquad \mathrm{Hom}(P, Q) \otimes \mathrm{Hom}(R, S) \to \mathrm{Hom}(P \otimes Q, R \otimes S)$$

that sends the tensor product $f \otimes g$ for $f \in \mathrm{Hom}(P, Q)$ and $g \in \mathrm{Hom}(R, S)$ to the corresponding homomorphism in $\mathrm{Hom}(P \otimes R, Q \otimes S)$ defines an isomorphism. Combining the two isomorphisms in (B.9.25) and (B.9.26) we obtain the assertion. $\qquad \square$

It follows from Proposition B.9.19 that for finite-dimensional vector spaces we can identify the tensor product of homomorphisms with the corresponding homomorphism between the tensor product of the vector spaces.

Example B.9.20. We modify Example B.8.20 and let $F, M_0, M_1, N_0, N_1 = \mathbb{Z}_3, f : \mathbb{Z}_3 \to \mathbb{Z}_3, v \mapsto 2v \bmod 4$. So, we have replaced the ring $\mathbb{Z}_4$ by the field $\mathbb{Z}_3$. We determine the homomorphism in $\mathrm{End}(\mathbb{Z}_3^{\otimes 2})$ associated with $f^{\otimes 2}$. It sends $x \otimes y \in \mathbb{Z}_3^{\otimes 2}$ to $(2x \bmod 3) \otimes (2y \bmod 3)$. This is the only representation of this map as a tensor product of endomorphisms of $\mathbb{Z}_3$.

B.9.4. Partial trace. Our next goal is to introduce the notion of the partial trace. In the discussion, we use direct products $\prod_{j \in I} M_j$ and tensor products $\bigotimes_{j \in I} M_j$ for subsets I of $\mathbb{Z}_m$. In these expressions, the indices are ordered by size: from smallest to largest.

First, we note that the following holds.

Proposition B.9.21. *For $0 \leq j < m$ let $f_j \in \mathrm{End}(V_j)$. Then we have*

$$(B.9.27) \qquad \mathrm{tr}\left(\bigotimes_{j=0}^{m-1} f_j\right) = \prod_{j=0}^{m-1} \mathrm{tr}\, f_j.$$

Exercise B.9.22. Prove Proposition B.9.21. Hint: Use induction on m and the formula (B.9.12) for the tensor product of matrices.

We introduce the partial trace.

Theorem B.9.23. *Let $J \subset \mathbb{Z}_m$. Then there is a uniquely determined linear map*

$$(B.9.28) \qquad \mathrm{tr}_J : \mathrm{End}\left(\bigotimes_{j \in \mathbb{Z}_m} V_j\right) \to \mathrm{End}\left(\bigotimes_{j \in \mathbb{Z}_m \setminus J} V_j\right)$$

that satisfies

$$(B.9.29) \qquad \mathrm{tr}_J\left(\bigotimes_{j \in \mathbb{Z}_m} f_j\right) = \prod_{j \in J} \mathrm{tr}\, f_j \bigotimes_{j \in \mathbb{Z}_m \setminus J} f_j$$

for all $(f_0, \ldots, f_{m-1}) \in \prod_{j=0}^{m-1} \mathrm{End}(V_j)$. It is called the partial trace *over the V_j, $j \in J$.*

Proof. Consider that map

$$(B.9.30) \qquad \prod_{j \in \mathbb{Z}_m} \mathrm{End}(V_j) \to \prod_{j \in J} \mathrm{tr}\, f_j \bigotimes_{j \in \mathbb{Z}_m \setminus J} f_j.$$

It is multilinear. Hence, it follows from Proposition B.8.9 that (B.9.29) defines the uniquely determined homomorphism (B.9.28). $\qquad \square$

Example B.9.24. Let $R, M_0, M_1 = \mathbb{Z}_3$, $f : \mathbb{Z}_3 \to \mathbb{Z}_3$, and $v \mapsto 2v \bmod 3$. The partial trace of $f^{\otimes 2}$ over M_0 is the map $(x, y) \mapsto (\mathrm{tr}\, f)f(y) = y$.

We show that the partial trace is trace-preserving.

Proposition B.9.25. *Let $J \subset \mathbb{Z}_m$ and let $f \in \bigotimes_{j=0}^{m-1} \mathrm{End}(V_j)$. Then we have*

$$(B.9.31) \qquad \mathrm{tr}(\mathrm{tr}_J(f)) = \mathrm{tr}(f).$$

Proof. We have $\mathrm{End}(\bigotimes_{j=0}^{m-1} V_j) = \bigotimes_{j=0}^{m-1} \mathrm{End}(V_j)$. Therefore, the linearity of the trace, Proposition B.9.21, and (B.9.29) imply the assertion. $\qquad \square$

Probability Theory

Quantum algorithms are probabilistic by nature. So their analysis requires some probability theory. This part of the appendix summarizes the concepts and results of probability theory that are required in the analyses of probabilistic and quantum algorithms in this book.

C.1. Basics

We begin with some basic definitions.

Definition C.1.1. A set S is called *countable* if it is finite or there is a bijection $\mathbb{N}_0 \to S$. Otherwise, S is called *uncountable*.

Exercise C.1.2. Show that the set $\mathbb{N} \times \mathbb{N}$ is countable.

Definition C.1.3. An infinite sum $\sum_{i=0}^{\infty} r_i$ with $r_i \in \mathbb{R}$ for all $i \in \mathbb{N}_0$ is called *absolute convergent* if $\sum_{i=0}^{\infty} |r_i|$ converges.

In the following, we need the Riemann Series Theorem which we state now.

Theorem C.1.4. *Consider an infinite sum $\sum_{i=0}^{\infty} r_i$ where $r_i \in \mathbb{R}$ for all $i \in \mathbb{N}_0$. Then the following statements are equivalent.*

(1) *The infinite sum $\sum_{i=0}^{\infty} r_i$ is absolute convergent.*

(2) *For all permutations $\pi : \mathbb{N}_0 \to \mathbb{N}_0$, the infinite sums $\sum_{i=0}^{\infty} r_{\pi(i)}$ are convergent and have the same limit.*

If the two statements hold, then we write $\sum_{r \in R} r$ for the limit of the infinite sum $\sum_{i=0}^{\infty} r_i$ where R represents any ordering of the sequence (r_i). If the elements of this sequence are pairwise distinct, then R is the set of these elements.

The proof of this theorem can be found in [**Rud76**, 3.55 Theorem].

Exercise C.1.5. Consider the following infinite series:

$$S = \sum_{n=1}^{\infty} \frac{(-1)^{n+1}}{n^2}.$$

(1) Show that the series S is absolute convergent.

(2) Calculate the sum of the series S using the original order of its terms.

(3) Now, consider a new series S' obtained by rearranging the terms of S as follows: First, take the positive terms in the order of increasing n, and then take the negative terms in the order of decreasing n. Prove that S' is equal to the sum calculated in step (2).

Definition C.1.6. (1) A *discrete probability space* is a pair (S, Pr), where S is a countable set, called a *sample space*. Its elements are called *samples* or *elementary events*. Also, Pr is a map

(C.1.1) $\mathrm{Pr} : S \to [0, 1]$

called a *probability distribution*, that satisfies

(C.1.2) $\sum_{s \in S} \mathrm{Pr}(s) = 1.$

We say that the probability distribution *assigns the probability* $\mathrm{Pr}(s)$ to each elementary event $s \in S$. The probability space is called *finite* if the sample space is finite. Otherwise, it is called *infinite*.

(2) The subsets of S are called *events*. The *probability* of an event $A \subset S$ is

(C.1.3) $\mathrm{Pr}(A) = \sum_{a \in A} \mathrm{Pr}(A).$

Note that by Theorem C.1.4, the condition (C.1.2) means that this sum converges to 1 for any ordering of the elements of S.

Example C.1.7. Consider the experiment of tossing a fair coin. The corresponding discrete probability space is $(\{0, 1\}, \mathrm{Pr})$ where 0 and 1 represent tails and heads, respectively, and Pr sends both 0 and 1 to $\frac{1}{2}$.

Example C.1.8. Consider the experiment of throwing a dice. The corresponding discrete probability space is $(\{1, \ldots, 6\}, \mathrm{Pr})$ where Pr sends all elements of $\{1, \ldots, 6\}$ to $\frac{1}{6}$.

Exercise C.1.9. Consider a fair coin, where the probability of getting heads is $\frac{1}{2}$ and the probability of getting tails is $\frac{1}{2}$. What is the probability of getting heads at least once when tossing the coin two times? Describe the corresponding probability space and event and use this to find the solution of the exercise.

Example C.1.10. Consider the experiment in which a dice is rolled until it shows 6. The sample space is the set of all finite sequences of length ≥ 1 where the last entry is 6 and all other entries are between 1 and 5. The probability distribution is

(C.1.4) $\mathrm{Pr} : S \to [0, 1], \quad s \mapsto \dfrac{5^{|s|-1}}{6^{|s|}}.$

This is a probability distribution because

(C.1.5) $$\sum_{s \in S} \Pr(s) = \sum_{i=1}^{\infty} \frac{5^{i-1}}{6^i} = \frac{1}{6} \sum_{i=1}^{\infty} \left(\frac{5}{6}\right)^{i-1} = 1.$$

Example C.1.11. We present another way to model the experiment of Example C.1.10. The sample space is $\mathbb{N}$. The sample or elementary event $s \in \mathbb{N}$ means that the experiment is successful after rolling the dice s times. The probability distribution is

(C.1.6) $$\Pr : \mathbb{N} \to [0,1], \quad s \mapsto \frac{5^{s-1}}{6^s}.$$

This is a probability distribution due to (C.1.5).

Exercise C.1.12. Consider the experiment in which a dice is rolled until an odd number occurs for the first time. Determine the corresponding discrete probability space as in Example C.1.11.

Definition C.1.13. A *random variable on a discrete probability space* $(S, \Pr)$ is a function

$$X : S \to \mathbb{R}.$$

The *expected value* or *expectation* of X is

(C.1.7) $$E[X] = \sum_{s \in S} \Pr(s) X(s)$$

if this sum is absolute convergent.

Example C.1.14. Use the notation of Example C.1.10 and define the random variable

(C.1.8) $$X : S \to \mathbb{R}, \quad s \mapsto |s|.$$

The expected value of this random variable is

(C.1.9) $$E[X] = \frac{1}{6} \sum_{n=1}^{\infty} n \left(\frac{5}{6}\right)^{n-1} = \frac{1}{6} \frac{1}{(1 - 5/6)^2} = 6.$$

This means that the expected number of times one needs to roll a dice until it shows a 6 is 6.

Exercise C.1.15. Calculate the expected number of rolls needed to achieve success in the experiment described in Exercise C.1.12.

Next, we show that the expectation of random variables has linearity properties.

Proposition C.1.16. *Let* $(S, \Pr)$ *be a discrete probability space and let* X *and* Y *be random variables on it such that the expectations* $E[X]$ *and* $E[Y]$ *are defined. Then the following hold.*

(1) $E[X] + E[Y] = E[X + Y]$.

(2) $E[rX] = rE[X]$.

Proof. The assertion follows from [**Rud76**, 3.47 Theorem].   □

We also require *Markov's inequality* which we state now.

Proposition C.1.17. *Let* $(S, \Pr)$ *be a discrete probability space, and let* $X : S \to \mathbb{R}_{\geq 0}$ *be a random variable on it such that* $E[X]$ *is defined. Let* $c \in \mathbb{R}_{>0}$ *and define the event* $X \geq cE[X]$ *to be the set of all elementary events* $s \in S$ *such* $X(s) \geq cE[X]$. *Then*

$$(C.1.10) \qquad\qquad \Pr(X \geq cE[X]) \leq \frac{1}{c}.$$

Proof. Let Y be the random variable satisfying $Y(s) = 0$ if $0 \leq X(s) < cE[X]$ and $Y(s) = cE[X]$ if $X(s) \geq cE[X]$ for all $s \in S$. Then we have

$$(C.1.11) \qquad\qquad E[X] \geq E[Y] = cE[X]\Pr(X \geq cE[X]).$$

This implies the assertion.   □

C.2. Bernoulli experiments

In this section, we discuss Bernoulli experiments that generalize Example C.1.11. Let $(S, \Pr)$ be a discrete probability distribution and let `success` and `failure` be two complementary events in S. Let p be the probability of `success`. Then the probability of `failure` is $1 - p$.

The corresponding *Bernoulli experiment* consists of repeating the above experiment until the event `success` happens for the first time. To model it, we define the discrete probability space $(\mathbb{N}, \Pr^*)$ as follows. An elementary event $i \in \mathbb{N}$ means that `success` occurs for the first time after i trials. So the probability distribution $\Pr^*$ is defined by

$$(C.2.1) \qquad\qquad \Pr^*(i) = (1 - p)^{i-1}p, \quad i \in \mathbb{N}.$$

Proposition C.2.1. *The pair* $(\mathbb{N}, \Pr^*)$ *is a discrete probability space.*

Exercise C.2.2. Prove Proposition C.2.1.

Example C.2.3. Consider the experiment of rolling a dice. So, we have $S = \{1, 2, \dots, 6\}$ and $\Pr(s) = \frac{1}{6}$ for all $s \in S$. We define a Bernoulli experiment by defining the outcome of 6 as a success and an outcome different from 6 as a failure. Therefore, we have

$$(C.2.2) \qquad\qquad \text{success} = \{6\}, \quad \text{failure} = \{1, 2, 3, 4, 5\}, \quad p = \frac{1}{6}.$$

The corresponding probability distribution $(\mathbb{N}, \Pr^*)$ was presented in Example C.1.11.

We are interested in the expected number of repetitions in the Bernoulli experiment required to be successful for the first time. So we consider the random variable

$$(C.2.3) \qquad\qquad X : \mathbb{N} \to \mathbb{N}, \quad i \mapsto i.$$

Its value is the number of trials required to be successful. Its expectation is now determined.

Proposition C.2.4. *The expected number of trials in the Bernoulli experiment is $\frac{1}{p}$.*

Proof. We have

(C.2.4) $$\sum_{i \in \mathbb{N}} i \Pr^*(i) = p \sum_{i=1}^{\infty} i(1-p)^{i-1} = \frac{p}{(1-(1-p))^2} = \frac{1}{p}. \qquad \Box$$

Example C.2.5. The expected number of rolls to obtain a 6 on a dice is 6. The expected number of coin tosses to obtain heads is 2.

Exercise C.2.6. Determine the expected number of rolls to obtain a number > 3 on a dice.

Solutions of
Selected Exercises

Solution of Exercise 1.1.8. Set $n = \lfloor \log_2 a \rfloor + 1$. Then we have $2^{n-1} \leq a < 2^n$. We prove the assertion by induction on n. If $n = 1$, we have $a = 1$ and that proves the assertion. Suppose that $n > 1$ and that the assertion holds for $n - 1$. Set $a' = a - 2^{n-1}$. Then we have $0 \leq a' < 2^{n-1}$. By the induction hypothesis, we can write $a' = \sum_{i=0}^{m-1} b_i' 2^{m-i-1}$ where $m < n$, $b_i \in \{0, 1\}$ for $0 \leq i < m$. If we set $b_0 = 1$, $b_1 = \cdots = b_{n-m-1} = 0$, and $b_{n-m} = b_0', \ldots, b_{n-1} = b_{m-1}'$, then $a = \sum_{i=0}^{n-1} b_i 2^{n-i-1}$. Also, two such representations of a give two representations of a' which proves the uniqueness. $\qquad\square$

Solution of Exercise 1.1.14. We have $0 \oplus 0 = 0 = 0 + 0 \bmod 2$, $0 \oplus 1 = 1 = 0 + 1 \bmod 2$, $1 \oplus 0 = 0 = 1 + 0 \bmod 2$, and $1 \oplus 1 = 0 = 1 + 1 \bmod 2$. $\qquad\square$

Solution of Exercise 1.1.20. Let $a = bc$ with two proper divisors b, c of a such that $1 < b \leq |c|$. Then we have $b^2 \leq |bc| = |a|$. This implies $1 < b \leq \sqrt{|a|}$. $\qquad\square$

Solution of Exercise 1.1.30. By assumption, we have $r_1 < r_0$. Also, we have $0 < r_{i+2} < r_{i+1}$ for all $i \in \mathbb{Z}_k$ since r_{i+2} is the remainder of the division of r_i by r_{i+1}. Hence, the sequence $(r_i)_{i \in \mathbb{Z}_{k+2}}$ is strictly decreasing. Let $i \in \mathbb{Z}_k$. Then we have $r_{i+2} < r_{i+1} < r_i$. If $r_{i+1} \leq r_i/2$, then $r_{i+2} < r_i/2$. Assume that

(D.1) $$r_{i+1} > r_i/2.$$

Now we have $r_i = qr_{i+1} + r_{i+2}$ with $q \in \mathbb{N}_0$ which implies $r_{i+2} = r_i - qr_{i+1}$. So $0 \leq r_{i+2} < r_{i+1}$ and (D.1) imply that $q = 1$ and $r_{i+2} < r_i/2$. Next, we see that $r_{i+2} < r_i/2$ implies $r_{2l} < r_0/2^l$ for all $l \in \mathbb{N}$ such that $2l \leq k + 1$. This implies $k = O(\log r_0) = O(\text{size}(r_0))$. $\qquad\square$

Solution of Exercise 1.2.7. Since a is composite, we can write $a = bc$ with $a, b \in \mathbb{N}$ and $1 < b \le \sqrt{a}$. Now we have $a < 2^{\text{bitLength}\,a}$. Hence, $b \le \sqrt{a} < 2^{(\text{bitLength}\,a)/2}$. This shows that $\text{bitLength}(b) \le \lceil (\text{bitLength}\,a)/2 \rceil = m(a)$. Since the binary expansion of a can be computed in polynomial time, the same is true for $m(a)$. □

Solution of Exercise 1.3.4. The set $\text{FRand}(A, a) \cup \{\infty\}$ is countable and by Lemma 1.3.2 $\Pr_{A,a}$ is a probability distribution on the sample space. If $\Pr_{A,a}(\infty) = 0$, then $\Pr$ is a probability distribution on the sample space $\text{FRand}(A, a)$. □

Solution of Exercise 1.3.18. Write $p = p_A(a)$ and $q = q_A(a)$ and denote by $q(a, k)$ the failure probability of $\text{repeat}_A(a, k)$. If $k \ge |\log \varepsilon|/p$, then it follows from (1.3.17) and from $0 < \varepsilon \le 1$ that

$$\text{(D.2)} \qquad\qquad q_A(a, k) \le e^{-kp} \le e^{-|\log \varepsilon|} = e^{\log \varepsilon} = \varepsilon.$$

Also, if $q_A(a, k) \le \varepsilon$, then (1.3.17) and $0 < \varepsilon \le 1$ imply

$$\text{(D.3)} \qquad\qquad \varepsilon \ge q_A(a, k) \ge e^{-kp/q}.$$

This implies

$$\text{(D.4)} \qquad\qquad \log \varepsilon \ge -kp/q$$

and thus

$$\text{(D.5)} \qquad\qquad k \ge |\log \varepsilon| p/q$$

as asserted. □

Solution of Exercise 1.4.23. The language is $L = \{(a, x) \ : \ a \in I, x \in \mathbb{R}_{>0}, a \text{ as a solution } b \text{ with size } b \le x\}$. □

Solution of Exercise 2.1.4. We have

$$\text{(D.6)} \qquad\qquad |0\rangle = \frac{|x_+\rangle + |x_-\rangle}{\sqrt{2}}, \quad |1\rangle = \frac{|x_+\rangle - |x_-\rangle}{\sqrt{2}}.$$

This proves the assertion. □

Solution of Exercise 2.2.10. Let $\vec{u} = (u_0, \ldots, u_{k-1})$, $\vec{v} = (v_0, \ldots, v_{k-1})$, $\vec{w} = (w_0, \ldots, w_{k-1}) \in \mathbb{C}^k$ and let $\alpha \in \mathbb{C}$. Denote by $\langle \cdot | \cdot \rangle$ the function defined in (2.2.9). Then we have

$$\langle \vec{u} | \vec{v} + \vec{w} \rangle = \sum_{i=0}^{k-1} \overline{u_i}(v_i + w_i) = \sum_{i=0}^{k-1} \overline{u_i} v_i + \sum_{i=0}^{k-1} \overline{u_i} w_i = \langle \vec{u} | \vec{v} \rangle + \langle \vec{u} | \vec{w} \rangle.$$

We also have

$$\langle \vec{v} | \alpha \vec{w} \rangle = \sum_{i=0}^{k-1} \overline{v_i}(\alpha w_i) = \alpha \sum_{i=0}^{k-1} \overline{v_i} w_i = \alpha \langle \vec{v} | \vec{w} \rangle.$$

This proves the linearity in the second argument. Next, we prove the conjugate symmetry:

$$\langle \vec{w}|\vec{v}\rangle = \sum_{i=0}^{k-1} \overline{w_i} v_i = \sum_{i=0}^{k-1} v_i \overline{w_i} = \overline{\sum_{i=0}^{k-1} \overline{v_i} w_i} = \overline{\langle \vec{v}|\vec{w}\rangle}.$$

Finally, we have

$$\langle \vec{v}|\vec{v}\rangle = \sum_{i=0}^{k-1} \overline{v_i} v_i = \sum_{i=0}^{k-1} |v_i|^2.$$

This implies the positive definiteness and concludes the proof of Theorem 2.2.9. □

Solution of Exercise 2.2.23. We prove that the map (2.2.20) is a norm on $\mathbb{C}$. We first prove the triangle inequality. Let $\alpha, \beta \in \mathbb{C}$. We apply the triangle inequality for the absolute value in $\mathbb{R}$ and obtain

$$
\begin{aligned}
|\alpha + \beta|^2 &= |\Re\alpha + \Re\beta|^2 + |\Im\alpha + \Im\beta|^2 \\
\text{(D.7)} \qquad &\leq (\Re\alpha)^2 + (\Re\beta)^2 + (\Im\alpha)^2 + (\Im\beta)^2 \\
&= |\alpha|^2 + |\beta|^2.
\end{aligned}
$$

The absolute homogeneity is seen as follows:

$$
\begin{aligned}
|\alpha\beta|^2 &= |(\Re\alpha + i\Im\alpha)(\Re\beta + i\Im\beta)|^2 \\
&= (\Re\alpha\Re\beta - \Im\alpha\Im\beta)^2 + (\Re\alpha\Im\beta + \Im\alpha\Re\beta)^2 \\
\text{(D.8)} \qquad &= (\Re\alpha\Re\beta)^2 + (\Im\alpha\Im\beta)^2 + (\Re\alpha\Im\beta)^2 + (\Im\alpha\Re\beta)^2 \\
&= ((\Re\alpha)^2 + (\Im\alpha)^2)^2((\Re\beta)^2 + (\Im\beta)^2)^2 \\
&= |\alpha|^2|\beta|^2.
\end{aligned}
$$

Finally, the positive definiteness follows directly from (2.2.20). □

Solution of Exercise 2.3.3. The matrix representation of Y with respect to B is

$$\text{(D.9)} \qquad \text{Mat}_B(Y) = \begin{pmatrix} 0 & -i \\ i & 0 \end{pmatrix}.$$

To determine the matrix representations of Y with respect to C we note that

$$\text{(D.10)} \qquad Y|x_+\rangle = \frac{Y|0\rangle + Y|1\rangle}{\sqrt{2}} = \frac{i|1\rangle - i|0\rangle}{\sqrt{2}} = -i|x_-\rangle$$

and

$$\text{(D.11)} \qquad Y|x_-\rangle = \frac{Y|0\rangle - Y|1\rangle}{\sqrt{2}} = \frac{i|1\rangle + i|0\rangle}{\sqrt{2}} = i|x_-\rangle.$$

Hence, we have

$$\text{(D.12)} \qquad \text{Mat}_C(Y) = \begin{pmatrix} 0 & i \\ -i & 0 \end{pmatrix}$$

which is equal to $-\text{Mat}_B(Y)$. Finally, to find $\text{Mat}_C(Z)$ we note that

$$\text{(D.13)} \qquad Z|x_+\rangle = \frac{Z|0\rangle + Z|1\rangle}{\sqrt{2}} = \frac{|0\rangle - |1\rangle}{\sqrt{2}} = |x_-\rangle$$

and

(D.14)
$$Z \left| x_- \right\rangle = \frac{Z \left| 0 \right\rangle - Z \left| 1 \right\rangle}{\sqrt{2}} = \frac{\left| 0 \right\rangle + \left| 1 \right\rangle}{\sqrt{2}} = \left| x_+ \right\rangle.$$

Hence, we have

(D.15)
$$\mathrm{Mat}_C(Z) = \begin{pmatrix} 0 & 1 \\ 1 & 0 \end{pmatrix}.$$

This matrix is equal to $\mathrm{Mat}_B(X)$. $\qquad\square$

Solution of Exercise 2.3.10. The identity (2.3.25) follows from the fact that transposition and conjugation of matrices are involutions. Next, we have $(A + B)^{\mathrm{T}} = A^{\mathrm{T}} + B^{\mathrm{T}}$ and $\overline{(A + B)} = \overline{A} + \overline{B}$ which implies (2.3.26). Also, $(\alpha A)^{\mathrm{T}} = \alpha A^{\mathrm{T}}$ and $\overline{\alpha A} = \overline{\alpha}\,\overline{A}$ imply (2.3.27).

Next, we prove (2.3.28). The rank r of A is the number of linearly independent column vectors of A. The conjugates of these column vectors are the row vectors of A^*. Since conjugation does not change linear dependence and independence, the number of linearly independent row vectors of A^* is also r. So, Proposition B.7.3 implies that A and A^* have the same rank.

Finally, equation (2.3.29) follows from the observations that $\overline{AB} = \overline{A}\,\overline{B}$ and $(AB)^{\mathrm{T}} = B^{\mathrm{T}}A^{\mathrm{T}}$. $\qquad\square$

Solution of Exercise 2.4.5. Assume that all eigenvalues of $A \in \mathbb{C}^{(k,k)}$ have algebraic multiplicity 1. It then follows from the definition of an eigenvalue and from Corollary B.7.26 that all eigenvalues also have geometric multiplicity 1. Since by Proposition 2.4.1 the characteristic polynomial $p_A(x)$ is a product of linear factors, Theorem B.7.28 implies the assertion. $\qquad\square$

Solution of Exercise 2.4.14. Let $A = (a_{i,j}) \in \mathbb{C}^{(k,k)}$. The diagonal elements of A are $a_{i,i}$ and the diagonal elements of A^* are $\overline{a_{i,i}}$. Since A is Hermitian, we have $A = A^*$ and therefore $a_{i,i} = \overline{a_{i,i}}$ for all $i \in \mathbb{Z}_k$. This proves the first assertion. The second assertion follows from Proposition 2.3.14. The remaining assertions can be deduced from Proposition 2.3.9 and the Hermitian property. $\qquad\square$

Solution of Exercise 2.4.42. Let P be a projection. Then by Proposition 2.4.39 also P^* is a projection. If P is Hermitian, then we have

$$
\begin{aligned}
\langle P\vec{v}, \vec{v} - P\vec{v} \rangle \\
= \langle P\vec{v}, \vec{v} \rangle - \langle P\vec{v}, P\vec{v} \rangle && \text{linearity of the inner product,} \\
= \langle P\vec{v}, \vec{v} \rangle - \langle P^*P\vec{v}, \vec{v} \rangle && \text{property of the adjoint,} \\
= \langle P\vec{v}, \vec{v} \rangle - \langle P^2\vec{v}, \vec{v} \rangle && P \text{ is Hermitian,} \\
= \langle P\vec{v}, \vec{v} \rangle - \langle P\vec{v}, \vec{v} \rangle = 0 && P \text{ is a projection.}
\end{aligned}
$$

So P is an orthogonal projection. Conversely, let P be orthogonal and let $\vec{v} \in \mathbb{C}^k$ or $\vec{v} \in V$. Then by Proposition 2.4.39 P^* is also an orthogonal projection and we have

$$\langle (P - P^*)\vec{v}, (P - P^*)\vec{v} \rangle$$

$$= \langle P\vec{v}, P\vec{v} \rangle - \langle P\vec{v}, P^*\vec{v} \rangle$$

$$\quad - \langle P^*\vec{v}, P\vec{v} \rangle + \langle P^*\vec{v}, P^*\vec{v} \rangle \qquad \text{linearity of inner product,}$$

$$= \langle P\vec{v}, P\vec{v} \rangle - \langle P^2\vec{v}, \vec{v} \rangle$$

$$\quad - \langle (P^*)^2\vec{v}, \vec{v} \rangle + \langle P^*\vec{v}, P^*\vec{v} \rangle \qquad \text{property of the adjoint,}$$

$$= \langle P\vec{v}, P\vec{v} \rangle - \langle P\vec{v}, \vec{v} \rangle$$

$$\quad - \langle P^*\vec{v}, \vec{v} \rangle + \langle P^*\vec{v}, P^*\vec{v} \rangle \qquad P \text{ and } P^* \text{ are projectors,}$$

$$= \langle P\vec{v}, P\vec{v} - \vec{v} \rangle + \langle P^*\vec{v}, P^*\vec{v} - \vec{v} \rangle \qquad \text{linearity of inner product,}$$

$$= 0 \qquad\qquad\qquad\qquad P \text{ and } P^* \text{ are orthogonal.}$$

Hence, $P = P^*$ which means that P is Hermitian. $\qquad\qquad\qquad\qquad\qquad$ $\square$

Solution of Exercise 2.4.50. If A is Hermitian, then $A = A^*$ and thus $A^*A = AA^*$. If A is unitary, then $A^*A = I_k = AA^*$. $\qquad\qquad\qquad\qquad\qquad\qquad\qquad\qquad\qquad$ $\square$

Solution of Exercise 3.1.10. Assume that $x, y \geq 0$. Since $x^2 + y^2 = 1$, we have $0 \leq x, y \leq 1$. Set $\gamma = \arcsin y$. Then γ is the uniquely determined number in $[0, \pi/2]$ such that $\sin \gamma = y$. Next, we have $x^2 = 1 - y^2 = 1 - \sin^2 \gamma = \cos^2 \gamma$. Since $\gamma \in [0, \pi/2]$ we have $\cos \gamma \geq 0$. Since $x \geq 0$ this implies that $x = \cos \gamma$. We claim that γ is uniquely determined in $[0, 2\pi[$. Let $\gamma' \in [0, 2\pi[$ such that $\cos \gamma' = x$ and $\sin \gamma' = y$. Since $y \geq 0$, it follows that $\gamma' \in [0, \pi]$ and because $x \geq 0$, it follows that $\gamma' \in [0, \pi/2]$. But γ is the only real number in $[0, \pi/2]$ with $x = \cos \gamma$. This implies $\gamma' = \gamma$. If $x > 0$ and $y < 0$, then we can replace γ with $2\pi - \gamma$ and use $\sin(2\pi - \gamma) = -\sin \gamma$ and $\cos(2\pi - \gamma) = \cos \gamma$. The other cases are treated analogously. $\qquad$ $\square$

Solution of Exercise 3.1.16. Since $|\alpha| = 1$, it follows that $(\Re\alpha)^2 + (\Im\alpha)^2 = 1$. By Lemma 3.1.9, there is $\gamma \in \mathbb{R}$ such that $\cos \gamma = \Re\alpha$ and $\sin \gamma = \Im\alpha$. Hence, we have $e^{i\gamma} = \cos \gamma + i \sin \gamma = \Re\alpha + i\Im\alpha = \alpha$. Conversely, if $\gamma \in \mathbb{R}$ with $\alpha = e^{i\gamma} = \cos \gamma + i \sin \gamma$, then $\Re\alpha = \cos \gamma$ and $\Im\alpha = \sin \gamma$. From Lemma 3.1.9 it follows that γ is uniquely determined modulo 2π. $\qquad\qquad\qquad\qquad\qquad\qquad\qquad\qquad\qquad\qquad\qquad$ $\square$

Solution of Exercise 3.1.21. We have $|x_+\rangle = \cos(\pi/4) |0\rangle + e^{i \cdot 0} \sin(\pi/4) |1\rangle$. Therefore, the spherical coordinates of the point on the Bloch sphere corresponding to $|x_+\rangle$ are $(\pi/2, 0)$. The Cartesian coordinates of this point are $(1, 0, 0)$. The proof for $|x_-\rangle$ is analogous.

Also, we have $|y_+\rangle = \cos(\pi/4) |0\rangle + e^{i \cdot \pi/2} \sin(\pi/4) |1\rangle$. Therefore, the spherical coordinates of the point on the Bloch sphere corresponding to $|y_+\rangle$ are $(\pi/2, \pi/2)$. The Cartesian coordinates of this point are $(0, 1, 0)$. The proof for $|y_-\rangle$ is analogous. $\qquad$ $\square$

Solution of Exercise 3.1.24. The relation R is reflexive, since for every $|\psi\rangle \in S$ we have $|\psi\rangle = e^{i\gamma} |\psi\rangle$ with $\gamma = 0$. If $|\varphi\rangle, |\psi\rangle \in S$ with $|\psi\rangle = e^{i\gamma} |\varphi\rangle$ for some $\gamma \in \mathbb{R}$, then we have $|\varphi\rangle = e^{i(-\gamma)}|\psi\rangle$. So R is symmetric. Finally, let $|\varphi\rangle, |\psi\rangle, |\xi\rangle \in S$ and let $\gamma, \delta \in \mathbb{R}$ such that $|\xi\rangle = e^{i\delta}|\psi\rangle$ and $|\psi\rangle = e^{i\gamma}|\varphi\rangle$. Then we have $|\xi\rangle = e^{i(\delta+\gamma)}|\varphi\rangle$. Therefore, R is transitive. $\qquad\square$

Solution of Exercise 3.6.8. Both O_A and I_B are Hermitian operators. Therefore, as noted in Section 2.5.5, it follows that $O_{AB} = O_A \otimes I_B$ is Hermitian and thus an observable of system AB. Also, it can be easily verified that the spectral decomposition of this observable is given by (3.6.11). The Measurement Postulate 3.6.5 implies that the eigenvalue λ is measured with probability

$$
\text{(D.16)} \quad
\begin{aligned}
\Pr(\lambda) &= \text{tr}((P_\lambda \otimes I_B)(\rho_A \otimes \rho_B)) = \text{tr}((P_\lambda\rho_A) \otimes (I_B\rho_B)) \\
&= \text{tr}((P_\lambda\rho_A) \otimes \rho_B) = \text{tr}(P_\lambda\rho_A)\,\text{tr}(\rho_B) = \text{tr}(P_\lambda\rho_A).
\end{aligned}
$$

Also, it follows from the Measurement Postulate 3.6.5 that if this outcome occurs, the state immediately after the measurement is

$$
\text{(D.17)} \quad \frac{(P_\lambda \otimes I_b)(\rho_A \otimes \rho_B)(P_\lambda \otimes I_b)}{\text{tr}(P_\lambda\rho_A)} = \frac{(P_\lambda\rho_A P_\lambda) \otimes \rho_B}{\text{tr}(P_\lambda\rho_A)}.
$$

Finally, the expectation value of $O \otimes I_b$ is

$$
\text{(D.18)} \quad \text{tr}((O \otimes I_b)(\rho_A \otimes \rho_B)) = \text{tr}(O\rho_A \otimes I_b\rho_B) = \text{tr}(O\rho_A)\,\text{tr}(\rho_B) = \text{tr}(O\rho_A). \quad\square
$$

Solution of Exercise 3.7.11. Proposition 2.4.27 implies

$$
\text{(D.19)} \quad \rho = |\xi\rangle\langle\xi| = \frac{1}{l}\sum_{i,j=0}^{l-1} |\varphi_i\rangle|\psi_i\rangle\langle\varphi_j|\langle\psi_j|.
$$

Since the sequence $(|\psi_i\rangle)$ is orthonormal, it follows that for all $i, j \in \mathbb{Z}_l$ we have

$$
\text{(D.20)} \quad \text{tr}_B |\varphi_i\rangle|\psi_i\rangle\langle\varphi_j|\langle\psi_j| = |\varphi_i\rangle\langle\varphi_j|\,\delta_{i,j}.
$$

Equations (D.19) and (D.20) imply

$$
\text{(D.21)} \quad \text{tr}_B |\xi\rangle\langle\xi| = \frac{1}{l}\sum_{i=0}^{l-1} |\varphi_i\rangle\langle\varphi_i|
$$

which proves the claim. $\qquad\square$

Solution of Exercise 4.1.5. We have $\text{tr}\,I^*X = \text{tr}\,X = 0$, $\text{tr}\,I^*Y = \text{tr}\,Y = 0$, $\text{tr}\,I^*Z = \text{tr}\,Z = 0$. Also Theorem 4.1.2 implies $\text{tr}\,X^*Y = \text{tr}\,XY = \text{tr}\,iZ = 0$, $\text{tr}\,Z^*X = \text{tr}\,ZX = \text{tr}\,iY = 0$, and $\text{tr}\,Y^*Z = \text{tr}\,YZ = \text{tr}\,iX = 0$. $\qquad\square$

Solution of Exercise 4.2.26. Let $B = (\hat{u}, \hat{v}, \hat{w}) \in \text{SO}(3)$. Then Proposition 4.2.25 implies $\text{Rot}_{\hat{w}}(\gamma) = B\,\text{Rot}_{\hat{z}}(\gamma)B^{-1}$. Choose $T \in \text{SO}(3)$ with $BT = (-\hat{u}, \hat{v}, -\hat{w})$. Then $T\,\text{Rot}_{\hat{z}}(\gamma)T^{-1} = \text{Rot}_{\hat{z}}(-\gamma)$. This implies

$$
\text{Rot}_{-\hat{w}}(\gamma) = BT\,\text{Rot}_{\hat{z}}(\gamma)T^{-1}B^{-1} = B\,\text{Rot}_{\hat{z}}(-\gamma)B^{-1} = \text{Rot}_{\hat{w}}(-\gamma). \quad\square
$$

Solution of Exercise 4.3.9. We have

$$
\text{(D.22)} \quad R_{\hat{x}}(\gamma)|0\rangle = \left(\cos\frac{\gamma}{2}I - i\sin\frac{\gamma}{2}X\right)|0\rangle = \cos\frac{\gamma}{2}|0\rangle - i\sin\frac{\gamma}{2}|1\rangle
$$

and

(D.23) $\qquad R_{\hat{x}}(\gamma)\,|1\rangle = \left(\cos\frac{\gamma}{2}I - i\sin\frac{\gamma}{2}X\right)|1\rangle = \cos\frac{\gamma}{2}\,|1\rangle - i\sin\frac{\gamma}{2}\,|0\rangle.$

This proves (4.3.8). We also have

(D.24) $\qquad R_{\hat{y}}(\gamma)\,|0\rangle = \left(\cos\frac{\gamma}{2}I - i\sin\frac{\gamma}{2}Y\right)|0\rangle = \cos\frac{\gamma}{2}\,|0\rangle + \sin\frac{\gamma}{2}\,|1\rangle$

and

(D.25) $\qquad R_{\hat{y}}(\gamma)\,|1\rangle = \left(\cos\frac{\gamma}{2}I - i\sin\frac{\gamma}{2}X\right)|1\rangle = \cos\frac{\gamma}{2}\,|1\rangle - \sin\frac{\gamma}{2}\,|0\rangle.$

This proves (4.3.9). Finally, we have

(D.26) $\qquad R_{\hat{z}}(\gamma)\,|0\rangle = \left(\cos\frac{\gamma}{2}I - i\sin\frac{\gamma}{2}Z\right)|0\rangle = \left(\cos\frac{\gamma}{2} - i\sin\frac{\gamma}{2}\right)|0\rangle\, e^{-i\gamma/2}\,|0\rangle$

and

(D.27) $\qquad R_{\hat{z}}(\gamma)\,|1\rangle = \left(\cos\frac{\gamma}{2}I - i\sin\frac{\gamma}{2}Z\right)|1\rangle = \left(\cos\frac{\gamma}{2} + i\sin\frac{\gamma}{2}\right)|0\rangle .\, e^{i\gamma/2}\,|0\rangle.$

This proves (4.3.10). $\qquad\square$

Solution of Exercise 4.3.13. Let $A \in \mathrm{su}(2)$,

(D.28) $$A = \begin{pmatrix} a & b \\ c & d \end{pmatrix}$$

with $a, b, c, d \in \mathbb{C}$. Since A is Hermitian, we have $a, d \in \mathbb{R}$ and $b = \bar{c}$. Since $\mathrm{tr}\,A = 0$, we have $d = -a$. Hence, A can be written as in the lemma. Conversely, if A has a representation as in the lemma, then A is Hermitian and has trace 0; that is, $A \in \mathrm{su}(2)$. $\qquad\square$

Solution of Exercise 4.3.18. Let $\hat{w} \in \mathbb{R}^3$ be a unit vector and let $\gamma \in \mathbb{R}$ such that $U = R_{\hat{w}}(\gamma)$. If $U \in \{\pm I\}$, then Theorem 4.3.15 implies $\gamma \equiv 0 \bmod 2\pi$. So, by Proposition 4.2.27 we have $\mathrm{Rot}(U) = I_3$. Assume that $U \neq \pm I$. Let $\hat{w}' \in \mathbb{R}^3$ be a unit vector and let $\gamma' \in \mathbb{R}$ such that $U = R_{\hat{w}'}(\gamma')$. Then Theorem 4.3.15 implies $\hat{w} = \hat{w}'$ and $\gamma \equiv \gamma' \bmod 2\pi$ or $\hat{w} = -\hat{w}'$ and $\gamma \equiv -\gamma' \bmod 2\pi$. So Proposition 4.2.27 implies that $\mathrm{Rot}_{\hat{w}}(\gamma) = \mathrm{Rot}_{\hat{w}'}(\gamma')$. $\qquad\square$

Solution of Exercise 4.3.23. Let $\tau = (\tau_1, \tau_2, \tau_3)$, $p = (p_1, p_2, p_3)$, and $B = (b_{i,j})$. Then we have

$$
\begin{aligned}
B\vec{p} \cdot \tau &= (b_{1,1}p_1 + b_{1,2}p_2 + b_{1,3}p_3)\tau_1 \\
&\quad + (b_{2,1}p_1 + b_{2,2}p_2 + b_{2,3}p_3)\tau_2 \\
&\quad + (b_{3,1}p_1 + b_{3,2}p_2 + b_{3,3}p_3)\tau_3 \\
&= p_1(b_{1,1}\tau_1 + b_{2,1}\tau_2 + b_{3,1}\tau_3) \\
&\quad + p_2(b_{1,2}\tau_1 + b_{2,2}\tau_2 + b_{3,2}\tau_3) \\
&\quad + p_3(b_{1,3}\tau_1 + b_{2,3}\tau_2 + b_{3,3}\tau_3) \\
&= \vec{p} \cdot (B \cdot \tau).
\end{aligned}
$$

(D.29) $\qquad\square$

Solution of Exercise 4.3.29. We have

$$(\vec{p} \cdot \sigma) |0\rangle = \cos \phi \sin \theta \, X |0\rangle + \sin \phi \sin \theta \, Y |0\rangle + \cos \theta \, Z |0\rangle$$

(D.30)
$$= \cos \phi \sin \theta \, |1\rangle + i \sin \phi \sin \theta \, |1\rangle + \cos \theta \, |0\rangle$$

$$= \cos \theta \, |0\rangle + \sin \theta \, e^{i\phi} |1\rangle.$$

We also have

$$(\vec{p} \cdot \sigma) |1\rangle = \cos \phi \sin \theta \, X |1\rangle + \sin \phi \sin \theta \, Y |1\rangle + \cos \theta \, Z |1\rangle$$

(D.31)
$$= \cos \phi \sin \theta \, |0\rangle - i \sin \phi \sin \theta \, |0\rangle - \cos \theta \, |1\rangle$$

$$= \sin \theta \, e^{-i\phi} |0\rangle - \cos \theta \, |1\rangle.$$

This proves the assertion. □

Solution of Exercise 4.4.3. Let $\bullet \in \{+, -\}$. Then we have

$$X |x_\bullet\rangle = X \frac{|0\rangle \bullet |1\rangle}{\sqrt{2}} = \frac{X |0\rangle \bullet X |1\rangle}{\sqrt{2}}$$

(D.32)
$$= \frac{|1\rangle \bullet |0\rangle}{\sqrt{2}} = \bullet \frac{|0\rangle \bullet |1\rangle}{\sqrt{2}}$$

$$= \bullet |x_\bullet\rangle.$$

On $\{+, -\}$ we define multiplication in the usual way:

(D.33)
$$+ \cdot + = - \cdot - = +, \quad + \cdot - = - \cdot + = -.$$

Then from (D.32) we obtain for all $\circ, \bullet \in \{+, -\}$

$$\text{CNOT} |x_\circ\rangle |x_\bullet\rangle = \text{CNOT} \frac{1}{2} (|0\rangle |x_\bullet\rangle \circ |1\rangle |x_\bullet\rangle)$$

$$= \frac{1}{\sqrt{2}} (\text{CNOT} |0\rangle |x_\bullet\rangle \circ \text{CNOT} |1\rangle |x_\bullet\rangle)$$

(D.34)
$$= \frac{1}{\sqrt{2}} (|0\rangle |x_\bullet\rangle \circ |1\rangle X |x_\bullet\rangle)$$

$$= \frac{1}{\sqrt{2}} (|0\rangle |x_\bullet\rangle \circ \cdot \bullet |1\rangle |x_\bullet\rangle)$$

$$= |x_{\circ \cdot \bullet}\rangle |x_\bullet\rangle.$$

This implies (4.4.4). □

Solution of Exercise 5.1.5. Since V is unitary and $|\psi\rangle$ is an eigenstate of V we have $V |\psi\rangle = e^{i\phi} |\psi\rangle$ with $\phi \in \mathbb{R}$. Also, we have

(D.35)
$$C(V) |x_+\rangle |\psi\rangle = \frac{|0\rangle + e^{i\phi} |1\rangle}{\sqrt{2}} |\psi\rangle.$$

The point on the Bloch sphere corresponding to $|x_+\rangle$ has the spherical coordinates $(1, \pi/2, 0)$. The point on the Bloch sphere corresponding to the first qubit of $C(V) |x_+\rangle |\psi\rangle$ has the spherical coordinates $(1, \pi/2, \phi)$. Finally, we have $U_f = (I \otimes X^{f(0)}) C(V_f)$. □

Solution of Exercise 5.5.8. We use (5.3.3) and obtain

$$H^{\otimes n} |\vec{z} \oplus S\rangle = \frac{1}{\sqrt{2^m}} \sum_{\vec{s} \in S} H^{\otimes n} |\vec{z} \oplus \vec{s}\rangle$$

(D.36)
$$= \frac{1}{\sqrt{2^{m+n}}} \sum_{\vec{s} \in S} \sum_{\vec{w} \in \{0,1\}^n} (-1)^{(\vec{z} \oplus \vec{s}) \cdot \vec{w}} |\vec{w}\rangle$$

$$= \frac{1}{\sqrt{2^{m+n}}} \sum_{\vec{w} \in \{0,1\}^n} (-1)^{\vec{z} \cdot \vec{w}} \left(\sum_{\vec{s} \in S} (-1)^{\vec{s} \cdot \vec{w}} \right) |\vec{w}\rangle.$$

We evaluate the inner sum of the last expression in (D.36). If $\vec{w} \in S^{\perp}$, we have

(D.37)
$$\sum_{\vec{s} \in S} (-1)^{\vec{s} \cdot \vec{w}} = \sum_{\vec{s} \in S} (-1)^0 = |S| = 2^m.$$

Let $\vec{w} \notin S^{\perp}$ and consider the map

(D.38)
$$S \to \{0,1\}, \quad \vec{s} \mapsto \vec{s} \cdot \vec{w}.$$

It is a surjective homomorphism of groups. By Theorem B.3.6, the kernel of this map contains $|S|/2 = 2^{m-1}$ elements. It follows that for half of the elements $\vec{s}$ of S we have $(-1)^{\vec{s} \cdot \vec{w}} = 1$ and for the other half we have $(-1)^{\vec{s} \cdot \vec{w}} = -1$. So, we have

(D.39)
$$\sum_{\vec{s} \in S} (-1)^{\vec{s} \cdot \vec{w}} = 0.$$

From (D.36), (D.37), and (D.39) we obtain

(D.40) $\quad H^{\otimes n} |\vec{z} \oplus S\rangle = \frac{2^m}{\sqrt{2^{m+n}}} \sum_{\vec{w} \in S^{\perp}} (-1)^{\vec{z} \cdot \vec{w}} |\vec{w}\rangle = \frac{1}{\sqrt{2^{n-m}}} \sum_{\vec{w} \in S^{\perp}} (-1)^{\vec{z} \cdot \vec{w}} |\vec{w}\rangle.$ $\quad\square$

Solution of Exercise 6.4.13. By definition, t_0 is the empty product times t which is t. Also, we have

(D.41) $\quad t_n \equiv \begin{cases} \prod_{l=0}^{n-1} a_{n-l-1}^{c_l} t \equiv a^{\sum_{l=0}^{n-1} c_l 2^{n-l-1}} t \equiv a^c t \bmod N & \text{if } t < N, \\ t & \text{if } t \geq N. \end{cases}$ $\quad\square$

Solution of Exercise 6.4.14. It suffices to show that the cardinality of the image of the map in (6.4.34) is 2^{2n}. The number of pairs $(x,y) \in \mathbb{Z}_{2^n}^2$ with $x \geq N$ is $k_1 = (2^n - N)2^n$. The number of pairs (x,y) with $x < N$ and $y \geq N$ is $k_2 = N(2^n - N)$. The number of pairs $(x,y) \in \mathbb{Z}_N^2$ with $\gcd(y,N) > 1$ is $k_3 = N(N - \varphi(N))$ where $\varphi(N)$ is the number of $y \in \mathbb{Z}_N$ with $\gcd(y,N) = 1$. Finally, if $y \in \mathbb{Z}_N$ with $\gcd(y,N) = 1$, then the map $\mathbb{Z}_N \to \mathbb{Z}_N, x \mapsto xy \bmod N$ is a bijection. Hence, the number of pairs $(x, xy \bmod N) \in \mathbb{Z}_N^2$ with $\gcd(y,N) = 1$ is $k_4 = N\varphi(N)$. So the cardinality of the image of the map in (6.4.34) is $k_1 + k_2 + k_3 + k_4 = (2^n - N)2^n + N(2^n - N) + N(N - \varphi(N)) + N\varphi(N) = 2^{2n} - 2^n N + 2^n N - N^2 + N^2 - N\varphi(N) + N\varphi(N) = 2^{2n}$. $\quad\square$

Solution of Exercise 7.1.8. Since the identity operator I_N and the projection $|s_1\rangle\langle s_1|$ are involutions, we have

(D.42)
$$\begin{aligned} U_1^2 &= (I_N - 2\,|s_1\rangle\langle s_1|)^2 \\ &= I_N^2 - 2I_N\,|s_1\rangle\langle s_1| - 2\,|s_1\rangle\langle s_1|\,I_N + 4(|s_1\rangle\langle s_1|)^2 \\ &= I_N - 2\,|s_1\rangle\langle s_1| - 2\,|s_1\rangle\langle s_1| + 4\,|s_1\rangle\langle s_1| \\ &= I_N. \end{aligned}$$

So U_1 is an involution. Also, since I_N and $|s_1\rangle\langle s_1|$ are Hermitian, it follows that U_1 is also Hermitian. It follows from Exercise 2.4.17 that U_1 is unitary. In the same way, it can be shown that U_s is a Hermitian, unitary involution. So $G = U_s U_1$ is unitary. $\quad\square$

Solution of Exercise 7.2.3. We have

$$\langle s_+|s_+\rangle = \langle s_-|s_-\rangle = \frac{1}{2}\left(\langle s_1|s_1\rangle + \langle s_0|s_0\rangle\right) = 1,$$

$$\langle s_+|s_-\rangle = \frac{1}{2}\left(\langle s_1|s_1\rangle - \langle s_0|s_0\rangle\right) = 0.$$

So $(|s_+\rangle, |s_-\rangle)$ is an orthonormal basis of P. Also, equation (7.2.3) implies

(D.43) $$G\,|s_0\rangle = G(\cos 0\,|s_0\rangle + \sin 0\,|s_1\rangle) = \cos 2\theta\,|s_0\rangle + \sin 2\theta\,|s_1\rangle$$

and

(D.44)
$$\begin{aligned} G\,|s_1\rangle &= G\left(\cos\frac{\pi}{2}\,|s_0\rangle + \sin\frac{\pi}{2}\,|s_1\rangle\right) \\ &= \cos\left(\frac{\pi}{2} + 2\theta\right)|s_0\rangle + \sin\left(\frac{\pi}{2} + 2\theta\right)|s_1\rangle \\ &= -\sin 2\theta\,|s_0\rangle + \cos 2\theta\,|s_1\rangle. \end{aligned}$$

From equations (D.43) and (D.44) we obtain

$$\begin{aligned} G\,|s_+\rangle &= \frac{1}{\sqrt{2}}\left(G\,|s_1\rangle + i\,|s_0\rangle\right) \\ &= \frac{1}{\sqrt{2}}\left(-\sin 2\theta\,|s_0\rangle + \cos 2\theta\,|s_1\rangle + i(\cos 2\theta\,|s_0\rangle + \sin 2\theta\,|s_1\rangle)\right) \\ &= \frac{1}{\sqrt{2}}\left((\cos 2\theta + i\sin 2\theta)\,|s_1\rangle + (i\cos 2\theta - \sin 2\theta)\,|s_0\rangle\right) \\ &= \frac{1}{\sqrt{2}}\left((\cos 2\theta - i\sin 2\theta)\,|s_0\rangle + i(\cos 2\theta + i\sin 2\theta)\,|s_1\rangle\right) \\ &= e^{2\theta}\,|s_+\rangle \end{aligned}$$

and

$$G\,|s_-\rangle = \frac{1}{\sqrt{2}}\,(G(|s_1\rangle - i\,|s_0\rangle)))$$

$$= \frac{1}{\sqrt{2}}\,(-\sin 2\theta\,|s_0\rangle + \cos 2\theta\,|s_1\rangle - i(\cos 2\theta\,|s_0\rangle + \sin 2\theta\,|s_1\rangle)))$$

$$= \frac{1}{\sqrt{2}}\,((\cos 2\theta - i\sin 2\theta)\,|s_1\rangle + (-i\cos 2\theta - \sin 2\theta)\,|s_0\rangle)$$

$$= \frac{1}{\sqrt{2}}\,((\cos 2\theta - i\sin 2\theta)\,|s_0\rangle - i(\cos 2\theta - i\sin 2\theta)\,|s_1\rangle)$$

$$= e^{-2\theta}\,|s_+\rangle.$$

This means that $|s_+\rangle$ and $|s_-\rangle$ are eigenstates of G associated with the eigenvales $e^{2\theta}$ and $e^{-2\theta}$, respectively. Finally, we prove (7.2.2). We have

$$\frac{-i}{\sqrt{2}}\left(e^{i\theta}\,|s_+\rangle - e^{-i\theta}\,|s_-\rangle\right)$$

$$= \frac{-i}{2}\left(e^{i\theta}(|s_1\rangle + i\,|s_0\rangle) - e^{-i\theta}(|s_1\rangle - i\,|s_0\rangle)\right)$$

(D.45)
$$= \frac{1}{2}\left(e^{i\theta} + e^{-i\theta}\right)|s_0\rangle + \frac{-i}{2}\left(e^{i\theta} - e^{-i\theta}\right)|s_1\rangle$$

$$= \cos\theta\,|s_0\rangle + \sin\theta\,|s_1\rangle = |s\rangle. \qquad \square$$

Solution of Exercise 8.1.1. We have $(A')^* = \begin{pmatrix} 0 & A \\ A^* & 0 \end{pmatrix}^* = A'$ and $A'\vec{x} = (A\vec{x}, \vec{0}) = (\vec{b}, \vec{0}) = \vec{b}'$. $\qquad \square$

Solution of Exercise A.1.11. Let $a, b \in S$ and assume that the equivalence classes of a and b have a common element c. Let $d \in S$. Also, let $(a, d) \in R$. Then $(a, c) \in R$, the symmetry, and the transitivity of R imply that $(b, d) \in R$. So the equivalence class of a is contained in the equivalence class of b and vice versa. Therefore, the equivalence classes are equal. $\qquad \square$

Solution of Exercise A.4.12. Both are abelian semigroups with identity elements 0 and 1, respectively. Also, $(\mathbb{Z}_k, +_k)$ is a group but $(\mathbb{Z}_k, \cdot_k)$ is not, since 0 has no inverse. The unit group of $(\mathbb{Z}_k, \cdot_k)$ is $(\mathbb{Z}_k^*, \cdot_k)$. $\qquad \square$

Solution of Exercise A.5.9. We use the trigonometric identities (A.5.1), (A.5.2), and (A.5.4) and obtain

$$\sin^2(x+y) - \sin^2 x$$
$$= (\sin x \cos y + \cos x \sin y)^2 - \sin^2 x$$
$$= \sin^2 x \cos^2 y + 2 \sin x \cos x \sin y \cos y + \cos^2 x \sin^2 y - \sin^2 x$$
$$= \sin^2 x(1 - \sin^2 y) + \sin x \cos x \sin(2y) + (1 - \sin^2 x) \sin^2 y - \sin^2 x$$
$$= \sin^2 x - \sin^2 x \sin^2 y + \sin x \cos x \sin(2y) + \sin^2 y - \sin^2 x \sin^2 y - \sin^2 x$$
$$= \sin x \cos x \sin(2y) + (1 - 2 \sin^2 x) \sin^2 y.$$

Likewise, we obtain from the trigonometric identities (A.5.1), (A.5.3), and (A.5.4)

$$\sin^2 x - \sin^2(x - y)$$
$$= \sin^2 x - (\sin x \cos y - \cos x \sin y)^2$$
$$= \sin^2 x - \sin^2 x \cos^2 y + 2 \sin x \cos x \sin y \cos y - \cos^2 x \sin^2 y$$
$$= \sin^2 x - \sin^2 x(1 - \sin^2 y) + \sin x \cos x \sin(2y) - (1 - \sin^2 x) \sin^2 y - \sin^2 x$$
$$= \sin^2 x - \sin^2 x + \sin^2 x \sin^2 y + \sin x \cos x \sin(2y) - \sin^2 y + \sin^2 x \sin^2 y$$
$$= \sin x \cos x \sin(2y) - (1 - 2 \sin^2 x) \sin^2 y. \qquad \square$$

Solution of Exercise C.1.9. The probability distribution is $(\{0, 1\}^2, \mathrm{Pr})$ where 0 and 1 represent tails and heads, respectively. Furthermore, Pr sends each pair $(a, b) \in \{0, 1\}^2$ to its probability $\frac{1}{4}$. The event "getting heads at least once" is $\{(0, 1), (1, 0), (1, 1)\}$. Its probability is $\frac{3}{4}$. $\qquad \square$

Solution of Exercise C.2.2. We must show that

$$(D.46) \qquad \qquad \sum_{i \in \mathbb{N}} \mathrm{Pr}^*(i) = 1.$$

In fact, we have

$$(D.47) \qquad \sum_{i \in \mathbb{N}} \mathrm{Pr}'(i) = p \sum_{i=0}^{\infty} (1-p)^i = \frac{p}{1 - (1-p)} = 1. \qquad \square$$

Bibliography

[AB09]   S. Arora and B. Barak, *Computational complexity: A modern approach*, Cambridge University Press, Cambridge, 2009, DOI 10.1017/CBO9780511804090. MR2500087

[Abr72]   M. Abramowitz (ed.), *Handbook of mathematical functions: with formulas, graphs, and mathematical tables*, 10th printing, with corrections, Applied mathematics series, no. 55, U. S. Government Printing Office, Washington, DC, 1972 (English).

[AGP94]   W. R. Alford, A. Granville, and C. Pomerance, *There are infinitely many Carmichael numbers*, Ann. of Math. (2) **139** (1994), no. 3, 703–722, DOI 10.2307/2118576. MR1283874

[AHU74]   A. V. Aho, J. E. Hopcroft, and J. D. Ullman, *The design and analysis of computer algorithms*, second printing, Addison-Wesley Series in Computer Science and Information Processing, Addison-Wesley Publishing Co., Reading, Mass.-London-Amsterdam, 1975. MR413592

[AKS04]   M. Agrawal, N. Kayal, and N. Saxena, *PRIMES is in P*, Ann. of Math. (2) **160** (2004), no. 2, 781–793, DOI 10.4007/annals.2004.160.781. MR2123939

[Aut23]   Wikipedia Authors, *Timeline of quantum computing and communication*, September 2023, Page Version ID: 1174829260.

[Ben80]   P. Benioff, *The computer as a physical system: a microscopic quantum mechanical Hamiltonian model of computers as represented by Turing machines*, J. Statist. Phys. **22** (1980), no. 5, 563–591, DOI 10.1007/BF01011339. MR574722

[BHT98]   G. Brassard, P. Høyer, and A. Tapp, *Quantum counting*, Automata, languages and programming (Aalborg, 1998), Lecture Notes in Comput. Sci., vol. 1443, Springer, Berlin, 1998, pp. 820–831, DOI 10.1007/BFb0055105. MR1683527

[BLM17]   J. Buchmann, K. E. Lauter, and M. Mosca (eds.), *Postquantum cryptography, part 1*, IEEE Security & Privacy, vol. 15, IEEE, 2017.

[BLM18]   J. Buchmann, K. E. Lauter, and M. Mosca (eds.), *Postquantum cryptography, part 2*, IEEE Security & Privacy, vol. 16, IEEE, 2018.

[BLP93]   J. P. Buhler, H. W. Lenstra Jr., and C. Pomerance, *Factoring integers with the number field sieve*, The development of the number field sieve, Lecture Notes in Math., vol. 1554, Springer, Berlin, 1993, pp. 50–94, DOI 10.1007/BFb0091539. MR1321221

[Buc04]   J. Buchmann, *Introduction to cryptography*, 2nd ed., Undergraduate Texts in Mathematics, Springer-Verlag, New York, 2004, DOI 10.1007/978-1-4419-9003-7. MR2075209

[BWP+17]   J. D. Biamonte, P. Wittek, N. Pancotti, P. Rebentrost, N. Wiebe, and S. Lloyd, *Quantum machine learning*, Nat. **549** (2017), no. 7671, 195–202.

[CEH+98]   R. Cleve, A. Ekert, L. Henderson, C. Macchiavello, and M. Mosca, *On quantum algorithms*, Complexity **4** (1998), no. 1, 33–42, DOI 10.1002/(SICI)1099-0526(199809/10)4:1<33::AID-CPLX10>3.0.CO;2-U. MR1653992

[Cle11]   R. Cleve, *Classical lower bounds for Simon's problem*, https://cs.uwaterloo.ca/~cleve/courses/F11CS667/SimonClassicalLB.pdf, 2011.

[CLRS22]   T. H. Cormen, C. E. Leiserson, R. L. Rivest, and C. Stein, *Introduction to algorithms*, 4th ed., The MIT Press, Cambridge, MA, 2022.

[Dav82]   M. Davis, *Computability & unsolvability*, Dover, New York, 1982.

[Deu85]   D. Deutsch, *Quantum theory, the Church-Turing principle and the universal quantum computer*, Proc. Roy. Soc. London Ser. A **400** (1985), no. 1818, 97–117. MR801665

[DGM⁺21] M. Dürmuth, M. Golla, P. Markert, A. May, and L. Schlieper, *Towards quantum large-scale password guessing on real-world distributions*, Cryptology and network security, Lecture Notes in Comput. Sci., vol. 13099, Springer, 2021, pp. 412–431, DOI 10.1007/978-3-030-92548-2_22. MR4460974

[DHM⁺18] D. Dervovic, M. Herbster, P. Mountney, S. Severini, N. Usher, and L. Wossnig, *Quantum linear systems algorithms: a primer*, CoRR **abs/1802.08227** (2018).

[DJ92] D. Deutsch and R. Jozsa, *Rapid solution of problems by quantum computation*, Proc. Roy. Soc. London Ser. A **439** (1992), no. 1907, 553–558, DOI 10.1098/rspa.1992.0167. MR1196433

[Fey82] R. P. Feynman, *Simulating physics with computers*, Physics of computation, Part II (Dedham, Mass., 1981), Internat. J. Theoret. Phys. **21** (1981/82), no. 6-7, 467–488, DOI 10.1007/BF02650179. MR658311

[FK03] J. B. Fraleigh and V. J. Katz, *A first course in abstract algebra*, 7th ed., Addison-Wesley, Boston, 2003.

[Fon12] F. Fontein, *The probability that two numbers are coprime*, https://math.fontein.de/2012/07/10/the-probability-that-two-numbers-are-coprime/, 2012.

[GLRS16] M. Grassl, B. Langenberg, M. Roetteler, and R. Steinwandt, *Applying Grover's algorithm to AES: quantum resource estimates*, Post-quantum cryptography, Lecture Notes in Comput. Sci., vol. 9606, Springer, 2016, pp. 29–43, DOI 10.1007/978-3-319-29360-8_3. MR3509727

[Gro96] L. K. Grover, *A fast quantum mechanical algorithm for database search*, Proceedings of the Twenty-eighth Annual ACM Symposium on the Theory of Computing (Philadelphia, PA, 1996), ACM, New York, 1996, pp. 212–219, DOI 10.1145/237814.237866. MR1427516

[HHL08] A. W. Harrow, A. Hassidim, and S. Lloyd, *Quantum algorithm for solving linear systems of equations*, 2008, cite arxiv:0811.3171, Comment: 15 pages; v2 is much longer, with errors fixed, run-time improved and a new BQP-completeness result added; v3 is the final published version and mostly adds clarifications and corrections to v2.

[HvdH21] D. Harvey and J. van der Hoeven, *Integer multiplication in time $O(n \log n)$*, Ann. of Math. (2) **193** (2021), no. 2, 563–617, DOI 10.4007/annals.2021.193.2.4. MR4224716

[IR10] K. Ireland and M. I. Rosen, *A classical introduction to modern number theory*, 2nd ed., 3rd printing, Graduate Texts in Mathematics, no. 84, Springer, New York, Berlin, Heidelberg, 2010 (English).

[Jor] S. Jordan, *Quantum algorithm zoo*, https://quantumalgorithmzoo.org/.

[KLM06] P. Kaye, R. Laflamme, and M. Mosca, *An introduction to quantum computing*, Oxford University Press, Oxford, 2007. MR2311153

[Knu82] D. E. Knuth, *The art of computer programming. 1: Fundamental algorithms*, 2nd ed., 7th printing, Addison-Wesley, Reading, MA, 1982.

[LP92] H. W. Lenstra Jr. and C. Pomerance, *A rigorous time bound for factoring integers*, J. Amer. Math. Soc. **5** (1992), no. 3, 483–516, DOI 10.2307/2152702. MR1137100

[LLMP93] A. K. Lenstra, H. W. Lenstra Jr., M. S. Manasse, and J. M. Pollard, *The factorization of the ninth Fermat number*, Math. Comp. **61** (1993), no. 203, 319–349, DOI 10.2307/2152957. MR1182953

[LP98] H. R. Lewis and C. H. Papadimitriou, *Elements of the theory of computation*, 2nd ed., Prentice-Hall, Upper Saddle River, N.J, 1998.

[Man80] Y. Manin, *Computable and noncomputable (Russian)*, Cybernetics, 1980.

[Man99] Y. I. Manin, *Classical computing, quantum computing, and Shor's factoring algorithm*, Astérisque **266** (2000), Exp. No. 862, 5, 375–404. Séminaire Bourbaki, Vol. 1998/99. MR1772680

[NC16] M. A. Nielsen and I. L. Chuang, *Quantum computation and quantum information*, Cambridge University Press, Cambridge, 2000. MR1796805

[RC18] R. Rines and I. L. Chuang, *High performance quantum modular multipliers*, arXiv:1801.01081 (2018).

[Rud76] W. Rudin, *Principles of mathematical analysis*, 3rd ed., International Series in Pure and Applied Mathematics, McGraw-Hill Book Co., New York-Auckland-Düsseldorf, 1976. MR385023

[Sho94] P. W. Shor, *Polynominal time algorithms for discrete logarithms and factoring on a quantum computer*, Algorithmic Number Theory, First International Symposium, ANTS-I, Ithaca, NY, USA, May 6–9, 1994, Proceedings (Leonard M. Adleman and Ming-Deh A. Huang, eds.), Lecture Notes in Computer Science, vol. 877, Springer, 1994, p. 289.

[Sim94] D. R. Simon, *On the power of quantum computation*, 35th Annual Symposium on Foundations of Computer Science (Santa Fe, NM, 1994), IEEE Comput. Soc. Press, Los Alamitos, CA, 1994, pp. 116–123, DOI 10.1109/SFCS.1994.365701. MR1489241

[Sim97] D. R. Simon, *On the power of quantum computation*, SIAM J. Comput. **26** (1997), no. 5, 1474–1483, DOI 10.1137/S0097539796298637. MR1471989

[Vol99] H. Vollmer, *Introduction to circuit complexity: A uniform approach*, Texts in Theoretical Computer Science. An EATCS Series, Springer-Verlag, Berlin, 1999, DOI 10.1007/978-3-662-03927-4. MR1704235

[Wan10] F. Wang, *The Hidden Subgroup Problem*, Publisher: arXiv Version Number: 1.

[Wat09] J. Watrous, *Quantum computational complexity*, Encyclopedia of Complexity and Systems Science (Robert A. Meyers, ed.), Springer, 2009, pp. 7174–7201.

Index

Selected Published Titles in This Series

For a complete list of titles in this series, visit the
AMS Bookstore at **www.ams.org/bookstore/amstextseries/**.